Cancer Sensitizing Agents for Chemotherapy

IMPROVING THE THERAPEUTIC RATIO IN HEAD AND NECK CANCER

VOLUME 6

Cancer Sensitizing Agents for Chemotherapy Series

Series Editor: Benjamin Bonavida, PhD

Volume 1: Nitric Oxide (Donor/Induced) in Chemosensitization
Edited by Benjamin Bonavida, PhD

Volume 2: Role of Nutraceuticals in Cancer Chemosensitization
Edited by Alok Chandra Bharti and Bharat Bhushan Aggarwal

Volume 3: Targeting Cell Survival Pathways to Enhance Response to Chemotherapy
Edited by Daniel M. Johnson, PhD

Volume 4: Protein Kinase Inhibitors as Sensitizing Agents for Chemotherapy
Edited by Zhe-Sheng Chen and Dong-Hua Yang

Volume 5: Breaking Tolerance to Pancreatic Cancer Unresponsiveness to Chemotherapy
Edited by Ganji Purnachandra Nagaraju, PhD, DSc

Volume 6: Improving the Therapeutic Ratio in Head and Neck Cancer
Edited by Randall J. Kimple, MD, PhD

Volume 7: Drug Efflux Pumps in Cancer Resistance Pathways: From Molecular Recognition and Characterization to Possible Inhibition Strategies in Chemotherapy
Edited by Alejandro Sosnik, PhD and Reina Bendayan, PharmD

Upcoming Volumes:

Terpenoids as Chemosensitizers in Cancer Therapy
Edited by Thomas Efferth

Drug Resistance in Colorectal Cancer: Molecular Mechanisms and Therapeutic Strategies
Edited by Chi Hin Cho, PhD and Tao Hu, PhD

Overcoming Ovarian Cancer Chemoresistance
Edited by Goli Samimi and Christina Annunziata

Breaking Tolerance to Unresponsiveness to Immunotherapy by Natural Killer Cells
Edited by Sandro Matosevic

Cancer Sensitizing Agents for Chemotherapy

IMPROVING THE THERAPEUTIC RATIO IN HEAD AND NECK CANCER

VOLUME 6

Edited by

RANDALL J. KIMPLE, MD, PhD
Department of Human Oncology, University of Wisconsin Carbone Cancer Center,
University of Wisconsin School of Medicine and Public Health,
University of Wisconsin, Madison, WI, United States

ELSEVIER

ACADEMIC PRESS
An imprint of Elsevier

Academic Press is an imprint of Elsevier
125 London Wall, London EC2Y 5AS, United Kingdom
525 B Street, Suite 1650, San Diego, CA 92101, United States
50 Hampshire Street, 5th Floor, Cambridge, MA 02139, United States
The Boulevard, Langford Lane, Kidlington, Oxford OX5 1GB, United Kingdom

Library of Congress Cataloging-in-Publication Data
A catalog record for this book is available from the Library of Congress

British Library Cataloguing-in-Publication Data
A catalogue record for this book is available from the British Library

ISSN: 2468-3183
ISBN: 978-0-12-817868-3

For information on all Academic Press publications
visit our website at https://www.elsevier.com/books-and-journals

Publisher: Stacy Masucci
Acquisition Editor: Rafael Teixeira
Editorial Project Manager: Samantha Allard
Production Project Manager: Punithavathy Govindaradjane
Cover Designer: Greg Harris

Typeset by SPi Global, India

Aims and Scope for Series "Cancer Sensitizing Agents for Chemotherapy"

Current cancer management strategies fail to adequately treat malignancies with chemotherapy, with multivariable does-restrictive factors such as systemic toxicity and multidrug resistance; hence, limiting therapeutic benefits, quality of life and complete long-term remission rates. The resistance of cancer cells to anti-cancer drugs is one of the major reasons for the failure of traditional cancer treatments. Cellular components and dysregulation of signaling pathways contribute to drug resistance. If modulated, such perturbations may restore the drug response and its efficacy. The recent understanding of the molecular mechanisms and targets that are implicated for cancer chemo-resistance have paved the way to develop a large battery of small molecules (sensitizing agents) that can target resistant factors and reduce the threshold of resistance and, thus, allowing their combination with chemotherapeutic drugs to be effective and to reverse the chemo-resistance. A large variety of chemo-therapy sensitizing agents has been developed and several have been shown to be effective in experimental models and in cancer patients.

The main objective of the proposed series "Cancer Sensitizing Agents for Chemotherapy" is to publish individualized and focused volumes whereby each volume is edited by an invited expert Editor(s). Each volume will dwell on specific chemosensitizing agents with similar targeting activities. The combination treatment of the sensitizing agent with chemotherapy may result in a synergistic/additive activity and the reversal of tumor cells resistance to drugs.

The Editor(s) will compile non-overlapping review chapters on reported findings in various cancers, both experimentally and clinically, with particular emphases on underlying biochemical, genetic, and molecular mechanisms of the sensitizing agent and the combination treatment.

The scope of the Series is to provide scientists and clinicians with updated and clinical information that will be valuable in their quest to investigate, develop, and apply novel combination therapies to reverse drug resistance and, thereby, prolong survival and even cure in cancer patients.

Dr. Benjamin Bonavida, PhD (Series Editor)

About the Series Editor

Dr. Benjamin Bonavida is currently a Distinguished Research Professor, Department of Microbiology, Immunology, and Molecular Genetics, David Geffen School of Medicine at the University of California, Los Angeles (UCLA). His research career, thus far, has focused on basic immunochemistry and cancer immunobiology. His research investigations have ranged from the mechanisms of cell-mediated killing and sensitization of resistant tumor cells to chemo/immunotherapies, characterization of resistant factors in cancer cells, cell-signaling pathways mediated by therapeutic anti-cancer antibodies, and characterization of each of the gene products in the dysregulated NF-κB/Snail/YY1/RKIP/PTEN loop in many cancers that regulate cell survival, proliferation, invasion, metastasis, and resistance. He has also investigated the role of nitric oxide in cancer and its potential anti-tumor activity. Many of the above studies are centered on the clinical challenging features of cancer patients' failure to respond to both conventional and targeted therapies. He has published extensively on the above research topics. He is currently Editor-in-Chief of *Critical Reviews in Oncogenesis*, the Series Editor for *Resistance to Targeted Anti-Cancer Therapeutics* published by Springer, and the Series Editor of the following three series: *Chemotherapy-Sensitizing Agents for Cancer*, *Breaking Cancer Resistance to Therapeutic Antibodies*, and *Breaking Tolerance to Anti-Cancer Immunotherapy*, published by Elsevier/Academic Press.

Acknowledgment: The editor wishes to acknowledge the excellent assistance of Ms. Kaiya Kozuma and Ms. Inesa Navasardyan who have worked diligently in the completion: both the editing, formatting of the various contributions of this volume. Ms. Kozuma and Ms. Navasardyan are undergraduate students at UCLA and are talented in this field The editors are indebted to their efforts and attention. I also wish to acknowledge both Sandra Harron and Samantha Allard, Editorial Project Managers, for their continuous assistance in the development of this volume.

About the Editor

Dr. Randall J. Kimple is an Associate Professor in the Departments of Human Oncology and Medical Physics at the University of Wisconsin School of Medicine and Public Health. He completed his MD, his PhD in Pharmacology, and his residency in Radiation Oncology at the University of North Carolina—Chapel Hill. His clinical practice focuses on the care of patients with cancers of the head and neck. He runs a translational research laboratory focused on understanding the molecular basis of therapeutic resistance in head and neck cancer. At Wisconsin he is a co-leader of the Imaging and Radiation Sciences Program within the UW Carbone Cancer Center, a member of the Multidisciplinary Head and Neck Program, and a project leader for the Wisconsin Head and Neck Cancer SPORE Grant. When away from the clinic and laboratory he enjoys spending time with his wife, son, and daughter and skiing or golfing.

Aims and Scope for "Improving the Therapeutic Ratio in Head and Neck Cancer"

Head and neck cancer affects over 600,000 people per year worldwide and is the sixth most common cancer in the United States. Radiation and chemotherapy play a key role in both the cure of patients with head and neck cancer and in the management of patients with incurable, metastatic disease. Unfortunately, over half of patients with locally advanced head and neck cancer will die of their disease or as a result of complications from treatment. Thus, resistance to current therapies remains a major concern for both patients and physicians. Strong data exist in head and neck cancer to support the use of concurrent chemotherapy and radiation therapy. This combination leads to improved disease control and cures more patients. Yet we lack strong data by which to identify those patients most likely to benefit from the toxic combination. In the metastatic setting, current cytotoxic chemotherapy results in a 20–30% response rate and the duration of response is often measured in months. Drugs targeting the immune system, while clearly having a role to play in head and neck cancer, still only result in responses in 20–30% of patients. Both intrinsic and acquired resistance are significant barriers to improving outcomes in these patients. The present volume titled *Improving the Therapeutic Ratio in Head and Neck Cancer* has been focused on novel approaches to overcome resistance and improve the therapeutic ratio for patients with head and neck cancer. Each chapter will include a summary of the underlying biology of the approach with a review of basic science identifying mechanisms of resistance and approaches to sensitization, a description of the translational studies that have been completed to date including a highlight of the potential next steps, and when appropriate will review and discuss the relevant clinical data and the opportunities available to improve treatments for patients with head and neck cancer.

Preface

Head and neck cancer remains the sixth most common cancer worldwide. Despite significant improvements in the multidisciplinary management of these complex cancer patients, many patients live with life-altering side effects of their cancer and/or treatment. These patients, seen every day in the clinics of many of the contributors to this volume, motivate us to do better; to cure more; to irradiate more precisely; to operative less invasively.

The selected contributors are an outstanding group of physicians and scientists who devote their time and energy to finding the next big discovery in head and neck cancer treatment. I am honored to consider them friends and colleagues. The contributors deserve my sincere gratitude for contributing their time, effort, and intellect to the chapters in this collection.

This volume includes 16 chapters that cover a wide spectrum of relevant topics spanning from an overview of head and neck cancer management to challenges in the design of clinical trials for these patients. Eleven of the chapters focus on specific signaling, metabolic, or microenvironmental pathways and how they can be manipulated to improve the therapeutic ratio. Two focus more generally on how multiple treatment approaches can be combined to optimize the therapeutic ratio. Briefly, the highlights of each chapter are briefly presented below.

The chapter by Li and Margalit titled "Overview of head and neck cancer management" provides an overview of the management of head and neck cancer, focusing on tumors of the oropharynx, nasopharynx, oral cavity, hypopharynx and larynx, salivary glands, and the challenge of managing recurrent/metastatic head and neck cancer. The authors discuss major clinical trials that inform current management as well as ongoing and future clinical trials to treat head and neck cancers while minimizing the acute and late effects of treatment.

The chapter by Harmon and Bruce titled "The role of molecularly targeted therapies and immunotherapy in head and neck cancer" discusses the landscape of molecularly targeted therapies and immunotherapy in advanced head and neck squamous cell carcinomas. The authors note that novel immunotherapy agents as well as new combinations of investigational agents with standard of care regimens are providing improvements in overall survival, but the full impact of these combinations are not yet known. This presents a major challenge for clinicians as they seek ways to maximize overall survival while minimizing toxicities.

The chapter by Ko et al. titled "Combining radiation and chemotherapy to improve the therapeutic ratio" reviews the role of radiation therapy in the treatment of head and neck cancer. The authors explain basic radiobiology principles and the uses of radiation therapy, surgical techniques, systemic agents and on-going efforts to combine these therapies to improve clinical outcomes while reducing treatment-related toxicities. They note that clinical trials focused on these aims have yielded disparate results, highlighting the importance of developing biomarkers to personalize treatment.

The chapter by Osman et al. titled "Targeting TP53 to augment therapeutic response in head and neck cancer" discusses TP53, a gene that is commonly mutated in patients with HPV-negative head and neck cancer. These mutations are significantly associated with poor survival as well as tumor resistance to radiation therapy and chemotherapy, suggesting the TP53 gene is a potential biomarker to predict treatment response as well as a possible treatment target. The authors discuss several treatment approaches, including TP53 gene transfer therapy, restoration of wild-type p53 confirmation and function, DNA damage and replication stress responses, selective ATM and ATR inhibitors, Chk1/2 dual inhibitors, specific Wee1 inhibitors, classical histone deacetylases (HDACs) inhibitors, and radiosensitization of HNSCC mutantTP53 cells through metabolic targeting.

The chapter by Galstya et al. titled "Radiosensitization of HNSCC mutant TP53 cells through metabolic targeting" reviews the role of the PI3K/Akt/mTOR signaling pathway in head and neck squamous cell carcinoma. The authors summarize preclinical and clinical investigations using PI3K pathway inhibitors and discuss the effect of radiation therapy on head and neck cancer models with distinct mutational profiles, which may help identify which patients should receive agents that target this pathway in conjunction with (or following) radiation.

The chapter by Ye et al. titled "Improving responses to immunotherapy in head and neck squamous cell carcinoma" discusses the use of immune checkpoint inhibitors targeting the programmed death pathway as an immunotherapy for head and neck squamous cell carcinoma. The authors note that patients who benefit from this form of immunotherapy have a durable response, but response rates are low. The authors review how anti-tumor immunity should develop, why it may be ineffective, how immune checkpoint inhibitors work, and how altering the immune microenvironment may enhance T cell immunity. They also summarize current studies aimed at enhancing response rates to immune checkpoint inhibitors in patients with advanced or relapsed head and neck cancer.

The chapter by Oklejas and Nör titled "Overcoming head and neck cancer stem cells" discusses how cancer stem cells function as a driver of tumor initiation, mediate therapeutic evasion, and promote recurrence and metastases of head and neck cancer. The authors note that cancer stem cells are highly resistant to conventional chemotherapy and radiation therapy. They suggest that targeting pathways that play a significant role in the maintenance of cancer stem cells could sensitize these cells to conventional therapies and result in better treatment outcomes.

The chapter by McDaniel et al. titled "TAM family proteins and therapy resistance" discusses the role the TAM family of receptor tyrosine kinases—Tyro, Axl, and MerTK—plays in cancer signaling, prognosis, and therapy resistance. The authors conclude that inhibiting these proteins could enhance a variety of treatment strategies. Specific to head and neck cancer, combination treatment of Axl and/or MerTK could improve radiation therapy, chemotherapy, and cetuximab therapy.

The chapter by Bhatia and Karam titled "Eph/ephrin family proteins and therapeutic resistance" discusses the functional role of Eph-ephrins in tumor development and progression and how their aberrant expression contributes toward therapeutic and radiation resistance. The authors provide an overview of the available agents in preclinical and clinical research aimed at targeting this gene family to counteract treatment resistance and enhance therapeutic sensitivity.

The chapter by Petronek and Allen titled "Manipulating redox active metal metabolism to enhance cancer therapy" explains the essential roles that redox active metals play in regulating cellular growth. By modulating the metabolism of these metals can provide a novel mechanism to improve the therapeutic ratio between cancer and normal tissues and can potentially enhance standard of care anti-cancer therapy. The authors describe current research on iron and manganese metabolism to improve head and neck cancer patient outcomes.

The chapter by SenthilKumar et al. titled "Fibroblast growth factor receptors as therapeutic targets in head and neck squamous cell carcinomas" discusses the role of fibroblast growth factor receptors (FGFR) as therapeutic targets and biomarkers in head and neck squamous cell carcinoma. The authors note that in head and neck cancer FGFR expression and FGFR1 amplification are associated with invasive disease, increased aggressiveness of disease, and overall poor survival. They describe in vitro studies that have demonstrated the efficacy of FGFR inhibition in head and neck cancer and the few clinical trials that suggest that FGFR inhibition can lead to tumor shrinkage.

The chapter by Francis et al. titled "The MET receptor as a therapeutic target in head and neck squamous cell carcinomas" discusses the role of the mesenchymal-epithelial transition factor (MET) in signaling pathways that drive head and neck squamous cell carcinoma. MET is a tyrosine kinase receptor that is overexpressed and dysregulated in head and neck cancer. MET activation leads to upregulation of downstream signaling through the PI3K/AKT and RAS/MAPK pathways, which in turn promotes cellular metabolism, growth, and proliferation. The authors write that MET inhibition represents a potential avenue for combination treatment protocols involving other molecular targeted agents in head and neck cancer.

The chapter by Lee et al. titled "Autophagy in head and neck cancer therapy" discusses the mechanisms by which chemotherapy and radiation therapy activate autophagy—an evolutionarily conserved cellular process that functions in the degradation and recycling of damaged cellular components in response to cell stress. The authors also discuss available drugs that can be used to modulate autophagy and how modulation of autophagy can aid in overcoming resistance to treatments commonly used in patients with head and neck cancer

The chapter by Nikolaev and Yang titled "Combining Chk1/2 inhibition with radiation in head and neck cancer" discusses the need for new targeted therapeutic approaches to improve the outcomes of patients with advanced high risk head and neck cancer. The authors highlight the emerging molecular therapeutics targeting DNA damage response and repair checkpoints and review the available data from pre-clinical studies and on-going clinical trials.

The chapter by Wu and Gutkind titled "GPCRs in head and neck squamous cell carcinoma" discussed a highly druggable class of transmembrane receptors, G-protein coupled receptors, and their emerging role in head and neck cancer. The authors highlight recent data demonstrating a role for GPCRs and alterations in heterotrimeric G-protein signaling in cancer and point to a potential route into the clinic for these widely studied receptors.

Finally, the chapter by Patel et al. titled "Clinical trial design in head and neck cancer-challenges and opportunities" discusses the current challenges in head and neck trials implementation and the opportunities available to ensure the execution of meaningful research. The authors offer suggestions on managing the changing landscape of tumor biology, accrual, site selection, trial design and defining appropriate endpoints.

It is clear from the above that this volume provides a diverse and wide-ranging review of the field of head and neck cancer research. The excellent contributors have provided suggestions and paths toward future research and the application of their science to overcome therapeutic resistance and improve the therapeutic ratio in head and neck cancer.

I would like to acknowledge the significant assistance in the preparation and editing of this volume provided by Robert Kelly and Kwang Nickel. I would also like to thank Katherine Baldwin for assistance in generating both the cover figure and figures in several additional chapters. Their assistance was extremely valuable in enabling the completion of this volume.

Randall J. Kimple, MD, PhD

Contents

6. Improving responses to immunotherapy in head and neck squamous cell carcinoma

Wenda Ye, Nicole C. Schmitt, Robert L. Ferris, Clint T. Allen

7. Overcoming head and neck cancer stem cells

Alexandra E. Oklejas, Jacques E. Nör

8. TAM family proteins and therapy resistance

Nellie K. McDaniel, Samantha R. Fischbach, Olivia J. Ondracek, Noah B. Welke, Mari Iida, Deric L. Wheeler

9. Eph/ephrin family proteins and therapeutic resistance

Shilpa Bhatia, Sana D. Karam

10. Manipulating redox active metal metabolism to enhance cancer therapy

M.S. Petronek, B.G. Allen

11. Fibroblast growth factor receptors as therapeutic targets in head and neck squamous cell carcinomas

Gopika Senthilkumar, David M. Francis, Andrew M. Baschnagel

12. The MET receptor as a therapeutic target in head and neck squamous cell carcinomas

David M. Francis, Gopika Senthilkumar, Huaising C. Ko, Andrew M. Baschnagel

13. Autophagy in head and neck cancer therapy

Yong-Syu Lee, Jaimee C. Eckers, Randall J. Kimple

14. Combining Chk1/2 inhibition with radiation in head and neck cancer

Anatoly Nikolaev, Eddy S. Yang

15. GPCRs in head and neck squamous cell carcinoma

Victoria H. Wu, J. Silvio Gutkind

Contributors

Clint T. Allen National Institute on Deafness and other Communication Disorders, National Institutes of Health; Department of Otolaryngology-Head and Neck Surgery, Johns Hopkins School of Medicine, Baltimore, MD, United States

B.G. Allen Department of Radiation Oncology, The University of Iowa, Iowa City, IA, United States

Andrew M. Baschnagel Department of Human Oncology, University of Wisconsin Carbone Cancer Center, University of Wisconsin School of Medicine and Public Health, University of Wisconsin, Madison, WI, United States

Shilpa Bhatia University of Colorado Denver, Anschutz Medical Campus, Aurora, CO, United States

Justine Yang Bruce Division of Hematology and Oncology, Department of Medicine, University of Wisconsin Carbone Cancer Center, University of Wisconsin School of Medicine and Public Health, University of Wisconsin, Madison, WI, United States

Janice Cho Department of Otolaryngology—Head and Neck Surgery, University of California at San Francisco, San Francisco, CA, United States

Pippa F. Cosper Department of Human Oncology, University of Wisconsin School of Medicine and Public Health, University of Wisconsin, Madison, WI, United States

Jaimee C. Eckers Department of Chemical & Biological Sciences, Rockford University, Rockford, IL, United States

Robert L. Ferris University of Pittsburgh Medical Center and Cancer Institute, Pittsburgh, PA, United States

Samantha R. Fischbach University of Wisconsin School of Medicine and Public Health, University of Wisconsin, Madison, WI, United States

David M. Francis Department of Human Oncology, University of Wisconsin Carbone Cancer Center, University of Wisconsin School of Medicine and Public Health, University of Wisconsin, Madison, WI, United States

Steven J. Frank Department of Radiation Oncology, The University of Texas MD Anderson Cancer Center, Houston, TX, United States

Arpine Galstyan Department of Otolaryngology—Head and Neck Surgery, University of California at San Francisco, San Francisco, CA, United States

Jennifer R. Grandis Department of Otolaryngology—Head and Neck Surgery, University of California at San Francisco, San Francisco, CA, United States

J. Silvio Gutkind Department of Pharmacology, UCSD Moores Cancer Center, La Jolla, CA, United States

Gil Harmon Division of Hematology and Oncology, Department of Medicine, University of Wisconsin Carbone Cancer Center, University of Wisconsin School of Medicine and Public Health, University of Wisconsin, Madison, WI, United States

Mari Iida University of Wisconsin School of Medicine and Public Health, University of Wisconsin, Madison, WI, United States

Daniel E. Johnson Department of Otolaryngology—Head and Neck Surgery, University of California at San Francisco, San Francisco, CA, United States

Sana D. Karam University of Colorado Denver, Anschutz Medical Campus, Aurora, CO, United States

Randall J. Kimple Department of Human Oncology, University of Wisconsin Carbone Cancer Center, University of Wisconsin School of Medicine and Public Health, University of Wisconsin, Madison, WI, United States

Huaising C. Ko Department of Human Oncology, University of Wisconsin Carbone Cancer Center, University of Wisconsin School of Medicine and Public Health, University of Wisconsin, Madison, WI, United States

Yong-Syu Lee Department of Human Oncology, University of Wisconsin School of Medicine and Public Health, University of Wisconsin, Madison, WI, United States

Puyao C. Li Harvard Radiation Oncology Program, Harvard Medical School, Boston, MA, United States

Antje Lindemann Department of Head and Neck Surgery, The University of Texas MD Anderson Cancer Center, Houston, TX, United States

Danielle N. Margalit Dana-Farber Cancer Institute/Brigham & Women's Hospital, Harvard Medical School, Boston, MA, United States

Nellie K. McDaniel University of Wisconsin School of Medicine and Public Health, University of Wisconsin, Madison, WI, United States

Jeffrey N. Myers Department of Head and Neck Surgery, The University of Texas MD Anderson Cancer Center, Houston, TX, United States

Anatoly Nikolaev Department of Radiation Oncology, Comprehensive Cancer Center, University of Alabama-Birmingham School of Medicine, Birmingham, AL, United States

Jacques E. Nör Department of Cariology, Restorative Sciences, Endodontics, University of Michigan School of Dentistry; Department of Otolaryngology, University of Michigan School of Medicine; University of Michigan Comprehensive Cancer Center; Department of Biomedical Engineering, University of Michigan College of Engineering, Ann Arbor, MI, United States

Alexandra E. Oklejas Department of Cariology, Restorative Sciences, Endodontics, University of Michigan School of Dentistry, Ann Arbor, MI, United States

Olivia J. Ondracek University of Wisconsin School of Medicine and Public Health, University of Wisconsin, Madison, WI, United States

Abdullah A. Osman Department of Head and Neck Surgery, The University of Texas MD Anderson Cancer Center, Houston, TX, United States

Malini M. Patel Division of Medical Oncology, Department of Medicine, Robert Wood Johnson Medical School; Rutgers Cancer Institute of New Jersey, New Brunswick, NJ, United States

M.S. Petronek Department of Radiation Oncology, The University of Iowa, Iowa City, IA, United States

Nicole C. Schmitt National Institute on Deafness and other Communication Disorders, National Institutes of Health; Department of Otolaryngology-Head and Neck Surgery, Johns Hopkins School of Medicine, Baltimore, MD, United States

Gopika Senthilkumar Department of Human Oncology, University of Wisconsin Carbone Cancer Center, University of Wisconsin School of Medicine and Public Health, University of Wisconsin, Madison, WI, United States

Li Wang Department of Radiation Oncology, The University of Texas MD Anderson Cancer Center, Houston, TX, United States

Noah B. Welke University of Wisconsin School of Medicine and Public Health, University of Wisconsin, Madison, WI, United States

Deric L. Wheeler University of Wisconsin School of Medicine and Public Health, University of Wisconsin, Madison, WI, United States

Matthew E. Witek Department of Human Oncology, University of Wisconsin School of Medicine and Public Health, University of Wisconsin, Madison, WI, United States

Victoria H. Wu Department of Pharmacology, UCSD Moores Cancer Center, La Jolla, CA, United States

Eddy S. Yang Department of Radiation Oncology, Comprehensive Cancer Center, University of Alabama-Birmingham School of Medicine, Birmingham, AL, United States

Wenda Ye National Institute on Deafness and other Communication Disorders, National Institutes of Health, Bethesda, MD, United States

CHAPTER

1

Overview of head and neck cancer management

Puyao C. Li[a], *Danielle N. Margalit*[b]

[a]**Harvard Radiation Oncology Program, Harvard Medical School, Boston, MA, United States**
[b]**Dana-Farber Cancer Institute/Brigham & Women's Hospital, Harvard Medical School, Boston, MA, United States**

Abstract

Head and neck cancers are a rich subject area for research—from basic science to clinical research. Recent clinical trials have focused on the role of viruses in cancer etiology and prognostication, individualizing treatment using de-intensification or intensification strategies based on tumor biology and patient risk factors, biomarker-driven treatment selection, targeted therapy with epidermal growth factor receptor (EGFR)-directed agents, and immune-modulating agents. The following chapter provides an overview of the management of head and neck cancer, focusing on tumors of the oropharynx, nasopharynx, oral cavity, hypopharynx and larynx, salivary glands, and the challenge of managing recurrent/metastatic head and neck cancer. In each section, major clinical trials that inform current management are discussed. Additionally, ongoing and future clinical trials are discussed in each section to emphasize current efforts to treat head and neck cancers while minimizing the acute and late effects of treatment.

Abbreviations

ACC	adenoid cystic carcinoma
AJCC	American Joint Committee on Cancer
CT	computed tomography
EBV	Epstein-Barr virus
ECOG	Eastern Cooperative Oncology Group
EORTC	European Organization for Research and Treatment of Cancer
ENE	extranodal extension
FDA	Food and Drug Administration
FDG	fluorodeoxyglucose
GORTEC	Groupe Oncologie Radiotherapie Tete et Cou
HER2	human epidermal growth factor receptor 2
HPV	human papillomavirus
IC	induction chemotherapy

Improving the Therapeutic Ratio in Head and Neck Cancer
https://doi.org/10.1016/B978-0-12-817868-3.00001-9

IHC	immunohistochemistry
IMRT	intensity-modulated radiation therapy
LRC	locoregional control
LRF	local-regional failure
LVI	lymphovascular invasion
MASC	mammary analog secretory carcinoma
MRI	magnetic resonance imaging
NCCN	National Comprehensive Cancer Network
PET	positron emission tomography
PCR	polymerase chain reaction
PFS	progression free survival
RBE	radiobiological effectiveness
RTOG	Radiation Therapy Oncology Group
SCC	squamous cell carcinoma
TLM	transoral laser microsurgery
TNM	tumor, node, metastasis
TORS	transoral robotic surgery
TPF	docetaxel, cisplatin, and fluorouracil
VA	Veterans Affairs
WHO	World Health Organization

Conflict of interest

No potential conflicts of interest were disclosed.

Introduction

Head and neck cancers are a heterogeneous group of neoplasms, each with distinct risk factors and epidemiology, biologic behavior, and treatment approaches. They are defined by their anatomic site of origin, which influences the patterns of spread, treatment approaches, and potential functional implications for the patient. Treatment of head and neck tumors can result in changes to appearance, and to speech, swallow, visual, and breathing functions. Therefore, treatment approaches consider both function and tumor control to optimize the therapeutic ratio. This chapter will provide an overview of head and neck cancer management, highlighting current topics of controversy and clinical research.

In general, treatment for head and neck cancers follows a multidisciplinary approach, often with complementary roles for surgery, radiotherapy, and systemic therapy including chemotherapy, biologics, and immune-modulators. Patients with early stage disease are generally managed with monotherapy, either surgery or radiotherapy alone, although high risk features may prompt combined modality therapy. More advanced stage disease typically requires multimodality therapy which is determined by the anatomic site of disease, risk factors for recurrence, and the estimated efficacy and morbidity of the therapy.

Anatomy

The major anatomic subsites of the head and neck are shown in Figs. 1 and 2 are described as follows:

- *Paranasal sinuses*: these include the four paired sinuses—the frontal, ethmoid, maxillary, and sphenoid sinuses.

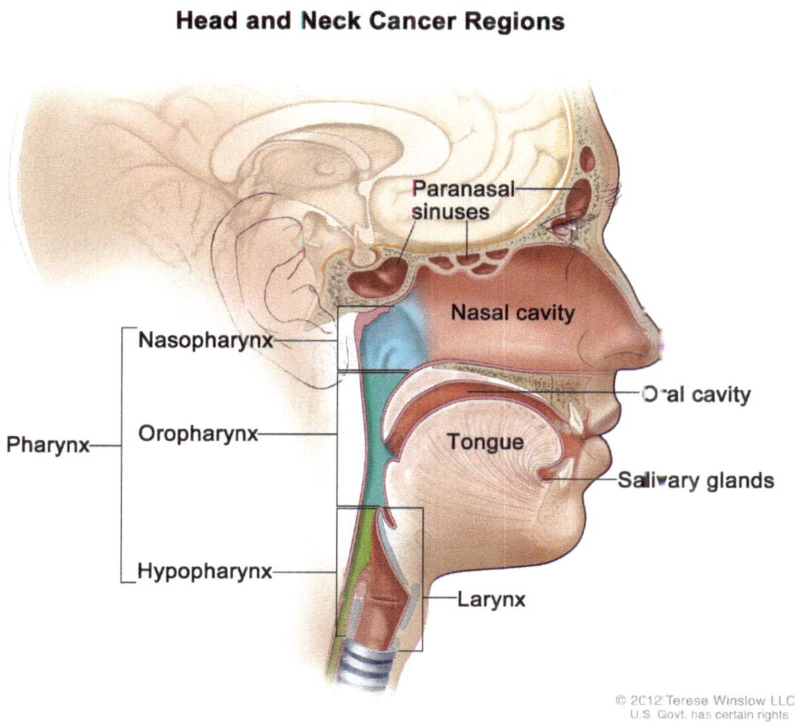

FIG. 1 The major anatomic regions of the head and neck are the paranasal sinuses and nasal cavity, nasopharynx, oropharynx, oral cavity, larynx, hypopharynx, and salivary glands.

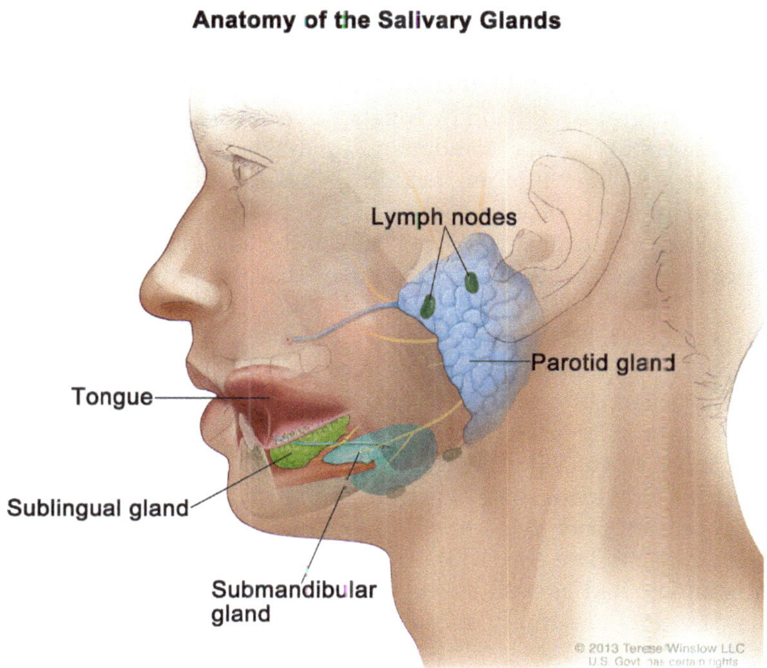

FIG. 2 The major salivary glands are the parotid, submandibular, and sublingual glands. Minor salivary glands are dispersed throughout the mucosa of the mouth and aerodigestive tract.

- *Nasopharynx*: bounded by the floor of the sphenoid superiorly, roof of the soft palate inferiorly, choanae anteriorly, and pharyngeal wall posteriorly.
- *Oropharynx*: contiguous with the nasopharynx superiorly, the larynx and hypopharynx inferiorly, and the oral cavity anteriorly. The oropharynx consists of four subsites: the tonsils, base of tongue, soft palate, and pharyngeal wall.
- *Oral cavity*: separated from the oropharynx by the junction of hard and soft palate and the circumvallate papillae. The oral cavity is comprised of eight subsites: the oral tongue, floor of mouth, retromolar trigone, alveolar ridge, hard palate, buccal mucosa, gingiva, and lips.
- *Larynx*: subdivided into the supraglottis, glottis, and subglottis. The supraglottis includes the false vocal folds, arytenoids, aryepiglottic folds, and epiglottis. The glottis extends from the true vocal folds to 0.5 cm inferiorly, and the subglottis extends inferiorly to the trachea.
- *Hypopharynx*: spans the area from the hyoid bone superiorly to the bottom of the cricoid cartilage inferiorly and includes the pyriform sinuses, the posterior pharyngeal wall, and the postcricoid region.
- *Salivary glands*: Shown in Fig. 2. Tumors may arise from the paired major salivary glands (the parotid, submandibular, or sublingual glands), or from minor salivary glands which are dispersed throughout the mucosa of the mouth and aerodigestive tract.

Clinical evaluation

Initial workup of patients with suspected head and neck cancer should include a complete head and neck examination with assessment of cranial nerves, particularly for nasopharyngeal carcinoma or tumors suspected to involve the skull base, to have perineural spread, or affect neurovascular bundles. With nasopharyngolaryngoscopy, contrast-enhanced computed tomography (CT) imaging of the head and neck and fluorodeoxyglucose (FDG)-positron emission tomography (PET) [1]. Magnetic resonance imaging (MRI) should also be performed for nasopharyngeal carcinoma and may also be helpful for soft tissue differentiation in a range of other tumors including those with perineural spread. A chest x-ray, or chest CT for more advanced disease, is performed to evaluate for distant metastases. When indicated, a direct laryngoscopy and exam under anesthesia is performed to biopsy the suspected primary and to assess the extent of disease and for synchronous primary cancers.

The diagnosis is determined through biopsy of the primary tumor and/or suspicious cervical lymph nodes. For oropharyngeal cancer, human papillomavirus (HPV) testing, including p16 immunohistochemistry and direct HPV assessment via Polymerase chain reaction (PCR)-based methods or HPV DNA in situ hybridization is recommended [2]. For suspected nasopharyngeal cancer, Epstein-Barr virus (EBV) testing is performed on the tissue (EBNA1 IHC) and serum (EBV viral load). Assessment by a dentist, nutritionist and speech and swallow therapist, is important prior to therapy for baseline assessment and support during treatment to minimize acute and late effects [3]. An audiology evaluation is also helpful especially when cisplatin is anticipated as well as the potential for Eustachian tube dysfunction and other auditory side effects of surgery, radiation, and chemotherapy.

Oropharynx

Epidemiology

Oropharynx cancer, of which the most common histology is squamous cell carcinoma (SCC), is relatively uncommon worldwide but is increasing in incidence. This increase reflects the growing proportion of oncogenic HPV-associated cancers, which now comprise an estimated 90% of cases in the United States. Meanwhile, alcohol and tobacco-related oropharynx cancers are declining [4, 5]. HPV-associated oropharyngeal cancers have a better prognosis than HPV-negative oropharyngeal cancers. Among the first studies to demonstrate this prognostic difference in a prospective clinical trial was the post-hoc analysis of Radiation Therapy Oncology Group (RTOG) 0129, which reported 3-year overall survival (OS) rates of 82.4% among patients with HPV-positive oropharynx cancer receiving chemoradiation compared to 57.1% among patients with HPV-negative cancer [6].

Pathology

The pathology and initial presentation of oropharyngeal SCCs differ based on HPV status. Compared with patients with HPV-negative disease, patients with HPV-positive disease are often younger and less likely to have alcohol or smoking exposures, but often have increased sexual exposures. Their disease is more likely to be poorly differentiated, nonkeratinizing, basaloid SCC involving the tonsil or base of tongue, with a higher likelihood of lower T-stage and more advanced N-stage [7]. Additionally, lymph nodes involved by HPV-positive SCC are more often cystic and appear to initially involute more quickly but have a more prolonged course to complete resolution compared to HPV-negative lymph nodes [8]. Posttreatment PET-CTs were shown to be a good strategy to monitor treatment response including nodal resolution [1].

The surrogate marker for HPV infection in oropharynx cancer, p16, is a cyclin-dependent kinase inhibitor and tumor suppressor gene induced by pRb inactivation by the HPV E7 oncoprotein. p16 is readily assayed using immunohistochemistry and, has been established as a surrogate marker for HPV infection in oropharynx cancer [6, 9]. The AJCC-8th edition staging system utilizes p16-status for staging; direct HPV assays are not required.

The disparate mutational landscapes of HPV-positive and HPV-negative tumors supports not only a clinical but biological difference between these diseases. HPV-positive tumors appear to have fewer gene mutations per tumor, have a lower frequency of tumor protein p53 mutations, and fewer epigenetic alterations [10, 11]. In addition, HPV-associated tumors have been reported to by more sensitive to radiation therapy [12, 13].

Anatomy and staging

The American Joint Committee on Cancer (AJCC) 8th edition tumor, node, metastasis (TNM) staging is shown in Tables 1 and 2. Compared to the previous edition, one of the major changes was the creation of separate staging systems for p16-positive (Table 1) and p16-negative (Table 2) oropharyngeal cancers [14, 15]. Additionally, unknown primary

TABLE 1 AJCC 8th edition tumor, node, metastasis (TNM) staging for HPV-positive oropharyngeal cancer

T stage	
T0	No evidence of primary tumor
T1	Tumor is ≤2 cm (cm) in greatest dimension
T2	Tumor is >2 cm but ≤4 cm
T3	Tumor >4 cm or extends to lingual surface of the epiglottis
T4	Moderately advanced local disease: tumor invades adjacent structure including the larynx, extrinsic muscle of tongue, medial pterygoid, hard palate, or mandible or beyond
Clinical N (cN) stage	
N0	No nodal involvement
N1	Involvement of ipsilateral lymph node(s), ≤6 cm in greatest dimension
N2	Involvement of contralateral or bilateral lymph nodes, ≤6 cm in greatest dimension
N3	Involvement of a lymph node >6 cm
Pathologic N (pN) stage	
N0	As above
N1	Involvement of ≤4 lymph nodes
N2	Involvement of >4 lymph nodes
M stage	
M0	No distant metastasis
M1	Distant metastasis
Clinical prognostic stage groups	
I	T0-2N0-1
II	T3 and/or N2
III	T4 and/or N3
IV	M1
Pathologic prognostic stage groups	
I	T0-2N0-1
II	T0-2N2 or T3-4N0-1
III	T3-4N2
IV	M1

TABLE 2 AJCC 8th edition tumor, node, metastasis (TNM) staging for HPV-negative oropharyngeal cancer

T stage	
T0	No evidence of primary tumor
T1	Tumor is ≤2 cm in greatest dimension
T2	Tumor is >2 cm but ≤4 cm
T3	Tumor >4 cm or extends to lingual surface of the epiglottis
T4a	Moderately advanced local disease: tumor invades adjacent structure including the larynx, extrinsic muscle of tongue, medial pterygoid, hard palate, or mandible
T4b	Very advanced local disease: tumor invades lateral pterygoid, pterygoid plates, lateral nasopharynx, or skull base, or encases carotid artery
Clinical N (cN) stage	
N0	No nodal involvement
N1	Involvement of single ipsilateral lymph node (≤3 cm in greatest dimension), without extranodal extension (ENE −)
N2a	Involvement of single ipsilateral lymph node (>3 cm but ≤6 cm) and ENE −
N2b	Involvement of multiple ipsilateral lymph nodes (≤6 cm) and ENE −
N2c	Involvement of bilateral or contralateral lymph nodes (≤6 cm) and ENE −
N3a	Involvement of a lymph node >6 cm and ENE −
N3b	Involvement of any lymph node(s) and clinically overt ENE +
Pathologic N (pN) stage	
N0	No nodal involvement
N1	Involvement of single ipsilateral lymph node (≤3 cm in greatest dimension) and ENE −
N2a	Involvement of single ipsilateral lymph node (>3 cm but ≤6 cm) and ENE − Or any lymph node ≤3 cm and ENE +
N2b	Involvement of multiple ipsilateral lymph nodes (≤6 cm) and ENE −
N2c	Involvement of bilateral or contralateral lymph nodes (≤6 cm) and ENE −
N3a	Involvement of a lymph node >6 cm and ENE −
N3b	Involvement of any lymph node(s) and clinically overt ENE +
M stage	
M0	No distant metastasis
M1	Distant metastasis
Prognostic stage groups	
I	T1N0M0
II	T2N0M0
III	T3N0 or T1-3N1
IVA	T4aN0-1 or T1-4aN2
IVB	T4b and/or N3
IVC	M1

tumors of the head and neck that are p16-positive are staged as per the p16-positive oropharyngeal cancer staging system.

General treatment approach

The management of oropharyngeal cancer is an area of active research as clinical trials evaluate treatments specifically tailored to the HPV-status of the patient. The theme of treatment *de-intensification* is germane to the topic of HPV-related oropharyngeal cancer. Treatment *intensification* is a theme relevant to HPV-negative disease or advanced HPV-positive disease in patients with a heavy tobacco-use history.

With the Food and Drug Administration (FDA)-approval of the use of robotic surgery for head and neck cancer in 2009, the ability to perform less-invasive surgery for oropharyngeal cancer using transoral robotic surgery (TORS) spurred a renewed interest in the surgical management of oropharyngeal cancer. Transoral laser microsurgery (TLM) is also an established surgical technique utilized in the surgical resection of oropharyngeal carcinoma.

In general, for early stage disease, the goal is unimodality treatment, and either surgery or definitive radiation may be chosen depending on the anatomic site and anticipated functional outcome. For patients with more advanced disease such as multiple involved nodes or clinical T3-4 tumors, definitive chemoradiation is often recommended in an effort to avoid the additional toxicity of trimodality treatment for HPV-related tumors. In some cases, particularly for T1-2 tumors and a low suspicion of extranodal extension, surgery followed by adjuvant therapy may be recommended with the goal of minimizing the acute toxicity of concurrent chemotherapy after surgery.

Surgery and radiotherapy have not been compared in a randomized clinical trial, but retrospective studies and metaanalyses of case series have reported similar survival outcomes [16, 17]. A rigorous comparison of outcomes associated with surgery versus radiation is currently underway in patients with early stage disease, in the ORATOR randomized clinical trial (NCT01590355), and in patients with HPV+ locally advanced disease in the ORATOR II randomized clinical trial (NCT03210103).

Initial surgery

For patients treated with initial surgery, the indications for adjuvant therapy are based on prospective studies performed during the pre-HPV era, prior to the epidemic of HPV-related oropharyngeal cancer. Whether the traditional risk factors that warrant adjuvant therapy are still relevant in the more favorable HPV-related oropharyngeal cancer, is an area of active research discussed below. While prospective studies have not reported results, the traditional indications for adjuvant therapy for locally advanced head and neck cancer are also utilized when considering adjuvant therapy for oropharyngeal cancer regardless of HPV-status. Specifically, radiotherapy is generally indicated for T3-4 disease, multiple positive lymph nodes or a lymph node >3cm, level IV to V lymph nodes, perineural invasion, and/or lymphovascular invasion [18]. Adjuvant chemoradiation is recommended for positive surgical margins and/or extranodal extension (ENE) [19–22]. Based on existing indications for adjuvant therapy, a large proportion of surgically-treated patients receive postoperative radiotherapy, and about one third of these patients ultimately received trimodality therapy

[16]. Trimodality therapy is generally considered overtreatment for HPV-related disease leading to increased cumulative toxicity of all modalities. Therefore, studies are evaluating redefining indications for adjuvant therapy for HPV-related oropharynx cancer.

Eastern Cooperative Oncology Group (ECOG) E3311 (NCT 01898494) is a highly anticipated phase II trial of patients with HPV-related disease undergoing primary surgical management with transoral robotic surgery, which not only evaluates the toxicity profile of surgical management, but also evaluates deintensification of adjuvant radiotherapy. Patients with negative margins, ≤4 positive nodes, and ≤1 mm extranodal extension were randomized to standard 60Gy adjuvant radiotherapy or a reduced dose of 50Gy. Patients with positive margins, >1 mm ENE, or >4 positive nodes were treated with standard adjuvant chemoradiation of 60Gy and concurrent weekly cisplatin.

A trial designed to identify the optimal therapy for HPV-negative oropharyngeal cancer, RTOG 1221 (NCT01953952), closed due to poor accrual. The goal was to compare surgery followed by risk-based adjuvant therapy to definitive chemoradiation.

Initial radiation-based therapy

For patients with locally advanced oropharyngeal cancer treated with definitive radiation-based therapy, the addition of systemic therapy improves OS compared to radiotherapy alone [23, 24]. The Intergroup study demonstrated that adding concurrent chemotherapy (bolus cisplatin) to definitive radiotherapy improved OS, 37% versus 23% ($P=.014$), and disease-free survival (DFS), 51% versus 33% ($P=.01$), at 3 years, for patients with locally advanced disease [25]. Similarly, the Groupe Oncologie Radiotherapie Tete et Cou (GORTEC) group demonstrated a substantial OS benefit (3-year OS 51% vs. 31%, $P=.02$) with the addition of carboplatin and 5-FU to radiotherapy for patients with advanced-stage oropharyngeal cancer [26]. A metaanalysis of randomized trials confirmed this benefit in the oropharynx, reporting an absolute OS benefit of 6.5% conferred by concurrent chemotherapy [23, 24].

Biologic therapy with cetuximab, a monoclonal antibody targeting the EGFR receptor, is the only FDA-approved targeted therapy for use in combination with radiotherapy. Combining concurrent cetuximab with radiotherapy improved OS and locoregional control (LRC) compared to radiotherapy alone in the definitive management of locally advanced head and neck cancer. Yet two recent studies, the RTOG 1016 and De-ESCALaTE trials, showed that concurrent cisplatin is more effective than concurrent cetuximab and should remain the standard for cisplatin-eligible patients [27–29]. Both trials compared radiotherapy with concurrent bolus cisplatin versus concurrent cetuximab in patients with HPV+ oropharyngeal cancer. The hope was that cetuximab would be noninferior to cisplatin and also less toxic and that this would constitute a strategy of de-intensification for HPV-association oropharyngeal cancer. Yet both studies showed that cetuximab was inferior to bolus cisplatin 100 mg/m^2 for OS and disease-free survival endpoints [30, 31]. Other efforts to de-intensify therapy for patients with HPV+ disease include the NRG-HN002 trial (NCT02254278) which compares reduced-dose radiotherapy to 60Gy and weekly cisplatin with slightly accelerated radiotherapy alone given 6 fractions a week to 60Gy.

Altered fractionation regimens when using radiotherapy alone have also improved outcomes compared to conventional daily fractionation of 1.8–2.0 Gy per day. For example, the RTOG 9003 trial demonstrated a local-regional control benefit to hyperfractionation and concomitant boost compared to conventional daily fractionation. The DAHANCA 6 & 7 studies also demonstrated a local control and DFS benefit when the overall treatment time was shortened with altered fractionation of 6 compared to 5 fractions a week, even in the setting of the hypoxic radiosensitizer nimorazole which most patients received concurrently. An OS benefit to hyperfractionation compared with conventional fractionation radiotherapy alone was confirmed in the MARCH metaanalysis [32–36]. However, in trials of chemoradiotherapy, including RTOG 0129 and GORTEC 99-02, no benefit was demonstrated for accelerated fractionation chemoradiotherapy over conventional chemoradiotherapy [6, 37].

The role of induction chemotherapy (IC) in head and neck cancer remains controversial but has been explored as a means of intensification for high-risk tumors and de-intensification strategies for favorable-risk patients such as HPV-associated oropharyngeal carcinoma. The TAX 323 [38] and TAX 324 [39, 40] studies established the optimal induction regimen of "TPF" (Taxotere, Platinol, fluorouracil) after demonstrating an improvement in OS when docetaxel was added to the cisplatin and fluorouracil backbone. Subsequent clinical trials compared sequential therapy (induction chemotherapy followed by concurrent chemoradiation) to concurrent chemoradiation. While one study from Italy demonstrated an OS and local-regional control benefit with the addition of induction chemotherapy to chemoradiation [41], the OS benefit was only seen in patients that received the cetuximab concurrent regimen, which has subsequently been shown to be inferior to cisplatin-based regimens [30, 31]. Both the PARADIGM [42] and DeCIDE [43] trials based in North America did not demonstrate an OS benefit to adding induction chemotherapy to concurrent chemoradiation. Even though a potential reduction in distant metastases is a rationale for using induction chemotherapy, the randomized trials of sequential versus concurrent chemoradiation have not shown a statistically significant reduction in distant metastases [41–44], with the use of IC.

The role of IC continues to be explored for organ preservation (e.g., larynx and hypopharynx cancer), improving oncologic outcomes for select patients, and response-adapted therapy whereby patients are selected for subsequent therapy based on induction response. For example, ECOG conducted a phase II study using induction chemotherapy to select patients with p16-positive oropharyngeal cancer for radiotherapy dose-deintensification with favorable results [45].

Key clinical research themes

- Optimizing treatment for patients with favorable-prognosis HPV-associated oropharyngeal cancer by de-intensifying curative treatment.
- Examining methods of intensification of treatment for patients with poor-prognosis HPV-negative oropharyngeal cancers.
- Exploring the efficacy of various strategies to alleviate acute and chronic side effects of therapy including dysphagia, xerostomia, musculoskeletal and other side effects of treatment.

Nasopharynx

Epidemiology

The epidemiology of nasopharyngeal carcinoma has substantial global variation. While relatively rare in North America and Europe, it is the most common head and neck cancer in Southern and Southeast Asia. The incidence is 1–2 per 100,000 people in the United States compared to up to 25 per 100,000 in Southern China [46].

The etiology of nasopharyngeal cancer is thought to be multifactorial. Several risk factors have been identified including EBV, HPV, smoking, diet, and genetic predisposition [47]. The role of EBV in the etiology of nasopharyngeal carcinoma was first elucidated by the discovery of EBV DNA in carcinoma tissue and an association between IgA and IgG antibodies against EBV and development of the malignancy [48, 49]. Recent studies have identified that circulating EBV DNA in plasma can be utilized as a screening tool for nasopharyngeal carcinoma [50]. Furthermore, plasma EBV DNA concentration can be used as a prognostic tool as both the pretreatment and posttreatment EBV DNA titers are prognostic for treated nasopharyngeal carcinoma and may have a role in monitoring response to therapy and for recurrence [50, 51].

Pathology

There are three histologic types of nasopharyngeal carcinoma defined by the World Health Organization (WHO): keratinizing squamous cell carcinoma (WHO type I) and nonkeratinizing carcinoma, which can be further subdivided into differentiated (WHO type II) and undifferentiated (WHO type III) forms, and basaloid squamous cell carcinoma. The keratinizing type is thought to be sporadic and associated with smoking, while the undifferentiated nonkeratinizing form is endemic in parts of Southern Asia and Northern Africa and, associated with Epstein-Barr virus (EBV) and better prognosis overall than the keratinizing type [52].

Anatomy and staging

Nasopharyngeal carcinoma can spread superiorly through the foramen lacerum to the cavernous sinus and posterolaterally through the jugular foramen and hypoglossal canal and can cause multiple cranial neuropathies. The disease is characterized by early lymph node spread, generally in a predictable, orderly fashion from retropharyngeal lymph nodes or level II cervical lymph nodes to lower level lymph nodes [53].

AJCC 8th edition staging is defined in Table 3 [14].

General treatment approach

Definitive radiotherapy is the mainstay of treatment for stage I disease and concurrent chemoradiation forms the basis of treatment for most cases of Stage II-IVA nasopharyngeal carcinoma [54, 55]. Due to anatomic constraints of the disease pattern of spread, upfront surgery is not generally performed. The benefit of adding concurrent cisplatin for locally advanced nasopharyngeal carcinoma was established by the Intergroup 0099 phase III study, which included 147 patients with stage III or IV nonmetastatic disease and compared radiation alone with concurrent chemoradiation. Radiation consisted of 70 Gray (Gy) delivered to the primary tumor and 50–70 Gy delivered to cervical lymph node stations and chemotherapy consisted of cisplatin at a dose of 100 mg/m^2 given every 3 weeks. Of note, patients in the chemoradiation arm also received adjuvant chemotherapy, which consisted of three cycles

TABLE 3 AJCC 8th edition tumor, node, metastasis (TNM) staging for nasopharyngeal cancer

T stage	
T0	EBV-positive cervical lymph node without a primary tumor identified
T1	Confined to nasopharynx, or extension to oropharynx and/or nasal cavity
T2	Parapharyngeal extension and/or involvement of adjacent soft tissue (medial pterygoid, lateral pterygoid, prevertebral muscles)
T3	Involvement of bony structures (base of skull, cervical vertebra, pterygoid structures) and/or paranasal sinuses
T4	Intracranial extension or cranial nerve involvement, or involvement of orbit, hypopharynx, parotid, beyond lateral pterygoid muscle
N stage	
N0	No nodal involvement
N1	Involvement of retropharyngeal node(s), unilateral cervical lymph nodes (above cricoid cartilage), ≤6cm
N2	Involvement of bilateral cervical nodes (above cricoid cartilage), ≤6cm
N3	Involvement of any cervical nodes >6cm or lymph nodes below cricoid cartilage
M stage	
M0	No distant metastasis
M1	Distant metastasis
Prognostic stage groups	
I	T1N0
II	T0-1N1 or T2N0-1
III	T0-1N2 or T2N2 or T3N0-2
IVA	T0-4N3 or T4N0-2
IVB	M1

of a regimen of cisplatin (80 mg/m^2) and fluorouracil (1000 mg/m^2) given every 3 weeks. The study demonstrated that adding chemotherapy improved progression free survival (PFS), 69% versus 24% ($P < .001$), and OS, 78% versus 47% ($P = .005$) at 3 years [56]. The study compromised a large proportion of patients with the nonendemic WHO I nasopharyngeal carcinoma yet subsequent studies exclusively including endemic nasopharyngeal carcinoma confirmed the benefit of concurrent chemotherapy both with or without adjuvant chemotherapy in patients with locally advanced nasopharyngeal carcinoma, including several metaanalyses [55, 57, 58].

For Stage II disease, there continues to be some controversy over which subgroup is likely to benefit the most from adding concurrent chemotherapy to radiation alone. Yet an OS advantage of concurrent weekly cisplatin (30 mg/m^2) over radiotherapy alone was demonstrated in patients with stage II disease in a phase III randomized clinical trial of patients in China [59].

Equipoise between weekly and bolus cisplatin in terms of survival outcomes has been established, but weekly cisplatin appears to be associated with improved patient quality of life [60]. Advances in radiotherapy delivery has also resulted in improvements in both patient-reported quality of life as well as observer-rated toxicity. Radiotherapy delivered via intensity-modulated radiation therapy (IMRT) allows for more conformal dose around the target and in general, reduces dose to organs at risk such as the parotid glands and has been shown to reduce xerostomia at least in the short term [61, 62]. Other studies have shown that modern-techniques with IMRT resulted in minimizing other toxicity such as hearing loss, temporal lobe neuropathy, cranial neuropathy, trismus, and neck fibrosis [63]. Proton therapy is another modality that may offer further reduction in acute and late effects yet prospective comparison with photon-IMRT are not yet published. Such trials are underway (for example, NCT01893307 and NCT00592501).

The role of adjuvant and/or induction chemotherapy is still somewhat unclear, but current National Comprehensive Cancer Network (NCCN) guidelines [64] recommend administration of one or the other in addition to concurrent chemoradiation, supported by a 2017 metaanalysis demonstrating the addition of either adjuvant or induction chemotherapy to be superior to concurrent chemoradiation alone, with adjuvant chemotherapy perhaps better for LRC and induction chemotherapy better for distant control [55]. The evidence for induction chemotherapy is stronger, with more recent studies demonstrating that regimens like docetaxel, cisplatin, and fluorouracil (TPF) are well tolerated at high-volume centers and provide a survival advantage [65, 66].

Key clinical research themes

- The role and timing of adjuvant versus sequential chemotherapy.
- Determining the optimal concurrent chemotherapy regimen (for example, NCT00896181).
- De-intensifying treatment by minimizing the extent of cervical nodal radiation in patients with cN0 stage disease.
- Intensifying treatment in patients with more advanced disease by delivering accelerated fractionation radiotherapy or through imaging markers of resistance.
- Lastly, clinical trials such as the NRG-HN001 (NCT02135042) are examining the utility of EBV as a biomarker for individualizing treatment. For example, patients with undetectable EBV titers after concurrent cisplatin-radiation are randomized to adjuvant cisplatin/5-FU versus observation. Such efforts highlight the potential translational role of biomarkers, particularly in virus-associated cancers, in determining the optimal treatment strategy for individual patients.

Oral cavity

Epidemiology and pathology

Cancers of the lip and oral cavity comprise 40% of head and neck cancers and are particularly common in Southeast Asia, an observation primarily attributed to the use of betel-quid, areca-nut chew, or tobacco [67]. The vast majority of these are SCC. Oral leukoplakia or other premalignant lesions can but does not always precede development of invasive SCC. Known risk factors for oral cavity SCC include exposure to tobacco, alcohol, and betel quid and areca nuts. Similar to oropharyngeal cancers, patients with tumors without p16 protein expression appear to have worse prognosis [68]. However, unlike oropharyngeal cancers, oral cavity cancers appear to be less frequently linked with HPV and p16 positivity was not found to be a good surrogate marker of HPV infection [69].

Anatomy and staging

AJCC 8th edition staging is defined in Table 4 [14].

TABLE 4 AJCC 8th edition tumor, node, metastasis (TNM) staging for oral cavity cancer

T stage	
T0	No evidence of primary tumor
Tis	Carcinoma in situ
T1	Tumor is ≤2cm in greatest dimension, ≤ 5mm depth of invasion (DOI)
T2	Tumor is ≤2cm and DOI is >5mm but ≤10mm or Tumor is >2cm but ≤4cm and DOI≤10mm
T3	Tumor >4cm or DOI>10mm
T4a	Moderately advanced local disease: tumor invades adjacent structures including cortical bone,[a] muscle of tongue, inferior alveolar nerve if lip tumor, floor of mouth, maxillary sinus, or skin of face
T4b	Very advanced local disease: tumor invades masticator space, pterygoid plates, skull base, and/or encases internal carotid artery
Clinical N (cN) stage	
N0	No nodal involvement
N1	Involvement of single ipsilateral lymph node (≤3cm in greatest dimension) and ENE−
N2a	Involvement of single ipsilateral lymph node (>3cm but ≤6cm) and ENE−
N2b	Involvement of multiple ipsilateral lymph nodes (≤6cm) and ENE−
N2c	Involvement of bilateral or contralateral lymph nodes (≤6cm) and ENE−
N3a	Involvement of a lymph node >6cm and ENE−
N3b	Involvement of any lymph node(s) and clinically overt ENE+
Pathologic N (pN) stage	
N0	No nodal involvement
N1	Involvement of single ipsilateral lymph node (≤3cm in greatest dimension) and ENE−
N2a	Involvement of single ipsilateral lymph node (>3cm but ≤6cm) and ENE− Or any lymph node ≤3cm and ENE+
N2b	Involvement of multiple ipsilateral lymph nodes (≤6cm) and ENE−

TABLE 4 AJCC 8th edition tumor, node, metastasis (TNM) staging for oral cavity cancer—Cont'd

Pathologic N (pN) stage	
N2c	Involvement of bilateral or contralateral lymph nodes (≤6 cm) and ENE−
N3a	Involvement of a lymph node >6 cm and ENE−
N3b	Single ipsilateral lymph node >3 cm and ENE+ Or multiple ipsilateral, contralateral, or bilateral lymph nodes and ENE+
M stage	
M0	No distant metastasis
M1	Distant metastasis
Prognostic stage groups	
0	TisN0M0
I	T1N0M0
II	T2N0M0
III	T3N0 or T1-3N1
IVA	T4a and/or N2
IVB	T4b and/or N3
IVC	M1

[a] *Gingival primary with superficial erosion of bone/tooth socket is not sufficient for T4 classification.*

General treatment approach

Surgery is the initial treatment for resectable oral cavity cancer. Patients with pathologic risk factors for recurrence subsequently receive adjuvant radiotherapy or chemoradiation. Patients who are not candidates for surgery may have definitive radiotherapy or chemoradiation; some patients may receive induction chemotherapy. Surgical management of the neck is always an important consideration. For example, an elective neck dissection is associated with improved overall survival for early stage tongue cancer [70].

Postoperative radiotherapy is recommended for patients with at least one intermediate risk factor for locoregional spread, including T3-4 disease, ≥N2 nodal stage, lymphovascular invasion (LVI), perineural invasion (PNI), and/or high-risk features of positive surgical margins or ENE [18, 71]. Indications for concurrent adjuvant chemotherapy include high risk features of positive surgical margins and/or ENE. Depth of invasion (DOI) is another important consideration when deciding upon initial neck management and adjuvant therapy, with ≥4 mm DOI linked to increased risk of recurrence and nodal involvement [72, 73]. DOI has been incorporated into the most recent AJCC staging system.

Retrospective and population-based studies have shown an OS benefit of postoperative radiotherapy [74, 75]. Pooled data from two randomized clinical trials, RTOG 9501 and European Organization for Research and Treatment of Cancer (EORTC) 22931, demonstrated improved LRC, DFS, and OS with the addition of concurrent bolus cisplatin, particularly among patients with positive margins and ENE. There was no statistically significant difference in the rate of distant metastases with chemotherapy [19–22]. Bolus cisplatin (100 mg/m^2) has been shown to be superior to weekly cisplatin (30 mg/m^2) in a randomized trial in providing better LRC, but comes at the expense of increased toxicity [76]. Also, weekly cisplatin at a dose of 30 mg/m^2 is lower than the starting dose of 40 mg/m^2 used concurrently with radiotherapy. A retrospective study suggested that cisplatin 40 mg/m^2 was similarly efficacious compared to cisplatin 100 mg/m^2 in the definitive management of locally advanced head and neck cancer [77].

In an effort to improve the efficacy and reduce toxicity of concurrent systemic therapy, cetuximab was explored in combination with radiation and weekly chemotherapy. A phase II clinical trial, RTOG 0234, demonstrated superior DFS with the addition of cetuximab to postoperative radiation and weekly cisplatin or weekly docetaxel when compared to historic controls extracted from RTOG 9501 with comparable toxicity [78]. A clinical trial, RTOG 1216 (NCT 01810913), is underway to determine the optimal adjuvant regimen for high-risk adjuvant therapy. It compares postoperative radiotherapy with concurrent cisplatin versus postoperative radiotherapy with concurrent docetaxel versus postoperative radiotherapy with concurrent docetaxel and cetuximab for patients with high risk head and neck SCC. Another trial, RTOG 0920 (NCT NCT00956007) is evaluating the benefit of concurrent cetuximab with postoperative radiotherapy compared to radiotherapy alone in patients with intermediate risk disease.

Like other head and neck SCCs, oral cancers may benefit from accelerated or hyperfractionation, with improved local control demonstrated in the RTOG 9003 [32, 33] and DAHANCA 6 & 7 studies [34] and the OS benefit demonstrated in the MARCH metaanalysis [35]. However, the oral cavity adjuvant therapy (OCAT) study (NCT 00193843) comparing postoperative conventional chemoradiation, concurrent chemoradiation, and accelerated radiotherapy for locally advanced oral cavity cancers did not report detecting a difference in LRC or survival [79]. Brachytherapy is an alternate radiotherapeutic option, and is used in institutions with expertise in brachytherapy to treat the primary site either in the adjuvant or definitive setting, often in combination with external beam radiotherapy, or as salvage treatment for patients with recurrent disease [80, 81].

Finally, for patients with unresectable disease, induction chemotherapy has been used [38, 39], especially when it is not anticipated to compromise the concurrent phase of treatment. Yet it is important to note that radiotherapy planning volumes should treat the preinduction volume of disease [82] and therefore it is essential to obtain imaging prior to initiation of induction chemotherapy.

Key clinical research themes

- The role of immunotherapy in treatment of oral cavity cancers (NCT02919683) and the utility of preoperative "window" trials to identify pathologic predictors of treatment response.
- Improving functional and cosmetic outcomes through novel surgical reconstruction techniques.
- Defining the optimal adjuvant therapy approach for intermediate- and high-risk SCC of the head and neck.

Larynx and hypopharynx

Epidemiology and staging

Worldwide, laryngeal cancer is relatively common while hypopharyngeal cancer is uncommon. AJCC 8th edition staging is defined in Table 5 [14].

Clinical and **pathologic nodal staging** and **prognostic stage groups** for larynx and hypopharynx are the same as for oral cavity above.

TABLE 5 AJCC 8th edition tumor (TNM) staging for larynx and hypopharynx cancer

Larynx			
T stage	**Supraglottis**	**Glottis**	**Subglottis**
Tis	Carcinoma in situ		
T1	Limited to one subsite with normal vocal cord (VC) mobility	T1a: Limited to one VC with normal moblity T1b: Involves both VCs with normal mobility	Limited to subglottis
T2	Invades mucosa of more than one adjacent subsite without larynx fixation	Extends to supraglottis, subglottis, and/or with impaired VC mobility	Extends to VC with normal or impaired mobility
T3	Limited to larynx with VC fixation and/or invades postcricoid, preepiglottic space, paraglottic space, and/or inner cortex of thyroid cartilage		
T4a	Moderately advanced local disease: invades outer cortex of thyroid cartilage and/or tissues beyond larynx		
T4b	Very advanced local disease: invades prevertebral space, encases carotid artery, or invades mediastinal structures		

Hypopharynx	
T stage	
T0	No evidence of primary tumor
Tis	Carcinoma in situ
T1	Tumor limited to one subsite and/or ≤2 cm
T2	Tumor invades multiple subsites or an adjacent site, or is >2 cm but ≤4 cm, without fixation of hemilarynx
T3	Tumor is >4 cm or with fixation of hemilarynx or extension to esophageal mucosa
T4a	Moderately advanced local disease: invades thyroid/cricoid cartilage, hyoid bone, thyroid gland, esophageal muscle, or central compartment soft tissue (i.e., prelaryngeal strap muscles and subcutaneous fat)
T4b	Very advanced local disease: invades prevertebral space, encases carotid artery, or invades mediastinal structures

Clinical and pathologic nodal staging and prognostic stage groups for larynx and hypopharynx are the same as for oral cavity in Table 4.

General treatment approach

Organ preservation is a primary consideration for these sites to maintain respiratory, swallowing, and vocal function. For early stage T1-2 laryngeal cancer, either endoscopic surgery or radiotherapy are options [83]. For early stage hypopharyngeal cancer, primary radiotherapy or surgery with partial laryngopharyngectomy are options. Yet due to the rich lymphatic drainage, neck management needs to be considered even in early-stage T1-2 hypopharyngeal lesions. For locally advanced hypopharyngeal or laryngeal cancer, organ preservation with chemoradiation, or induction chemotherapy followed by radiation, are preferred options in nonbulky primary tumors or when there is adequate baseline laryngeal function. Patients with bulky T3 or T4 lesions or those with functional impairment in respiratory, swallow, and speech function, may benefit from an upfront total laryngectomy or laryngopharyngectomy with adjuvant radiation or chemoradiation depending on pathologic risk factors [19–22].

For early stage T1N0 laryngeal cancer, accelerated fractionation was demonstrated to be superior to standard fractionation radiotherapy, with better local control for patients receiving 2.25 Gy per fraction compared to 2 Gy per fraction [84].

For treatment of advanced laryngeal cancer with functional baseline swallowing, voice, and respiratory function, nonsurgical organ preservation approaches are appropriate. Such "organ-preservation" approaches were initially established in the Veterans Affairs (VA) laryngeal cancer study [85]. The study was the first to demonstrate that using an initial nonsurgical approach, and salvage laryngectomy for failures or chemotherapy nonresponders, OS was not compromised and a substantial proportion of patients were able to preserve their larynx (64%). The study established treatment with induction chemotherapy, consisting of bolus cisplatin and fluorouracil, followed by radiotherapy-alone to be an alternative to upfront laryngectomy and adjuvant radiotherapy. The RTOG 9111 trial later explored the optimal nonsurgical approach [86] and compared radiotherapy-alone, concurrent radiotherapy-cisplatin, and induction chemotherapy followed by radiotherapy-alone as used in the VA larynx study. The study demonstrated that concurrent chemoradiation (bolus cisplatin alone), had comparable OS rates compared with the sequential VA laryngeal study regimen and a higher rate of larynx preservation. Yet the long-term follow-up of RTOG 9111 [87] suggested a potential higher death rate among the concurrent chemoradiation arm such that the composite endpoint of laryngectomy-free survival (i.e., alive with an intact larynx) was similar in the concurrent chemoradiation arm and the induction chemotherapy arm. The reason for the increased mortality rate was not known but was postulated to potentially reflect late effects of concurrent chemoradiation such as aspiration events. Sequential therapy (Induction chemotherapy followed by definitive chemoradiation) has not been shown to improve survival compared with concurrent chemoradiation alone [88]. For a discussion of this, please see the oropharynx section where induction chemotherapy is discussed in more detail.

For advanced hypopharyngeal cancer, EORTC 24891 demonstrated no difference in OS or LRC for induction chemotherapy, consisting of bolus cisplatin and fluorouracil (PF), followed by radiotherapy compared with surgery followed by adjuvant radiotherapy [89, 90]. The GORTEC 2000–01 study later demonstrated better response and laryngeal preservation with docetaxel, cisplatin, and fluorouracil (TPF) compared with PF induction chemotherapy, but no difference in survival rates for patients with advanced laryngeal or hypopharyngeal cancer at 3 years [91]. The Pignon metaanalysis of randomized trials showed that for hypopharyngeal cancers, a 3.9% OS benefit was conferred by concurrent chemotherapy, compared with sequential chemotherapy followed by radiotherapy at 5 years [23, 24].

Key clinical research themes

Ongoing clinical research efforts in laryngeal and hypopharyngeal cancers focus on both the early-stage and advanced-stage setting:

- For early-stage larynx cancers, clinical research topics include optimizing focused partial laryngeal irradiation rather than traditional techniques that treat the entire larynx (for example, NCT03759431) and hypofractionated radioablative techniques to make treatment shorter and more convenient (for example, NCT03114462).
- In the advanced-stage setting clinical studies are exploring treatment-modification using predictive biomarkers such as p53 mutation (NCT02734537), and investigating novel systemic agents to improve outcomes for advanced disease for which local-regional and distant failures are still common (NCT00956007).

Salivary gland tumors

Epidemiology and pathology

Salivary gland tumors are a heterogeneous group of neoplasms with unique histologic, genetic, and immunohistochemical features, and biologic behavior. This family of tumors are an example of how identification of cytogenetic and protein expression patterns has enabled further characterization of tumors that may appear histologically similar. For example, mammary analog secretory carcinoma (MASC) was recently included as a distinct tumor in the 4th edition of the WHO classification of salivary tumors. MASC was formerly grouped under acinic cell carcinoma yet was identified as a distinct entity based on the *ETV6-NTRK3* gene fusion [92]. Table 6 lists the most recent WHO classification of salivary gland tumors [93]. Pleomorphic adenoma is the most common benign histology, and accounts for approximately half of all salivary gland tumors. Mucoepidermoid carcinoma and adenoid cystic carcinoma (ACC) are the most common malignant salivary tumors and account for a quarter to a third of cases [94].

Many salivary neoplasms have targetable overexpression of several oncogenes and high levels of human epidermal growth factor receptor 2 (HER2), androgen receptors, c-kit tyrosine kinase receptor, and EGFR identified on immunohistochemical staining [95]. This observation has led to prospective trials and clinical approaches that vary by institution, aimed to tailor systemic therapy. For example, a translocation involving (11;19)(q21;p13), producing the fusion protein, MECT1-MAML2, occurs commonly in mucoepidermoid carcinomas and EGFR and HER2 may be overexpressed in many cases [96]. However, targeted agents like cetuximab and trastuzumab have met with little success in early phase clinical trials [97, 98].

TABLE 6 WHO classification of tumors of the salivary gland

Malignant tumors
Mucoepidermoid carcinoma
Adenoid cystic carcinoma
Acinic cell carcinoma
Polymorphous adenocarcinoma
Clear cell carcinoma
Basal cell carcinoma
Intraductal carcinoma
Adenocarcinoma, NOS
Salivary duct carcinoma
Myoepithelial carcinoma
Epithelial-myoepithelial carcinoma
Carcinoma ex pleomorphic adenoma
Secretory carcinoma
Sebaceous adenocarcinoma
Carcinosarcoma
Poorly differentiated carcinoma (undifferentiated carcinoma, large cell neuroendocrine carcinoma, small cell neuroendocrine carcinoma)
Lymphoepithelial carcinoma
Squamous cell carcinoma
Oncocytic carcinoma
Uncertain malignant potential (sialoblastoma)
Benign tumors
Pleomorphic adenoma
Myoepithelioma
Basal cell adenoma
Warthin tumor
Oncocytoma
Lymphadenoma
Cystadenoma
Sialadenoma
Ductal papillomas
Sebaceous adenoma
Canalicular adenoma and other ductal adenomas

TABLE 6 WHO classification of tumors of the salivary gland—Cont'd

Nonneoplastic epithelial lesions
Sclerosing polycystic adenosis
Nodular oncocytic hyperplasia
Lymphoepithelial sialadenitis
Intercalated duct hyperplasia
Benign soft tissue lesions
Hemangioma
Lipoma/sialolipoma
Nodular fasciitis
Hematolymphoid tumors
Extranodal marginal zone lymphoma of mucosa-associated lymphoid tissue (MALT lymphoma)

ACCs are locally aggressive tumors with a high risk of recurrence, frequently via perineural spread. Solid growth pattern and p53 expression have been identified as prognostic for this histology [99]. The vast majority of ACCs overexpress c-kit, but targeted therapy using c-kit inhibitors has not been successful thus far [100–102]. ACCs also characteristically harbor a translocation t(6;9) leading to the fusion gene *MYB-NFIB* [103] and overexpression of the oncogene MYB. MYB has been useful as a diagnostic marker and being evaluated as a potential therapeutic target [104].

Salivary duct carcinomas and adenocarcinomas are rare histologic subtypes that characteristically express androgen receptor. In a small trial, androgen deprivation had a 50% clinical benefit ratio (patients had partial response or stable disease) [105]. The efficacy and safety of this treatment is now under further investigation in a larger phase II trial (NCT01969578).

Anatomy and staging

AJCC 8th edition staging is defined in the Table 7 [14].

Clinical and **pathologic nodal staging** and **prognostic stage groups** for salivary gland tumors are the same as for oral cavity above.

General treatment approach and key clinical research themes

Primary treatment for salivary gland tumors is typically surgical resection of the primary with neck management considerations for intermediate and high-grade salivary tumors. While there are no randomized trials providing level one evidence, retrospective studies suggest a benefit for adjuvant radiotherapy or chemoradiation for risk factors for recurrence. These include close or positive margins, gross residual disease, lymph node involvement,

TABLE 7 AJCC 8th edition tumor (TNM) staging for salivary gland cancer

T stage	
T0	No evidence of primary tumor
Tis	Carcinoma in situ
T1	Tumor is $\leq$2cm without extraparenchymal extension
T2	Tumor is >2cm but $\leq$4cm, without extraparenchymal extension
T3	Tumor is >4cm and/or with extraparenchymal extension
T4a	Moderately advanced local disease: invades skin, mandible, ear canal, and/or facial nerve
T4b	Very advanced local disease: invades skull base and/or pterygoid plates and/or encases carotid artery

Clinical and pathologic nodal staging and prognostic stage groups for are the same as for oral cavity in Table 4.

adenoid cystic histology, intermediate or high-grade histology, perineural invasion, lymphovascular invasion, and T3-4 stage disease [106–108].

The benefit of adding chemotherapy to adjuvant radiotherapy for salivary gland tumors is under investigation in several trials, including RTOG 1008 (NCT01220583). In addition to the previous efforts to treat salivary gland tumors with targeted agents described above, there are multiple ongoing studies of molecular and genetic tumor profiling of salivary tumors, particularly in the recurrent and metastatic setting, and deliver tailored targeted therapy (for example, NCT02069730, NCT03602079, and NCT02576431).

Inoperable tumors may be treated with definitive radiotherapy or chemoradiation. In small trials, neutron therapy and carbon-ion therapy have yielded favorable outcomes for unresectable salivary tumors compared to historical series yet with an increased risk of toxicity due to the high radiobiological effectiveness (RBE) values of these particles [109, 110].

Recurrent and metastatic squamous cell carcinoma of the head and neck

Despite advances in treatment that have improved the prognosis of many H&N cancers over time, recurrence is still a challenge for which clinical trials are needed. For most head and neck cancers, local-regional failure (LRF) is the most common pattern of failure. One of the exceptions to this pattern of failure, is endemic nasopharyngeal carcinoma for which distant metastasis is frequently the most common site of first failure [65]. In the RTOG 0129 and 0522 trials, the 3-year LRF rate was 14–17% in p16-postive oropharyngeal tumors but 33–35% in p16-negative oropharyngeal tumors. Distant recurrence, either isolated or with LRF constituted 43–46% of all failures [6, 29]. In another modern prospective trial from Spain in which patients with locally advanced head and neck cancer received either sequential therapy or concurrent chemoradiation, the 2-year LRF rate ranged from 47 to 51% by treatment arm [44]. Clearly, LRF remains a problem in head and neck SCC management.

When LRF is surgically resectable, salvage surgery is associated with improvement in OS [111, 112]. Following salvage surgery, adjuvant re-irradiation and concurrent chemotherapy is associated with improved LRC compared to observation yet the improvement in LRC did not translate to an OS benefit [113]. The lack of OS benefit may reflect an increase in treatment-related deaths, distant metastases, or second primary tumors in the adjuvant re-irradiation arm. This has led some investigators to prospectively assess the role of adjuvant systemic therapy with immune checkpoint inhibitors and to forego adjuvant reirradiation altogether (NCT03341936 and NCT02769520).

For unresectable local-regional recurrences, reirradiation and systemic therapy is an option for select patients and can result in prolonged disease control and even cure in select patients [114, 115]. However, improvements in local-regional control must be balanced with the significant risks of treatment-related toxicity [116]. Current trials are exploring the addition of immune checkpoint inhibitors in combination with definitive long-course reirradiation (NCT02289209) as well as the role of stereotactic body radiation therapy (SBRT) in combination with adjuvant immune checkpoint inhibitors (NCT03546582). Additional clinical studies are needed to compare long-course re-irradiation over several weeks to shorter-course SBRT and to identify which patients will benefit from each modality. For example, patients with recurrent disease <2 years from their initial radiation as well as organ dysfunction such as feeding tube or tracheostomy have a median OS of only 8 months and may best be treated with palliative systemic therapy or short-course SBRT rather than several weeks of reirradiation.

Whether recurrence is distant or an unresectable local-regional recurrence, advances in systemic therapy have improved the OS of patients with recurrent/metastatic disease. The EXTREME trial demonstrated an OS benefit when cetuximab was added to platinum-based chemotherapy and fluorouracil in the first-line setting. The median OS benefit improved from 7.4 to 10.1 months in the cetuximab group. The preplanned interim analysis of the Keynote-048 trial (presented in abstract form Burtness et al. *ESMO* 2018), was the first prospective data showing the benefit of first-line immune checkpoint inhibitors compared to first line chemotherapy. It also represents one of few prospective trials that indicates biomarkers may be useful to select specific therapy. The study compared the EXTREME regimen to first-line pembrolizumab-alone or pembrolizumab and chemotherapy (platinum chemotherapy +5-FU). The study also included a biomarker analysis of combined PD-L1 score (percent PD-L1 staining of tumor and immune cells/total cells). The biomarker-select patients with a PD-L1 CPS $\geq$20 had improved OS (median 14.9 months vs. 10.7 months) when treated with pembrolizumab alone compared to EXTREME. Similarly, patients with a PD-L1 CPS $\geq$1 had improved median OS when treated with pembrolizumab compared to EXTREME (12.3 vs. 10.3 months). In the total population, without biomarker selection, pembrolizumab and chemotherapy was superior to EXTREME (median OS 13.0 vs. 10.7 months).

In the second line setting, pembrolizumab and nivolumab were approved after clinical trials showed a benefit compared to standard therapy [117, 118]. The Checkmate 141 trial showed a median OS benefit with nivolumab compared to investigator's choice of single-agent systemic therapy (MTX, docetaxel, or cetuximab) with a median OS 7.5 vs. 5.1 months. [118]. Cytotoxic chemotherapy such as methotrexate or docetaxel, and targeted therapy such as cetuximab and small molecule tyrosine kinase inhibitors such as afatinib [119], are also options beyond first-line management.

Lastly, there is growing evidence that patients with oligometastatic disease and an otherwise favorable prognosis, may benefit from intensive local management. For example, patients with p16-positive oropharyngeal cancer have improved survival even after recurrence compared to p16-neagtive patients with a median OS of 2.6 compared to 0.8 years [120]. Such patients may benefit from local therapy such as metastasectomy or radiation with SBRT. It is not clear if systemic therapy should be added after treatment of a metastatic lesion, in the absence of measurable disease; this is a potential area of investigation.

Conclusion

This chapter has discussed the clinical trials and studies that underpin treatment advances in head and neck cancer management. It is clear that with the discovery of the diverse etiologic factors for each tumor type, there is a need to tailor therapy based on anatomic site, biomarkers status including viral and immunologic markers, and based on patient shared-decision making. Improvements in radiation therapy technique such as IMRT, surgical innovation and reconstruction techniques, as well as the introduction of biologic and immune-modulatory therapy, have decreased adverse effects of head and neck cancer treatments. Yet there is still tremendous opportunity for improving the therapeutic ratio via translation of bench discovery to the clinic.

References

[1] Mehanna H, Wong WL, McConkey CC, Rahman JK, Robinson M, Hartley AG, Nutting C, Powell N, Al-Booz H, Robinson M, Junor E, Rizwanullah M, von Zeidler SV, Wieshmann H, Hulme C, Smith AF, Hall P, Dunn J, Group P-NTM. PET-CT surveillance versus neck dissection in advanced head and neck cancer. N Engl J Med 2016;374(15):1444–54.
[2] Lewis Jr. JS, Beadle B, Bishop JA, Chernock RD, Colasacco C, Lacchetti C, Moncur JT, Rocco JW, Schwartz MR, Seethala RR, Thomas NE, Westra WH, Faquin WC. Human papillomavirus testing in head and neck carcinomas: guideline from the college of American pathologists. Arch Pathol Lab Med 2018;142(5):559–97.
[3] NCCN. Clinical practice guidelines in oncology; 2017. Version 2.
[4] Gillison ML, D'Souza G, Westra W, Sugar E, Xiao W, Begum S, Viscidi R. Distinct risk factor profiles for human papillomavirus type 16-positive and human papillomavirus type 16-negative head and neck cancers. J Natl Cancer Inst 2008;100(6):407–20.
[5] Chaturvedi AK, Engels EA, Pfeiffer RM, Hernandez BY, Xiao W, Kim E, Jiang B, Goodman MT, Sibug-Saber M, Cozen W, Liu L, Lynch CF, Wentzensen N, Jordan RC, Altekruse S, Anderson WF, Rosenberg PS, Gillison ML. Human papillomavirus and rising oropharyngeal cancer incidence in the United States. J Clin Oncol 2011;29(32):4294–301.
[6] Ang KK, Harris J, Wheeler R, Weber R, Rosenthal DI, Nguyen-Tan PF, Westra WH, Chung CH, Jordan RC, Lu C, Kim H, Axelrod R, Silverman CC, Redmond KP, Gillison ML. Human papillomavirus and survival of patients with oropharyngeal cancer. N Engl J Med 2010;363(1):24–35.
[7] Marur S, D'Souza G, Westra WH, Forastiere AA. HPV-associated head and neck cancer: a virus-related cancer epidemic. Lancet Oncol 2010;11(8):781–9.
[8] Huang SH, O'Sullivan B, Xu W, Zhao H, Chen DD, Ringash J, Hope A, Razak A, Gilbert R, Irish J, Kim J, Dawson LA, Bayley A, Cho BC, Goldstein D, Gullane P, Yu E, Perez-Ordonez B, Weinreb I, Waldron J. Temporal nodal regression and regional control after primary radiation therapy for N2-N3 head-and-neck cancer stratified by HPV status. Int J Radiat Oncol Biol Phys 2013;87(5):1078–85.
[9] Nguyen-Tan PF, Zhang Q, Ang KK, Weber RS, Rosenthal DI, Soulieres D, Kim H, Silverman C, Raben A, Galloway TJ, Fortin A, Gore E, Westra WH, Chung CH, Jordan RC, Gillison ML, List M, Le QT. Randomized phase III trial to test accelerated versus standard fractionation in combination with concurrent cisplatin for head and neck carcinomas in the radiation therapy oncology group 0129 trial: long-term report of efficacy and toxicity. J Clin Oncol 2014;32(34):3858–66.

[10] Agrawal N, Frederick MJ, Pickering CR, Bettegowda C, Chang K, Li RJ, Fakhry C, Xie TX, Zhang J, Wang J, Zhang N, El-Naggar AK, Jasser SA, Weinstein JN, Trevino L, Drummond JA, Muzny DM, Wu Y, Wood LD, Hruban RH, Westra WH, Koch WM, Califano JA, Gibbs RA, Sidransky D, Vogelstein B, Velculescu VE, Papadopoulos N, Wheeler DA, Kinzler KW, Myers JN. Exome sequencing of head and neck squamous cell carcinoma reveals inactivating mutations in NOTCH1. Science 2011;333(6046):1154–7.

[11] Stransky N, Egloff AM, Tward AD, Kostic AD, Cibulskis K, Sivachenko A, Kryukov GV, Lawrence MS, Sougnez C, McKenna A, Shefler E, Ramos AH, Stojanov P, Carter SL, Voet D, Cortes ML, Auclair D, Berger MF, Saksena G, Guiducci C, Onofrio RC, Parkin M, Romkes M, Weissfeld JL, Seethala RR, Wang L, Rangel-Escareno C, Fernandez-Lopez JC, Hidalgo-Miranda A, Melendez-Zajgla J, Winckler W, Ardlie K, Gabriel SB, Meyerson M, Lander ES, Getz G, Golub TR, Garraway LA, Grandis JR. The mutational landscape of head and neck squamous cell carcinoma. Science 2011;333(6046):1157–60.

[12] Rieckmann T, Tribius S, Grob TJ, Meyer F, Busch CJ, Petersen C, Dikomey E, Kriegs M. HNSCC cell lines positive for HPV and p16 possess higher cellular radiosensitivity due to an impaired DSB repair capacity. Radiother Oncol 2013;107(2):242–6.

[13] Kimple RJ, Smith MA, Blitzer GC, Torres AD, Martin JA, Yang RZ, Peet CR, Lorenz LD, Nickel KP, Klingelhutz AJ, Lambert PF, Harari PM. Enhanced radiation sensitivity in HPV-positive head and neck cancer. Cancer Res 2013;73(15):4791–800.

[14] Amin MB, Edge SB, Greene FL, Byrd DR, Brookland RK, Washington MK, Gershenwald JE. AJCC Cancer staging manual. 8th ed. Springer International Publishing: American Joint Commission on Cancer; 2017.

[15] O'Sullivan B, Huang SH, Su J, Garden AS, Sturgis EM, Dahlstrom K, Lee N, Riaz N, Pei X, Koyfman SA, Adelstein D, Burkey BB, Friborg J, Kristensen CA, Gothelf AB, Hoebers F, Kremer B, Speel EJ, Bowles DW, Raben D, Karam SD, Yu E, Xu W. Development and validation of a staging system for HPV-related oropharyngeal cancer by the international collaboration on oropharyngeal cancer network for staging (ICON-S): a multicentre cohort study. Lancet Oncol 2016;17(4):440–51.

[16] de Almeida JR, Byrd JK, Wu R, Stucken CL, Duvvuri U, Goldstein DP, Miles BA, Teng MS, Gupta V, Genden EM. A systematic review of transoral robotic surgery and radiotherapy for early oropharynx cancer: a systematic review. Laryngoscope 2014;124(9):2096–102.

[17] Morisod B, Simon C. Meta-analysis on survival of patients treated with transoral surgery versus radiotherapy for early-stage squamous cell carcinoma of the oropharynx. Head Neck 2016;38(Suppl 1):E2143–50.

[18] Ang KK, Trotti A, Brown BW, Garden AS, Foote RL, Morrison WH, Geara FB, Klotch DW, Goepfert H, Peters LJ. Randomized trial addressing risk features and time factors of surgery plus radiotherapy in advanced head-and-neck cancer. Int J Radiat Oncol Biol Phys 2001;51(3):571–8.

[19] Bernier J, Domenge C, Ozsahin M, Matuszewska K, Lefebvre JL, Greiner RH, Giralt J, Maingon P, Rolland F, Bolla M, Cognetti F, Bourhis J, Kirkpatrick A, van Glabbeke M, European Organization for R, Treatment of Cancer T. Postoperative irradiation with or without concomitant chemotherapy for locally advanced head and neck cancer. N Engl J Med 2004;350(19):1945–52.

[20] Bernier J, Cooper JS, Pajak TF, van Glabbeke M, Bourhis J, Forastiere A, Ozsahin EM, Jacobs JR, Jassem J, Ang KK, Lefebvre JL. Defining risk levels in locally advanced head and neck cancers: a comparative analysis of concurrent postoperative radiation plus chemotherapy trials of the EORTC (#22931) and RTOG (# 9501). Head Neck 2005;27(10):843–50.

[21] Cooper JS, Pajak TF, Forastiere AA, Jacobs J, Campbell BH, Saxman SB, Kish JA, Kim HE, Cmelak AJ, Rotman M, Machtay M, Ensley JF, Chao KS, Schultz CJ, Lee N, Fu KK, Radiation Therapy Oncology Group I. Postoperative concurrent radiotherapy and chemotherapy for high-risk squamous-cell carcinoma of the head and neck. N Engl J Med 2004;350(19):1937–44.

[22] Cooper JS, Zhang Q, Pajak TF, Forastiere AA, Jacobs J, Saxman SB, Kish JA, Kim HE, Cmelak AJ, Rotman M, Lustig R, Ensley JF, Thorstad W, Schultz CJ, Yom SS, Ang KK. Long-term follow-up of the RTOG 9501/intergroup phase III trial: postoperative concurrent radiation therapy and chemotherapy in high-risk squamous cell carcinoma of the head and neck. Int J Radiat Oncol Biol Phys 2012;84(5):1198–205.

[23] Pignon JP, Bourhis J, Domenge C, Designe L. Chemotherapy added to locoregional treatment for head and neck squamous-cell carcinoma: three meta-analyses of updated individual data. MACH-NC collaborative group. Meta-analysis of chemotherapy on head and neck cancer. Lancet 2000;355(9208):949–55.

[24] Pignon JP, le Maitre A, Maillard E, Bourhis J, Group M-NC. Meta-analysis of chemotherapy in head and neck cancer (MACH-NC): an update on 93 randomised trials and 17,346 patients. Radiother Oncol 2009;92(1):4–14.

[25] McHam SA, Adelstein DJ, Rybicki LA, Lavertu P, Esclamado RM, Wood BG, Strome M, Carroll MA. Who merits a neck dissection after definitive chemoradiotherapy for N2-N3 squamous cell head and neck cancer? Head Neck 2003;25(10):791–8.
[26] Calais G, Alfonsi M, Bardet E, Sire C, Germain T, Bergerot P, Rhein B, Tortochaux J, Oudinot P, Bertrand P. Randomized trial of radiation therapy versus concomitant chemotherapy and radiation therapy for advanced-stage oropharynx carcinoma. J Natl Cancer Inst 1999;91(24):2081–6.
[27] Bonner JA, Harari PM, Giralt J, Azarnia N, Shin DM, Cohen RB, Jones CU, Sur R, Raben D, Jassem J, Ove R, Kies MS, Baselga J, Youssoufian H, Amellal N, Rowinsky EK, Ang KK. Radiotherapy plus cetuximab for squamous-cell carcinoma of the head and neck. N Engl J Med 2006;354(6):567–78.
[28] Bonner JA, Harari PM, Giralt J, Cohen RB, Jones CU, Sur RK, Raben D, Baselga J, Spencer SA, Zhu J, Youssoufian H, Rowinsky EK, Ang KK. Radiotherapy plus cetuximab for locoregionally advanced head and neck cancer: 5-year survival data from a phase 3 randomised trial, and relation between cetuximab-induced rash and survival. Lancet Oncol 2010;11(1):21–8.
[29] Ang KK, Zhang Q, Rosenthal DI, Nguyen-Tan PF, Sherman EJ, Weber RS, Galvin JM, Bonner JA, Harris J, El-Naggar AK, Gillison ML, Jordan RC, Konski AA, Thorstad WL, Trotti A, Beitler JJ, Garden AS, Spanos WJ, Yom SS, Axelrod RS. Randomized phase III trial of concurrent accelerated radiation plus cisplatin with or without cetuximab for stage III to IV head and neck carcinoma: RTOG 0522. J Clin Oncol 2014;32(27):2940–50.
[30] Gillison ML, Trotti AM, Harris J, Eisbruch A, Harari PM, Adelstein DJ, Sturgis EM, Burtness B, Ridge JA, Ringash J, Galvin J, Yao M, Koyfman SA, Blakaj DM, Razaq MA, Colevas AD, Beitler JJ, Jones CU, Dunlap NE, Seaward SA, Spencer S, Galloway TJ, Phan J, Dignam JJ, Le QT. Radiotherapy plus cetuximab or cisplatin in human papillomavirus-positive oropharyngeal cancer (NRG oncology RTOG 1016): a randomised, multicentre, non-inferiority trial. Lancet 2018;.
[31] Mehanna H, Robinson M, Hartley A, Kong A, Foran B, Fulton-Lieuw T, Dalby M, Mistry P, Sen M, O'Toole L, Al Booz H, Dyker K, Moleron R, Whitaker S, Brennan S, Cook A, Griffin M, Aynsley E, Rolles M, De Winton E, Chan A, Srinivasan D, Nixon I, Grumett J, Leemans CR, Buter J, Henderson J, Harrington K, McConkey C, Gray A, Dunn J, De EHPVTG. Radiotherapy plus cisplatin or cetuximab in low-risk human papillomavirus-positive oropharyngeal cancer (De-ESCALaTE HPV): an open-label randomised controlled phase 3 trial. Lancet 2019;393(10166):51–60.
[32] Fu KK, Pajak TF, Trotti A, Jones CU, Spencer SA, Phillips TL, Garden AS, Ridge JA, Cooper JS, Ang KK. A radiation therapy oncology group (RTOG) phase III randomized study to compare hyperfractionation and two variants of accelerated fractionation to standard fractionation radiotherapy for head and neck squamous cell carcinomas: first report of RTOG 9003. Int J Radiat Oncol Biol Phys 2000;48(1):7–16.
[33] Beitler JJ, Zhang Q, Fu KK, Trotti A, Spencer SA, Jones CU, Garden AS, Shenouda G, Harris J, Ang KK. Final results of local-regional control and late toxicity of RTOG 9003: a randomized trial of altered fractionation radiation for locally advanced head and neck cancer. Int J Radiat Oncol Biol Phys 2014;89(1):13–20.
[34] Overgaard J, Hansen HS, Specht L, Overgaard M, Grau C, Andersen E, Bentzen J, Bastholt L, Hansen O, Johansen J, Andersen L, Evensen JF. Five compared with six fractions per week of conventional radiotherapy of squamous-cell carcinoma of head and neck: DAHANCA 6 and 7 randomised controlled trial. Lancet 2003;362(9388):933–40.
[35] Bourhis J, Overgaard J, Audry H, Ang KK, Saunders M, Bernier J, Horiot JC, Le Maitre A, Pajak TF, Poulsen MG, O'Sullivan B, Dobrowsky W, Hliniak A, Skladowski K, Hay JH, Pinto LH, Fallai C, Fu KK, Sylvester R, Pignon JP. Hyperfractionated or accelerated radiotherapy in head and neck cancer: a meta-analysis. Lancet 2006;368(9538):843–54.
[36] Lacas B, Bourhis J, Overgaard J, Zhang Q, Gregoire V, Nankivell M, Zackrisson B, Szutkowski Z, Suwinski R, Poulsen M, O'Sullivan B, Corvo R, Laskar SG, Fallai C, Yamazaki H, Dobrowsky W, Cho KH, Beadle B, Langendijk JA, Viegas CMP, Hay J, Lotayef M, Parmar MKB, Auperin A, van Herpen C, Maingon P, Trotti AM, Grau C, Pignon JP, Blanchard P, Group MC. Role of radiotherapy fractionation in head and neck cancers (MARCH): an updated meta-analysis. Lancet Oncol 2017;18(9):1221–37.
[37] Bourhis J, Sire C, Graff P, Gregoire V, Maingon P, Calais G, Gery B, Martin L, Alfonsi M, Desprez P, Pignon T, Bardet E, Rives M, Geoffrois L, Daly-Schveitzer N, Sen S, Tuchais C, Dupuis O, Guerif S, Lapeyre M, Favrel V, Hamoir M, Lusinchi A, Temam S, Pinna A, Tao YG, Blanchard P, Auperin A. Concomitant chemoradiotherapy versus acceleration of radiotherapy with or without concomitant chemotherapy in locally

advanced head and neck carcinoma (GORTEC 99–02): an open-label phase 3 randomised trial. Lancet Oncol 2012;13(2):145–53.

[38] Vermorken JB, Remenar E, van Herpen C, Gorlia T, Mesia R, Degardin M, Stewart JS, Jelic S, Betka J, Preiss JH, van den Weyngaert D, Awada A, Cupissol D, Kienzer HR, Rey A, Desaunois I, Bernier J, Lefebvre JL. Cisplatin, fluorouracil, and docetaxel in unresectable head and neck cancer. N Engl J Med 2007;357(17):1695–704.

[39] Posner MR, Hershock DM, Blajman CR, Mickiewicz E, Winquist E, Gorbounova V, Tjulandin S, Shin DM, Cullen K, Ervin TJ, Murphy BA, Raez LE, Cohen RB, Spaulding M, Tishler RB, Roth B, Viroglio Rdel C, Venkatesan V, Romanov I, Agarwala S, Harter KW, Dugan M, Cmelak A, Markoe AM, Read PW, Steinbrenner L, Colevas AD, Norris Jr. CM, Haddad RI. Cisplatin and fluorouracil alone or with docetaxel in head and neck cancer. N Engl J Med 2007;357(17):1705–15.

[40] Lorch JH, Goloubeva O, Haddad RI, Cullen K, Sarlis N, Tishler R, Tan M, Fasciano J, Sammartino DE, Posner MR, Group TAXS. Induction chemotherapy with cisplatin and fluorouracil alone or in combination with docetaxel in locally advanced squamous-cell cancer of the head and neck: long-term results of the TAX 324 randomised phase 3 trial. Lancet Oncol 2011;12(2):153–9.

[41] Ghi MG, Paccagnella A, Ferrari D, Foa P, Alterio D, Codeca C, Nole F, Verri E, Orecchia R, Morelli F, Parisi S, Mastromauro C, Mione CA, Rossetto C, Polsinelli M, Koussis H, Loreggian L, Bonetti A, Campostrini F, Azzarello G, D'Ambrosio C, Bertoni F, Casanova C, Emiliani E, Guaraldi M, Bunkheila F, Bidoli P, Niespolo RM, Gava A, Massa E, Frattegiani A, Valduga F, Pieri G, Cipani T, Da Corte D, Chiappa F, Rulli E, Group GIS. Induction TPF followed by concomitant treatment versus concomitant treatment alone in locally advanced head and neck cancer. A phase II-III trial. Ann Oncol 2017;28(9):2206–12.

[42] Haddad R, O'Neill A, Rabinowits G, Tishler R, Khuri F, Adkins D, Clark J, Sarlis N, Lorch J, Beitler JJ, Limaye S, Riley S, Posner M. Induction chemotherapy followed by concurrent chemoradiotherapy (sequential chemoradiotherapy) versus concurrent chemoradiotherapy alone in locally advanced head and neck cancer (PARADIGM): a randomised phase 3 trial. Lancet Oncol 2013;14(3):257–64.

[43] Cohen EE, Karrison TG, Kocherginsky M, Mueller J, Egan R, Huang CH, Brockstein BE, Agulnik MB, Mittal BB, Yunus F, Samant S, Raez LE, Mehra R, Kumar P, Ondrey F, Marchand P, Braegas B, Seiwert TY, Villaflor VM, Haraf DJ, Vokes EE. Phase III randomized trial of induction chemotherapy in patients with N2 or N3 locally advanced head and neck cancer. J Clin Oncol 2014;32(25):2735–43.

[44] Hitt R, Grau JJ, Lopez-Pousa A, Berrocal A, Garcia-Giron C, Irigoyen A, Sastre J, Martinez-Trufero J, Brandariz Castelo JA, Verger E, Cruz-Hernandez JJ, Spanish H, Neck Cancer Cooperative G. A randomized phase III trial comparing induction chemotherapy followed by chemoradiotherapy versus chemoradiotherapy alone as treatment of unresectable head and neck cancer. Annal Oncol 2014;25(1):216–25.

[45] Marur S, Li S, Cmelak AJ, Gillison ML, Zhao WJ, Ferris RL, Westra WH, Gilbert J, Bauman JE, Wagner LI, Trevarthen DR, Balkrishna J, Murphy BA, Agrawal N, Colevas AD, Chung CH, Burtness B. E1308: phase II trial of induction chemotherapy followed by reduced-dose radiation and weekly cetuximab in patients with HPV-associated resectable squamous cell carcinoma of the oropharynx—ECOG-ACRIN Cancer Research Group. J Clin Oncol 2017;35(5):490–7.

[46] Ferlay J, Soerjomataram I, Dikshit R, Eser S, Mathers C, Rebelo M, Parkin DM, Forman D, Bray F. Cancer incidence and mortality worldwide: sources, methods and major patterns in GLOBOCAN 2012. Int J Cancer 2015;136(5):E359–86.

[47] Chua MLK, Wee JTS, Hui EP, Chan ATC. Nasopharyngeal carcinoma. Lancet 2016;387(10022):1012–24.

[48] Pathmanathan R, Prasad U, Sadler R, Flynn K, Raab-Traub N. Clonal proliferations of cells infected with Epstein-Barr virus in preinvasive lesions related to nasopharyngeal carcinoma. N Engl J Med 1995;333(11):693–8.

[49] Chien YC, Chen JY, Liu MY, Yang HI, Hsu MM, Chen CJ, Yang CS. Serologic markers of Epstein-Barr virus infection and nasopharyngeal carcinoma in Taiwanese men. N Engl J Med 2001;345(26):1877–82.

[50] Chan KCA, Woo JKS, King A, Zee BCY, Lam WKJ, Chan SL, Chu SWI, Mak C, Tse IOL, Leung SYM, Chan G, Hui EP, Ma BBY, Chiu RWK, Leung SF, van Hasselt AC, Chan ATC, Lo YMD. Analysis of plasma Epstein-Barr virus DNA to screen for nasopharyngeal Cancer. N Engl J Med 2017;377(6):513–22.

[51] Lin JC, Wang WY, Chen KY, Wei YH, Liang WM, Jan JS, Jiang RS. Quantification of plasma Epstein-Barr virus DNA in patients with advanced nasopharyngeal carcinoma. N Engl J Med 2004;350(24):2461–70.

[52] Stelow EB, Wenig BM. Update from the 4th edition of the World Health Organization classification of head and neck tumours: nasopharynx. Head Neck Pathol 2017;11(1):16–22.

[53] Tang L, Mao Y, Liu L, Liang S, Chen Y, Sun Y, Liao X, Lin A, Liu M, Li L, Ma J. The volume to be irradiated during selective neck irradiation in nasopharyngeal carcinoma: analysis of the spread patterns in lymph nodes by magnetic resonance imaging. Cancer 2009;115(3):680–8.

[54] Langendijk JA, Leemans CR, Buter J, Berkhof J, Slotman BJ. The additional value of chemotherapy to radiotherapy in locally advanced nasopharyngeal carcinoma: a meta-analysis of the published literature. J Clin Oncol 2004;22(22):4604–12.

[55] Ribassin-Majed L, Marguet S, Lee AWM, Ng WT, Ma J, Chan ATC, Huang PY, Zhu G, Chua DTT, Chen Y, Mai HQ, Kwong DLW, Cheah SL, Moon J, Tung Y, Chi KH, Fountzilas G, Bourhis J, Pignon JP, Blanchard P. What is the best treatment of locally advanced nasopharyngeal carcinoma? An individual patient data network meta-analysis. J Clin Oncol 2017;35(5):498–505.

[56] Al-Sarraf M, LeBlanc M, Giri PG, Fu KK, Cooper J, Vuong T, Forastiere AA, Adams G, Sakr WA, Schuller DE, Ensley JF. Chemoradiotherapy versus radiotherapy in patients with advanced nasopharyngeal cancer: phase III randomized intergroup study 0099. J Clin Oncol Off J Am Soc Clin Oncol 1998;16(4):1310–7.

[57] Wee J, Tan EH, Tai BC, Wong HB, Leong SS, Tan T, Chua ET, Yang E, Lee KM, Fong KW, Tan HS, Lee KS, Loong S, Sethi V, Chua EJ, Machin D. Randomized trial of radiotherapy versus concurrent chemoradiotherapy followed by adjuvant chemotherapy in patients with American joint committee on cancer/international union against cancer stage III and IV nasopharyngeal cancer of the endemic variety. J Clin Oncol 2005;23(27):6730–8.

[58] Blanchard P, Lee A, Marguet S, Leclercq J, Ng WT, Ma J, Chan AT, Huang PY, Benhamou E, Zhu G, Chua DT, Chen Y, Mai HQ, Kwong DL, Cheah SL, Moon J, Tung Y, Chi KH, Fountzilas G, Zhang L, Hui EP, Lu TX, Bourhis J, Pignon JP, Group M-NC. Chemotherapy and radiotherapy in nasopharyngeal carcinoma: an update of the MAC-NPC meta-analysis. Lancet Oncol 2015;16(6):645–55.

[59] Chen QY, Wen YF, Guo L, Liu H, Huang PY, Mo HY, Li NW, Xiang YQ, Luo DH, Qiu F, Sun R, Deng MQ, Chen MY, Hua YJ, Guo X, Cao KJ, Hong MH, Qian CN, Mai HQ. Concurrent chemoradiotherapy vs radiotherapy alone in stage II nasopharyngeal carcinoma: phase III randomized trial. J Natl Cancer Inst 2011;103(23): 1761–70.

[60] Lee JY, Sun JM, Oh DR, Lim SH, Goo J, Lee SH, Kim SB, Park KU, Kim HK, Hong DS, Kim JS, Kim SG, Yi SY, Yun HJ, Hyun MS, Kim HJ, Jung SH, Park K, Ahn YC, Ahn MJ. Comparison of weekly versus triweekly cisplatin delivered concurrently with radiation therapy in patients with locally advanced nasopharyngeal cancer: a multicenter randomized phase II trial (KCSG-HN10-02). Radiother Oncol 2016;118(2):244–50.

[61] Kam MK, Leung SF, Zee B, Chau RM, Suen JJ, Mo F, Lai M, Ho R, Cheung KY, Yu BK, Chiu SK, Choi PH, Teo PM, Kwan WH, Chan AT. Prospective randomized study of intensity-modulated radiotherapy on salivary gland function in early-stage nasopharyngeal carcinoma patients. J Clin Oncol 2007;25(31):4873–9.

[62] Pow EH, Kwong DL, McMillan AS, Wong MC, Sham JS, Leung LH, Leung WK. Xerostomia and quality of life after intensity-modulated radiotherapy vs. conventional radiotherapy for early-stage nasopharyngeal carcinoma: initial report on a randomized controlled clinical trial. Int J Radiat Oncol Biol Phys 2006;66(4):981–91.

[63] Peng G, Wang T, Yang KY, Zhang S, Zhang T, Li Q, Han J, Wu G. A prospective, randomized study comparing outcomes and toxicities of intensity-modulated radiotherapy vs. conventional two-dimensional radiotherapy for the treatment of nasopharyngeal carcinoma. Radiother Oncol 2012;104(3):286–93.

[64] National Comprehensive Cancer Network. Head and neck cancer, Version 2.2018: Available from: https://www.nccn.org/professionals/physician_gls/pdf/head-and-neck.pdf; 2018. (Accessed 14 December 2018).

[65] Sun Y, Li WF, Chen NY, Zhang N, Hu GQ, Xie FY, Sun Y, Chen XZ, Li JG, Zhu XD, Hu CS, Xu XY, Chen YY, Hu WH, Guo L, Mo HY, Chen L, Mao YP, Sun R, Ai P, Liang SB, Long GX, Zheng BM, Feng XL, Gong XC, Li L, Shen CY, Xu JY, Guo Y, Chen YM, Zhang F, Lin L, Tang LL, Liu MZ, Ma J. Induction chemotherapy plus concurrent chemoradiotherapy versus concurrent chemoradiotherapy alone in locoregionally advanced nasopharyngeal carcinoma: a phase 3, multicentre, randomised controlled trial. Lancet Oncol 2016;17(11):1509–20.

[66] Frikha M, Auperin A, Tao Y, Elloumi F, Toumi N, Blanchard P, Lang P, Sun S, Racadot S, Thariat J, Alfonsi M, Tuchais C, Cornely A, Moussa A, Guigay J, Daoud J, Bourhis J. Gortec. A randomized trial of induction docetaxel-cisplatin-5FU followed by concomitant cisplatin-RT versus concomitant cisplatin-RT in nasopharyngeal carcinoma (GORTEC 2006-02). Annal Oncol 2018;29(3):731–6.

[67] IARC. Betel-quid and Areca-nut chewing and some Areca-nut-dervied nitrosamines. Lyon, France: International Agency for resarch on Cancer; 2004.

[68] Chung CH, Zhang Q, Kong CS, Harris J, Fertig EJ, Harari PM, Wang D, Redmond KP, Shenouda G, Trotti A, Raben D, Gillison ML, Jordan RC, Le QT. p16 protein expression and human papillomavirus status as

prognostic biomarkers of nonoropharyngeal head and neck squamous cell carcinoma. J Clin Oncol 2014;32(35): 3930–8.

[69] Lingen MW, Xiao W, Schmitt A, Jiang B, Pickard R, Kreinbrink P, Perez-Ordonez B, Jordan RC, Gillison ML. Low etiologic fraction for high-risk human papillomavirus in oral cavity squamous cell carcinomas. Oral Oncol 2013;49(1):1–8.

[70] D'Cruz AK, Vaish R, Kapre N, Dandekar M, Gupta S, Hawaldar R, Agarwal JP, Pantvaidya G, Chaukar D, Deshmukh A, Kane S, Arya S, Ghosh-Laskar S, Chaturvedi P, Pai P, Nair S, Nair D, Badwe R, Head and Neck Disease Management Group. Elective versus therapeutic neck dissection in node-negative oral cancer. N Engl J Med 2015;373(6):521–9.

[71] Hinerman RW, Mendenhall WM, Morris CG, Amdur RJ, Werning JW, Villaret DB. Postoperative irradiation for squamous cell carcinoma of the oral cavity: 35-year experience. Head Neck 2004;26(11):984–94.

[72] Ganly I, Goldstein D, Carlson DL, Patel SG, O'Sullivan B, Lee N, Gullane P, Shah JP. Long-term regional control and survival in patients with "low-risk," early stage oral tongue cancer managed by partial glossectomy and neck dissection without postoperative radiation: the importance of tumor thickness. Cancer 2013;119(6): 1168–76.

[73] Huang SH, Hwang D, Lockwood G, Goldstein DP, O'Sullivan B. Predictive value of tumor thickness for cervical lymph-node involvement in squamous cell carcinoma of the oral cavity: a meta-analysis of reported studies. Cancer 2009;115(7):1489–97.

[74] Lavaf A, Genden EM, Cesaretti JA, Packer S, Kao J. Adjuvant radiotherapy improves overall survival for patients with lymph node-positive head and neck squamous cell carcinoma. Cancer 2008;112(3):535–43.

[75] Kao J, Lavaf A, Teng MS, Huang D, Genden EM. Adjuvant radiotherapy and survival for patients with node-positive head and neck cancer: an analysis by primary site and nodal stage. Int J Radiat Oncol Biol Phys 2008;71 (2):362–70.

[76] Noronha V, Joshi A, Patil VM, Agarwal J, Ghosh-Laskar S, Budrukkar A, Murthy V, Gupta T, D'Cruz AK, Banavali S, Pai PS, Chaturvedi P, Chaukar D, Pande N, Chandrasekharan A, Talreja V, Vallathol DH, Mathrudev V, Manjrekar A, Maske K, Bhelekar AS, Nawale K, Kannan S, Gota V, Bhattacharjee A, Kane S, Juvekar SL, Prabhash K. Once-a-week versus once-every-3-weeks cisplatin chemoradiation for locally advanced head and neck cancer: a phase III randomized noninferiority trial. J Clin Oncol 2018;36(11):1064–72.

[77] Bauml JM, Vinnakota R, Anna Park YH, Bates SE, Fojo T, Aggarwal C, Limaye S, Damjanov N, Di Stefano J, Ciunci C, Genden EM, Wisnivesky JP, Ferrandino R, Mamtani R, Langer CJ, Cohen RB, Sigel K. Cisplatin every 3 weeks versus weekly with definitive concurrent radiotherapy for squamous cell carcinoma of the head and neck. J Natl Cancer Inst 2018;111(5):490–7.

[78] Harari PM, Harris J, Kies MS, Myers JN, Jordan RC, Gillison ML, Foote RL, Machtay M, Rotman M, Khuntia D, Straube W, Zhang Q, Ang K. Postoperative chemoradiotherapy and cetuximab for high-risk squamous cell carcinoma of the head and neck: radiation therapy oncology group RTOG-0234. J Clin Oncol 2014;32(23): 2486–95.

[79] Laskar SG, Chaukar D, Deshpande M, Chatterjee A, Hawaldar R, Chakraborty S, Sharma S. Phase III randomized trial of surgery followed by conventional radiotherapy (5 fr/Wk) (Arm A) vs concurrent chemoradiotherapy (Arm B) vs accelerated radiotherapy (6fr/Wk) (Arm C) in locally advanced, stage III and IV, resectable, squamous cell carcinoma of oral cavity-oral cavity adjuvant therapy (OCAT): final results (NCT00193843). J Clin Oncol 2016;34:6004.

[80] Grabenbauer GG, Rodel C, Brunner T, Schulze-Mosgau S, Strnad V, Muller RG, Iro H, Sauer R. Interstitial brachytherapy with Ir-192 low-dose-rate in the treatment of primary and recurrent cancer of the oral cavity and oropharynx. Review of 318 patients treated between 1985 and 1997. Strahlenther Onkol 2001;177(7):338–44.

[81] Melzner WJ, Lotter M, Sauer R, Strnad V. Quality of interstitial PDR-brachytherapy-implants of head-and-neck-cancers: predictive factors for local control and late toxicity? Radiother Oncol 2007;82(2):167–73.

[82] Salama JK, Haddad RI, Kies MS, Busse PM, Dong L, Brizel DM, Eisbruch A, Tishler RB, Trotti AM, Garden AS. Clinical practice guidance for radiotherapy planning after induction chemotherapy in locoregionally advanced head-and-neck cancer. Int J Radiat Oncol Biol Phys 2009;75(3):725–33.

[83] Forastiere AA, Ismaila N, Lewin JS, Nathan CA, Adelstein DJ, Eisbruch A, Fass G, Fisher SG, Laurie SA, Le QT, O'Malley B, Mendenhall WM, Patel S, Pfister DG, Provenzano AF, Weber R, Weinstein GS, Wolf GT. Use of larynx-preservation strategies in the treatment of laryngeal cancer: American Society of Clinical Oncology clinical practice guideline update. J Clin Oncol 2018;36(11):1143–69.

[84] Yamazaki H, Nishiyama K, Tanaka E, Koizumi M, Chatani M. Radiotherapy for early glottic carcinoma (T1N0M0): results of prospective randomized study of radiation fraction size and overall treatment time. Int J Radiat Oncol Biol Phys 2006;64(1):77–82.
[85] Wolf G. Induction chemotherapy plus radiation compared with surgery plus radiation in patients with advanced laryngeal cancer. The Department of Veterans Affairs Laryngeal Cancer Study Group. N Engl J Med 1991;324(24):1685–90.
[86] Forastiere AA, Goepfert H, Maor M, Pajak TF, Weber R, Morrison W, Glisson B, Trotti A, Ridge JA, Chao C, Peters G, Lee DJ, Leaf A, Ensley J, Cooper J. Concurrent chemotherapy and radiotherapy for organ preservation in advanced laryngeal cancer. N Engl J Med 2003;349(22):2091–8.
[87] Forastiere AA, Zhang Q, Weber RS, Maor MH, Goepfert H, Pajak TF, Morrison W, Glisson B, Trotti A, Ridge JA, Thorstad W, Wagner H, Ensley JF, Cooper JS. Long-term results of RTOG 91-11: a comparison of three nonsurgical treatment strategies to preserve the larynx in patients with locally advanced larynx cancer. J Clin Oncol 2013;31(7):845–52.
[88] Lefebvre JL, Pointreau Y, Rolland F, Alfonsi M, Baudoux A, Sire C, de Raucourt D, Malard O, Degardin M, Tuchais C, Blot E, Rives M, Reyt E, Tourani JM, Geoffrois L, Peyrade F, Guichard F, Chevalier D, Babin E, Lang P, Janot F, Calais G, Garaud P, Bardet E. Induction chemotherapy followed by either chemoradiotherapy or bioradiotherapy for larynx preservation: the TREMPLIN randomized phase II study. J Clin Oncol 2013;31(7):853–9.
[89] Lefebvre JL, Chevalier D, Luboinski B, Kirkpatrick A, Collette L, Sahmoud T. Larynx preservation in pyriform sinus cancer: preliminary results of a European Organization for Research and Treatment of Cancer phase III trial. EORTC Head and Neck Cancer Cooperative Group. J Natl Cancer Inst 1996;88(13):890–9.
[90] Lefebvre JL, Andry G, Chevalier D, Luboinski B, Collette L, Traissac L, de Raucourt D, Langendijk JA, Head E, Neck Cancer G. Laryngeal preservation with induction chemotherapy for hypopharyngeal squamous cell carcinoma: 10-year results of EORTC trial 24891. Ann Oncol 2012;23(10):2708–14.
[91] Pointreau Y, Garaud P, Chapet S, Sire C, Tuchais C, Tortochaux J, Faivre S, Guerrif S, Alfonsi M, Calais G. Randomized trial of induction chemotherapy with cisplatin and 5-fluorouracil with or without docetaxel for larynx preservation. J Natl Cancer Inst 2009;101(7):498–506.
[92] Seethala RR, Stenman G. Update from the 4th edition of the World Health Organization classification of head and neck tumours: tumors of the salivary gland. Head Neck Pathol 2017;11(1):55–67.
[93] El-Naggar AK, Chan JKC, Grandis JR, Takata T, Slootweg PJ, World Health Organizagion (WHO). Classification of head and neck tumours. Lyon: International Agency for Research on Cancer (IARC) Press; 2017.
[94] Jones AV, Craig GT, Speight PM, Franklin CD. The range and demographics of salivary gland tumours diagnosed in a UK population. Oral Oncol 2008;44(4):407–17.
[95] Cros J, Sbidian E, Hans S, Roussel H, Scotte F, Tartour E, Brasnu D, Laurent-Puig P, Bruneval P, Blons H, Badoual C. Expression and mutational status of treatment-relevant targets and key oncogenes in 123 malignant salivary gland tumours. Ann Oncol 2013;24(10):2624–9.
[96] Jaehne M, Roeser K, Jaekel T, Schepers JD, Albert N, Loning T. Clinical and immunohistologic typing of salivary duct carcinoma: a report of 50 cases. Cancer 2005;103(12):2526–33.
[97] Haddad R, Colevas AD, Krane JF, Cooper D, Glisson B, Amrein PC, Weeks L, Costello R, Posner M. Herceptin in patients with advanced or metastatic salivary gland carcinomas. A phase II study. Oral Oncol 2003;39(7):724–7.
[98] Locati LD, Bossi P, Perrone F, Potepan P, Crippa F, Mariani L, Casieri P, Orsenigo M, Losa M, Bergamini C, Liberatoscioli C, Quattrone P, Calderone RG, Rinaldi G, Pilotti S, Licitra L. Cetuximab in recurrent and/or metastatic salivary gland carcinomas: a phase II study. Oral Oncol 2009;45(7):574–8.
[99] da Cruz Perez DE, de Abreu AF, Nobuko Nishimoto I, de Almeida OP, Kowalski LP. Prognostic factors in head and neck adenoid cystic carcinoma. Oral Oncol 2006;42(2):139–46.
[100] Laurie SA, Ho AL, Fury MG, Sherman E, Pfister DG. Systemic therapy in the management of metastatic or locally recurrent adenoid cystic carcinoma of the salivary glands: a systematic review. Lancet Oncol 2011;12(8):815–24.
[101] Pfeffer MR, Talmi Y, Catane R, Symon Z, Yosepovitch A, Levitt M. A phase II study of Imatinib for advanced adenoid cystic carcinoma of head and neck salivary glands. Oral Oncol 2007;43(1):33–6.
[102] Wong SJ, Karrison T, Hayes DN, Kies MS, Cullen KJ, Tanvetyanon T, Argiris A, Takebe N, Lim D, Saba NF, Worden FP, Gilbert J, Lenz HJ, Razak AR, Roberts JD, Vokes EE, Cohen EE. Phase II trial of dasatinib for

recurrent or metastatic c-KIT expressing adenoid cystic carcinoma and for nonadenoid cystic malignant salivary tumors. Annal Oncol 2016;27(2):318–23.

[103] Persson M, Andren Y, Mark J, Horlings HM, Persson F, Stenman G. Recurrent fusion of MYB and NFIB transcription factor genes in carcinomas of the breast and head and neck. Proc Natl Acad Sci U S A 2009;106(44):18740–4.

[104] Andersson MK, Afshari MK, Andren Y, Wick MJ, Stenman G. Targeting the oncogenic transcriptional regulator MYB in adenoid cystic carcinoma by inhibition of IGF1R/AKT signaling. J Natl Cancer Inst 2017;109(9).

[105] Boon E, van Boxtel W, Buter J, Baatenburg de Jong RJ, van Es RJJ, Bel M, Fiets E, Oosting SF, Slingerland M, Hoeben A, MET T, Jonker MA, Flucke UE, Nationwide N, Registry of H, Cytopathology G, van der Graaf WTA, van Herpen CML. Androgen deprivation therapy for androgen receptor-positive advanced salivary duct carcinoma: a nationwide case series of 35 patients in the Netherlands. Head Neck 2018;40(3):605–13.

[106] Chen AM, Granchi PJ, Garcia J, Bucci MK, Fu KK, Eisele DW. Local-regional recurrence after surgery without postoperative irradiation for carcinomas of the major salivary glands: implications for adjuvant therapy. Int J Radiat Oncol Biol Phys 2007;67(4):982–7.

[107] Terhaard CH, Lubsen H, Rasch CR, Levendag PC, Kaanders HH, Tjho-Heslinga RE, van Den Ende PL, Burlage F, Dutch H, Dutch Head and Neck Oncology Cooperative Group. The role of radiotherapy in the treatment of malignant salivary gland tumors. Int J Radiat Oncol Biol Phys 2005;61(1):103–11.

[108] Mendenhall WM, Morris CG, Amdur RJ, Werning JW, Hinerman RW, Villaret DB. Radiotherapy alone or combined with surgery for adenoid cystic carcinoma of the head and neck. Head Neck 2004;26(2):154–62.

[109] Laramore GE, Krall JM, Griffin TW, Duncan W, Richter MP, Saroja KR, Maor MH, Davis LW. Neutron versus photon irradiation for unresectable salivary gland tumors: final report of an RTOG-MRC randomized clinical trial. Radiation therapy oncology group. Medical Research Council. Int J Radiat Oncol Biol Phys 1993;27(2): 235–40.

[110] Jensen AD, Nikoghosyan AV, Lossner K, Haberer T, Jakel O, Munter MW, Debus J. COSMIC: a regimen of intensity modulated radiation therapy plus dose-escalated, raster-scanned carbon ion boost for malignant salivary gland tumors: results of the prospective phase 2 trial. Int J Radiat Oncol Biol Phys 2015;93(1):37–46.

[111] Awan MJ, Nedzi L, Wang D, Tumati V, Sumer B, Xie XJ, Smith I, Truelson J, Hughes R, Myers LL, Lavertu P, Wong S, Yao M. Final results of a multi-institutional phase II trial of reirradiation with concurrent weekly cisplatin and cetuximab for recurrent or second primary squamous cell carcinoma of the head and neck. Ann Oncol 2018;29(4):998–1003.

[112] Ward MC, Riaz N, Caudell JJ, Dunlap NE, Isrow D, Zakem SJ, Dault J, Awan MJ, Vargo JA, Heron DE, Higgins KA, Beitler JJ, Marcrom S, Boggs DH, Hassanzadeh C, Reddy CA, Borner JA, Yao M, Machtay M, Siddiqui F, Trotti AM, Lee NY, Koyfman SA. Multi-Institution Reirradiation C. Refining patient selection for Reirradiation of head and neck squamous carcinoma in the IMRT era: a multi-institution cohort study by the MIRI collaborative. Int J Radiat Oncol Biol Phys 2018;100(3):586–94.

[113] Janot F, de Raucourt D, Benhamou E, Ferron C, Dolivet G, Bensadoun RJ, Hamoir M, Gery B, Julieron M, Castaing M, Bardet E, Gregoire V, Bourhis J. Randomized trial of postoperative reirradiation combined with chemotherapy after salvage surgery compared with salvage surgery alone in head and neck carcinoma. J Clin Oncol 2008;26(34):5518–23.

[114] Spencer SA, Harris J, Wheeler RH, Machtay M, Schultz C, Spanos W, Rotman M, Meredith R, Ang KK. Final report of RTOG 9610, a multi-institutional trial of reirradiation and chemotherapy for unresectable recurrent squamous cell carcinoma of the head and neck. Head Neck 2008;30(3):281–8.

[115] Langer CJ, Harris J, Horwitz EM, Nicolaou N, Kies M, Curran W, Wong S, Ang K. Phase II study of low-dose paclitaxel and cisplatin in combination with split-course concomitant twice-daily reirradiation in recurrent squamous cell carcinoma of the head and neck: results of radiation therapy oncology group protocol 9911. J Clin Oncol 2007;25(30):4800–5.

[116] Margalit DN, Schoenfeld JD, Rawal B, Haddad RI, Catalano PJ, Goguen LA, Chau NG, Rabinowits G, Lorch JH, Annino DJ, Tishler RB. Patient-oriented toxicity endpoints after head and neck reirradiation with intensity modulated radiation therapy. Oral Oncol 2017;73:160–5.

[117] Seiwert TY, Burtness B, Mehra R, Weiss J, Berger R, Eder JP, Heath K, McClanahan T, Lunceford J, Gause C, Cheng JD, Chow LQ. Safety and clinical activity of pembrolizumab for treatment of recurrent or metastatic squamous cell carcinoma of the head and neck (KEYNOTE-012): an open-label, multicentre, phase 1b trial. Lancet Oncol 2016;17(7):956–65.

[118] Ferris RL, Blumenschein Jr. G, Fayette J, Guigay J, Colevas AD, Licitra L, Harrington K, Kasper S, Vokes EE, Even C, Worden F, Saba NF, Iglesias Docampo LC, Haddad R, Rordorf T, Kiyota N, Tahara M, Monga M, Lynch M, Geese WJ, Kopit J, Shaw JW, Gillison ML. Nivolumab for recurrent squamous-cell carcinoma of the head and neck. N Engl J Med 2016;375:1856–67.
[119] Seiwert TY, Fayette J, Cupissol D, Del Campo JM, Clement PM, Hitt R, Degardin M, Zhang W, Blackman A, Ehrnrooth E, Cohen EE. A randomized, phase II study of afatinib versus cetuximab in metastatic or recurrent squamous cell carcinoma of the head and neck. Ann Oncol 2014;25(9):1813–20.
[120] Fakhry C, Zhang Q, Nguyen-Tan PF, Rosenthal D, El-Naggar A, Garden AS, Soulieres D, Trotti A, Avizonis V, Ridge JA, Harris J, Le QT, Gillison M. Human papillomavirus and overall survival after progression of oropharyngeal squamous cell carcinoma. J Clin Oncol 2014;32(30):3365–73.

CHAPTER

2

The role of molecularly targeted therapies and immunotherapy in head and neck cancer

Gil Harmon, Justine Yang Bruce

Division of Hematology and Oncology, Department of Medicine, University of Wisconsin Carbone Cancer Center, University of Wisconsin School of Medicine and Public Health, University of Wisconsin, Madison, WI, United States

Abstract

Treatment of recurrent or metastatic squamous cell carcinoma of the head and neck has grown increasingly more complex with the development of new immunotherapy agents as well as novel combinations of immunotherapy with existing cytotoxic chemotherapy and targeted agents. With the advent of new immunotherapy agents, the overall survival of patients with metastatic disease continues to improve although there is still an unmet need for patients with continued disease progression. Immunochemotherapy may offer additional benefits with increased response rate as well as durable response. The combination of both immunotherapy and cytotoxic chemotherapy may also offer an improved toxicity profile while improving efficacy, although clinical trials are ongoing, and the data is still maturing. Further investigations into biomarker testing and new scoring systems of PD-L1 expression levels are still being evaluated. Traditionally, the tumor positive score (TPS) had been utilized as a benchmark for defining patient populations; however, emerging data using a combined positive score (CPS) may offer additional insight regarding the benefit of PD-1 inhibitors.

Abbreviations

CPS	combined positive score
CTLA-1	anti-cytotoxic T-lymphocyte-associated antigen 4
EGFR	epidermal growth factor receptor
EXTREME	ErbituX in first-line Treatment of Recurrent or MEtastatic head and neck cancer
HNSCC	squamous cell carcinoma of the head and neck
HPV	human papillomavirus
ORR	overall response rate
PD-1	programmed death receptor-1
PD-L1	programmed death ligand-1
RTOG	Radiation Therapy Oncology Group

Improving the Therapeutic Ratio in Head and Neck Cancer
https://doi.org/10.1016/B978-0-12-817868-3.00002-0

Conflict of interest

No potential conflicts of interest were disclosed.

Current therapies for recurrent/metastatic disease

Despite aggressive local treatments, local or regional disease recurrences are seen in 30% of head and neck cancer patients, and distant metastases appear in 25% of these patients resulting in a 5 year overall survival rate of approximately 40% [1]. Although a proportion of those patients with recurrent disease remain potentially curable with further local treatment approaches (e.g., surgery or radiation), retreatment is technically challenging and can be accompanied by a significant risk of irreversible damage to normal tissues. Thus, palliative systemic treatment is often utilized in an effort to improve overall survival and palliate symptoms associated with the disease.

The current standard of care for patients with recurrent/metastatic head and neck cancer typically incorporates cytotoxic systemic chemotherapy in those patients with adequate performance status [2]. While overall survival rates for patients with metastatic head and neck cancer have not changed much in the last 20 years, in 2011, the use of EGFR targeted therapy with cetuximab was found to improve overall survival compared to traditional cytotoxic therapy alone [3]. More recently, novel agents capable of modulating the immune system have been approved for use in head and neck and other advanced cancers [4–7]. It is hoped that these advances will lead to improved survival in patients with advanced disease. Despite the expansion of immunotherapy, both as monotherapy or in combination with cytotoxic chemotherapy, there is still an unmet need for patients who continue to have disease progression despite treatment with the currently FDA approved agents. Here we discuss the landscape of molecularly targeted therapies and immunotherapy in head and neck squamous cell carcinomas.

Early work with cytotoxic agents

Initial chemotherapy for head and neck cancer followed the approaches used in most other cancers: cytotoxic chemotherapies. Due to non-overlapping toxicities, combination regimens of cytotoxic chemotherapy are typically recommended for treatment of patients with unresectable or metastatic head and neck cancers [8–11]. Although not a preferred initial regimen, single agent chemotherapy can be utilized for those with borderline performance status or for those patients who have progressed through multiple lines of standard chemotherapy.

Platinum agents have been established as the most active agents against head and neck cancer. Cisplatin has become the standard agent in the treatment of both locoregional and metastatic disease. Cisplatin monotherapy was first investigated as an active agent for head and neck squamous cell carcinomas in the 1980s, when it was noted to prolong overall survival by up to 10 weeks [12]. Since that time, cisplatin has been compared to other monotherapy agents, and no other cytotoxic chemotherapy agent has been shown to improve survival compared to cisplatin [13].

Based on the activity of cisplatin monotherapy, combination regimens utilizing platinum doublets were developed. Several of these combinations demonstrated an improvement in overall response rates. In a phase III randomized trial, cisplatin and 5-fluorouracil were evaluated both as single agents and in combination for the treatment of recurrent/metastatic head and neck cancer [9]. The combination arm demonstrated a statistically significant improvement in overall response rate (ORR) of 32% versus 17% and 13% for platinum and fluorouracil monotherapy, respectively. Although there was no significant difference in median overall survival (5.7 months) among the groups, (monotherapy arms versus a platinum combination regimen), the improvement in overall response rate prompted the adoption of a platinum based doublet as the early first line standard treatment for advanced disease. Up until 2008, no randomized phase III trial had demonstrated superior overall survival (averaging about 6 months) of any platinum-based cytotoxic therapy combination over single agent regimens, despite the improved response rates with combination regimens [8, 10, 11]. Thus, selection of a specific cytotoxic chemotherapy regimen for each patient was personalized based on the toxicity profile of the combination, the performance status of the patient, the burden of disease, and the requirements for a rapid response.

Emergence of targeted therapy against the epidermal growth factor receptor (EGFR)

The majority of squamous cell carcinomas of the head and neck overexpress the epidermal growth factor receptor (EGFR), and in fact, EGFR overexpression (as an independent prognostic factor) adversely impacts overall survival [14]. As combinations of various cytotoxic agents failed to incrementally advance overall survival, interest increased in targeting the EGFR [8, 10, 11] (Table 1).

Cetuximab is a chimeric monoclonal antibody which binds with high affinity to the extracellular domain of the human EGFR. The high affinity of cetuximab to EGFR results in inhibition of the receptor function. Cetuximab initially demonstrated activity in a phase 1b clinical trial that included 12 patients with recurrent or metastatic squamous cell cancer of the head and neck [15]. Patients received cisplatin in combination with cetuximab, and though the primary endpoints were receptor saturation and dose determination, six out of nine evaluable patients demonstrated a clinical response, including patients with prior cisplatin exposure. Importantly, the combination of cisplatin and cetuximab therapy did not demonstrate overlapping toxicities in this study. This study established the recommended phase II dose for future trials to be 400 mg/m^2 as a loading dose followed by a maintenance dose of 250 mg/m^2. This dose regimen achieved a high percentage of saturation of EGFR in tumor tissues.

Given the use of radiotherapy as the principal treatment for locoregionally advanced HNSCC, cetuximab was also evaluated in combination with definitive radiation therapy [16, 17]. Since radiation increases the expression of EGFR in cancer cells, the study hypothesized that inhibition of EGFR signaling might sensitize cells to the effects of radiation [18, 19]. In a multinational, randomized study, radiotherapy alone was compared to radiotherapy plus cetuximab in the treatment of locoregionally advanced squamous cell carcinoma of the head and neck [17]. Patients were randomized to high dose radiotherapy alone or high dose radiotherapy plus weekly cetuximab. The trial's primary end point was duration of control of locoregional disease. The median duration of locoregional control was found to be significantly longer with concurrent cetuximab at 24.4 months compared to 14.9 months for patients receiving radiotherapy alone. This study firmly established

TABLE 1 Selected phase 2 interventional trials for recurrent/metastatic squamous cell carcinoma of the head and neck

Title and treatment regimen	Interventions	NCT number
Pembrolizumab plus cetuximab in HNSCC	Pembrolizumab Cetuximab	NCT03082534
Durvalumab, tremelimumab, and stereotactic body radiation therapy in recurrent/metastatic disease	Durvalumab Tremelimumab Stereotactic body radiation therapy	NCT03522584
Ficlatuzumab w/wo cetuximab in patients with cetuximab resistant disease	Cetuximab	NCT03422536
Cetuximab plus nivolumab	Cetuximab Nivolumab	NCT03370276
Durvalumab plus cetuximab	Durvalumab Cetuximab	NCT03691714
MEDI0457 plus durvalumab in HPV associated recurrent/metastatic disease	MEDI0457 Durvalumab	NCT03162224
Intratumoral injection of SD-101 plus pembrolizumab	SD-101 Pembrolizumab	NCT02521870
LN-145 autologous tumor infiltrating lymphocytes	LN-145	NCT03083873
Bimiralisib in HNSCC patients harboring NOTCH1 loss of function mutations	Bimiralisib	NCT03740100
Monalizumab and cetuximab	Monalizumab Cetuximab	NCT02643550
Docetaxel and capecitabine	Docetaxel Capecitabine	NCT02524275
MEDI4736 with either AZD9150 or AZD5069	MEDI4736 AZD9150 AZD5069	NCT02499328
Expansion trial for axitinib	Axitinib	NCT02762513
Trigriluzole with nivolumab and pembrolizumab	Trigriluzole Nivolumab Pembrolizumab	NCT03229278
Dual inhibition of EGFR with afatinib and cetuximab	Afatinib Cetuximab	NCT02979977

cetuximab as a radiosensitizer that could improve locoregional control without have an excess of severe toxic effects that are commonly seen in chemoradiation therapy.

Given the improved toxicity profile, cetuximab was also evaluated as a potential comparable radiosensitizer in the treatment of human papillomavirus (HPV)-positive oropharyngeal squamous cell carcinomas. Prognosis for patients with HPV-positive oropharyngeal carcinoma is better when compared with HPV-negative oropharyngeal carcinoma when treated with radiotherapy plus high dose cisplatin [20]. Given the concern regarding late treatment-related toxicities (severe dysphagia, feeding tube dependence, or death) for patients with HPV-positive oropharyngeal carcinoma treated with cisplatin chemotherapy, the RTOG 1016 was established as a randomized clinical trial with a non-inferiority design [21]. This clinical trial directly compared the overall survival for patients with HPV-positive oropharyngeal carcinoma who were treated with radiotherapy plus cetuximab versus radiotherapy plus cisplatin [21–23]. Despite the favorable prognosis of HPV-positive oropharyngeal squamous cell carcinomas, this study demonstrated that overall survival was significantly worse with cetuximab with an estimated 5 year overall survival of 78% in the cetuximab group compared to 85% in the cisplatin group. In addition, the progression free survival was also significantly lower in the cetuximab group compared with the cisplatin arm (5-year progression free survival of 67% in the cetuximab arm versus 78% in the cisplatin arm). RTOG 1016 was the first prospective randomized study to investigate treatment de-intensification for patients with HPV-positive oropharyngeal carcinoma, and demonstrated that radiotherapy plus cisplatin should remain as the standard of care for eligible patients with HPV-positive oropharyngeal carcinoma.

Cetuximab was also evaluated in the metastatic platinum-refractory setting, as published by Vermorken et al. [24]. In this open-label, multicenter phase II study, patients with recurrent/metastatic head and neck squamous cell carcinoma who were previously treated with a platinum based treatment were eligible to receive cetuximab monotherapy. In the single agent phase, the overall response rate was noted to be 13% with a median duration of response of 5.8 months. The results of these trials led to the FDA approval in 2006 for cetuximab as monotherapy in patients who had platinum refractory disease.

Results of these two landmark studies (as monotherapy in the platinum refractory setting and in combination with definitive radiotherapy) subsequently led to additional trials that focused on a broader range of cetuximab use, primarily in combination with cytotoxic chemotherapy.

Cetuximab plus chemotherapy in the first line metastatic setting

Prior to the emergence of targeted therapy and immunotherapy, the overall survival for patients with recurrent/metastatic head and neck squamous cell carcinoma was approximately 6 months, despite improvements in the overall response rate with a combination of 5-fluorouracil with a platinum agent [3, 8, 25]. In the metastatic setting, doublet therapy with 5-fluorouracil plus cisplatin was considered the standard of care chemotherapy regimen in the first line setting. Given the improvement of cetuximab as a monotherapy and the known overexpression of EGFR in head and neck squamous cell carcinomas, Vermorken et al. conducted a phase III randomized trial (the EXTREME trial) comparing fluorouracil plus a platinum agent (either cisplatin or carboplatin), either alone or in combination with cetuximab [3]. Patients were randomly assigned in a 1:1 ratio, with the primary end point being overall survival: defined as the time from randomization to death. Patients were to complete 6 cycles of combination therapy followed by maintenance weekly cetuximab so long as therapy was tolerable and disease response was stable disease or better (imaging was performed every 6 weeks).

The median overall survival for the cetuximab-containing group was 10.1 months compared to 7.4 months in the group without cetuximab. The cetuximab group was also noted to have an improvement in progression free survival (5.6 months vs 3.3 months). The cetuximab containing regimen was tolerable as grade 3 and 4 adverse events, when taken in aggregate, were not significantly different between the two treatment arms. The addition of cetuximab to platinum based chemotherapy resulted in a significant reduction in pain and difficulties with eating and speech, which positively impacted social functioning and quality of life for these patients [26]. The EXTREME trial was the first trial to demonstrate a benefit in overall survival by adding cetuximab to a platinum-based chemotherapy. Thus, this landmark trial led to the FDA approval of a third indication for cetuximab use: in combination with platinum based therapy for recurrent locoregional disease or metastatic squamous cell carcinoma of the head and neck.

Panitumumab as an alternative targeted agent for HNSCC

With the approval of cetuximab in the first line metastatic setting, another trial (SPECTRUM) was also conducted using panitumumab, a fully human anti-EGFR monoclonal antibody that is approved for the treatment of metastatic colorectal cancer. The SPECTRUM trial was an open label randomized phase III trial evaluating patients with locoregionally recurrent or distant metastatic HNSCC [27]. Patients were randomized to receive 5-fluorouracil plus cisplatin, alone, or in combination with panitumumab. Treatment was continued for a maximum of six every three-week cycles, and continuation of panitumumab was optional after the completion of cytotoxic chemotherapy until disease progression, intolerable toxicity, or study withdrawal. The SPECTRUM trial did not meet its primary endpoint of overall survival (defined as time from randomization to death) as the panitumumab group had a median overall survival of 11.1 months compared to 9 months in the control group. Median progression free survival was 5.8 months for the panitumumab group and 4.6 months in the control group. The results of the SPECTRUM study were surprising when compared to the EXTREME trial, but multiple rationalizations for the differences have been put forth. For example, the SPECTRUM trial had a more robust patient population since patients were required to receive treatment with cisplatin (compared to either carboplatin or cisplatin in the EXTREME study), thus increasing the overall survival for the control group. Second, patients in the SPECTRUM study were not required to continue with panitumumab monotherapy (compared to a required cetuximab monotherapy after six cycles of chemotherapy in the EXTREME trial). Lastly, the population accrued in the SPECTRUM was from a global population compared to the Western Europe population studied in EXTREME. The inclusion of patients from the Asia-Pacific region (who have fairly long overall survival) may account for the extended overall survival in the control group and a failure to identify an improvement. Nevertheless, the results of the EXTREME trial supported the use of targeted agents given the improvement in overall survival, even if the lack of ability to demonstrate this effect across multiple drugs in the EGFR-inhibitor class somewhat tempers enthusiasm.

Practical approach to first line systemic treatment

The only combination chemotherapy regimen that has demonstrated an improvement in overall survival to date is a platinum, 5-fluorouracil, and cetuximab regimen [3]. For fit patients who desire aggressive therapy, this regimen should be utilized. For those patients who require a relatively rapid response, a platinum doublet should be considered [2, 8, 9]. In our clinical practice, we typically use a platinum agent combined with a taxane. This choice is based on the E1395 trial which compared cisplatin plus 5-fluorouracil against cisplatin plus paclitaxel [28]. In this trial, the overall response rate (25%) and overall survival (8–9 months) were similar between both groups. The platinum plus taxane treatment option avoids the need for 4 days of infusional therapy with 5-fluorouracil, and for that reason, we often favor platinum plus taxane given the improved patient convenience. In an effort to mitigate the side effect profile, we often consider replacing cisplatin with carboplatin. Though evidence for the combination of carboplatin plus taxane in advanced head and neck cancer is not as strong [29, 30], many investigators consider them to be relatively interchangeable (a further discussion of the interchangeability of carboplatin and cisplatin in the metastatic setting is beyond the scope of this chapter) [8, 31–34].

Dawn of an immunotherapy trend

Pembrolizumab

As in many other tumor types, checkpoint inhibition has changed the landscape of head and neck cancer research and treatment. Though currently only having FDA approval in the recurrent/metastatic setting after platinum chemotherapy, checkpoint inhibition is quickly moving into multiple treatment paradigms [5–7, 35, 36]. To date, pembrolizumab and nivolumab have both received FDA approval as second line monotherapies in the recurrent/metastatic setting.

Pembrolizumab is a programmed death receptor-1 (PD-1) blocking antibody that was evaluated in a multicenter phase 1b clinical trial (KEYNOTE-012) for patients with PD-L1 positive recurrent or metastatic squamous cell carcinoma of the head and neck [6, 7, 37]. Out of the 104 patients who were screened for the study, 81 patients were PD-L1 positive. Of these, 60 patients were treated with pembrolizumab. Patients with HNSCC tolerated pembrolizumab well, with an overall response rate of 18% [6, 7, 37]. The proportion of patients with an overall response was higher in HPV positive than HPV negative patients (25% vs 19%), but regardless of HPV status, patients still responded to pembrolizumab. The median time to response was 8 weeks, and the median duration of response was 53 weeks. Based on the early data from this trial, the FDA granted accelerated approval for pembrolizumab.

Subsequent to the positive activity noted in the phase I trial, pembrolizumab was then evaluated in a single arm, phase II study (KEYNOTE-055) that targeted patients who previously had received prior platinum and cetuximab treatment [5]. Although the combination regimen of 5-fluorouracil, platinum, and cetuximab was a standard first line standard treatment (according to the EXTREME trial), there was no approved treatment options in the second line setting. Patients enrolled in KEYNOTE-055 were required to have disease progression within 6 months of platinum and cetuximab treatment, and were treated with pembrolizumab 200 mg every 3 weeks. This single arm study met its primary endpoint of overall response rate at 16% with the response rates being similar regardless of HPV status. In addition, durable responses were noted, with responses noted to be >12 months in some patients. In comparison, the historical response rate to single agent cetuximab, afatinib, and methotrexate in patients with recurrent/metastatic HNSCC that progressed after platinum treatment were 10%, 10%, and 6% [38, 39].

Although PD-L1 expression was evaluated in these studies, testing of PD-L1 levels was not implemented as a standard of care laboratory test in clinical practice [5–7, 37]. A growing number of trials note that patients whose tumors express PD-L1 are more likely to respond to treatment with PD-1 pathway inhibitors [40–42]. However, meaningful improvements in response rates were noted in patients who were PD-L1 negative [6, 35, 43, 44]. Thus, the data suggest that therapeutic benefit extends to patients regardless of their PD-L1 status.

An international, randomized, phase 3, open-label study (KEYNOTE-040) specifically compared pembrolizumab to investigator's choice of standard doses of methotrexate, docetaxel, or cetuximab (considered to be the standard of care group) in patients with recurrent/metastatic HNSCC who had disease progression within 3–6 months of previous platinum-based therapy [36]. Within the study, patients were evenly matched in each arm. The study did not meet is primary endpoint of overall survival, although the median overall survival was 8.4 months with pembrolizumab compared to 6.9 months with standard of care chemotherapy (methotrexate, docetaxel, or cetuximab). Patients who received pembrolizumab experienced fewer grade 3 or worse treatment related adverse events (13% with pembrolizumab compared to 36% with standard of care). The negative results from the study were surprising given the improved toxicity profile and tolerability, and the announcement did not impact the approval status of the compound.

Most recently, results from an open label, randomized phase 3 study (KEYNOTE-048) of pembrolizumab, pembrolizumab plus chemotherapy, and the standard of care first line treatment of fluorouracil, platinum, and cetuximab (EXTREME regimen) noted an improvement in overall survival with upfront pembrolizumab monotherapy in patients who had a PD-L1 combined positive score ≥20 [35]. The promising results reinvigorated interest in pembrolizumab as a viable agent for the treatment of head and neck cancer. This international study enrolled 882 patients in the first line setting for recurrent or metastatic squamous cell carcinoma of the head and neck. Given that prior studies noted no significant difference in survival with based on HPV status, further evaluation in the PD-L1 levels were assessed. Patients were tested for the PD-L1 combined positive score (CPS) with a cutoff of 20 (CPS ≥20 vs CPS ≥1). For those patients who had a CPS ≥20, overall survival was improved for patients treated with pembrolizumab monotherapy compared to standard chemotherapy (EXTREME) at 14.9 versus 10.7 months. For patients who had a CPS ≥1, the overall survival also was improved with pembrolizumab monotherapy (12.3 vs 10.3 months), although this demonstrated a smaller improvement. In the overall population, pembrolizumab was deemed to be non-inferior in overall survival when compared to standard of care chemotherapy.

This study also was the first to evaluate the combination of pembrolizumab with cytotoxic chemotherapy, and noted that the overall survival was not statistically superior when compared to standard of care chemotherapy in either population of CPS ≥20 or CPS ≥1. However, the investigators noted that the combination chemotherapy plus pembrolizumab may offer a more favorable safety profile than standard of care chemotherapy as patients had a lower incidence of treatment related adverse events that led to treatment discontinuation. However, further analysis are still ongoing.

Another notable point from KEYNOTE-048 is that there was no difference in progression free survival or overall response rate in patients who received combined chemotherapy plus pembrolizumab compared to the EXTREME regimen of 5-fluorouracil, platinum, plus cetuximab [35]. However, responses to the pembrolizumab/chemotherapy combination were more durable. It is unclear what the time to response is with chemotherapy plus pembrolizumab, but for those patients who require a more rapid response due to the burden of the disease, a combination regimen of chemoimmunotherapy could potentially offer similar response rates as standard chemotherapy with a more durable response due to the inclusion of immunotherapy. The final analysis of this study for overall survival is still ongoing.

Nivolumab

Nivolumab is a fully human IgG4 anti-programmed death (PD-1) monoclonal antibody that has been approved in other tumor types, such as lung cancer. Nivolumab was evaluated in a randomized, open-label, phase 3 trial for patients with recurrent/metastatic HNSCC whose disease had progressed within 6 months after platinum-based chemotherapy [4]. Patients were assigned, in a 2:1 ratio, to nivolumab or investigator's choice of single agent standard chemotherapy (methotrexate, docetaxel, or cetuximab). The trial met its primary endpoint of overall survival with an improvement of 7.5 months with nivolumab compared to 5.1 months with standard of care treatment. In addition, this study also evaluated physical and social functioning, and found that these qualities were stable in the nivolumab group, whereas they were meaningfully worse in the standard therapy group [4]. This positive study led to the FDA approval of nivolumab in 2016.

Although checkpoint inhibition has led to significant improvements in overall survival for patients with advanced disease, there remains an unmet need for patients who continue to have disease progression. Thus, combination immunotherapy regimens have been designed to target other immunomodulating pathways. For instance, ipilimumab, an anti-cytotoxic T-lymphocyte-associated antigen 4 (CTLA-4) fully human monoclonal antibody acts as a critical negative regulator of the antitumor T cell response [45, 46]. When combined with nivolumab, the combination regimen has been demonstrated (in treatment of malignant melanoma) to result in longer progression free survival and a higher rate of objective responses compared to single agent nivolumab or single agent ipilimimumab [47]. Thus, the use of combination immunotherapy regimens are currently being evaluated in head and neck cancers as well [48, 49].

Durvalumab

Durvalumab is an anti-PD-L1 monoclonal antibody that has been examined in the immunotherapy treatment of head and neck cancers. Given that the agent is primarily targeted against the programmed cell death ligand-1, the early trials of durvalumab distinguished HNSCC patients with high or low PD-L1 expression.

The phase II CONDOR trial evaluated heavily pretreated patients with recurrent or metastatic head and neck cancer specifically with low or negative PD-L1 expression [50]. These patients were randomized to three arms: durvalumab monotherapy, tremelimumab monotherapy, and the combination of both durvalumab plus tremelimumab. Durvalumab monotherapy had an overall response rate of 9.2% and the combination therapy had an overall response rate of 7.8%. Tremelimumab monotherapy had a 1.6% overall response rate. On the other hand, the companion phase II HAWK study evaluated patients with HNSCC that expressed high levels of PD-L1, who were treated with single agent durvalumab. Patients enjoyed an overall response rate of 16.2% with a median progression free survival of 2.1 months and a median overall survival of 7.1 months.

The promising results of durvalumab in the early studies prompted the phase III EAGLE trial evaluating durvalumab, durvalumab plus the anti-CTLA4 antibody tremelimumab, and standard of care chemotherapy after platinum based chemotherapy. Unfortunately, initial reports suggest that the study did not meet its primary endpoint of improving overall survival compared to standard of care chemotherapy [51]. Final results remain pending at this time. An ongoing phase III study (KESTREL trial) of durvalumab plus tremelimumab in recurrent/metastatic patients who are chemotherapy naïve is being conducted globally, with the results anticipated to be released in 2019 [52].

Ongoing trials and future directions

Novel immunotherapy agents are constantly being developed, both targeting PD-1 as well as PD-L1. In addition, other immunotherapy targets are being developed, such as OX-40 or GITR, both in combination with checkpoint inhibitors, as well as monotherapy. Early data that demonstrated that combination regimens of multiple immunotherapy agents as well as chemoimmunotherapy likely would offer improved response rate, progression free survival, and overall survival. These studies likely will offer additional evidence and data for an improvement in this patient population.

Another avenue of intense research is the incorporation of immunotherapy in combination with radiation therapy for definitive treatment of intermediate and high risk locoregionally advanced head and neck cancers. RTOG 3504 is a cooperative group study evaluating nivolumab, nivolumab plus cetuximab, or nivolumab plus cisplatin given concurrently with definitive radiation therapy, followed by maintenance immunotherapy for 1 year. This multicenter trial has completed accrual with analysis forthcoming. Initial results from this study demonstrated the safety and feasibility of adding immunotherapy to curative intent treatment with cisplatin and radiation therapy without adding additional significant toxicities and compromising the delivery of radiation therapy.

Future directions

A unique opportunity for additional treatment in head and neck squamous cell carcinoma is the incorporation of vaccines in combination with immunotherapy. One approach is to use intratumoral injection of investigational agents in order to activate a complementary part of the immune system aside. For example, SD-101, is a synthetic CpG-oligodeoxynucleotide toll-like receptor 9 agonist that can stimulate dendritic cells to release interferon alpha and subsequently active downstream T cell antitumor responses [53]. In combination with other immunotherapy agents, such as the approved checkpoint inhibitors, the combination has demonstrated promising improvement in response rate [54]. Talimogene laherparepvec (T-VEC) has also been evaluated as an oncolytic immunotherapy that enhances systemic antitumor immune responses when given in combination with pembrolizumab [55].

Treatments for advanced head and neck cancer is ever-changing, with the advent of novel immunotherapy agents as well as new combinations of investigational agents with standard of care regimens. Immunotherapy has provided promising improvements in overall survival with tolerable toxicities, although the extent of its impact has not been fully delineated as new data and analysis remain pending. Given the breadth of treatments that are becoming more available, clinicians will have the daunting task of deciphering the optimal order of treatment regimens in an effort to maximize overall survival while minimizing toxicities. Finally, given the relatively modest response rates seen even with combination therapies, approaches to overcome both intrinsic and acquired therapy resistance are desperately needed.

Acknowledgments

This project was supported in part by the Specialized Program of Research Excellence (SPORE) program, through the NIH National Institute for Dental and Craniofacial Research (NIDCR) and National Cancer Institute (NCI), grant P50DE026787. The content is solely the responsibility of the authors and does not necessarily represent the official views of the NIH.

References

[1] Laramore GE, Scott CB, al-Sarraf M, Haselow RE, Ervin TJ, Wheeler R, Jacobs JE, Schuller DE, Gahbauer RA, Schwade JG, et al. Adjuvant chemotherapy for resectable squamous cell carcinomas of the head and neck: report on Intergroup Study 0034. Int J Radiat Oncol Biol Phys 1992;23(4):705–13.

[2] Sacco AG, Cohen EE. Current treatment options for recurrent or metastatic head and neck squamous cell carcinoma. J Clin Oncol Off J Am Soc Clin Oncol 2015;33(29):3305–13.

[3] Vermorken JB, Mesia R, Rivera F, Remenar E, Kawecki A, Rottey S, Erfan J, Zabolotnyy D, Kienzer HR, Cupissol D, Peyrade F, Benasso M, Vynnychenko I, De Raucourt D, Bokemeyer C, Schueler A, Amellal N, Hitt R. Platinum-based chemotherapy plus cetuximab in head and neck cancer. N Engl J Med 2008; 359(11):1116–27.

[4] Ferris RL, Blumenschein Jr G, Fayette J, Guigay J, Colevas AD, Licitra L, Harrington K, Kasper S, Vokes EE, Even C, Worden F, Saba NF, Iglesias Docampo LC, Haddad R, Rordorf T, Kiyota N, Tahara M, Monga M, Lynch M, Geese WJ, Kopit J, Shaw JW, Gillison ML. Nivolumab for recurrent squamous-cell carcinoma of the head and neck. N Engl J Med 2016;375:1856–67.

[5] Bauml J, Seiwert TY, Pfister DG, Worden F, Liu SV, Gilbert J, Saba NF, Weiss J, Wirth L, Sukari A, Kang H, Gibson MK, Massarelli E, Powell S, Meister A, Shu X, Cheng JD, Haddad R. Pembrolizumab for platinum- and Cetuximab-refractory head and neck cancer: results from a single-arm, phase II study. J Clin Oncol 2017;35(14):1542–9.

[6] Chow LQ, Haddad R, Gupta S, Mahipal A, Mehra R, Tahara M, Berger R, Eder JP, Burtness B, Lee SH, Keam B, Kang H, Muro K, Weiss J, Geva R, Lin CC, Chung HC, Meister A, Dolled-Filhart M, Pathiraja K, Cheng JD, Seiwert TY. Antitumor activity of Pembrolizumab in biomarker-unselected patients with recurrent and/or metastatic head and neck squamous cell carcinoma: results from the phase Ib KEYNOTE-012 expansion cohort. J Clin Oncol 2016;34:3838–45.

[7] Seiwert TY, Burtness B, Mehra R, Weiss J, Berger R, Eder JP, Heath K, McClanahan T, Lunceford J, Gause C, Cheng JD, Chow LQ. Safety and clinical activity of Pembrolizumab for treatment of recurrent or metastatic squamous cell carcinoma of the head and neck (KEYNOTE-012): an open-label, multicentre, phase 1b trial. Lancet Oncol 2016;17(7):956–65.

[8] Forastiere AA, Metch B, Schuller DE, Ensley JF, Hutchins LF, Triozzi P, Kish JA, McClure S, Von Feldt E, Williamson SK, et al. Randomized comparison of cisplatin plus fluorouracil and carboplatin plus fluorouracil versus methotrexate in advanced squamous-cell carcinoma of the head and neck: a Southwest Oncology Group study. J Clin Oncol Off J Am Soc Clin Oncol 1992;10(8):1245–51.
[9] Jacobs C, Lyman G, Velez-Garcia E, Sridhar KS, Knight W, Hochster H, Goodnough LT, Mortimer JE, Einhorn LH, Schacter L, et al. A phase III randomized study comparing cisplatin and fluorouracil as single agents and in combination for advanced squamous cell carcinoma of the head and neck. J Clin Oncol Off J Am Soc Clin Oncol 1992;10(2):257–63.
[10] Clavel M, Vermorken JB, Cognetti F, Cappelaere P, de Mulder PH, Schornagel JH, Tueni EA, Verweij J, Wildiers J, Clerico M, et al. Randomized comparison of cisplatin, methotrexate, bleomycin and vincristine (CABO) versus cisplatin and 5-fluorouracil (CF) versus cisplatin (C) in recurrent or metastatic squamous cell carcinoma of the head and neck. A phase III study of the EORTC Head and Neck Cancer Cooperative Group. Ann Oncol 1994;5(6):521–6.
[11] Price KA, Cohen EE. Current treatment options for metastatic head and neck cancer. Curr Treat Options Oncol 2012;13(1):35–46.
[12] Morton RP, Rugman F, Dorman EB, Stoney PJ, Wilson JA, McCormick M, Veevers A, Stell PM. Cisplatinum and bleomycin for advanced or recurrent squamous cell carcinoma of the head and neck: a randomised factorial phase III controlled trial. Cancer Chemother Pharmacol 1985;15(3):283–9.
[13] A phase III randomised trial of cistplatinum, methotrextate, cisplatinum + methotrexate and cisplatinum + 5-FU in end stage squamous carcinoma of the head and neck. Liverpool Head and Neck Oncology Group. Br J Cancer 1990;61(2):311–5.
[14] Ang KK, Berkey BA, Tu XY, Zhang HZ, Katz R, Hammond EH, Fu KK, Milas L. Impact of epidermal growth factor receptor expression on survival and pattern of relapse in patients with advanced head and neck carcinoma. Cancer Res 2002;62(24):7350–6.
[15] Shin DM, Donato NJ, Perez-Soler R, Shin HJ, Wu JY, Zhang P, Lawhorn K, Khuri FR, Glisson BS, Myers J, Clayman G, Pfister D, Falcey J, Waksal H, Mendelsohn J, Hong WK. Epidermal growth factor receptor-targeted therapy with C225 and cisplatin in patients with head and neck cancer. Clin Cancer Res 2001;7(5):1204–13.
[16] Bonner JA, Harari PM, Giralt J, Azarnia N, Shin DM, Cohen RB, Jones CU, Sur R, Raben D, Jassem J, Ove R, Kies MS, Baselga J, Youssoufian H, Amellal N, Rowinsky EK, Ang KK. Radiotherapy plus cetuximab for squamous-cell carcinoma of the head and neck. N Engl J Med 2006;354(6):567–78.
[17] Bonner JA, Harari PM, Giralt J, Cohen RB, Jones CU, Sur RK, Raben D, Baselga J, Spencer SA, Zhu J, Youssoufian H, Rowinsky EK, Ang KK. Radiotherapy plus cetuximab for locoregionally advanced head and neck cancer: 5-year survival data from a phase 3 randomised trial, and relation between cetuximab-induced rash and survival. Lancet Oncol 2010;11(1):21–8.
[18] Liang K, Ang KK, Milas L, Hunter N, Fan Z. The epidermal growth factor receptor mediates radioresistance. Int J Radiat Oncol Biol Phys 2003;57(1) advanced head and:246–54.
[19] Bonner JA, Maihle NJ, Folven BR, Christianson TJ, Spain K. The interaction of epidermal growth factor and radiation in human head and neck squamous cell carcinoma cell lines with vastly different radiosensitivities. Int J Radiat Oncol Biol Phys 1994;29(2):243–7.
[20] Ang KK, Harris J, Wheeler R, Weber R, Rosenthal DI, Nguyen-Tan PF, Westra WH, Chung CH, Jordan RC, Lu C, Kim H, Axelrod R, Silverman CC, Redmond KP, Gillison ML. Human papillomavirus and survival of patients with oropharyngeal cancer. N Engl J Med 2010;363(1):24–35.
[21] Gillison ML, Trotti AM, Harris J, Eisbruch A, Harari PM, Adelstein DJ, Sturgis EM, Burtness B, Ridge JA, Ringash J, Galvin J, Yao M, Koyfman SA, Blakaj DM, Razaq MA, Colevas AD, Beitler JJ, Jones CU, Dunlap NE, Seaward SA, Spencer S, Galloway TJ, Phan J, Dignam JJ, Le QT. Radiotherapy plus cetuximab or cisplatin in human papillomavirus-positive oropharyngeal cancer (NRG Oncology RTOG 1016): a randomised, multicentre, non-inferiority trial. Lancet 2019;393(10166):40–50.
[22] Adelstein DJ, Li Y, Adams GL, Wagner Jr H, Kish JA, Ensley JF, Schuller DE, Forastiere AA. An intergroup phase III comparison of standard radiation therapy and two schedules of concurrent chemoradiotherapy in patients with unresectable squamous cell head and neck cancer. J Clin Oncol 2003;21(1):92–8.
[23] Machtay M, Moughan J, Trotti A, Garden AS, Weber RS, Cooper JS, Forastiere A, Ang KK. Factors associated with severe late toxicity after concurrent chemoradiation for locally advanced head and neck cancer: an RTOG analysis. J Clin Oncol 2008;26(21):3582–9.

[24] Vermorken JB, Trigo J, Hitt R, Koralewski P, Diaz-Rubio E, Rolland F, Knecht R, Amellal N, Schueler A, Baselga J. Open-label, uncontrolled, multicenter phase II study to evaluate the efficacy and toxicity of cetuximab as a single agent in patients with recurrent and/or metastatic squamous cell carcinoma of the head and neck who failed to respond to platinum-based therapy. J Clin Oncol 2007;25(16):2171–7.

[25] Adamo V, Ferraro G, Pergolizzi S, Sergi C, Laudani A, Settineri N, Alafaci E, Scimone A, Spano F, Spitaleri G. Paclitaxel and cisplatin in patients with recurrent and metastatic head and neck squamous cell carcinoma. Oral Oncol 2004;40(5):525–31.

[26] Rivera F, Garcia-Castano A, Vega N, Vega-Villegas ME, Gutierrez-Sanz L. Cetuximab in metastatic or recurrent head and neck cancer: the EXTREME trial. Expert Rev Anticancer Ther 2009;9(10):1421–8.

[27] Vermorken JB, Stohlmacher-Williams J, Davidenko I, Licitra L, Winquist E, Villanueva C, Foa P, Rottey S, Skladowski K, Tahara M, Pai VR, Faivre S, Blajman CR, Forastiere AA, Stein BN, Oliner KS, Pan Z, Bach BA, SPECTRUM Investigators. Cisplatin and fluorouracil with or without panitumumab in patients with recurrent or metastatic squamous-cell carcinoma of the head and neck (SPECTRUM): an open-label phase 3 randomised trial. Lancet Oncol 2013;14(8):697–710.

[28] Gibson MK, Li Y, Murphy B, Hussain MH, DeConti RC, Ensley J, Forastiere AA, Eastern Cooperative Oncology Group. Randomized phase III evaluation of cisplatin plus fluorouracil versus cisplatin plus paclitaxel in advanced head and neck cancer (E1395): an intergroup trial of the Eastern Cooperative Oncology Group. J Clin Oncol Off J Am Soc Clin Oncol 2005;23(15):3562–7.

[29] Clark JI, Hofmeister C, Choudhury A, Matz G, Collins S, Bastian R, Melian E, Emami B, Petruzzelli G. Phase II evaluation of paclitaxel in combination with carboplatin in advanced head and neck carcinoma. Cancer 2001; 92(9):2334–40.

[30] Samlowski WE, Moon J, Kuebler JP, Nichols CR, Gandara DR, Ozer H, Williamson SK, Atkins JN, Schuller DE, Ensley JF. Evaluation of the combination of docetaxel/carboplatin in patients with metastatic or recurrent squamous cell carcinoma of the head and neck (SCCHN): a Southwest Oncology Group Phase II study. Cancer Invest 2007;25(3):182–8.

[31] Al-Sarraf M, Metch B, Kish J, Ensley J, Rinehart JJ, Schuller DE, Coltman Jr CA. Platinum analogs in recurrent and advanced head and neck cancer: a Southwest Oncology Group and Wayne State University Study. Cancer Treat Rep 1987;71(7–8):723–6.

[32] De Andres L, Brunet J, Lopez-Pousa A, Burgues J, Vega M, Tabernero JM, Mesia R, Lopez JJ. Randomized trial of neoadjuvant cisplatin and fluorouracil versus carboplatin and fluorouracil in patients with stage IV-M0 head and neck cancer. J Clin Oncol Off J Am Soc Clin Oncol 1995;13(6):1493–500.

[33] Welkoborsky HJ, Bleier R, Wissen-Siegert I. Antineoplastic effectiveness and unwanted side effects of polychemotherapy of extensive oro- and hypopharyngeal cancers—results of a prospective therapy study with 5-FU/cisplatin versus 5-FU/carboplatin. Laryngorhinootologie 1992;71(5):261–6.

[34] Ebeling O, Eckel HE, Volling P, Heitmann K, Vossing M. Cisplatin/5-FU versus carboplatin/5-FU. 5 year follow-up. HNO 1994;42(10):629–35.

[35] Burtness B, Harrington K, Greil R, Soulieres D, Tahara M, de Castro Jr G, Psyrri A, Baste Rotllan N, Neupane PC, Bratland A, Fuereder T, BGM H, Mesia R, Ngamphaiboon N, Rordorf T, Wan Ishak WZ, Roy A, Cheng J, Jin F, Rischin D. KEYNOTE-048: phase 3 study of first-line pembrolizumab (P) for recurrent/metastatic head and neck squamous cell carcinoma (R/M HNSCC). In: ESMO 2018 congress; 2018.

[36] Cohen EEW, Soulieres D, Le Tourneau C, Dinis J, Licitra L, Ahn MJ, Soria A, Machiels JP, Mach N, Mehra R, Burtness B, Zhang P, Cheng J, Swaby RF, Harrington KJ, KEYNOTE-040 Investigators. Pembrolizumab versus methotrexate, docetaxel, or cetuximab for recurrent or metastatic head-and-neck squamous cell carcinoma (KEYNOTE-040): a randomised, open-label, phase 3 study. Lancet 2019;393(10167):156–67.

[37] Mehra R, Seiwert TY, Gupta S, Weiss J, Gluck I, Eder JP, Burtness B, Tahara M, Keam B, Kang H, Muro K, Geva R, Chung HC, Lin CC, Aurora-Garg D, Ray A, Pathiraja K, Cheng J, Chow LQM, Haddad R. Efficacy and safety of pembrolizumab in recurrent/metastatic head and neck squamous cell carcinoma: pooled analyses after long-term follow-up in KEYNOTE-012. Br J Cancer 2018;119(2):153–9.

[38] Machiels JP, Haddad RI, Fayette J, Licitra LF, Tahara M, Vermorken JB, Clement PM, Gauler T, Cupissol D, Grau JJ, Guigay J, Caponigro F, de Castro Jr G, de Souza Viana L, Keilholz U, Del Campo JM, Cong XJ, Ehrnrooth E, Cohen EE, LUX-H&N 1 Investigators. Afatinib versus methotrexate as second-line treatment in patients with recurrent or metastatic squamous-cell carcinoma of the head and neck progressing on or after platinum-based therapy (LUX-Head & Neck 1): an open-label, randomised phase 3 trial. Lancet Oncol 2015;16(5):583–94.

[39] Seiwert TY, Fayette J, Cupissol D, Del Campo JM, Clement PM, Hitt R, Degardin M, Zhang W, Blackman A, Ehrnrooth E, Cohen EE. A randomized, phase II study of afatinib versus cetuximab in metastatic or recurrent squamous cell carcinoma of the head and neck. Ann Oncol 2014;25(9):1813–20.

[40] Herbst RS, Baas P, Kim DW, Felip E, Perez-Gracia JL, Han JY, Molina J, Kim JH, Arvis CD, Ahn MJ, Majem M, Fidler MJ, de Castro Jr G, Garrido M, Lubiniecki GM, Shentu Y, Im E, Dolled-Filhart M, Garon EB. Pembrolizumab versus docetaxel for previously treated, PD-L1-positive, advanced non-small-cell lung cancer (KEYNOTE-010): a randomised controlled trial. Lancet 2016;387(10027):1540–50.

[41] Reck M, Rodriguez-Abreu D, Robinson AG, Hui R, Csoszi T, Fulop A, Gottfried M, Peled N, Tafreshi A, Cuffe S, O'Brien M, Rao S, Hotta K, Vandormael K, Riccio A, Yang J, Pietanza MC, Brahmer JR. Updated analysis of KEYNOTE-024: pembrolizumab versus platinum-based chemotherapy for advanced non-small-cell lung cancer with PD-L1 tumor proportion score of 50% or greater. J Clin Oncol Off J Am Soc Clin Oncol 2019;.

[42] Reck M, Rodriguez-Abreu D, Robinson AG, Hui R, Csoszi T, Fulop A, Gottfried M, Peled N, Tafreshi A, Cuffe S, O'Brien M, Rao S, Hotta K, Leiby MA, Lubiniecki GM, Shentu Y, Rangwala R, Brahmer JR, KEYNOTE-024 Investigators. Pembrolizumab versus chemotherapy for PD-L1-positive non-small-cell lung cancer. N Engl J Med 2016;375(19):1823–33.

[43] Paz-Ares L, Luft A, Vicente D, Tafreshi A, Gumus M, Mazieres J, Hermes B, Cay Senler F, Csoszi T, Fulop A, Rodriguez-Cid J, Wilson J, Sugawara S, Kato T, Lee KH, Cheng Y, Novello S, Halmos B, Li X, Lubiniecki GM, Piperdi B, Kowalski DM, KEYNOTE-407 Investigators. Pembrolizumab plus chemotherapy for squamous non-small-cell lung cancer. N Engl J Med 2018;379(21):2040–51.

[44] Gandhi L, Rodriguez-Abreu D, Gadgeel S, Esteban E, Felip E, De Angelis F, Domine M, Clingan P, Hochmair MJ, Powell SF, Cheng SY, Bischoff HG, Peled N, Grossi F, Jennens RR, Reck M, Hui R, Garon EB, Boyer M, Rubio-Viqueira B, Novello S, Kurata T, Gray JE, Vida J, Wei Z, Yang J, Raftopoulos H, Pietanza MC, Garassino MC, KEYNOTE-189 Investigators. Pembrolizumab plus chemotherapy in metastatic non-small-cell lung cancer. N Engl J Med 2018;378(22):2078–92.

[45] Tarhini AA, Kirkwood JM. CTLA-4-blocking immunotherapy with ipilimumab for advanced melanoma. Oncology (Williston Park) 2010;24(14):1302. 1304.

[46] Tarhini A, Lo E, Minor DR. Releasing the brake on the immune system: ipilimumab in melanoma and other tumors. Cancer Biother Radiopharm 2010;25(6):601–13.

[47] Wolchok JD, Chiarion-Sileni V, Gonzalez R, Rutkowski P, Grob JJ, Cowey CL, Lao CD, Wagstaff J, Schadendorf D, Ferrucci PF, Smylie M, Dummer R, Hill A, Hogg D, Haanen J, Carlino MS, Bechter O, Maio M, Marquez-Rodas I, Guidoboni M, McArthur G, Lebbe C, Ascierto PA, Long GV, Cebon J, Sosman J, Postow MA, Callahan MK, Walker D, Rollin L, Bhore R, Hodi FS, Larkin J. Overall survival with combined Nivolumab and Ipilimumab in advanced melanoma. N Engl J Med 2017;377(14):1345–56.

[48] Haddad R, Gillison M, Ferris RL, Harrington K, Monga M, Baudelet C, Geese WJ, Argiris A. Double-blind, two-arm, phase 2 study of nivolumab (nivo) in combination with ipilimumab (ipi) versus nivo and ipi-placebo (PBO) as first-line (1L) therapy in patients (pts) with recurrent or metastatic squamous cell carcinoma of the head and neck (R/M SCCHN)—CheckMate 714. Ann Oncol 2016;27 [suppl_6, 1 October 2016].

[49] Argiris A, Gillison M, Ferris RL, Harrington K, Sanchez TK, Baudelet C, Geese WJ, Shaw JW, Haddad R. A randomized, open-label, phase 3 study of nivolumab in combination with ipilimumab vs extreme regimen (cetuximab + cisplatin/carboplatin + fluorouracil) as first-line therapy in patients with recurrent or metastatic squamous cell carcinoma of the head and neck—CheckMate 651. Ann Oncol 2016;27 [suppl_6, 1 October 2016].

[50] Siu LL, Even C, Mesia R, Remenar E, Daste A, Delord JP, Krauss J, Saba NF, Nabell L, Ready NE, Brana I, Kotecki N, Zandberg DP, Gilbert J, Mehanna H, Bonomi M, Jarkowski A, Melillo G, Armstrong JM, Wildsmith S, Fayette J. Safety and efficacy of Durvalumab with or without Tremelimumab in patients with PD-L1-low/negative recurrent or metastatic HNSCC: the phase 2 CONDOR randomized clinical trial. JAMA Oncol 2019;5:193–203.

[51] Ferris RL, Even C, Haddad R, Tahara M, Goswami T, Franks A, Emeribe U, Jarkowski A, Melillo G, Licitra L. Phase III, randomized, open-label study of durvalumab (MEDI4736) monotherapy, or durvalumab + tremelimumab, versus standard of care (SoC), in recurrent or metastatic (R/M) squamous cell carcinoma of the head and neck (SCCHN): eagle. J Immunother Cancer 2015;3(Suppl 2):P150.

[52] Schwarz JK, Payton JE, Rashmi R, Xiang T, Jia Y, Huettner P, Rogers BE, Yang Q, Watson M, Rader JS, Grigsby PW. Pathway-specific analysis of gene expression data identifies the PI3K/Akt pathway as a novel therapeutic target in cervical cancer. Clin Cancer Res 2012;18(5):1464–71.

[53] Ribas A, Medina T, Kummar S, Amin A, Kalbasi A, Drabick JJ, Barve M, Daniels GA, Wong DJ, Schmidt EV, Candia AF, Coffman RL, Leung ACF, Janssen RS. SD-101 in combination with Pembrolizumab in advanced melanoma: results of a phase Ib, multicenter study. Cancer Discov 2018;8(10):1250–7.
[54] Cohen E, Bishnoi S, Laux DE, Wong D, Amin A, Nabell L, Schmidt EV, Xing B, Leung AC, Janssen R. Phase Ib/II, open label, multicenter study of intratumoral SD-101 in combination with pembrolizumab in anti-PD-1 treatment naive patients with recurrent or metastatic head and neck squamous cell carcinoma (HNSCC). In: Proceedings: AACR annual meeting. vol. 78; 2018 [13].
[55] Harrington KJ, Kong AH, Mach N, Rordorf T, Jaime JC, Espeli V, Treichel S, Gumuscu B, Kim JJ, Chesney JA. Safety and preliminary efficacy of talimogene laherparepvec (T-VEC) in combination with pembrolizumab in patients with recurrent or metastatic squamous cell carcinoma of the head and neck: A multicenter, phase 1b study (MASTERKEY-232). J Clin Oncol 2018;36(suppl):6036.

CHAPTER

3

Combining radiation and chemotherapy to improve the therapeutic ratio

Huaising C. Ko[a], *Pippa F. Cosper*[b], *Randall J. Kimple*[a], *Matthew E. Witek*[b]

[a]Department of Human Oncology, University of Wisconsin Carbone Cancer Center, University of Wisconsin School of Medicine and Public Health, University of Wisconsin, Madison, WI, United States [b]Department of Human Oncology, University of Wisconsin School of Medicine and Public Health, University of Wisconsin, Madison, WI, United States

Abstract

Clinical outcomes and general treatment paradigms for patients with oral cavity, laryngeal, hypopharyngeal, and human papillomavirus (HPV) negative oropharyngeal squamous cell carcinoma (OPSCC) have remained relatively stable over the past several decades. Since the early 2000s it has become clear that patients with HPV+ OPSCC have significantly improved outcomes compared to those with HPV− disease. This has led to considerable efforts being focused on defining the best approach to maximize the therapeutic ratio in patients with HPV+ OPSCC in order to minimize the toxicities associated with treatment. Simultaneously, for patients with HPV− disease, attempts to intensify treatment have been taken in order to improve outcomes in patients with a significantly worse prognosis. In this chapter we review the role of radiation therapy in the treatment of head and neck cancer by focusing on the basics of radiobiology juxtaposed with the intricacies of modern radiotherapy, surgical techniques, systemic agents, and the on-going efforts aimed at combining their interactions to improve clinical outcomes while reducing treatment-related toxicities.

Abbreviations

DNA	deoxyribonucleic acid
ECE	extracapsular extension
Gy	Gray
HPV	human papillomavirus
LD	lethal damage

https://doi.org/10.1016/B978-0-12-817868-3.00003-2

OPSCC oropharyngeal squamous cell carcinoma
PLC potentially lethal damage
PNI perineural invasion
SLD sublethal damage
TLM transoral laser microsurgery
TORS transoral robotic surgery

Conflict of interest

No potential conflicts of interest were disclosed.

Basic principles of radiobiology

Classical radiobiology teaching states that radiotherapy exploits the reduced ability of tumor cells to repair DNA damage compared to normal cells. Conventionally, the effectiveness of radiotherapy depends of what is referred to as, "The Four R's of Radiobiology," which include DNA repair, reassortment of the cell cycle, reoxygenation, and repopulation [1].

Two-thirds of double-stranded DNA breaks result indirectly from the formation of oxygen radicals that act upon DNA while one-third are caused directly by "hitting" the DNA. The damage caused by radiotherapy can be categorized as either lethal damage (LD), potentially lethal damage (PLD), or sublethal damage (SLD). LD is irreversible and leads to cell death. PLD can be repaired under suitable environmental conditions. SLD is similar to PLD, in which it can be repaired, unless additional SLD occurs prior to repair. SLD is considered the major contributor of DNA damage caused by radiotherapy-mediated death and is a fundamental reason for giving daily radiotherapy treatment fractions.

In addition to SLD providing a rationale for treating tumors with small daily treatments, the role of cell cycle reassortment also contributes to this approach. Cells live and propagate through several phases including the G0/G1, S-phase, G2, and the M-phase. Each phase of the cell cycle exhibits different levels of sensitivity to radiotherapy. Cells in the G2 phase a typically the most sensitive while those in S-phase are the most resistant. By delivering radiotherapy in daily fractions radiation oncologists can exploit the properties of SLD and cell cycle reassortment to improve the therapeutic ratio.

Radiation, by killing a proportion of tumor cells, can permit tumors to continue to divide by removing nutrient and oxygen depriving competitors. This response is termed repopulation. The rate of repopulation increases (accelerated repopulation) as the time course of treatment is prolonged. When this occurs, the total dose needed to achieve cure needs to be increased. In vitro and clinical studies have shown that local control decreases by 1.4% per day of the prolonged overall treatment time [2, 3].

The presence of oxygen is key to "fixing" or making permanent an otherwise transient event of DNA damage induced by oxygen radical species. Hypoxia, conversely, is radioprotective to cells. For this reason, reoxygenation is another phenomenon that places fractionated radiotherapy at an advantage over delivering the total dose in a single treatment as fractionation allows for the hypoxic cells to become reoxygenated between doses of radiation thus becoming more vulnerable to the next fraction [4, 5].

The importance of these principles and their relationship to fractionation in the management of head and neck cancers have been demonstrated by multiple prospective clinical trials (e.g., DAHANCA-6, DAHANCA-7, and RTOG 90-03) and have been further analyzed via metaanalyses [6–10].

Basic principles of radiation-induced cell death

Radiotherapy induces cell death through mitotic catastrophe, apoptosis, and engagement of the immune system. Radiotherapy-induced mitotic catastrophe results following cell division(s) with chromosomal abnormalities that include dicentric chromosomes, acentric chromosomal fragments, ring chromosomes, and anaphase bridges. Cell death following formation of the above-mentioned chromosomal abnormalities may occur rapidly but can also be seen after several rounds of replication spanning days to weeks. Radiotherapy has the potential to induce cellular apoptosis, which occurs through a rapidly occurring chain of cellular events involving endonuclease activation, chromatin condensation, cellular shrinkage, and cellular fragmentation [4].

Unlike mitotic catastrophe and apoptosis, which are well understood, immunogenic cell death is a complicated intertwined mechanism involving numerous cellular systems that continue to be unraveled. One demonstrated pathway includes the release of tumor antigens following irradiation allowing the uptake of the tumor antigens by dendritic cells. The immune system is then capable of stimulating tumor-antigen specific cytotoxic T-lymphocytes that can act both at the site of the irradiated tumor and distant untreated sites, a phenomenon termed the abscopal effect [11, 12].

Basic principles of radiotherapy in head and neck cancer management

Radiotherapy serves a key role in the treatment of patients with head and neck cancer in the definitive nonsurgical setting, as an adjuvant to surgery, or for palliation of tumor-related symptoms in incurable cases [13]. The clinical scenario dictates the radiotherapy target volume, total dose, and fraction number. Factors such as the cancer stage or postoperative pathologic feathers are used as indicators of the benefit of systemic agents.

Several common themes underlie radiotherapy decisions in the clinic. Target volumes are typically classified as high-, intermediate-, and low-risk depending on their likelihood of containing gross or subclinical disease (Fig. 1). In the definitive setting high-risk regions include the primary tumor or pathologic lymph nodes and usually receive doses of 66–72 Gray (Gy) in fractions sizes ranging from 2 to 2.2 Gy if given daily or 1.2–1.4 Gy if given twice daily. In the adjuvant setting, high-risk regions include resection cavities with positive margins or lymph nodes with extracapsular extension and are commonly treated to doses ranging from 60 to 66 Gy in 30–33 daily fraction. Intermediate-risk volumes that commonly receive 60–63 Gy in 1.8–2 Gy daily fractions include areas at considerable risk for containing microscopic disease such as areas adjacent to gross disease, resection-margin negative tumor beds, or dissected nodal levels that contained pathologic lymph nodes. Low-risk regions that are usually treated to doses of 50–56 Gy in daily fraction sizes of 1.6–1.8 Gy are areas that are

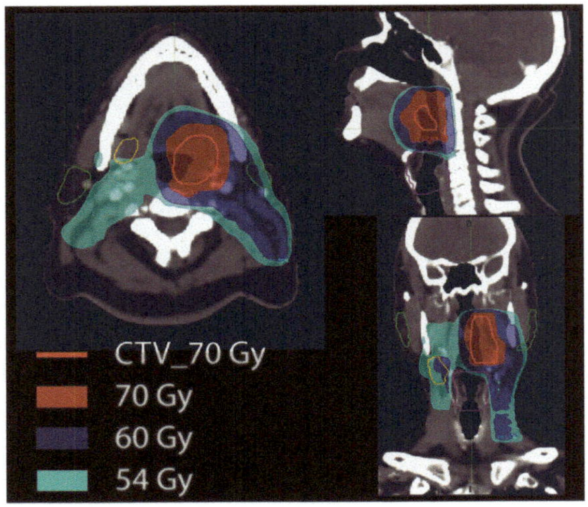

FIG. 1 Head and neck cancer radiation plan. Dose distribution *(shaded areas)*, high-risk clinical target volume (CTV_70), larynx *(purple)*, parotid glands *(green)*, and submandibular gland *(yellow)* from a patient with a T2N1M0 HPV-related squamous cell carcinoma of the left tonsil.

of lower concern for containing microscopic disease but enough to warrant treatment such as a contralateral lymph node level from a primary tumor that approaches or crosses midline.

Standards of care for palliation are more variable and range from doses of 20 Gy in 5 fractions in cases of advanced disease in patients with a limited life expectancy; to near curative doses in patients with metastatic disease that exhibit slow progression distantly but might exhibit severe local symptoms if uncontrolled (e.g., adenoid cystic carcinoma). Systemic therapy, which have thus far consisted of cytotoxic chemotherapy such as cisplatin, carboplatin, and taxanes or biologics such as the epidermal growth factor receptor inhibitors, are commonly employed in the definitive setting for advanced clinical disease stages (e.g., T3-4N+) [14–16] or in the adjuvant setting when a positive surgical margin or extracapsular extension is identified on final pathology [17, 18].

Basic principles of toxicities of head and neck cancer management

Similar to the myriad radiotherapy options for management of tumors, the resultant toxicities are also numerous and vary according to the volume and dose of radiotherapy a particular tissue receives [19, 20].

Radiotherapy toxicities are typically categorized as acute or late. Acute toxicities are those that occur during and immediately (i.e., <3 months) following treatment. Late toxicities are those that occur months to years following the completion of radiation. Acute toxicities are seen in most patients receiving head and neck cancer radiotherapy and can include dysgeusia, xerostomia, dermatitis, dysphagia, mucositis, pharyngitis, laryngitis, and fatigue. These are typically temporary and the majority are self-limited and resolve within weeks of finished radiation. Dysgeusia and xerostomia are two that often have life-long effects. Late toxicities are much less common but can be quite severe. Examples of late toxicities can include soft tissue fibrosis, osteoradionecrosis of the mandible, laryngeal chondronecrosis, hypopharyngeal stricture, carotid artery rupture, pharyngeal constrictor dysfunction, spinal cord myelopathy, dental caries, trismus, telangiectasias, lymphedema, and radiation-induced malignancies [21–23]. The incidence of late toxicities has decreased with modern radiation delivery approaches.

High quality data on toxicities is biased because most trials utilize "worst grade" methodologies in which only the presence or absence of a toxicity is noted [24]. This approach systematically excludes a high percentage of total reported grade 3 and grade 4 acute adverse event leading to progressive bias and favoring higher toxicity regimens by not collecting data on the duration and timing of events. Therefore, the TAME (short-term (acute) Toxicity (T), Adverse long-term (late) effects (A), and Mortality risk (M) generated by a treatment program (E = End results)) system was developed to enable a uniform way of comparing relative risks among different treatment groups [25].

Toxicities of systemic therapy depend on the specific drug used. The most common chemotherapies given concurrently with radiation for head and neck cancer are cisplatin and cetuximab. Recently, trials have begun investigating the role of taxanes such as doxetaxel given concurrently with radiation. In a few centers, 5-fluorouracil is combined with radiation for head and neck cancer. Cisplatin is commonly associated with nephrotoxicity, ototoxicity, peripheral neuropathy, myelosuppression, nausea and vomiting. Carboplatin is mainly associated with myelosuppression. Taxanes, commonly docetaxel, cause peripheral neuropathy. 5-fluorouracil usually given in combination with a taxanes or platins is most commonly known to cause diarrhea and mucositis. Cetuximab can cause an acneiform rash, mucositis, and occasionally anaphylaxis. The role of immunotherapy given concurrent with radiation is an area of active investigation. At this time it is most commonly used for patients with metastatic or recurrent cancers. The toxicities of immune modulators differ from that typically seen with cytotoxic agents and reflect an autoimmune phenotype and manifest with toxicities such as hypophysitis, colitis, hyperthyroidism, hypothyroidism, nephritis, and pneumonitis.

Surgical toxicities are dependent on the initial health of the patient, extent of tumor, type of surgery, and skill of the surgeon and can also be appreciated in the acute and late time frames. Acutely, there is a risk of bleeding during the surgery as well as anesthetic risks. In the postoperative setting patients are at risk of pain, infection and wound dehiscence, and failure of a viable blood supply for grafted tissue. Given structural changes following surgery, patients can be faced with dysarthria, dysphagia, loss of natural speech, shoulder pain and weakness, and changes in appearance.

Widening the therapeutic window

Multiple clinical trials have focused on decreasing the toxicity associated with head and neck cancer therapy while maintaining or improving upon expected clinical outcomes. The majority of these trials focus on HPV-positive OPSCC given the favorable outcomes seen in this group of patients. However, work is also focused on widening the therapeutic window in patients with non-HPV-related head and neck cancer. For convenience and ease of comparison, the different trials have been grouped below into similar approaches and assumed to be inclusive of HPV-/p16-positive disease unless otherwise noted.

Radiotherapy dose/volume reduction in the absence of a surgical component or induction chemotherapy

A radiotherapy-based standard of care for patients with locally advanced HPV+ OPSCC is 70 Gy of radiotherapy to gross disease with differential dosing of 60 Gy and 54–56 Gy to areas of intermediate- and low-risk, respectively given with concurrent cisplatin at 100 mg/m^2 every 3 weeks. Prospective analysis of 44 patients treated with a reduced dose of radiotherapy to gross disease of 60 Gy and chemotherapy to cisplatin given weekly at 30 mg/m^2 demonstrated 3-year local control, regional control, cause-specific survival, distant metastasis-free survival, and overall survival rates were 100%, 100%, 100%, 100%, and 95%, respectively [26]. Toxicities were thoroughly recorded in this trial and reported as improved compared to those associated with standard therapy [26]. This significantly reduced approach and excellent results need to be interpreted with caution as the patient cohort was very favorable. No patient had T4 disease and 82% were never smokers.

HN002 (Table 1) is a phase II controlled trial that randomized favorable risk patients that report ≤10 pack-years and present with clinical stages T1-T2N1-N2b or T3, N0-N2b to either 60 Gy of radiotherapy with weekly cisplatin in 6 weeks versus accelerated radiotherapy of 60 Gy in 5 weeks [27].

The aforementioned trials evaluated the efficacy of reducing dose in aggregate, primary tumor and nodal volumes. In addition to global dose reduction, there have been multiple trials that have investigated selectively reducing the radiotherapy dose to elective nodal volumes. A single arm prospective analysis of 54 patients with locally advanced head and neck squamous cell carcinoma with 57% being HPV+OPSCC evaluated the elective nodal failure rate after treating gross disease to 70 Gy and 36 Gy to elective nodal regions with concurrent cisplatin at 35 mg/m^2 [28]. Toxicity profiles were evaluated by the FACT-HN and QOL-RTI surveys. Toxicity assessments were favorable and there were no failures in the elective nodal levels treated with 36 Gy. A caveat that needs to be taken into consideration regarding the 36 Gy volume is that the treatment planning was done sequentially therefore the boost dose to the primary had the potential to increase dose to first echelon nodes. Indeed, post-hoc analysis suggested that the first echelon nodes received a mean and median dose of 45.5 Gy and 44.9 Gy, respectively, whereas the 38.6 Gy and 36.7 Gy were the mean and median dose to second echelon noes, respectively [28]. A recently published multicenter randomized phase III clinical trial demonstrated similar regional control for patients with locally advanced head and neck squamous cell carcinoma treated with either 70 Gy to gross disease and either 40 Gy or 50 Gy to elective nodal stations with or without chemotherapy [29]. Regional failure in elective nodal stations was 13.0% in the 40 Gy experimental arm and 6% in the 50 Gy control arm ($P=.08$). Patients in the experimental 40 Gy arm exhibited a significant reduction of any salivary gland toxicity at 6 and 18 months and a trend of decreased dysphagia [29]. The difference in elective nodal failure rates between these two similar trials likely reflects the patient population treated as the latter included approximately 40% oropharyngeal patients of which 30% were HPV-positive.

More aggressive radiotherapy de-escalation approaches include reduction of dose to and volume of elective nodal stations. INFIELD (Table 1) is a prospective phase II study of patients with oropharyngeal or laryngeal SCC that treats only lymph node levels with >5% risk of

TABLE 1 Clinical trials for various treatment regimens for head and neck cancer

Title and treatment regimen	Interventions	NCT number
HN002—Reduced-dose intensity-modulated radiation therapy with or without cisplatin in treating patients with advanced oropharyngeal cancer	Arm A) 60 Gy (6 weeks) + weekly cisplatin Arm B) 60 Gy (5 weeks)	NCT02254278
Involved field elective volume de-intensification radiation therapy for head and neck cancer (INFIELD)	RT (40 Gy to low risk node) +/− concurrent chemotherapy	NCT03067610
Quarterback 2: A phase II clinical trial of sequential therapy and de-intensified chemoradiotherapy for locally advanced HPV positive oropharynx cancer	Induction chemotherapy followed by radiotherapy (56 Gy in 28 fractions)	NCT02945631
Postoperative adjuvant therapy de-intensification trial for human papillomavirus-related, p16+ oropharynx cancer (ADEPT)	60 Gy +/− chemotherapy	NCT01687413
Postoperative adjuvant treatment for HPV-positive tumors (PATHOS) (PATHOS)	Arm A) no adjuvant treatment Arm B1) 60 Gy in 30 fractions Arm B2) 50 Gy in 25 fractions Arm C1) 60 Gy in 30 fractions with cisplatin Arm C2) 60 Gy in 30 fractions	NCT02215265
Transoral surgery followed by low-dose or standard-dose radiation therapy with or without chemotherapy in treating patients with HPV positive stage III-IVA oropharyngeal cancer	Arm A) no adjuvant treatment Arm B) 50 Gy in 5 weeks Arm C) 60 Gy in 6 weeks Arm D) 60–66 Gy + chemotherapy	NCT01398494
Definitive chemo-radiotherapy for regionally advanced head and neck cancer with or without upfront neck dissection (UPFRONT-NECK)	Arm A) 70 Gy in 35 fractison + cisplatin Arm B) Neck dissection followed by CRT	NCT02918955
Randomized trial of intensity-modulated proton beam therapy (IMPT) versus intensity-modulated photon therapy (IMRT) for the treatment of oropharyngeal cancer of the head and neck	A) IMRT (Photons) B) IMPT (Protons)	NCT01893307
The quarterback trial: reduced dose radiotherapy for HPV + oropharynx cancer	Arm A) 70 Gy + q3w carboplatin x 2cc Arm B) 56 Gy + q3w carboplatin	NCT01706939
ECOG 3311 transoral surgery followed by low-dose or standard-dose radiation therapy with or without chemotherapy in treating patients with HPV positive stage III-IVA oropharyngeal cancer	Arm A) TORS Arm Arm B1) TORS + 50 Gy in 25 fractions Arm Arm B2) TORS + 60 Gy in 30 fractions Arm Arm C) TORS + 66 Gy in 33 fractions + qw cisplatin	NCT01898494
Uniform FDG-PET guided gradient dose prescription to reduce late radiation	Arm A) 45 Gy to the elective neck; no intermediate dose level Arm B) 35 Gy to the	NCT02442375

Continued

TABLE 1 Clinical trials for various treatment regimens for head and neck cancer—Cont'd

Title and treatment regimen	Interventions	NCT number
toxicity (UPGRADE-RT): a randomized controlled trial with dose reduction to the elective neck in head and neck squamous cell carcinoma	elective neck; 60 Gy intermediate dose level	
A randomized placebo-controlled trial of synchronous nimorazole versus radiotherapy alone in patients with locally advanced head and neck squamous cell carcinoma not suitable for synchronous chemotherapy or cetuximab	Arm A) Radiotherapy Arm B) Radiotherarpy + Nimorazole	NCT01950689

harboring disease to a dose of 40 Gy with or without systemic therapy [30]. The primary objective is 2-year solitary elective volume recurrence and secondary objectives include patient reported outcomes and toxicities. The study has accrued but data has yet to be reported.

Radiotherapy dose/volume reduction following induction chemotherapy

Induction chemotherapy followed by concurrent chemoradiotherapy has been studied extensively and been shown to generate similar outcomes compared to concurrent chemoradiotherapy without induction [31, 32]. Utilizing chemotherapy as a way of selecting patients that will respond well to definitive radiotherapy has been supported in the larynx cancer literature [33]. Given these data as a foundation, several trials have evaluated the use of chemoselection in patients with HPV+OPSCC to determine those that may benefit from radiotherapy de-intensification. E1308 was a phase II trial of induction chemotherapy followed by reduced-dose radiotherapy with weekly cetuximab in patients that exhibited a clinical complete response at the primary site [34]. Of the 80 patients evaluable 70% achieved a primary tumor clinical complete response and proceeded with concurrent cetuximab and radiotherapy to 54 Gy. With a median follow-up of 35.4 months, 2-year progression-free survival and overall survival rates were 80% and 94%, respectively. Two-year progression free survival and overall survival for patients with favorable risk disease ($n = 27$, $<$ T4, $<$ N2c, and $\leq$ 10 pack-years of cigarette smoking) was 96% and 96%, respectively. Toxicity for patients treated with 54 Gy was significantly improved as they experienced less difficulty with swallowing solids and manifesting impaired nutrition. A single-arm phase II study performed at two academic US medical centers treated p16-positive patients with induction paclitaxel and carboplatin followed by concurrent paclitaxel and radiotherapy [35]. Patients that achieved a partial or complete response to induction therapy received 54 Gy whereas those that received less than a partial response received 60 Gy. Fifty-five percent of patients achieved a partial or complete response. Two-year progression-free survival was 92% for the entire cohort with 7% developing locoregional recurrences and 2% distant metastases. A favorable toxicity profile was achieved with no Grade 4 event reported and no patients were gastrostomy tube-dependent by 6 months. These data although promising have to be interpreted in context of the very favorable patient population that consisted of nearly

70% reported to never have smoked cigarettes and with 77% having T1-T2 tumors [35]. Yet to be reported are data from a larger phase III randomized trial which evaluated induction therapy followed by reduced dose radiotherapy. The Quarterback trial is evaluating induction with 3 cycles of docetaxel, cisplatin, and 5-fluorouracil [36]. Patients with a clinical or radiographic partial or complete response to induction were randomized in a 2:1 fashion to either 70 Gy with concurrent carboplatin or the experimental arm of 56 Gy with concurrent carboplatin. Toxicity data have been collected, but mature data have not been reported. This group has opened a follow-up study (Quarterback 2, Table 1) of sequential chemotherapy followed by reduced dose radiotherapy (56 Gy) [37].

In addition to dose reduction following a response to induction therapy, investigators have evaluated limiting radiotherapy treatment volumes as well. Following induction therapy with cisplatin, paclitaxel, and cetuximab, patients (71% oropharynx, HPV-status not recorded) with a good response, defined by a $\geq$50% reduction in the sum of all target lesions, received radiotherapy to a dose of 75 Gy in 1.5 Gy twice per day treatments with concurrent systemic therapy consisting of paclitaxel, fluorouracil and hydroxyurea to the area of the gross disease with a 1.5 cm expansion, whereas patients that were not good responders received the same dose to the primary as well as an elective 45 Gy volume to account for subclinical disease [38].

Systemic therapy toxicity studies without radiotherapy or surgical components

There is an estimated absolute benefit of 8% in 5-year overall survival for patients with head and neck squamous cell carcinoma treated with the addition of platinum-based chemotherapy to radiotherapy [39]. This benefit is similar when specifically assessing patients with oropharyngeal SCC [40]. This advantage, however, is beset with increased moderate to severe acute toxicities [14] and likely more long-term dysphagia and feeding tube dependence [21, 23].

Cetuximab, a monoclonal antibody targeting the epidermal growth factor receptor was shown to improve survival without increasing toxicity in a phase III randomized trial comparing radiotherapy versus radiotherapy plus cetuximab [15, 41]. The suggested improvement in the therapeutic index of cetuximab over cisplatin in combination with radiotherapy led to two phase III randomized trials, RTOG 1016 and De-ESCALaTE [42, 43]. RTOG 1016 was a prospective phase III noninferiority study of 849 patients with HPV+OPSCC randomized to accelerated radiotherapy of 70 Gy with either cetuximab (given as a loading dose of 400 mg/m^2 followed by seven weekly doses of 250 mg/m^2) or cisplatin (two cycles at 100 mg/m^2 give on days 1 and 22). At a median follow-up of 4.5 years, noninferiority was not met for overall survival. Estimated 5-year overall survival was 77.9% (95% CI 73.4–82.5) in the cetuximab cohort and 84.6% (95% CI 80.6–88.6) in the cetuximab cohort. Locoregional failure was significantly higher in the cetuximab arm while progression free survival was lower. Patients receiving cetuximab were more likely to develop acneiform rash while patients receiving cisplatin developed more myelosuppression, nausea, anemia, vomiting, dehydration, hyponatremia, kidney injury, anorexia, and hearing impairment [43]. De-ESCALaTE was a randomized phase III trial of 304 patients with HPV+OPSCC designed to detect a reduction of >25% in the overall number of Grade 3-5 acute and late toxicities between patients receiving concurrent cetuximab or cisplatin. The trial did not meet its primary goal as there was no significant difference in acute and late Grade 3-5 toxicities. However, similar to RTOG 1016, at 2-years overall survival for patients receiving cisplatin was 97.5% compared to 89.4% for those receiving cetuximab (HR 5.0; 95% CI 1.7–14.7; $P = .0007$) [42]. These data suggest that concurrent cisplatin should remain the standard of care for patients with HPV+OPSCC. However, given phase III data demonstrating improved overall with survival with cetuximab given concurrent with radiotherapy compared to radiotherapy alone, cetuximab should be considered in patients who are not able to receive concurrent cisplatin. Finally, data from both of these trials should provide caution in rapidly adopting novel approaches for patients with head and neck squamous cell carcinoma until they can be confirmed through randomized phase III data.

Radiotherapy dose/volume reductions with a surgical component

OPSCC had been predominately managed with radiotherapy with or without systemic therapy as the preponderance of data suggested similar outcomes with significantly lower morbidity [44]. However, as surgical techniques, such as transoral laser microsurgery (TLM) and transoral robotic surgery (TORS) have been developed and exhibit reduced toxicity compared with open approaches, they are more commonly used in the management of patients with OPSCC and provide an opportunity to decrease the need for and dose of radiotherapy and/or chemotherapy, which may decrease acute and long-term toxicities. For example, postoperative doses of radiotherapy are typically 10 Gy less than definitive doses (70 Gy versus 60 Gy), which has been shown to be important in long-term dysphagia as patients receiving 60 Gy to at least 70% of their cricopharyngeus inlet exhibit a 50% risk of developing stricture [21]. Further, reduction in neck dissection toxicity has been demonstrated in more contemporary series as patients are undergoing more selective neck dissections [45].

There are several on-going or maturing trials evaluating the utility of surgical in the de-escalation approach in patients with HPV+ OPSCC. ADEPT is a prospective phase III trial that evaluates patients with HPV+ OPSCC, T-stages 1-4a, who underwent TLM or TORS and were found to have extracapsular extension on final pathology [46]. Patients are randomized or allowed to choose the treatment arm of either 60 Gy with or without concurrent weekly cisplatin therapy. The primary outcome is disease-free survival. However, secondary outcomes include the cumulative incidence of acute toxicity at 4.5 months and function and quality of life at 2 years. PATHOS is a phase III trial that is evaluating if swallowing function of patients with HPV+OPSCC can be improved following TORS and reduced intensity adjuvant treatment regimens [47]. Multiple adjuvant therapy protocols are included in this protocol (Table 1) based on final pathology findings such as T- and N-stage and the presence or absence of adverse features such as positive margins (<1 mm), extranodal extension, perineural invasion, and vascular invasion. The primary outcome measure is the MD Anderson Dysphagia Inventory score at 12 months after completion of therapy. ECOG 3311 is a phase II trial examining the role of transoral surgery in treatment de-intensification for stage III/IV HPV-associated OPC with adjuvant strategies based on pathologic findings [48]. Following surgery with neck dissection, patients are classified into risk groups based upon pathological criteria. Patients classified as low risk will have T1-2N0-1 disease and negative surgical margins, no ECE or PNI, and these patients will be observed following surgery. Patients with close margins, minimal ECE, N2a-N2b disease or perineural invasion will compose the intermediate risk group and will be randomized to receive 50 Gy or 60 Gy of adjuvant radiation. Finally, patients with positive margins, extensive ECE, or >5 positive lymph nodes will be high risk and will be treated with standard of care chemoradiation therapy with 66 Gy radiation concurrent with cisplatin. However, unlike PATHOS, the primary outcome is progression-free survival while toxicity measures are secondary outcomes.

Risk stratification based upon surgical pathology is also being studied in a phase 3 trial based in Bern, Switzerland. UPFRONT-NECK is investigating whether upfront neck dissection prior to chemoradiation therapy results in decreased treatment toxicity by allowing the surgical pathology results to tailor the adjuvant radiation volume in patients with head and neck squamous cell carcinoma of the oropharynx, hypopharynx or larynx [49]. This study will compare acute and subacute toxicity between patients receiving upfront neck dissection and tailored adjuvant (chemo)radiation therapy versus those receiving definitive (chemo)radiation therapy with standard neck volumes and doses.

Radiotherapy protectors

There have been numerous compounds evaluated for their ability to reduce radiotherapy toxicity, specifically agents to reduce xerostomia and mucositis [50–52]. To date, there are no agents that are commonly used in the clinical setting. A promising agent that deserves mentioning is the compound that is a first-in-class molecular enzyme mimetic that selectively converts superoxide generated by radiotherapy into hydrogen peroxide, which is more toxic to tumor than normal cells. Initial clinical data support the ability of GC4419 to reduce the duration, incidence, and severity of severe oral mucositis [52]. The compound is currently undergoing in a phase II randomized, double-blind, placebo-controlled, multicenter trial [53]. The primary outcome measure is duration of radiation induced severe oral mucositis as defined by the WHO criteria.

Proton therapy

The clinical benefits of proton therapy, which theoretically should minimize toxicities in patients with head and neck cancer given theoretical dosimetric advantages, has never been demonstrated to be superior in a randomized trial [54]. A phase II/III randomized trial of intensity-modulated proton beam therapy versus intensity-modulated photon therapy for patients with oropharynx squamous cell carcinoma [55]. The primary objective is late grade 3-5 toxicity from 90 days to 2 years post therapy according to the Common Terminology Criteria for Adverse Events (CTCAE) version 4. The secondary outcome is progression-free survival.

Conclusion

Over the last 20 years there have been numerous advances in therapies available to manage patients with head and neck malignancies. Significantly improved clinical outcomes of patients with HPV+OPSCC have led to myriad clinical trials evaluating regimens that are aimed at maintaining the expected outcomes but reduce both acute and long-term toxicities. To date, several studies have been reported with some demonstrating successful treatment de-intensification, while others have not. A major component of these disparate results is likely a reflection of differing patient populations. This highlights the importance of improved markers to personalize therapy for each individual patient.

Acknowledgments

None.

References

[1] Hall EJ, Giaccia AJ. Radiobiology for the radiologist. Philadelphia: Wolters Kluwer Health/Lippincott Williams & Wilkins; 2019.
[2] Fowler JF, Lindstrom MJ. Loss of local control with prolongation in radiotherapy. Int J Radiat Oncol Biol Phys 1992;23(2):457–67.
[3] Withers HR, Taylor JM, Maciejewski B. The hazard of accelerated tumor clonogen repopulation during radiotherapy. Acta Oncol 1988;27(2):131–46.
[4] Hall E. Radiobiology for the radiologist. 6th ed. Philadelphia: Lippincott Williams & Wilkins; 2006.
[5] Withers R. The four Rs of radiotherapy. Adv Radiat Biol 1975;5:241–71.
[6] Bourhis J, Overgaard J, Audry H, Ang KK, Saunders M, Bernier J, Horiot JC, Le Maitre A, Pajak TF, Poulsen MG, O'Sullivan B, Dobrowsky W, Hliniak A, Skladowski K, Hay JH, Pinto LH, Fallai C, Fu KK, Sylvester R, Pignon JP. Meta-Analysis of Radiotherapy in Carcinomas of Head and neck (MARCH) Collaborative Group. Hyperfractionated or accelerated radiotherapy in head and neck cancer: a meta-analysis. Lancet 2006;368(9538):843–54.
[7] Fu KK, Pajak TF, Trotti A, Jones CU, Spencer SA, Phillips TL, Garden AS, Ridge JA, Cooper JS, Ang KK. A radiation therapy Oncology group (RTOG) phase III randomized study to compare hyperfractionation and two variants of accelerated fractionation to standard fractionation radiotherapy for head and neck squamous cell carcinomas: first report of RTOG 9003. Int J Radiat Oncol Biol Phys 2000;48(1):7–16.
[8] Lacas B, Bourhis J, Overgaard J, Zhang Q, Gregoire V, Nankivell M, Zackrisson B, Szutkowski Z, Suwinski R, Poulsen M, O'Sullivan B, Corvo R, Laskar SG, Fallai C, Yamazaki H, Dobrowsky W, Cho KH, Beadle B, Langendijk JA, Viegas CMP, Hay J, Lotayef M, Parmar MKB, Auperin A, van Herpen C, Maingon P, Trotti AM, Grau C, Pignon JP, Blanchard P, Group MC. Role of radiotherapy fractionation in head and neck cancers (MARCH): an updated meta-analysis. Lancet Oncol 2017;18(9):1221–37.
[9] Lyhne NM, Primdahl H, Kristensen CA, Andersen E, Johansen J, Andersen LJ, Evensen J, Mortensen HR, Overgaard J. The DAHANCA 6 randomized trial: Effect of 6 vs 5 weekly fractions of radiotherapy in patients with glottic squamous cell carcinoma. Radiother Oncol 2015;117(1):91–8.
[10] Overgaard J, Hansen HS, Specht L, Overgaard M, Grau C, Andersen E, Bentzen J, Bastholt L, Hansen O, Johansen J, Andersen L, Evensen JF. Five compared with six fractions per week of conventional radiotherapy of squamous-cell carcinoma of head and neck: DAHANCA 6 and 7 randomised controlled trial. Lancet 2003;362(9388):933–40.
[11] Abuodeh Y, Venkat P, Kim S. Systematic review of case reports on the abscopal effect. Curr Probl Cancer 2016;40(1):25–37.
[12] Postow MA, Callahan MK, Barker CA, Yamada Y, Yuan J, Kitano S, Mu Z, Rasalan T, Adamow M, Ritter E, Sedrak C, Jungbluth AA, Chua R, Yang AS, Roman RA, Rosner S, Benson B, Allison JP, Lesokhin AM, Gnjatic S, Wolchok JD. Immunologic correlates of the abscopal effect in a patient with melanoma. N Engl J Med 2012;366(10):925–31.
[13] Pfister DGSS, Adelstein D, Adkins D, Brizel DM, Burtness B, Busse PM, Caudell JJ, Cmelak AJ, Eisele DW, Fenton M, Foote RL, Gilbert J, Gillison ML, Haddad RI, Hicks WL, Hitchcock YJ, Jimeno A, Leizman D, Maghami E, Mell LK, Mittal BB, Pinto HA, Ridge JA, Rocco J, Rodriguez CP, Shah JP, Weber RS, Witek M,

Worden F, Zhen W, Burns JL, Darlow SD. NCCN guideline insights: head and neck cancers, version 2.2018. J Natl Compr Canc Netw 2018;16(5):479–90.
[14] Adelstein DJ, Li Y, Adams GL, Wagner Jr H, Kish JA, Ensley JF, Schuller DE, Forastiere AA. An intergroup phase III comparison of standard radiation therapy and two schedules of concurrent chemoradiotherapy in patients with unresectable squamous cell head and neck cancer. J Clin Oncol 2003;21(1):92–8.
[15] Bonner JA, Harari PM, Giralt J, Azarnia N, Shin DM, Cohen RB, Jones CU, Sur E, Raben D, Jassem J, Ove R, Kies MS, Baselga J, Youssoufian H, Amellal N, Rowinsky EK, Ang KK. Radiotherapy plus cetuximab for squamous-cell carcinoma of the head and neck. N Engl J Med 2006;354(6):567–78.
[16] Calais G, Alfonsi M, Bardet E, Sire C, Germain T, Bergerot P, Rhein B, Tortochaux J, Oudinot P, Bertrand P. Randomized trial of radiation therapy versus concomitant chemotherapy and radiation therapy for advanced-stage oropharynx carcinoma. J Natl Cancer Inst 1999;91(24):2081–6.
[17] Bernier J, Domenge C, Ozsahin M, Matuszewska K, Lefebvre JL, Greiner RH, Giralt J, Maingon P, Rolland F, Bolla M, Cognetti F, Bourhis J, Kirkpatrick A, van Glabbeke M, European Organization for Research and Treatment of Cancer Trial 22931. Postoperative irradiation with or without concomitant chemotherapy for locally advanced head and neck cancer. N Engl J Med 2004;350(19):1945–52.
[18] Cooper JS, Pajak TF, Forastiere AA, Jacobs J, Campbell BH, Saxman SB, Kish JA, Kim HE, Cmelak AJ, Rotman M, Machtay M, Ensley JF, Chao KS, Schultz CJ, Lee N, Fu KK, Radiation Therapy Oncology Group I. Postoperative concurrent radiotherapy and chemotherapy for high-risk squamous-cell carcinoma of the head and neck. N Engl J Med 2004;350(19):1937–44.
[19] Emami B, Lyman J, Brown A, Coia L, Goitein M, Munzenrider JE, Shank B, Solin LJ, Wesson M. Tolerance of normal tissue to therapeutic irradiation. Int J Radiat Oncol Biol Phys 1991;21(1):109–22.
[20] Marks LB, Yorke ED, Jackson A, Ten Haken RK, Constine LS, Eisbruch A, Bentzen SM, Nam J, Deasy JO. Use of normal tissue complication probability models in the clinic. Int J Radiat Oncol Biol Phys 2010;76(3 Suppl) S10–9.
[21] Chen AM, Li BQ, Jennelle RL, Lau DH, Yang CC, Courquin J, Vijayakumar S, Purdy JA. Late esophageal toxicity after radiation therapy for head and neck cancer. Head Neck 2010;32(2):178–83.
[22] Lee MK, Nalliah RP, Kim MK, Elangovan S, Allareddy V, Kumar-Gajendrareddy P, Allareddy V. Prevalence and impact of complications on outcomes in patients hospitalized for oral and oropharyngeal cancer treatment. Oral Surg Oral Med Oral Pathol Oral Radiol Endod 2011;112(5):581–91.
[23] Machtay M, Moughan J, Trotti A, Garden AS, Weber RS, Cooper JS, Forastiere A, Ang KK. Factors associated with severe late toxicity after concurrent chemoradiation for locally advanced head and neck cancer: an RTOG analysis. J Clin Oncol 2008;26(21):3582–9.
[24] Thanarajasingam G, Hubbard JM, Sloan JA, Grothey A. The imperative for a new approach to toxicity analysis in oncology clinical trials. J Natl Cancer Inst 2015;107(10).
[25] Trotti A, Pajak TF, Gwede CK, Paulus R, Cooper J, Forastiere A, Ridge JA, Watkins-Bruner D, Garden AS, Ang KK, Curran W. TAME: development of a new method for summarising adverse events of cancer treatment by the Radiation Therapy Oncology Group. Lancet Oncol 2007;8(7):613–24.
[26] Chera BS, Amdur RJ, Tepper JE, Tan X, Weiss J, Grilley-Olson JE, Hayes DN, Zanation A, Hackman TG, Patel S, Sheets N, Weissler MC, Mendenhall WM. Mature results of a prospective study of deintensified chemoradiotherapy for low-risk human papillomavirus-associated oropharyngeal squamous cell carcinoma. Cancer 2018;124(11):2347–54.
[27] Oncology N, Institute NC. Reduced-dose intensity-modulated radiation therapy with or without cisplatin in treating patients with advanced oropharyngeal cancer, https://ClinicalTrials.gov/show/NCT02254278; 2014.
[28] Maguire PD, Neal CR, Hardy SM, Schreiber AM. Single-arm phase 2 trial of elective nodal dose reduction for patients with locoregionally advanced squamous cell carcinoma of the head and neck. Int J Radiat Oncol Biol Phys 2018;100(5):1210–6.
[29] Nevens D, Duprez F, Daisne JF, Dok R, Belmans A, Voordeckers M, Van den Weyngaert D, De Neve W, Nuyts S. Reduction of the dose of radiotherapy to the elective neck in head and neck squamous cell carcinoma; a randomized clinical trial. Effect on late toxicity and tumor control. Radiother Oncol 2017;122(2):171–7.
[30] Center UoTSM. Involved field elective volume De-intensification radiation therapy for head and neck cancer, https://ClinicalTrials.gov/show/NCT03067610; 2017.
[31] Iocca O, Farcomeni A, Di Rocco A, Di Maio P, Golusinski P, Pardinas Lopez S, Savo A, Pellini R, Spriano G. Locally advanced squamous cell carcinoma of the head and neck: a systematic review and Bayesian network meta-analysis of the currently available treatment options. Oral Oncol 2018;80:40–51.

[32] Budach W, Bolke E, Kammers K, Gerber PA, Orth K, Gripp S, Matuschek C. Induction chemotherapy followed by concurrent radio-chemotherapy versus concurrent radio-chemotherapy alone as treatment of locally advanced squamous cell carcinoma of the head and neck (HNSCC): a meta-analysis of randomized trials. Radiother Oncol 2016;118(2):238–43.
[33] Kiong KL, de Souza NN, Sultana R, Iyer NG. Meta-analysis of induction chemotherapy as a selection marker for chemoradiation in the head and neck. Laryngoscope 2018;128(7):1594–601.
[34] Marur S, Li S, Cmelak AJ, Gillison ML, Zhao WJ, Ferris RL, Westra WH, Gilbert J, Bauman JE, Wagner LI, Trevarthen DR, Balkrishna J, Murphy BA, Agrawal N, Colevas AD, Chung CH, Burtness B. E1308: Phase II trial of induction chemotherapy followed by reduced-dose radiation and weekly cetuximab in patients with HPV-associated resectable squamous cell carcinoma of the oropharynx—ECOG-ACRIN Cancer Research Group. J Clin Oncol 2017;35(5):490–7.
[35] Chen AM, Felix C, Wang PC, Hsu S, Basehart V, Garst J, Beron P, Wong D, Rosove MH, Rao S, Melanson H, Kim E, Palmer D, Qi L, Kelly K, Steinberg ML, Kupelian PA, Daly ME. Reduced-dose radiotherapy for human papillomavirus-associated squamous-cell carcinoma of the oropharynx: a single-arm, phase 2 study. Lancet Oncol 2017;18(6):803–11.
[36] Sinai ISoMaM, Institute TB, University AS. The quarterback trial: reduced dose radiotherapy for HPV + oropharynx cancer, https://ClinicalTrials.gov/show/NCT01706939; 2012.
[37] Sinai ISoMaM. Quarterback 2—sequential therapy with reduced dose chemoradiotherapy for HPV oropharynx cancer, https://ClinicalTrials.gov/show/NCT02945631; 2016.
[38] Villaflor VM, Melotek JM, Karrison TG, Brisson RJ, Blair EA, Portugal L, De Souza JA, Ginat DT, Stenson KM, Langerman A, Kocherginsky M, Spiotto MT, Hannigan N, Seiwert TY, Cohen EE, Vokes EE, Haraf DJ. Response-adapted volume de-escalation (RAVD) in locally advanced head and neck cancer. Ann Oncol 2016;27(5):908–13.
[39] Pignon JP, le Maitre A, Maillard E, Bourhis J, Group M-NC. Meta-analysis of chemotherapy in head and neck cancer (MACH-NC): an update on 93 randomised trials and 17,346 patients. Radiother Oncol 2009;92(1):4–14.
[40] Blanchard P, Baujat B, Holostenco V, Bourredjem A, Baey C, Bourhis J, Pignon JP, Group M-CC. Meta-analysis of chemotherapy in head and neck cancer (MACH-NC): a comprehensive analysis by tumour site. Radiother Oncol 2011;100(1):33–40.
[41] Bonner JA, Harari PM, Giralt J, Cohen RB, Jones CU, Sur RK, Raben D, Baselga J, Spencer SA, Zhu J, Youssoufian H, Rowinsky EK, Ang KK. Radiotherapy plus cetuximab for locoregionally advanced head and neck cancer: 5-year survival data from a phase 3 randomised trial, and relation between cetuximab-induced rash and survival. Lancet Oncol 2010;11(1):21–8.
[42] Mehanna H, Robinson M, Hartley A, Kong A, Foran B, Fulton-Lieuw T, Dalby M, Mistry P, Sen M, O'Toole L, Al Booz H, Dyker K, Moleron R, Whitaker S, Brennan S, Cook A, Griffin M, Aynsley E, Rolles M, De Winton E, Chan A, Srinivasan D, Nixon I, Grumett J, Leemans CR, Buter J, Henderson J, Harrington K, McConkey C, Gray A, Dunn J, De EHPVTG. Radiotherapy plus cisplatin or cetuximab in low-risk human papillomavirus-positive oropharyngeal cancer (De-ESCALaTE HPV): an open-label randomised controlled phase 3 trial. Lancet 2019;393(10166):51–60.
[43] Gillison ML, Trotti AM, Harris J, Eisbruch A, Harari PM, Adelstein DJ, Sturgis EM, Burtness B, Ridge JA, Ringash J, Galvin J, Yao M, Koyfman SA, Blakaj DM, Razaq MA, Colevas AD, Beitler JJ, Jones CU, Dunlap NE, Seaward SA, Spencer S, Galloway TJ, Phan J, Dignam JJ, Le QT. Radiotherapy plus cetuximab or cisplatin in human papillomavirus-positive oropharyngeal cancer (NRG Oncology RTOG 1016): a randomised, multicentre, non-inferiority trial. Lancet 2019;393(10166):40–50.
[44] Parsons JT, Mendenhall WM, Stringer SP, Amdur RJ, Hinerman RW, Villaret DB, Moore-Higgs GJ, Greene BD, Speer TW, Cassisi NJ, Million RR. Squamous cell carcinoma of the oropharynx: surgery, radiation therapy, or both. Cancer 2002;94(11):2967–80.
[45] Adelstein DJ, Ridge JA, Brizel DM, Holsinger FC, Haughey BH, O'Sullivan B, Genden EM, Beitler JJ, Weinstein GS, Quon H, Chepeha DB, Ferris RL, Weber RS, Movsas B, Waldron J, Lowe V, Ramsey S, Manola J, Yueh B, Carey TE, Bekelman JE, Konski AA, Moore E, Forastiere A, Schuller DE, Lynn J, Ullmann CD. Transoral resection of pharyngeal cancer: summary of a National Cancer Institute head and neck cancer steering committee clinical trials planning meeting, November 6-7, 2011, Arlington, Virginia. Head Neck 2012;34(12):1681–703.
[46] Medicine WUSo. Post operative adjuvant therapy De-intensification trial for human papillomavirus-related, p16+ oropharynx cancer, https://ClinicalTrials.gov/show/NCT01687413; 2013.

[47] Nixon L, Trust VN. Post-operative adjuvant treatment for HPV-positive Tumours (PATHOS), https://ClinicalTrials.gov/show/NCT02215265; 2015.

[48] Group ECO, Institute NC. Transoral surgery followed by low-dose or standard-dose radiation therapy with or without chemotherapy in treating patients with HPV positive stage III-IVA oropharyngeal cancer, https://ClinicalTrials.gov/show/NCT01898494; 2013.

[49] University Hospital Inselspital B, Bern Uo. Definitive chemo-radiotherapy for regionally advanced head and neck Cancer with or without up-front neck dissection, https://ClinicalTrials.gov/show/NCT02918955; 2016.

[50] Lee MG, Freeman AR, Roos DE, Milner AD, Borg MF. Randomized double-blind trial of amifostine versus placebo for radiation-induced xerostomia in patients with head and neck cancer. J Med Imaging Radiat Oncol 2019;63(1):142–50.

[51] Brizel DM, Wasserman TH, Henke M, Strnad V, Rudat V, Monnier A, Eschwege F, Zhang J, Russell L, Oster W, Sauer R. Phase III randomized trial of amifostine as a radioprotector in head and neck cancer. J Clin Oncol 2000;18(19):3339–45.

[52] Anderson CM, Sonis ST, Lee CM, Adkins D, Allen BG, Sun W, Agarwala SS, Venigalla ML, Chen Y, Zhen W, Mould DR, Holmlund JT, Brill JM, Buatti JM. Phase 1b/2a trial of the superoxide dismutase mimetic GC4419 to reduce chemoradiotherapy-induced oral mucositis in patients with oral cavity or oropharyngeal carcinoma. Int J Radiat Oncol Biol Phys 2018;100(2):427–35.

[53] Galera Therapeutics, Inc. A study to investigate the effects of GC4419 on radiation induced Oral mucositis in patients with head/neck cancer, https://ClinicalTrials.gov/show/NCT03689712; 2018.

[54] Blanchard P, Gunn GB, Lin A, Foote RL, Lee NY, Frank SJ. Proton therapy for head and neck cancers. Semin Radiat Oncol 2018;28(1):53–63.

[55] Center MDAC, Institute NC, Dental NIo Research C, Hospital MG. Randomized trial of intensity-modulated proton beam therapy (IMPT) versus intensity-modulated photon therapy (IMRT) for the treatment of oropharyngeal cancer of the head and neck, https://ClinicalTrials.gov/show/NCT01893307; 2013.

CHAPTER

4

Targeting *TP53* to augment therapeutic response in head and neck cancer

Abdullah A. Osman[a], *Antje Lindemann*[a], *Li Wang*[b], *Steven J. Frank*[b], *Jeffrey N. Myers*[a]

[a]Department of Head and Neck Surgery, The University of Texas MD Anderson Cancer Center, Houston, TX, United States [b]Department of Radiation Oncology, The University of Texas MD Anderson Cancer Center, Houston, TX, United States

Abstract

Despite recent advances in treatment modalities, patient survival for non-viral mediated oropharyngeal cancer patients continues to remain poor over the past years. Whole-exome sequencing studies reveal that *TP53* is commonly mutated in human papillomavirus-negative HNSCC patients. Clinically, *TP53* mutations are significantly associated with poor survival and tumor resistance to radiotherapy and chemotherapy in HNSCC patients, suggesting a potential utilization of the *TP53* mutational status as a useful molecular prognostic marker to predict clinical response. Several approaches for restoration of wild-type p53 function in HNSCC tumors have been developed using gene transfer therapy *via* adenovirus and nanoparticles delivery. These approaches have shown relative success inducing cell-cycle arrest and apoptosis. Recently, small molecules that specifically reactivate mutant p53 and trigger mutant p53-dependent apoptosis have been developed and tested in clinical trials and show great promise. Targeting HNSCC cells harboring mutant p53 with ATR, Chk1/2 and Wee1 inhibitors alone or in combination with chemotherapy and/or radiation has become an intensive area of investigation. Numerous studies have revealed a novel function of p53 in regulation of cellular metabolism. A compelling evidence for a novel gain-of-function (GOF) of mutant p53 proteins in promoting metabolic changes has been recently demonstrated in HNSCC. We have demonstrated that *TP53* mutations are associated with poor clinical outcome and increased radioresistance in HNSCC. We have also provided evidence that metformin acts as a radiosensitizer in vitro in *TP53* mutant HNSCC by increasing expression of p21 and reactive oxygen species (ROS) levels, leading to induction of cellular senescence. Collectively, these novel strategies raise hopes for more effective therapies for head and neck cancers.

Improving the Therapeutic Ratio in Head and Neck Cancer
https://doi.org/10.1016/B978-0-12-817868-3.00004-4

Abbreviations

ATM	ataxia-telangiectasia mutated
ATR	ATM-Rad3 related
CDK	cyclin dependent kinase
Chk1/2	checkpoint kinase-1/2
DDR	DNA damage response
EAp53	evolutionary action score of *TP53*
EGFR	epidermal growth factor receptor
GOF	gain-of-function
HDAC	histone deacetylase
HNSCC	head and neck squamous cell carcinoma
HPV	human papillomavirus
LRR	locoregional recurrence
PIKK	phosphatidylinositol 3-kinase-related kinase
ROS	reactive oxygen species
RPA	replication protein A
SEER	the surveillance, epidemiology, and end results
TCGA	The Cancer Genome Atlas
TME	tumor microenvironment
wt	wild-type

Conflict of interest

No potential conflicts of interest were disclosed.

Introduction

Head and neck squamous cell Carcinoma (HNSCC) encompasses a heterogeneous group of squamous epithelial malignancies arising from the mucosal lining of the upper aerodigestive region that includes cancer of the sinonasal tract, oral cavity, nasopharynx, oropharynx, hypopharynx, and larynx. With more than 680,000 estimated cases diagnosed annually, HNSCC is the seventh leading cause of cancer related deaths [1, 2]. The major risk factors for the development of HNSCC include tobacco, alcohol consumption, viral infections (Epstein–Barr virus, human papillomavirus (HPV), and herpes simplex virus), chewing betel nut, occupational exposure to carcinogens, immunodeficiency, irradiation, diet, and genetic predisposition [3, 4]. As a whole, the prevalence of HNSCC has been declining in the United States over recent years, as a result of declining rates of tobacco use [5]. However, incidence rates are significantly increasing for certain subsets of HNSCC. Most notably, of these, is the rise in squamous cell carcinoma of the oropharynx, attributed to oncogenic HPV infection [6–8]. Despite trends towards intensification of treatment, modest survival outcomes in latest patient cohorts have been observed as the result of improved superior survival of HPV-associated malignancies compared to those related to tobacco use [9]. Nevertheless, the overall survival for HNSCC across all stages has not improved significantly over the past decades and still remains in the range of 60% at 5 years (The Surveillance, Epidemiology, and End Results [SEER] Program https://seer.cancer.gov/).It is clear that there is an unmet need for novel therapeutic modalities and/or active combinations approaches to improve survival for patients with HNSCC. Multiple studies have demonstrated that *TP53*, the gene that encodes the p53 protein, is by far the most commonly mutated gene found in head and neck cancer and is, in turn, associated with treatment failure and poor survival outcomes [10, 11]. The p53 protein is a critical component of cellular mechanisms that respond to certain stresses, such as DNA damage, hypoxia, and oncogenic stress to preserve genomic integrity. The Cellular DNA damage often leads to stabilization and accumulation of wildtype p53, which in turn leads to enhanced transcription of several essential downstream target genes including, p21, Puma, and Noxa and subsequently cell-cycle arrest, apoptosis, and senescence [12].

Current treatment for head and neck squamous cell carcinoma (HNSCC)

Because of the remarkable differences in anatomical subsites and molecular pathogenesis of HNSCC, treatment of this disease often requires multiple modality therapy delivered by a multidisciplinary team of surgeons and clinical oncologists. Early-stage disease (stage I/II) is treated fairly well with single-modality therapy which most often entails surgery or radiotherapy. However, patients presenting with locoregionally advanced disease (stage III/IV), are often treated with local therapy with surgery and/or radiotherapy (i.e., intensity-modulated radiotherapy, IMRT), and frequently combined with systemic therapy. For metastatic disease and unresectable locoregional recurrence, following previous radiotherapy, the goal of treatment shifts to achieve palliation. As such, addition of cetuximab, an anti-epidermal growth factor receptor (EGFR) monoclonal antibody, to first-line chemotherapy improves the survival of recurrent or metastatic HNSCC patients [13]. In spite of recent advances in radiation, imaging, surgical techniques, and current systemic treatments, patients overall survival rates have not changed significantly over the past three decades. Thus, HNSCC remains a major formidable clinical challenge. and currently there are no useful biomarkers to guide treatment decisions. Therefore, identification of reliable biomarkers and novel molecular targets to classify patients for precision and personalized treatment plans are urgently needed. Such approach can be more effective and less toxic than current standard of-care therapy including platinum-based chemotherapy and radiation. Because *TP53* is highly mutated in head and neck cancers and associated with poor outcome, attempts have recently been made to stratify patients for therapy based on the *TP53* status. Currently, an ongoing phase II randomized clinical trial of adjuvant radiotherapy with or without cisplatin is treating patients with stage III-IVA HNSCC who have undergone surgery. In this trial, wildtype p53 and mutant p53 statuses are employed as predictive biomarkers of survival benefit given post-operative concurrent radiation and cisplatin (NCT02734537).

Mutations in *TP53* are common deleterious genetic alterations in HNSCC

Recent whole exome sequencing studies confirmed that *TP53* is the most commonly mutated tumor suppressor gene in HNSCCs, with a mutation frequency of approximately 80% in non-HPV-associated HNSCC cases [14–16]. An updated The Cancer Genome Atlas (TCGA) analysis results of 520 HNSCC cases demonstrated that 68% of tumors have *TP53* mutations (Table 1).

Moreover, tumors of the larynx and hypopharynx have the highest *TP53* mutation rate (79.4%), tumors of the tongue and oral cavity have a *TP53* mutation rate of 73.6%, and tumors of the oropharynx, including the tonsils, and base of the tongue have the lowest *TP53*

TABLE 1 Prevalence of *TP53* mutations in HNSCC TCGA data ($n = 520$)

Anatomical sites	Cases (*n*)	HPV+ (%)	HPV+ & WT *TP53* (%)	*TP53* mutation (%)	*TP53* mutation/ HPV− (%)
Larynx and hypopharynx	126	2.4 (3/126)	2.4 (3/126)	79.4 (100/126)	81.3 (100/123)
Oral cavity and tongue	314	5.1 (16/314)	4.8 (15/314)	73.6 (231/314)	77.2 (230/298)
Oropharynx, tonsil and base of tongue	80	65 (52/80)	65 (52/80)	28.8 (23/80)	82.1 (23/28)
Total number of cases	520	13.7 (71/520)	13.5 (70/520)	68.1 (354/520)	78.4 (353/450)

mutation rate (28.8%). Not surprisingly, tumors of the oropharynx and tonsils have the lowest *TP53* mutation rate, since the majority of oropharyngeal cancers are HPV-associated with *TP53* that is almost commonly wildtype in status. However, in most cases, the p53 protein is inactivated by the HPV oncogene-E6 through ubiquitination-dependent degradation pathways [17].

Tumor suppressive functions of *TP53*

The p53 protein consists of functionally specialized domains, that work coordinately, in a context-dependent manner, to successfully attain DNA binding and transactivation. These include a transactivation domain (TAD, amino acid residues 1–61), the proline rich domain (PR, residues 64–92), the central DNA-binding domain (DBD, residues 100–300), the oligomerization domain (OD, residues 323–335), and the C-terminal domain (CTD, residues 364–393). Within the unstructured C-terminal domain, p53 can form a dimer with another p53 molecule, which in turn combines with a second p53 dimer, leading to the formation of a final functional tetrameric complex [18]. Consistent with findings regarding many other tumor types, most *TP53* mutations in HNSCC are missense mutations, commonly occurring in the region of the protein that functions as the p53 DNA-binding domain (Fig. 1). Recently updated HNSCC TCGA data revealed that although the spectrum of *TP53* missense mutations is vast, some of these mutations, including those located at codons R248, R273, G245, R175, R282, and H179, have been shown to be the most prevalent hotspot mutations in HNSCCs. While TCGA results confirmed that HNSCC develops through multistep process of carcinogenesis, it seems that genomic alterations of *TP53* occur sometimes early during progression, since they can be easily detected in premalignant tumors [19]. Currently, the exact mechanisms of p53-mediated tumor suppression are not fully understood. The wildtype (wt) p53 acts as a transcription factor that suppresses tumor growth through regulation of many downstream target genes involved in cell-cycle arrest, apoptosis, senescence, DNA repair, and metabolism [20]. The p53 protein has a short half-life and is tightly kept at a very low level under normal physiology, as a result of degradation by its E3 ubiquitin ligase, MDM2, pirh2, and COP1 [21]. When cells are exposed to different genotoxic stresses, p53 undergoes post-translational modification through phosphorylation, acetylation, and other modifications resulting in its stabilization. Then, its level increases dramatically, causing

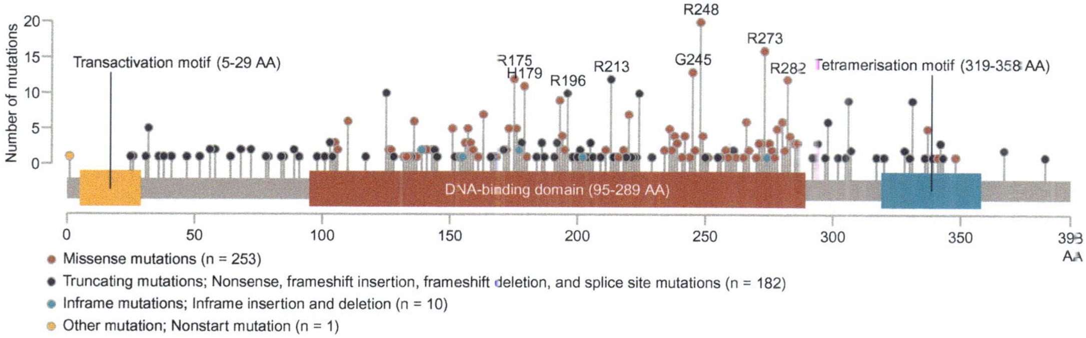

FIG. 1 Histogram of number of cases with *TP53* mutations by p53 codon position and the top 8 most frequently mutated p53 codons located in the DNA binding domain (DBD) in HNSCC-TCGA ($n = 446$). *Adapted from Lindemann A, Takahashi H, Patel AA, Osman AA, Myers JN. Targeting the DNA damage response in OSCC with TP53 mutations. J Dent Res 2018;97(6):635–44.*

transcription and activation of hundreds of canonical genes including p21, Noxa, and Puma that trigger cell-cycle arrest, senescence, apoptosis, and inhibit cell differentiation [12]. Consequently, these p53 activities ensure that cells with excessive DNA damage do not transmit this damage through proliferation. Therefore, when p53 loses its activity as a result of gene deletion or mutation, normal cells will be unable to control the DNA damage through repair or death, causing genomic instability and ultimately leading to cancer development [22]. In addition to loss of tumor-suppressive activity of wild-type p53 (wtp53), the stabilized mutant p53 proteins can gain new oncogenic properties called gain-of-function (GOF), mainly through specific binding to other transcriptional factors [23]. Recently, much attention has been devoted towards understanding the mechanism(s) of GOF activity of *TP53* mutations and their effects on tumor development, resistance to therapies and poor survival outcome. For instance, mutant p53 typically fails to regulate the expression of the wildtype p53-induced target genes due to loss of its ability to interact with canonical p53-responsive elements (p53REs) on the promoters of p53's target genes. Several studies have also demonstrated that mutant p53s are capable of binding to novel promoter sequences in target genes in a DNA structure-specific manner to activate them. On the other hand, the mutant p53 can act as enhancer or repressor of expression of some target genes through binding and cooperation with other transcription factors, such as p63, p73, nuclear factor-κB, NF-Y, E2F1, VDR, SP1, SREBP, and ETS. [22, 24]. Gaiddon and colleagues demonstrated that mutant p53 proteins can abolish the tumor-suppressive activities of the p53 family members, p63 and p73 and contribute to the invasive characteristics of these tumor cells [25]. Recent studies showed that mutant p53 epigenetically regulated chromatin remodeling through the SWI/SNF complex [26], and chromatin regulatory proteins methyltransferase MLL1/2 and acetyltransferase MOZ [27]. We have also shown that high-risk mutant p53 significantly decreases the p21 promoter activity and inhibits apoptosis and accelerated cellular senescence following treatment with cisplatin in HNSCC cells [28, 29]. Our recent report has also demonstrated that GOF *TP53* mutations can bind to and inhibit the activity of the master metabolic regulatory protein, AMPK, which leads to gains of oncogenic functions in HNSCC cells [30], indicating that a transcription-independent mechanism is also involved in the GOF properties of mutant p53. Additionally, we have recently presented evidence that GOF mutant p53 promotes the oncogenic potential of HNSCC cells by targeting the transcription factors FOXO3a and FOXM1 [31]

TP53 mutations as a prognostic and predictive biomarker for poor outcomes in HNSCC

The *TP53* gene is by far the most frequently mutated gene in HNSCC, indicating a strong selective pressure for *TP53* inactivation during tumor development. The prognostic implication of *TP53* mutations has been evaluated in numerous studies [10, 11]. Poeta and colleagues demonstrated that the 'disruptive' *TP53* mutations (defined as those mutations leading to nonsense mutation or an amino acid change that significantly changes the DNA binding domain of p53), were associated with worse disease-free and overall survival [29]. A study by Lindenbergh and colleagues found that truncating *TP53* mutations had poor outcomes compared to the disruptive *TP53* mutations [30]. Given the high frequency of *TP53* mutations in HNSCC and the lack of a unified and reliable classification system for *TP53*, it is unclear which group of patients would benefit most from clinical decisions and prognostication based on *TP53* status. In an effort to accurately predict the clinical impact of *TP53* mutations, we established and validated a novel algorithm called evolutionary action score of *TP53* (EAp53, http://mammoth.bcm.tmc.edu/EAp53/). In this algorithm, each *TP53* missense mutation is assigned to a score based on the functional sensitivity of p53 to amino acid sequence differences and classified as either high- or low-risk [32]. Sequence positions with high scores are typically associated with larger phylogenetic divergences and the least conservative amino acid changes than those sequence positions with lower scores. This classification system accurately separated patients with *TP53* mutations (high-risk) that have poor outcome, from those with an outcome similar to patients with wildtype *TP53* [32]. In addition, EAp53 revealed that HNSCC tumor cells expressing high-risk *TP53* mutations are highly resistant to platinum based therapy when compared to low-risk *TP53* mutations or wildtype *TP53* and associated with worse survival [28]. While the EAp53 algorithm did not take into account the nonsense/indel *TP53* mutations into classification, these mutations were included in the high-risk groups. This inclusion was made because we showed that these mutations were quite resistant to cisplatin and associated with decreased clonogenic survival in HNSCC (unpublished observation). These results were consistent with a previous study which demonstrated that truncated mutations were associated with poor outcomes in HNSCC [11]. With the current advances in novel therapies and genomics, it is important to identify head and neck cancer patients who will benefit from specific treatments and the identification of these patients may be accomplished using the EAp53 classification system for *TP53* mutations.

TP53 gene transfer therapy for HNSCC

Direct expression of wildtype p53 to replace missing or mutant p53 in cancer cells by means of gene transfer strategies have been conducted in the past. Consequently, the delivered wildtype p53 has successfully induced cell-cycle arrest and apoptosis leading to cancer cell killing. Liu and colleagues demonstrated previously that overexpression of the wtp53 or the transcription factor, E2F-1 genes in squamous cell carcinoma of the head and neck cell lines *via* adenovirus-mediated gene transfer suppressed tumor growth both in vitro and in vivo, as well as in actual clinical trials [33, 34]. Using another approach in preclinical studies, in vivo systemic administration of SGT-53, a nanocomplex that contains a normal human wtp53 cDNA, have significantly enhanced antitumor response to docetaxel [35]. Furthermore, a variety of advanced solid tumors including HNSCC treated with SGT-53 in several clinical trials have been shown to be sensitized to conventional radiation/chemotherapy, resulting in tumor growth inhibition and long-term regression [35, 36]. In a recent phase I study, Senzer and colleagues have shown that not only are systemically delivered SGT-53 nanoparticles well tolerated but they also exhibit anticancer activity specifically in metastatic lesions from advanced solid tumors [37]. An ongoing phase II clinical trials is currently investigating the combination of Adenoviral p53 (Ad-p53) administered intra-tumorally with Nivolumab in patients with advanced, recurrent head and neck squamous cell carcinoma and favorable wildtype p53 status (NCT03544723). No results have been reported yet from this trial. While the adenoviral and nanoparticle delivery approaches of p53 have demonstrated impressive responses, these approaches have not been adopted for widespread use due to a number of issues including low transduction efficiency and safety concerns. Hopefully, with advances in cutting-edge delivery technologies, p53 gene therapy is poised to become amenable for routine first line therapy for head and neck cancers.

Restoration of wild-type p53 confirmation and function in HNSCC as a novel therapeutic approach

The wtp53 can induce apoptosis and cellular senescence when expressed in tumor cells, and therefore much attention has been paid to restore wild-type properties in mutant p53. However, attempts have not been fruitful because wtp53 lacks enzymatic activities and rather works mainly as a DNA sequence-specific transcription factor. Additionally, restoring functional activity of a defective tumor suppressor protein is extremely difficult compared to abrogating the activity of a hyperactive oncoprotein [38]. Nonetheless, several compounds that have the ability to restore wtp53 function have been developed and investigated in recent years (Fig. 2). A small molecule PRIMA-1, which is able to reactivate mutant p53 [39], was later modified to a new derivative, PRIMA-1-met (also known as APR-246). This newly derived specific p53 reactivator can bind to several mutant p53 proteins *via* the DNA-binding domain, thereby promoting correct folding of mutant proteins and restoration of some wtp53 functions [39]. Recently, in addition to its impressive efficacy in preclinical models of various

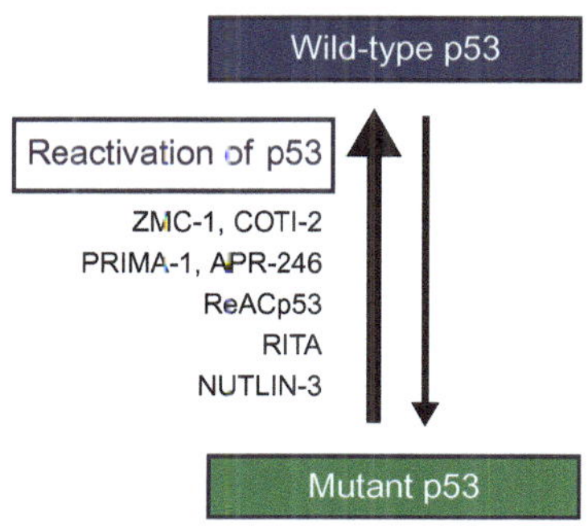

FIG. 2 Current approach to reactivate mutant p53 in HNSCC cells using novel small molecules and synthetic peptides.

types of cancers, PRIMA-1-met has entered a Phase I/II clinical trial in patients with solid tumors and no data have been yet released (Table 2) [40]. Restoring normal p53 function has also been investigated *via* the use of RITA (reactivation of p53 and induction of tumor cell apoptosis), a small molecule that can induce a conformational change in p53 to inhibit p53-HMD-2 interaction, leading to activation of its downstream targets [41]. RITA is thought to act primarily *via* the induction of apoptosis, and indeed RITA, alone or in combination with cisplatin, can induce apoptosis in many HNSCC cell lines [42]. However, this effect is not universal and our recently published work has demonstrated that in vitro RITA treatment induces senescence in HNSCC cells and synergizes with radiation [43]. Another class of small molecules, nutlins, was identified from a diverse chemical library to displace the binding of wild-type p53 from MDM2 with high potency and selectivity [44, 45]. Nutlins bind to the p53-binding pocket of MDM2 to stabilize p53 resulting in p53 pathway activation in wild-type p53, but not in mutant p53, carcinoma cells [45]. Nutlin-3 as single agent and when combined with cisplatin caused more apoptosis in wild-type p53 HNSCC cells than in mutant p53 HNSCC cells [46]. Both RITA and Nutlin-3 are currently under early preclinical testing and their evaluation in clinical trials has yet to be seen. Although metal zinc (Zn^{2+}) ions are also critical for the stabilization of the proper folding of the DNA binding domain of wtp53, many mutant p53 in cancer cells can bind to zinc with much lower affinity compared to wtp53 and therefore have a propensity to misfold and aggregate ZMC1 (NSC319726), a small molecule and belongs to the family of the thiosemicarbazone metal ion chelators, was shown to transport Zn^{2+} to the DNA binding domain of mutant p53, thereby facilitating its proper folding

TABLE 2 Overview of mutant-p53 targeting agents in clinical trials

Compound name	Main target	Phase of development	Clinical trials: Details	Patients with	Identifier
LY2606368 (Prexasertib)	Chk1/2	Phase 2	Alone	Ovarian, breast, prostate cancer	NCT02203513
		Phase 2	Alone	Extensive stage disease small cell lung cancer (ED-SCLC)	NCT02735980
		Phase 2	Alone	Advanced cancer	NCT02873975
		Phase 2	Alone	Ovarian cancer	NCT03414047
		Phase 1	Combination with PD-L1 inhibitor LY3300054	Advanced solid tumor	NCT03495323
		Phase 1	Combination with olaparib	Advanced solid tumor	NCT03057145
		Phase 1	Alone	Recurrent or refractory solid tumors	NCT02808650
		Phase 1			NCT02860780

TABLE 2 Overview of mutant-p53 targeting agents in clinical trials—Cont'd

Compound name	Main target	Phase of development	Clinical trials: Details	Patients with	Identifier
			Combination with ralimetinib	Advanced or metastatic cancer	
		Phase 1	Combination with chemotherapy (cisplatin or cetuximab) and radiation	Head and neck cancer	NCT02555644
		Phase 1	Alone	Advanced cancer	NCT02778126
		Phase 1	Alone	Advanced cancer	NCT02514603
		Phase 1	Alone	Advanced cancer	NCT01115790
		Phase 1	Combination with anti-cancer drugs (cisplatin, cetuximab, pemetrexed, fluorouracil or LY3023414)	Advanced cancer	NCT02124148
GDC-0425	Chk1	Phase 1	Combination with gemcitabine	Refractory solid tumors	NCT01359696
AZD1775 (MK-1775)[a]	Wee1	Phase 2	Combination with cisplatin	Recurrent or metastatic head and neck cancer	NCT02196168
		Phase 1	Combination with cisplatin and radiotherapy	Head and neck cancer	NCT03028766
		Phase 1	Combination with cisplatin and radiotherapy	Intermediate/high risk HNSCC	NCT02585973
		Phase 1	Combination with cisplatin and docetaxel	Stage III-IVB head and neck squamous cell carcinoma (HNSCC)	NCT02508246
AZD0156	ATM	Phase 1	Combination with olaparib, irinotecan, fluorouracil, folinic acid	Advanced solid tumors	NCT02588105
AZD6738	ATR	Phase 2	Combination with durvalumab	Non-small cell lung cancer (NSCLC)	NCT03334617
		Phase 2	Combination with olaparib	Metastatic triple negative breast cancer	NCT03330847

Continued

TABLE 2 Overview of mutant-p53 targeting agents in clinical trials—Cont'd

Compound name	Main target	Phase of development	Clinical trials: Details	Patients with	Identifier
		Phase 2	Combination with olaparib	Small cell lung cancer	NCT03428607
		Phase 2	Combination with olaparib	Recurrent ovarian cancer	NCT03462342
		Phase 2	Combination with AZD2281, AZD5363, AZD1775	Cancer	NCT02576444
		Phase 2	Combination with olaparib	Platinum refractory extensive-stage small cell lung carcinoma	NCT02937818
		Phase 1 Phase 2	Combination with carboplatin, olaparib and/or MEDI4736	Advanced/ metastatic solid malignancies	NCT02264678
		Phase 1	Combination with paclitaxel	Refractory cancer	NCT02630199
		Phase 1	Combination with radiotherapy	Solid tumor	NCT02223923
		Phase I	Biomarker study, combination with olaparib	HNSCC	NCT03022409
BAY1895344	ATR	Phase 1	Alone	Advanced solid tumors and lymphomas	NCT03188965
VX-970 (M6620, VE-822)	ATR	Phase 2	Combination with gemcitabine hydrochloride	Recurrent ovarian, primary peritoneal, or fallopian tube cancer	NCT02595892
		Phase 2	Combination with cisplatin and gemcitabine hydrochloride	Metastatic urothelial cancer	NCT02567409
		Phase 2	Combination with docetaxel	Metastatic castration-resistant prostate cancer	NCT03517969
		Phase 2	Combination with irinotecan	Gastric or gastroesophageal junction cancer	NCT03641313
		Phase 2	Combination with gemcitabine, cisplatin, etoposide,	Advanced solid tumor	NCT02157792

TABLE 2 Overview of mutant-p53 targeting agents in clinical trials—Cont'd

Compound name	Main target	Phase of development	Clinical trials: Details	Patients with	Identifier
			carboplatin, irinotecan		
		Phase 1 Phase 2	Combination with topotecan	Small cell cancers	NCT02487095
		Phase 1	Combination with veliparib and cisplatin	Refractory solid tumors	NCT02723864
		Phase 1	Combination with irinotecan hydrochloride	Metastatic and unresectable solid tumors	NCT02595931
		Phase 1	Combination with whole brain radiation therapy	Brain metastases from non-small cell lung cancer, small cell lung cancer, or neuroendocrine tumors	NCT02589522
		Phase 1	Combination with cisplatin and radiotherapy	Locally advanced HPV-negative HNSCC	NCT02567422
Vorinostat	Histone deacetylases (HDAC)	Phase 2	Combination with capecitabine	Recurrent and/or metastatic head and neck cancer	NCT01267240
		Phase 1 Phase 2	Combination with pembrolizumab	Recurrent squamous cell head and neck cancer or salivary gland cancer	NCT02538510
		Phase 1	Combination with cisplatin or radiation	Advanced staged oropharyngeal squamous cell carcinoma	NCT01064921
COTI-2	P53	Phase 1	Alone	Gynecologic malignancies and head and neck squamous cell carcinoma	NCT02433626
APR-246 (PRIMA-1-met)	P53	Phase 1 Phase 2	Alone	Esophageal cancer	NCT02999893
		Phase 1 Phase 2	Combination with dabrafenib	Melanoma	NCT03391050

Continued

TABLE 2 Overview of mutant-p53 targeting agents in clinical trials—Cont'd

Compound name	Main target	Phase of development	Clinical trials: Details	Patients with	Identifier
	•	Phase 2	Combination with Pegylated liposomal doxorubicin hydrochloride (PLD)	Ovarian cancer	NCT03268382
		Phase 1	Alone	Refractory hematologic cancer or prostate cancer	NCT00900614
		Phase 1 Phase 2	Combination with carboplatin and Pegylated liposomal doxorubicin hydrochloride (PLD)	Recurrent high grade serous ovarian cancer	NCT02098343

[a] *Currently, there are many clinical trials with AZD1775 or Vorinostat, therefore only trials in HNSCCs are listed.*

and restoring wtp53 like function in a small number of mutant p53 expressing tumor cell lines.. Unfortunately, ZMC1 can cause unpleasant adverse effects in patients since its chemical activity is broad and not p53-specific [47]. COTI-2 is a novel third generation thiosemicarbazone derivative which was discovered using a prediction model through a proprietary computational platform called CHEMSAS. COTI-2 displayed significant antitumor effects in vitro as single agent and inhibited cell proliferation, leading to death through apoptosis in many cancer cells [48]. Furthermore, COTI-2 seemed to work by restoring wild-type-like p53 function and perhaps by negatively regulating the PI3K/AKT/mTOR pathway, which is a classical prosurvival pathway constitutively activated in many cancers [48]. Recent study has shown that COTI-2 inhibits tumor growth and it is well-tolerated in vivo in xenograft mouse models [48]. We have demonstrated that COTI-2 shows single agent activity and significantly decreases cell growth in vitro at nanomolar concentration in HNSCC cells irrespective of the p53 status, and inhibits tumor growth in vivo in orthotopic mouse models of HNSCC (unpublished data). Additionally, COTI-2 synergizes with cisplatin and potentiates response to radiation in vitro and in vivo in HNSCC expressing high-risk p53 mutations (unpublished data). Mechanistically, our study has revealed that COTI-2 appears to reactivate p53 downstream target genes. This compound is currently being tested in a Phase I clinical trial of patients with refractory gynecologic and head and neck cancers (Table 2).

Using a rational design approach, Soragni and colleagues have recently identified a peptide dubbed ReACp53, which is capable of blocking the amyloid-like aggregation of mutant p53 proteins containing either R248Q or R175H mutations [49]. The ReACp53 peptide inhibited aggregation and restored wtp53 functions determined by induction of its downstream target genes and apoptosis. Furthermore, treatment of mice bearing ovarian cancer cells expressing aggregation-prone p53 mutations with conjugated ReACp53 inhibited cell proliferation and caused significant tumor shrinkage [49]. The spectrum and types of *TP53* mutations targeted by this unique small peptide remains to be elucidated. Despite their remarkable target specificity, ineffective delivery, short half-life and poor cell permeability are major drawbacks of the *TP53* reactivation peptides that could potentially limit their clinical applications.

DNA damage and replication stress responses as therapeutic targets in *TP53* mutant HNSCC

Number of cellular stresses, including DNA damage, hypoxia and inadequate nucleotide supply can activate p53. For example, DNA damage triggered by ionizing radiation and chemotherapy can activate p53 through phosphorylation and acetylation cascades to preserve cellular genomic integrity. Therefore, inhibitors of critical proteins that regulate DNA damage response and cell cycle progression have great potential to sensitize HNSCC with mutant *TP53* to chemotherapy and radiation. The DNA damage normally stabilizes and activates p53 in part through a DNA damage-induced signaling cascade that includes the sensor kinases, Ataxia-telangiectasia mutated (ATM) and ATM-Rad3 related (ATR) and effector kinases, such as checkpoint kinase-1/2 (Chk1/2) and Wee1, leading to transcriptional activation of many genes involved in cell cycle regulation and apoptosis [50]. High levels of DNA damage resulting in generation of DNA single- or-double strand breaks and premature mitosis can interfere with DNA replication and delay its progression during the cell cycle [51]. If nucleotides or components of the replication machinery are inadequate or not properly assembled to ensure completion of DNA replication, the cells become "stressed." Cells with high levels of replication stress are often characterized by stalling of replication forks, and increased binding of the replication protein A (RPA) to the single strand DNA segments [52]. It is well documented that loss of essential mitotic and S-phase cell cycle inhibitors leads to stalled replication forks in cancer cells [53]. This is in turn, followed by the activation of the DNA sensors such as ATR, ATM and DNA dependent protein kinase (DNA-PK). These sensors recruit their immediate downstream target proteins including Chk1, p53, Wee1, in order to suppress new origins of replication from firing, stabilize advancing replication forks, prevent mitosis, and causes cell cycle arrest. This activation step occurs to ensures that replication errors can be faithfully repaired through excision of double stranded DNA breaks or Rad51-mediated homologous recombination [54]. If the mechanisms that deal with faithful replication become defective, more lethal DNA breaks occur and slowly accumulate leading to genomic instability, replication stress [53, 54]. The ATR-ATM-Chk1-Wee1 signaling pathway is a critical component of the general surveillance mechanism that monitors the DNA replication and is activated at low thresholds even during unperturbed S-phase [55, 56]. Therefore, Inhibition of this pathway at different levels can stimulate mutant *TP53* HNSCC cells experiencing replication stress to die through senescence or apoptosis. Numerous studies demonstrated that treatment with the histone deacetylase (HDAC) inhibitor, vorinostat or knockdown of HDAC8 showed preferential cytotoxicity for mutant *TP53* rather than wildtype or -null *TP53* tumor cells, providing the scientific premise for using vorinostat-based regimens for the treatment of mutant *TP53*-specific tumors [57, 58]. HDAC inhibitors can destabilize the mutant p53 by preventing the interaction of HDAC6 with heat shock protein 90 (Hsp90) [58] and inhibit mutant p53 transcription through inactivation of HDAC8 [57]. These findings suggest that HDACs are potential therapeutic targets in HNSCC harboring *TP53* mutations. Inhibitors of the DNA damage repair and histone deacetylation and their mechanism of actions are discussed in details in the following sections.

Selective ATM and ATR inhibitors

ATM and ATR are serine/threonine kinases belong to a structurally unique family of the phosphatidylinositol 3-kinase-related kinase (PIKK) [55]. These kinases activate their downstream targets and function as essential mediators of the DNA damage response (DDR), which induce DNA repair machinery proteins and cause cell cycle arrest. Resistance to genotoxic treatments, such radiation or chemotherapy is often associated with increased levels of DDR. Many cancers including HNSCC with mutant p53 have defects in some components of the DDR, and as such become highly dependent on the remaining DNA repair pathways for survival. Therefore, targeting the DDR has become an attractive approach in cancer treatment. Recent preclinical data have provided a strong evidence supporting clinical evaluation of ATR and ATM inhibitors either alone or in combination with chemotherapy, radiation, and molecularly targeted therapies to treat tumors with deficiencies in certain DDR components. Somatic mutations in ATR and ATM are seen in various frequencies in HNSCC tumors with ATR (4–10%) and ATM (1–16%) respectively making them more vulnerable to inhibitors of these kinases [59]. Selective and potent oral inhibitors of ATR, AZD6738 and VX-970 (previously known as VE-822 or berzosertib) have been developed and currently being tested in Phase I/II clinical trials in solid tumors including HNSCC (NCT02264678; NCT02567422). The majority of the ATM inhibitors are still undergoing an early preclinical investigation and only AZD0156 has entered Phase I clinical trial as monotherapy or in combination with the PARP inhibitor olaparib and other cytotoxic agents in patients with advanced solid tumors (NCT02588105) (Table 2).

Chk1/2 dual inhibitors

The Chk1 and Chk2 are structurally distinct serine/threonine kinases with an overlapping catalytic functions. In response to different genotoxic insults, Chk1 and Chk2 become activated through phosphorylation by ATR and ATM respectively. These two kinases are major regulators of fundamental cellular functions including DNA replication, cell cycle progression, chromatin remodeling and apoptosis [60]. In cells lacking functional p53, cell cycle arrest and DNA repair depend mainly on the function of Chk1/2 for faithful DNA repair and replication. We have shown that inhibition of Chk1/2 in HNSCC cells expressing mutant p53 with the small molecule Chk1/2 inhibitor, AZD7762 sensitizes these cells to cisplatin through induction of mitotic catastrophe [29], a type of cell death which is typically characterized by presence of multinucleated cells and polyploidy. Recently, several Chk inhibitors have been evaluated and showed significant synergistic effects on inhibiting cancer cell growth when combined with specific inhibitors targeting critical cell cycle regulators such as Wee1 kinase [61]. Remarkably, in some in vitro and in vivo cancer models including HNSCC, we and others have demonstrated that Chk1 inhibitors can exert single agent activity in cells undergoing replication stress [62–64]. In our study, we showed that CDKN2A/p16 deletion in

HNSCC cells is associated with cyclin dependent kinase 2 (CDK2) activation and replication stress, making these cells vulnerable to CHK1 inhibition with siRNAs or volasertib [64]. Taken together, results from our study suggest a biomarker-driven strategy for selecting HNSCC patients who may benefit the most from CHK inhibitor therapy. Several clinical trials are currently evaluating some specific Chk1/2 inhibitors in combination with various chemotherapeutic agents in HNSCC (Table 2).

Specific Wee1 inhibitors

Wee1 is a serine/threonine kinase which responds to single or double strand breaks generated by DNA damage, and causes G2-M cell cycle arrest by inactivating CDK1 (also known as CDC2) through phosphorylation of the Tyr15 residue [65]. While Wee1 is necessary to prevent cells with DNA damage from entering mitosis prematurely, and therefore inhibition of its activity can override the G2 cell cycle arrest resulting in accumulation of cells with extensive DNA damage which subsequently die through mitotic catastrophe [66, 67]. Tumor cells carrying wtp53 protein arrest at the G1 checkpoint of the cell cycle to repair damaged DNA, before DNA replication. However, tumor cells with defective p53 pathway, as for those carrying mutant p53 proteins, rely mainly on DNA repair at the G2 checkpoint [68]. These findings support the mitotic lethality rationale, and that abrogation of the G2-M checkpoint by Wee1 inhibition sensitizes tumor cells with defective p53 to DNA-damaging agents. Numerous studies have been conducted with AZD1775 (formerly known as MK-1775), a specific inhibitor of Wee1, and demonstrate that Wee1 inhibition abrogates the G2 checkpoint and sensitizes tumor cells lacking functional p53 to various DNA-damaging agents and radiotherapy [69–73]. Based on these results, AZD1775 was evaluated in Phase I clinical trial as a chemosensitizer in patients with advanced solid tumors and showed good tolerability with minimal adverse side effects, and further advanced to Phase II clinical trials in solid tumors (Table 2) [68, 74]. We also demonstrated that AZD1775 synergizes with cisplatin in HNSCC cells with high-risk mutant p53 through mitotic arrest followed by senescence rather than apoptosis both in vitro and in vivo [75]. Interestingly, AZD1775 shows single agent activity and significantly enhances the response of the HPV-positive HNSCC cells to cisplatin treatment both in vitro and in vivo [76]. However, unlike HPV-negative HNSCC cells, AZD1775 induces significant apoptosis initiated by selective degradation (or cleavage) of the antiapoptotic proteins MCl-1 and XIAP in HPV-positive HNSCC cells. Taken together, our data provides compelling evidence that HNSCC cancer cells experiencing replication stress can be successfully targeted with Wee1 kinase inhibitor in combination with chemotherapy. Mendez and colleagues have recently conducted a phase I clinical trial (NCT02508246) testing AZD1775 in combination with neoadjuvant weekly docetaxel and cisplatin before definitive therapy in HNSCC [77]. This trial has revealed that the triplet combination of AZD1775, cisplatin, and docetaxel is safe and tolerable. Furthermore, pharmacodynamic analysis has revealed a reduction in phosphorylation of CDK1 (pY15) and increases in the DNA damage and replication stress markers, γH2AX, and phospho-RPA32 respectively in responders *versus* nonresponders [77]. These data are consistent with evidence of replication stress and G2-M abrogation in preclinical models of HNSCC [75, 76]. Preliminary results from the trial show encouraging antitumor efficacy in advanced HNSCC, warranting further investigation at the recommended phase II dose (Table 2).

Classical histone deacetylases (HDACs) inhibitors

Modification of the acetylation status of histones and transcription factors is crucial step in regulating gene expression [78, 79]. Histone acetylation is normally associated with an enhanced transcription, whereas deacetylated histones are often linked to repressed transcription [80]. Acetylation is tightly controlled by opposing actions of two large families of catalytic enzymes which include histone acetyltransferases (HATs) and histone deacetylases (HDACs). HDAC inhibitors function enzymatically to facilitate the removal of the acetyl group from hyperacetylated histones and suppress gene transcription [80]. Increased activities of HDACs associated with poor outcome has been reported in a number of human cancers including HNSCC [78, 81, 82]. These results indicate that HDACs may serve as an ideal therapeutic targets in HNSCC. In fact, numerous studies demonstrated that HDAC inhibitors induced growth arrest, differentiation, and apoptosis in various cancer cell lines in vitro and inhibited tumor growth in vivo in mouse xenograft models including HNSCC [78, 83, 84]. Vorinostat, a small-molecule HDAC inhibitor (formerly known as suberoylanide hydroxamic acid, SAHA), showed preferential cytotoxicity in vitro and in vivo in cancer cells expressing *TP53* mutations [57, 58, 85]. Although HDAC inhibitors lack single agent activity, they are now emerging as an attractive novel classes of antitumor agents being evaluated clinically in combination with various conventional chemotherapies or molecularly targeted agents [86]. Recent study demonstrated that addition of the HDAC inhibitor, trichostatin A (TSA), significantly reduced cisplatin resistance in HNSCC [87]. Another report also provided evidence that, treatment with pan-HDAC inhibitor SAHA synergized with cisplatin in vitro and significantly inhibited tumor growth and metastasis in vivo in HNSCC [88]. Another recent study has revealed that acetylation-mediated histone post-translational modification plays critical roles in DDR in lung cancer cells [89]. This study shows that suppression of sirtuin 1 (sirt1), a histone deacetylase enzyme, causes DNA damage and sensitizes lung cancer cells to Wee1 inhibition with AZD1775 leading to apoptosis [89]. Consistent with this, our work has revealed that vorinostat synergizes with AZD1775 in HNSCC cells with mutant p53 in vitro and in vivo through induction of replication stress associated with decreased Rad51-mediated homologous recombination, resulting in apoptotic cell death within the S-phase of the cell cycle [90]. We further demonstrated that the vorinostat-induced replication stress is associated with significant CDK1 activation and reduced Chk1 phosphorylation in these cells. Thus, our study provides a rationale for initiation of clinical trials using combination of HDAC and Wee1 inhibitors, particularly for patients with advanced, recurrent, and/or metastatic HNSCC. Tumors of these patients harbor high incidence of *TP53* mutations and they are likely to become more sensitive to such combination regimens. A recent phase I clinical trial has reported that vorinostat (Zolinza) in combination with concurrent chemoradiation therapy is a safe and highly effective treatment regimen in advanced stage HNSCC (Table 2).

Radiosensitization of HNSCC mutant *TP53* cells through metabolic targeting

Recent studies have revealed a novel function of p53 in regulation of cellular metabolism [91–93]. In addition to its role as a transcriptional factor, p53 regulates mitochondrial oxidative phosphorylation, glycolysis, glutamine metabolism, lipid metabolism, and antioxidant defense [91–95]. Through the regulation of these metabolic processes, p53 maintains the homeostasis of cellular metabolism and redox balance in cells, which contributes significantly to the role of p53 as a tumor suppressor. Inactivation of p53 can lead to metabolic disturbances in cancer cells. Recent studies have reported compelling evidence for a novel GOF of mutant p53 proteins in promoting metabolic changes [30, 96]. These studies suggest that metabolic pathways normally regulated by wt p53 are altered by mutant p53 to facilitate the supply of metabolites required for proliferation. Recent data from our laboratory has demonstrated that disruptive *TP53* mutations are associated with poor clinical outcome in HNSCC and increased radioresistance [97]. Parallel work in our laboratory has shown that cells with disruptive *TP53* exhibit a therapeutically exploitable decrease in metabolic flexibility [98]. One agent that can target this process is metformin. Recently, it has been shown that patients taking metformin have improved responses following neoadjuvant chemotherapy [99]. Furthermore, metformin acts as a radiosensitizer in vitro by increasing expression of p21 and reactive oxygen species (ROS) levels, leading to induction of cellular senescence [100]. Interestingly, when we examined the effects of metformin in vitro, radiation-induced senescence and toxicity were potentiated only in cells expressing disruptive *TP53* mutations, an effect that was modulated by altering p53 expression (Fig. 3). This dramatic effect on the efficacy of radiation was also seen in our in vivo orthotopic model, as well as in HNSCC patients taking metformin at the time of PORT (predictive of poor response to a specific therapy) [97]. Although this only included 10 patients in the metformin group, the reduction in locoregional recurrence (LRR) was profound and argues for the possibility of a dramatic clinical benefit with targeting of radiation-induced senescence. Currently, there are two active Phase I/II clinical trials

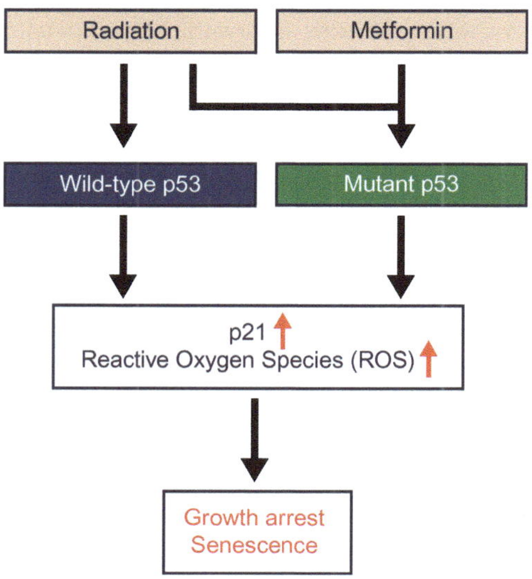

FIG. 3 Proposed mechanism of action of radiation and metformin in HNSCC based on p53 mutational status.

investigating efficacy of metformin alone (NCT02083692) or in combination with cisplatin and radiation in patients with head and neck cancer (NCT02949700) and no results have been released yet.

Summary and concluding remark

Head and neck squamous cell carcinoma (HNSCC) persists as one of the most common and deadly types of cancer worldwide and in the United States. In the clinic, *TP53* mutations are significantly associated with poor survival and tumor resistance to chemotherapy and radiation in HNSCC patients. This suggests that *TP53* mutation can be utilized as a potential prognostic and predictive marker of clinical response. *TP53* mutations are diverse and often act through a complex and intricate network of multiple proteins. Therefore, discovering novel drug-based approaches to safely and efficiently targeting mutant p53 in HNSCC remains challenging. Realizing this goal will require considerable knowledge and better understanding of the biology of mutant p53, such as identification of its interacting partners and their downstream signaling pathways. Since p53 activation causes cell-cycle arrest and initiates of DNA repair in response to DNA damage, inhibitors of critical proteins that regulate DNA damage response and induction of replicative stress during cell cycle progression have the potential to sensitize HNSCC cells expressing mutant p53 to chemotherapy and radiation. Recently, this therapeutic strategy has been tested using small molecule inhibitors of Chk1/2, Wee1 and HDAC in preclinical models of HNSCC. Additionally, promising results have emerged from recent clinical trials which have demonstrated antitumor efficacy and good tolerability with the use of AZD1775 in combination with chemotherapy in head and neck cancer patients. With better understanding of the mechanisms of DNA repair and replication we expect that the development of novel cancer therapies in this field will have a significant impact on HNSCC patients. Reactivation of mutant p53 in HNSCC cells with specific small molecules and synthetic peptides is another therapeutic approach and intense research efforts have been initiated in recent years in different cancer types using this novel strategy. Although, there are some limitations in restoring the activity of a defective p53 and effectively delivering the *TP53* reactivation peptides into the tumor cells, this approach will possibly become impactful on the treatment of cancer patients with tumor harboring mutant p53. There is growing evidence that tumor microenvironment (TME) supplies tumor cells with nutrients and metabolites to survive and grow, a known process called metabolic dependency. This process often leads to an inappropriate metabolic reprograming and an increase in oncometabolites favoring uncontrollable tumor development and progression. Although, many agents that target tumor metabolism have been evaluated in the past, their success was limited more by toxicity than by their ability to kill cancer cells. New approaches to evaluate metabolism and accurately assess metabolic dependency in living tumors in human and mice may improve our ability to understand how metabolism is regulated and which altered pathways can be therapeutically targeted to improve care of cancer patients. In conclusion, identification of reliable biomarkers and novel molecular targets is critical to stratify and select NSCC patients for precision and personalized treatment plans.

Acknowledgments

This work was supported by the NIH/NIDCR R01DE024601 (A.A. Osman and J.N Myers) and the University of Texas MD Anderson Cancer Center Stiefel Oropharyngeal Cancer Program (A.A. Osman and J.N. Myers). The authors declare no potential conflicts of interest with respect to the authorship and/or publication of this article.

References

[1] Ferlay J, Shin HR, Bray F, Forman D, Mathers C, Parkin DM. Estimates of worldwide burden of cancer in 2008: GLOBOCAN 2008. Int J Cancer 2010;127(12):2893–917.

[2] Haddad RI, Shin DM. Recent advances in head and neck cancer. N Engl J Med 2008;359(11):1143–54.

[3] Argiris A, Karamouzis MV, Raben D, Ferris RL. Head and neck cancer. Lancet 2008;371(9625):1695–709.

[4] Hashibe M, Brennan P, Chuang SC, Boccia S, Castellsague X, Chen C, Curado MP, Dal Maso L, Daudt AW, Fabianova E, Fernandez L, Wunsch-Filho V, Franceschi S, Hayes RB, Herrero R, Kelsey K, Koifman S, La Vecchia C, Lazarus P, Levi F, Lence JJ, Mates D, Matos E, Menezes A, MD MC, Muscat J, Eluf-Neto J, Olshan AF, Purdue M, Rudnai P, Schwartz SM, Smith E, Sturgis EM, Szeszenia-Dabrowska N, Talamini R, Wei Q, Winn DM, Shangina O, Pilarska A, Zhang ZF, Ferro G, Berthiller J, Boffetta P. Interaction between tobacco and alcohol use and the risk of head and neck cancer: pooled analysis in the International Head and Neck Cancer Epidemiology Consortium. Cancer Epidemiol Biomarkers Prev 2009;18(2):541–50.

[5] Sturgis EM, Ang KK. The epidemic of HPV-associated oropharyngeal cancer is here: is it time to change our treatment paradigms? J Natl Compr Canc Netw 2011;9(6):665–73.

[6] Chaturvedi AK, Engels EA, Pfeiffer RM, Hernandez BY, Xiao WH, Kim E, Jiang B, Goodman MT, Sibug-Saber M, Cozen W, Liu LH, Lynch CF, Wentzensen N, Jordan RC, Altekruse S, Anderson WF, Rosenberg PS, Gillison ML. Human papillomavirus and rising oropharyngeal cancer incidence in the United States. J Clin Oncol 2011;29(32):4294–301.

[7] Gillison ML, Koch WM, Capone RB, Spafford M, Westra WH, Wu L, Zahurak ML, Daniel RW, Viglione M, Symer DE, Shah KV, Sidransky D. Evidence for a causal association between human papillomavirus and a subset of head and neck cancers. J Natl Cancer Inst 2000;92(9):709–20.

[8] Hammarstedt L, Lindquist D, Dahlstrand H, Romanitan M, Onelov L, Joneberg J, Creson N, Lindholm J, Ye WM, Dalianis T, Munck-Wikland E. Human papillomavirus as a risk factor for the increase in incidence of tonsillar cancer. Int J Cancer 2006;119(11):2620–3.

[9] Ang KK, Harris J, Wheeler R, Weber R, Rosenthal DI, Nguyen-Tan PF, Westra WH, Chung CH, Jordan RC, Lu C, Kim H, Axelrod R, Silverman CC, Redmond KP, Gillison ML. Human papillomavirus and survival of patients with oropharyngeal cancer. New Engl J Med 2010;363(1):24–35.

[10] Poeta ML, Manola J, Goldwasser MA, Forastiere A, Benoit N, Califano JA, Ridge JA, Goodwin J, Kenady D, Saunders J, Westra W, Sidransky D, Koch WM. TP53 mutations and survival in squamous-cell carcinoma of the head and neck. N Engl J Med 2007;357(25):2552–61.

[11] Lindenbergh-van der Plas M, Brakenhoff RH, Kuik DJ, Buijze M, Bloemena E, Snijders PJ, Leemans CR, Braakhuis BJ. Prognostic significance of truncating TP53 mutations in head and neck squamous cell carcinoma. Clin Cancer Res 2011;17(11):3733–41.

[12] Vousden KH, Prives C. Blinded by the light: the growing complexity of p53. Cell 2009;137(3):413–31.

[13] Vermorken JB, Mesia R, Rivera F, Remenar E, Kawecki A, Rottey S, Erfan J, Zabolotnyy D, Kienzer HR, Cupissol D, Peyrade F, Benasso M, Vynnychenko I, De Raucourt D, Bokemeyer C, Schueler A, Amellal N, Hitt R. Platinum-based chemotherapy plus cetuximab in head and neck cancer. New Engl J Med 2008;359(11):1116–27.

[14] Agrawal N, Frederick MJ, Pickering CR, Bettegowda C, Chang K, Li RJ, Fakhry C, Xie TX, Zhang J, Wang J, Zhang N, El-Naggar AK, Jasser SA, Weinstein JN, Trevino L, Drummond JA, Muzny DM, Wu Y, Wood LD, Hruban RH, Westra WH, Koch WM, Califano JA, Gibbs RA, Sidransky D, Vogelstein B, Velculescu VE, Papadopoulos N, Wheeler DA, Kinzler KW, Myers JN. Exome sequencing of head and neck squamous cell carcinoma reveals inactivating mutations in NOTCH1. Science 2011;333(6046):1154–7.

[15] Cancer Genome Atlas Network. Comprehensive genomic characterization of head and neck squamous cell carcinomas. Nature 2015;517(7536):576–82.

[16] Pickering CR, Zhang J, Yoo SY, Bengtsson L, Moorthy S, Neskey DM, Zhao M, Ortega Alves MV, Chang K, Drummond J, Cortez E, Xie TX, Zhang D, Chung W, Issa JP, Zweidler-McKay PA, Wu X, El-Naggar AK, Weinstein JN, Wang J, Muzny DM, Gibbs RA, Wheeler DA, Myers JN, Frederick MJ. Integrative genomic characterization of oral squamous cell carcinoma identifies frequent somatic drivers. Cancer Discov 2013;3 (7):770–81.

[17] Rautava J, Syrjanen S. Biology of human papillomavirus infections in head and neck carcinogenesis. Head Neck Pathol 2012;6(Suppl 1):S3–S15.

[18] Okorokov AL, Orlova EV. Structural biology of the p53 tumour suppressor. Curr Opin Struct Biol 2009;19 (2):197–202.

[19] Boyle JO, Hakim J, Koch W, van der Riet P, Hruban RH, Roa RA, Correo R, Eby YJ, Ruppert JM, Sidransky D. The incidence of p53 mutations increases with progression of head and neck cancer. Cancer Res 1993;53 (19):4477–80.

[20] Lane D, Levine A. p53 research: the past thirty years and the next thirty years. Cold Spring Harb Perspect Biol 2010;2(12):a000893.

[21] Pant V, Lozano G. Limiting the power of p53 through the ubiquitin proteasome pathway. Genes Dev 2014;28 (16):1739–51.

[22] Muller PA, Vousden KH. p53 mutations in cancer. Nat Cell Biol 2013;15(1):2–8.

[23] Kim MP, Lozano G. Mutant p53 partners in crime. Cell Death Differ 2018;25(1):161–8.

[24] Muller PAJ, Vousden KH. Mutant p53 in cancer: new functions and therapeutic opportunities. Cancer Cell 2014;25(3):304–17.

[25] Gaiddon C, Lokshin M, Ahn J, Zhang T, Prives C. A subset of tumor-derived mutant forms of p53 down-regulate p63 and p73 through a direct interaction with the p53 core domain. Mol Cell Biol 2001;21(5):1874–87.

[26] Pfister NT, Fomin V, Regunath K, Zhou JY, Zhou W, Silwal-Pandit L, Freed-Pastor WA, Laptenko O, Neo SP, Bargonetti J, Hoque M, Tian B, Gunaratne J, Engebraaten O, Manley JL, Borresen-Dale AL, Neilsen PM, Prives C. Mutant p53 cooperates with the SWI/SNF chromatin remodeling complex to regulate VEGFR2 in breast cancer cells. Genes Dev 2015;29(12):1298–315.

[27] Zhu JJ, Sammons MA, Donahue G, Dou ZX, Vedadi M, Getlik M, Barsyte-Lovejoy D, Al-Awar R, Katona BW, Shilatifard A, Huang J, Hua XX, Arrowsmith CH, Berger SL. Gain-of-function p53 mutants co-opt chromatin pathways to drive cancer growth. Nature 2015;525(7568):206.

[28] Osman AA, Neskey DM, Katsonis P, Patel AA, Ward AM, Hsu TK, Hicks SC, McDonald TO, Ow TJ, Alves MO, Pickering CR, Skinner HD, Zhao M, Sturgis EM, Kies MS, El-Naggar A, Perrone F, Licitra L, Bossi P, Kimmel M, Frederick MJ, Lichtarge O, Myers JN. Evolutionary action score of TP53 coding variants is predictive of platinum response in head and neck Cancer patients. Cancer Res 2015;75(7):1205–15.

[29] Gadhikar MA, Sciuto MR, Alves MV, Pickering CR, Osman AA, Neskey DM, Zhao M, Fitzgerald AL, Myers JN, Frederick MJ. Chk1/2 inhibition overcomes the cisplatin resistance of head and neck cancer cells secondary to the loss of functional p53. Mol Cancer Ther 2013;12(9):1860–73.

[30] Zhou G, Wang JP, Zhao M, Xie TX, Tanaka N, Sano D, Patel AA, Ward AM, Sandulache VC, Jasser SA, Skinner HD, Fitzgerald AL, Osman AA, Wei YK, Xia XF, Zhou SY, Mills GB, Hung MC, Caulin C, Liang JY, Myers JN. Gain-of-function mutant p53 promotes cell growth and cancer cell metabolism via inhibition of AMPK activation. Mol Cell 2014;54(6):960–74.

[31] Tanaka N, Zhao M, Zhang JX, Wang J, Zhou G, Myers JN. Gain-of-function mutant p53 promotes the oncogenic potential of head and neck squamous cell carcinoma cell by targeting forkhead transcription factors FOXO3a and FOXM1. Clin Cancer Res 2017;23(23).

[32] Neskey DM, Osman AA, Ow TJ, Katsonis P, McDonald T, Hicks SC, Hsu TK, Pickering CR, Ward A, Patel A, Yordy JS, Skinner HD, Giri U, Sano D, Story MD, Beadle BM, El-Naggar AK, Kies MS, William WN, Caulin C, Frederick M, Kimmel M, Myers JN, Lichtarge O. Evolutionary action score of TP53 identifies high-risk mutations associated with decreased survival and increased distant metastases in head and neck cancer. Cancer Res 2015;75(7):1527–36.

[33] Clayman GL, Elnaggar AK, Roth JA, Zhang WW, Goepfert H, Taylor DL, Liu TJ. In-vivo molecular therapy with P53 adenovirus for microscopic residual head and neck squamous carcinoma. Cancer Res 1995;55(1):1–6.

[34] Liu TJ, Wang M, Breau RL, Henderson Y, El-Naggar AK, Steck KD, Sicard MW, Clayman GL. Apoptosis induction by E2F-1 via adenoviral-mediated gene transfer results in growth suppression of head and neck squamous cell carcinoma cell lines. Cancer Gene Ther 1999;6(2):163–71.

[35] Pirollo KF, Nemunaitis J, Leung PK, Nunan R, Adams J, Chang EH. Safety and efficacy in advanced solid tumors of a targeted nanocomplex carrying the p53 gene used in combination with docetaxel: a phase 1b study. Mol Ther 2016;24(9):1697–706.

[36] Xu L, Huang CC, Huang WQ, Tang WH, Rait A, Yin YZ, Cruz I, Xiang LM, Pirollo KF, Chang EH. Systemic tumor-targeted gene delivery by anti-transferrin receptor scFv-immunoliposomes. Mol Cancer Ther 2002;1(5):337–46.

[37] Senzer N, Nemunaitis J, Nemunaitis D, Bedell C, Edelman G, Barve M, Nunan R, Pirollo KF, Rait A, Chang EH. Phase I study of a systemically delivered p53 nanoparticle in advanced solid tumors. Mol Ther 2013;21(5):1096–103.

[38] Oren M, Tal P, Rotter V. Targeting mutant p53 for cancer therapy. Aging (Albany NY) 2016;8(6):1159–60.

[39] Bykov VJ, Zache N, Stridh H, Westman J, Bergman J, Selivanova G, Wiman KG. PRIMA-1(MET) synergizes with cisplatin to induce tumor cell apoptosis. Oncogene 2005;24(21):3484–91.

[40] Bykov VJ, Zhang Q, Zhang M, Ceder S, Abrahmsen L, Wiman KG. Targeting of mutant p53 and the cellular redox balance by APR-246 as a strategy for efficient cancer therapy. Front Oncol 2016;6:21.

[41] Issaeva N, Bozko P, Enge M, Protopopova M, Verhoef LGGC, Masucci M, Pramanik A, Selivanova G. Small molecule RITA binds to p53, blocks p53-HDM-2 interaction and activates p53 function in tumors. Nat Med 2004;10(12):1321–8.

[42] Roh JL, Ko JH, Moon SJ, Ryu CH, Choi JY, Koch WM. The p53-reactivating small-molecule RITA enhances cisplatin-induced cytotoxicity and apoptosis in head and neck cancer. Cancer Lett 2012;325(1):35–41.

[43] Chuang HC, Yang LP, Fitzgerald AL, Osman A, Woo SH, Myers JN, Skinner HD. The p53-reactivating small molecule RITA induces senescence in head and neck cancer cells. Plos One 2014;9(8).

[44] Shen H, Maki CG. Pharmacologic activation of p53 by small-molecule MDM2 antagonists. Curr Pharm Design 2011;17(6):560–8.

[45] Vassilev LT, Vu BT, Graves B, Carvajal D, Podlaski F, Filipovic Z, Kong N, Kammlott U, Lukacs C, Klein C, Fotouhi N, Liu EA. In vivo activation of the p53 pathway by small-molecule antagonists of MDM2. Science 2004;303(5659):844–8.

[46] Roh JL, Kang SK, Minn I, Califano JA, Sidransky D, Koch WM. p53-reactivating small molecules induce apoptosis and enhance chemotherapeutic cytotoxicity in head and neck squamous cell carcinoma. Oral Oncol 2011;47(1):8–15.

[47] Yu X, Vazquez A, Levine AJ, Carpizo DR. Allele-specific p53 mutant reactivation. Cancer Cell 2012;21(5):614–25.

[48] Salim KY, Maleki Vareki S, Danter WR, Koropatnick J. COTI-2, a novel small molecule that is active against multiple human cancer cell lines in vitro and in vivo. Oncotarget 2016;7(27):41363–79.

[49] Soragni A, Janzen DM, Johnson LM, Lindgren AG, Thai-Quynh Nguyen A, Tiourin E, Soriaga AB, Lu J, Jiang L, Faull KF, Pellegrini M, Memarzadeh S, Eisenberg DS. A designed inhibitor of p53 aggregation rescues p53 tumor suppression in ovarian carcinomas. Cancer Cell 2016;29(1):90–103.

[50] Coutts AS, La Thangue N. The p53 response during DNA damage: impact of transcriptional cofactors. Biochem Soc Symp 2006;73:181–9.

[51] Allen C, Ashley AK, Hromas R, Nickoloff JA. More forks on the road to replication stress recovery. J Mol Cell Biol 2011;3(1):4–12.

[52] Dobbelstein M, Sorensen CS. Exploiting replicative stress to treat cancer. Nat Rev Drug Discov 2015;14(6):405–23.

[53] Suram A, Herbig U. The replicometer is broken: telomeres activate cellular senescence in response to genotoxic stresses. Aging Cell 2014;13(5):780–6.

[54] Pfister SX, Markkanen E, Jiang Y, Sarkar S, Woodcock M, Orlando G, Mavrommati I, Pai CC, Zalmas LP, Drobnitzky N, Dianov GL, Verrill C, Macaulay VM, Ying S, La Thangue NB, D'Angiolella V, Ryan AJ, Humphrey TC. Inhibiting WEE1 selectively kills histone H3K36me3-deficient cancers by dNTP starvation. Cancer Cell 2015;28(5):557–68.

[55] Shechter D, Costanzo V, Gautier J. ATR and ATM regulate the timing of DNA replication origin firing. Nat Cell Biol 2004;6(7):648–55.

[56] Syljuasen RG, Sorensen CS, Hansen LT, Fugger K, Lundin C, Johansson F, Helleday T, Sehested M, Lukas J, Bartek J. Inhibition of human Chk1 causes increased initiation of DNA replication, phosphorylation of ATR targets, and DNA breakage. Mol Cell Biol 2005;25(9):3553–62.

[57] Yan W, Liu S, Xu E, Zhang J, Zhang Y, Chen X, Chen X. Histone deacetylase inhibitors suppress mutant p53 transcription via histone deacetylase 8. Oncogene 2013;32(5):599–609.
[58] Li D, Marchenko ND, Moll UM. SAHA shows preferential cytotoxicity in mutant p53 cancer cells by destabilizing mutant p53 through inhibition of the HDAC6-Hsp90 chaperone axis. Cell Death Differ 2011;18 (12):1904–13.
[59] Seiwert TY, Zuo Z, Keck MK, Khattri A, Pedamallu CS, Stricker T, Brown C, Pugh TJ, Stojanov P, Cho J, Lawrence MS, Getz G, Bragelmann J, DeBoer R, Weichselbaum RR, Langerman A, Portugal L, Blair E, Stenson K, Lingen MW, Cohen EE, Vokes EE, White KP, Hammerman PS. Integrative and comparative genomic analysis of HPV-positive and HPV-negative head and neck squamous cell carcinomas. Clin Cancer Res 2015;21 (3):632–41.
[60] Bartek J, Lukas J. Chk1 and Chk2 kinases in checkpoint control and cancer. Cancer Cell 2003;3(5):421–9.
[61] Guertin AD, Martin MM, Roberts B, Hurd M, Qu X, Miselis NR, Liu Y, Li J, Feldman I, Benita Y, Bloecher A, Toniatti C, Shumway SD. Unique functions of CHK1 and WEE1 underlie synergistic anti-tumor activity upon pharmacologic inhibition. Cancer Cell Int 2012;12(1):45.
[62] Chen CC, Kennedy RD, Sidi S, Look AT, D'Andrea A. CHK1 inhibition as a strategy for targeting Fanconi Anemia (FA) DNA repair pathway deficient tumors. Mol Cancer 2009;8:24.
[63] Ferrao PT, Bukczynska EP, Johnstone RW, McArthur GA. Efficacy of CHK inhibitors as single agents in MYC-driven lymphoma cells. Oncogene 2012;31(13):1661–72.
[64] Gadhikar MA, Zhang JX, Shen L, Rao XY, Wang J, Zhao M, Kalu NN, Johnson FM, Byers LA, Heymach J, Hittelman WN, Udayakumar D, Pandita RK, Pandita TK, Pickering CR, Redwood AB, Piwnica-Worms H, Schlacher K, Frederick MJ, Myers JN. CDKN2A/p16 deletion in head and neck cancer cells is associated with CDK2 activation, replication stress, and vulnerability to CHK1 inhibition. Cancer Res 2018;78 (3):781–97.
[65] Squire CJ, Dickson JM, Ivanovic I, Baker EN. Structure and inhibition of the human cell cycle checkpoint kinase, Wee1A kinase: an atypical tyrosine kinase with a key role in CDK1 regulation. Structure 2005;13(4):541–50.
[66] De Witt Hamer PC, Mir SE, Noske D, Van Noorden CJ, Wurdinger T. WEE1 kinase targeting combined with DNA-damaging cancer therapy catalyzes mitotic catastrophe. Clin Cancer Res 2011;17(13):4200–7.
[67] Portugal J, Mansilla S, Bataller M. Mechanisms of drug-induced mitotic catastrophe in cancer cells. Curr Pharm Des 2010;16(1):69–78.
[68] Leijen S, Schellens JH, Shapiro G, Pavlick AC, Tibes R, Demuth T, Viscusi J, Cheng JD, Xu Y, Oza AM. A phase I pharmacological and pharmacodynamic study of MK-1775, a Weel tyrosine kinase inhibitor, in monotherapy and combination with gemcitabine, cisplatin, or carboplatin in patients with advanced solid tumors. J Clin Oncol 2010;28(15).
[69] Hirai H, Arai T, Okada M, Nishibata T, Kobayashi M, Sakai N, Imagaki K, Ohtani J, Sakai T, Yoshizumi T, Mizuarai S, Iwasawa Y, Kotani H. MK-1775, a small molecule Weel inhibitor, enhances anti-tumor efficacy of various DNA-damaging agents, including 5-fluorouracil. Cancer Biol Ther 2010;9(7):514–22.
[70] Hirai H, Iwasawa Y, Okada M, Arai T, Nishibata T, Kobayashi M, Kimura T, Kaneko N, Ohtani J, Yamanaka K, Itadani H, Takahashi-Suzuki I, Fukasawa K, Oki H, Nambu T, Jiang J, Sakai T, Arakawa H, Sakamoto T, Sagara T, Yoshizumi T, Mizuarai S, Kotani H. Small-molecule inhibition of Weel kinase by MK-1775 selectively sensitizes p53-deficient tumor cells to DNA-damaging agents. Mol Cancer Ther 2009;8(11):2992–3000.
[71] Rajeshkumar NV, De Oliveira E, Ottenhof N, Watters J, Brooks D, Demuth T, Shumway SD, Mizuarai S, Hirai H, Maitra A, Hidalgo M. MK-1775, a potent Weel inhibitor, synergizes with gemcitabine to achieve tumor regressions, selectively in p53-deficient pancreatic cancer xenografts. Clin Cancer Res 2011;17(9):2799–806.
[72] Wang Y, Decker SJ, Sebolt-Leopold J. Knockdown of Chk1, Weel and Myt1 by RNA interference abrogates G2 checkpoint and induces apoptosis. Cancer Biol Ther 2004;3(3):305–13.
[73] Bridges KA, Hirai H, Buser CA, Brooks C, Liu H, Buchholz TA, Molkentine JM, Mason KA, Meyn RE. MK-1775, a novel Weel kinase inhibitor, radiosensitizes p53-defective human tumor cells. Clin Cancer Res 2011;17 (17):5638–48.
[74] Leijen S, van Geel RM, Sonke GS, de Jong D, Rosenberg EH, Marchetti S, Pluim D, van Werkhoven E, Rose S, Lee MA, Freshwater T, Beijnen JH, Schellens JH. Phase II study of WEE1 inhibitor AZD1775 plus carboplatin in patients with TP53-mutated ovarian cancer refractory or resistant to first-line therapy within 3 months. J Clin Oncol 2016;34(36):4354–61.
[75] Osman AA, Monroe MM, Ortega Alves MV, Patel AA, Katsonis P, Fitzgerald AL, Neskey DM, Frederick MJ, Woo SH, Caulin C, Hsu TK, McDonald TO, Kimmel M, Meyn RE, Lichtarge O, Myers JN. Wee-1 kinase

inhibition overcomes cisplatin resistance associated with high-risk TP53 mutations in head and neck cancer through mitotic arrest followed by senescence. Mol Cancer Ther 2015;14(2):608–19.

[76] Tanaka N, Patel AA, Wang J, Frederick MJ, Kalu NN, Zhao M, Fitzgerald AL, Xie TX, Silver NL, Caulin C, Zhou G, Skinner HD, Johnson FM, Myers JN, Osman AA. Wee-1 kinase inhibition sensitizes high-risk HPV + HNSCC to apoptosis accompanied by downregulation of MCl-1 and XIAP antiapoptotic proteins. Clin Cancer Res 2015;21(21):4831–44.

[77] Mendez E, Rodriguez CP, Kao MC, Raju S, Diab A, Harbison RA, Konnick EQ, Mugundu GM, Santana-Davila R, Martins R, Futran ND, Chow LQM. A phase I clinical trial of AZD1775 in combination with neoadjuvant weekly docetaxel and cisplatin before definitive therapy in head and neck squamous cell carcinoma. Clin Cancer Res 2018;24(12):2740–8.

[78] Roth SY, Denu JM, Allis CD. Histone acetyltransferases. Annu Rev Biochem 2001;70:81–120.

[79] Gillenwater AM, Zhong M, Lotan R. Histone deacetylase inhibitor suberoylanilide hydroxamic acid induces apoptosis through both mitochondrial and Fas (Cd95) signaling in head and neck squamous carcinoma cells. Mol Cancer Ther 2007;6(11):2967–75.

[80] Struhl K. Histone acetylation and transcriptional regulatory mechanisms. Genes Dev 1998;12(5):599–606.

[81] Chang HH, Chiang CP, Hung HC, Lin CY, Deng YT, Kuo MY. Histone deacetylase 2 expression predicts poorer prognosis in oral cancer patients. Oral Oncol 2009;45(7):610–4.

[82] Weichert W. HDAC expression and clinical prognosis in human malignancies. Cancer Lett 2009;280(2):168–76.

[83] Prystowsky MB, Adomako A, Smith RV, Kawachi N, McKimpson W, Atadja P, Chen Q, Schlecht NF, Parish JL, Childs G, Belbin TJ. The histone deacetylase inhibitor LBH589 inhibits expression of mitotic genes causing G2/M arrest and cell death in head and neck squamous cell carcinoma cell lines. J Pathol 2009;218(4):467–77.

[84] Iglesias-Linares A, Yanez-Vico RM, Gonzalez-Moles MA. Potential role of HDAC inhibitors in cancer therapy: insights into oral squamous cell carcinoma. Oral Oncol 2010;46(5):323–9.

[85] Blagosklonny MV, Trostel S, Kayastha G, Demidenko ZN, Vassilev LT, Romanova LY, Bates S, Fojo T. Depletion of mutant p53 and cytotoxicity of histone deacetylase inhibitors. Cancer Res 2005;65(16):7386–92.

[86] Lane AA, Chabner BA. Histone deacetylase inhibitors in cancer therapy. J Clin Oncol 2009;27(32):5459–68.

[87] Almeida LO, Abrahao AC, Rosselli-Murai LK, Giudice FS, Zagni C, Leopoldino AM, Squarize CH, Castilho RM. NF kappa B mediates cisplatin resistance through histone modifications in head and neck squamous cell carcinoma (HNSCC). Febs Open Bio 2014;4:96–104.

[88] Kumar B, Yadav A, Lang JC, Teknos TN, Kumar P. Suberoylanilide hydroxamic acid (SAHA) reverses chemoresistance in head and neck cancer cells by targeting cancer stem cells via the downregulation of nanog. Genes Cancer 2015;6(3–4):169–81.

[89] Chen G, Zhang B, Xu H, Sun Y, Shi Y, Luo Y, Jia H, Wang F. Suppression of Sirt1 sensitizes lung cancer cells to WEE1 inhibitor MK-1775-induced DNA damage and apoptosis. Oncogene 2017;36(50):6863–72.

[90] Tanaka N, Patel AA, Tang L, Silver NL, Lindemann A, Takahashi H, Jaksik R, Rao X, Kalu NN, Chen TC, Wang J, Frederick MJ, Johnson F, Gleber-Netto FO, Fu S, Kimmel M, Wang J, Hittelman WN, Pickering CR, Myers JN, Osman AA. Replication stress leading to apoptosis within the S-phase contributes to synergism between Vorinostat and AZD1775 in HNSCC harboring high-risk TP53 mutation. Clin Cancer Res 2017;23(21):6541–54.

[91] Bensaad K, Tsuruta A, Selak MA, Vidal MNC, Nakano K, Bartrons R, Gottlieb E, Vousden KH. TIGAR, a p53-inducible regulator of glycolysis and apoptosis. Cell 2006;126(1):107–20.

[92] Liang YJ, Liu J, Feng ZH. The regulation of cellular metabolism by tumor suppressor p53. Cell Biosci 2013;3.

[93] Matoba S, Kang JG, Patino WD, Wragg A, Boehm M, Gavrilova O, Hurley PJ, Bunz F, Hwang PM. p53 regulates mitochondrial respiration. Cardiovasc Drug Ther 2006;20(6):415.

[94] Fitzgerald AL, Osman AA, Xie TX, Patel A, Skinner H, Sandulache V, Myers JN. Reactive oxygen species and p21Waf1/Cip1 are both essential for p53-mediated senescence of head and neck cancer cells. Cell Death Dis 2015;6.

[95] Lam-ubol A, Fitzgerald AL, Ritdej A, Phonyiam T, Zhang H, Myers JN, Huang P, Trachootham D. Sensory acceptable equivalent doses of ss-phenylethyl isothiocyanate (PEITC) induce cell cycle arrest and retard the growth of p53 mutated oral cancer in vitro and in vivo. Food Funct 2018;9(7):3640–56.

[96] Zhang C, Liu J, Liang YJ, Wu R, Zhao YH, Hong XH, Lin MH, Yu HY, Liu LX, Levine AJ, Hu WW, Feng ZH. Tumour-associated mutant p53 drives the Warburg effect. Nat Commun 2013;4.

[97] Skinner HD, Sandulache VC, Ow TJ, Meyn RE, Yordy JS, Beadle BM, Fitzgerald AL, Giri U, Ang KK, Myers JN. TP53 disruptive mutations Lead to head and neck cancer treatment failure through inhibition of radiation-induced senescence. Clin Cancer Res 2012;18(1):290–300.

[98] Sandulache VC, Skinner HD, Ow TJ, Zhang AJ, Xia XF, Luchak JM, Wong LJC, Pickering CR, Zhou G, Myers JN. Individualizing antimetabolic treatment strategies for head and neck squamous cell carcinoma based on TP53 mutational status. Cancer-Am Cancer Soc 2012;118(3):711–21.

[99] Jiralerspong S, Palla SL, Giordano SH, Meric-Bernstam F, Liedtke C, Barnett CM, Hsu LM, Hung MC, Hortobagyi GN, Gonzalez-Angulo AM. Metformin and pathologic complete responses to neoadjuvant chemotherapy in diabetic patients with breast cancer. J Clin Oncol 2009;27(20):3297–302.

[100] Sanli T, Rashid A, Liu CQ, Harding S, Bristow RG, Cutz JC, Singh G, Wright J, Tsakiridis T. Ionizing radiation activates amp-activated kinase (Ampk): a target for radiosensitization of human cancer cells. Int J Radiat Oncol 2010;78(1):221–9.

CHAPTER

5

Modulation of the PI3K/mTOR pathways

Arpine Galstyan, Janice Cho, Daniel E. Johnson, Jennifer R. Grandis

Department of Otolaryngology—Head and Neck Surgery, University of California at San Francisco, San Francisco, CA, United States

Abstract

Activation of the PI3K/Akt/mTOR signaling pathway has been associated with tumor progression and treatment resistance in head and neck squamous cell carcinoma (HNSCC). In the tumor microenvironment, PI3K/Akt/mTOR signaling confers a survival advantage by contributing to cellular growth, proliferation, and metabolism. Agents that target nodes within this pathway have demonstrated both tumor suppressive and anti-metastatic potential. Novel inhibitors of the PI3K/Akt/mTOR signaling pathway are currently being investigated in HNSCC clinical trials, with the goal of optimizing anticancer treatment and bypassing radio-resistance. Here, we review the role of the PI3K/Akt/mTOR signaling pathway in HNSCC with a focus on published preclinical and clinical investigations using PI3K pathway inhibitors. We also discuss approaches to elucidate the effect of radiation therapy on HNSCC models with distinct mutational profiles.

Abbreviations

4E-BP1	eukaryotic initiation factor 4E-binding protein 1
Akt	protein kinase B
ECOG-PS	Eastern cooperative oncology group performance status
GPCRs	G-protein-coupled receptors
HNSCC	head and neck squamous cell carcinoma
JNK	c-Jun NH-terminal kinase
mTORC1	mammalian target of rapamycin complex 1
mTORC2	mammalian target of rapamycin complex 2
PBMCs	peripheral blood, mononuclear cells
PDK1	phosphoinositide-dependent protein kinase-1
PDXs	patient-derived xenografts
PI3K	phosphoinositide 3-kinase
PIP_2	phosphatidylinositol 4,5-bisphosphate

Improving the Therapeutic Ratio in Head and Neck Cancer
https://doi.org/10.1016/B978-0-12-817868-3.00005-6

PIP_3	phosphatidylinositol (3,4,5)-trisphosphate
pS6	ribosomal protein S6
RGD	Arg-Gly-Asp
RTKs	receptor tyrosine kinases
TCGA	the cancer genome atlas
TKI	tyrosine kinase inhibitor
TRAIL	tumor necrosis factor-related apoptosis-inducing ligand

Conflicts of interest

Dr. Grandis receives research support from Incyte Pharmaceuticals and has patents that have optioned by STAT3 Therapeutics.

The PI3K/Akt/mTOR pathway

The phosphoinositide 3-kinase (PI3K) signaling pathway is the most frequently altered pathway in HNSCC and plays an important role in cell proliferation, growth, survival, and metabolism [1–4]. PI3Ks are a family of lipid kinases integral for mediating intracellular signaling pathways in response to extracellular stimuli. There are three classes of PI3K enzymes: class I, II, and III, based on their substrate structures and specificities [5]. Class I PI3Ks are heterodimers, consisting of a catalytic (p110) subunit and a regulatory (p85) subunit, and are further subdivided into class IA and class IB [5]. Class IA PI3Ks include three different p110 catalytic isoforms (p110α, p110β, and p110δ encoded by the genes *PIK3CA*, *PIK3CB*, and *PIK3CD*, respectively) that interact with five different p85 regulatory isoforms (p85α and its splicing variants p55α and p50α encoded by *PIK3R1* gene; p85β encoded by *PIK3R2*; p55γ encoded by *PIK3R3*) [5]. Class IB PI3Ks are heterodimers of the p110γ catalytic isoform and two regulatory isoforms (p87 and p101). Class IA and IB PI3Ks phosphorylate phosphatidylinositol 4,5-bisphosphate (PIP_2) and class III phosphorylates phosphatidylinositol [5]. Class II PI3Ks can be activated by various stimuli, including cytokines and Ca^{2+} [5] yet remain incompletely understood. Further information on the various classes of PI3Ks and their subunits can be found in Thorpe et al. [5].

In the absence of an activating signal, class IA PI3Ks exist as heterodimers, with p85 serving to inhibit the catalytic kinase activity of the p110 subunit [2–4]. However, the binding of growth factors and cytokines to cell surface G-protein-coupled receptors (GPCRs) or receptor tyrosine kinases (RTKs) results in the recruitment of class I PI3Ks to the plasma membrane [2, 5]. This leads to the dissociation of p110 from p85 and activation of p110 catalytic activity. The activated p110 then phosphorylates PIP_2 to generate phosphatidylinositol (3,4,5)-trisphosphate (PIP_3) [2, 3, 5]. PIP_3 is a lipid second messenger that recruits phosphoinositide-dependent protein kinase-1 (PDK1), leading to PDK1-mediated phosphorylation/activation of the serine/threonine kinase Akt (protein kinase B) on Thr 308 and Ser 473 residues [2, 3, 5]. Akt-dependent signaling pathways influence a number of key cellular activities, including growth, motility, and survival. Of particular interest is Akt regulation of mTOR, a serine/threonine kinase that is the catalytic subunit of two complexes—mammalian target of rapamycin complex 1 (mTORC1) and mTORC2, which are important for cell metabolism and growth [2, 4, 5]. The mTORC1 complex activates ribosomal protein S6 (pS6) kinase and eukaryotic initiation factor 4E-binding protein 1 (4E-BP1), which play key roles in cell growth and proliferation by increasing mRNA translation, lipid and nucleotide synthesis while decreasing autophagy [2, 4]. The PTEN lipid phosphatase is a negative regulator of PI3K signaling pathway, which exerts its effects by removing the 3' phosphate from PIP_3, thereby inactivating the PI3K/Akt/mTOR signaling.

In class III PI3Ks the VPS34-VPS15 heterodimer is activated by extracellular signals, including glucose and amino acids [2, 5]. Upon activation, it phosphorylates phosphatidylinositol generating phosphatidylinositol 3-phosphate, which plays a pivotal role in facilitating endosomal trafficking, phagocytosis, and autophagy [5]. Class II PI3Ks also serve crucial roles in cellular function and activity. Though poorly understood, class II PI3Ks have been shown to be essential in glucose transport, cell migration, and endocytosis [5]. In this chapter, we will primarily focus on class IA PI3Ks.

Dysregulation of the PI3K/Akt/mTOR pathway

Surgery, chemotherapy, and radiation offer limited survival benefits to a significant subset of HNSCC patients underscoring the need for improved treatments based on biology. One approach is to identify aberrant signaling pathways and to target critical nodes in these pathways using molecular-based treatments. Such studies to date in HNSCC have generally been carried out in unselected patient populations and failed to significantly improve clinical outcomes. The majority of the genetic and epigenetic alterations associated with HNSCC lesions result in constitutive activation of the PI3K/Akt/mTOR signaling axis. Therefore, the PI3K/Akt/mTOR pathway can be exploited for therapeutic purposes. By improving our understanding of how specific genetic alterations in this pathway respond to targeted therapy, we can tailor medical treatment to increase the likelihood that patients will experience a therapeutic benefit.

PIK3CA, encoding the p110α subunit of PI3K, is commonly mutated in HNSCC. In The Cancer Genome Atlas (TCGA), *PIK3CA* mutation and/or amplification and PTEN loss was observed in 35% and 30% of 504 HNSCC cases, respectively [6]. Whole exome sequencing of 151 HNSCC tumors revealed the following canonical mutation sites in *PIK3CA*: H1047R/L, E545K/G, and E542K/G [7]. These canonical mutations result in activation of the PI3K/Akt/mTOR signaling pathway and account for 64% of *PIK3CA* mutations in HNSCC The remaining 36% of mutations occur at noncanonical sites. The functional consequences of noncanonical *PIK3CA* mutations in HNSCC is incompletely understood. Previously unreported *PIK3CA* mutations were also discovered (R115L, G363A, C971A, C971R, and R975S), all of which conferred a survival advantage relative to wild-type [7]. In addition, 4% of HNSCC tumors had PI3K-γ mutations, and mutations were found in four of the PI3K regulatory subunits, each occurring at a rate of 2% [7].

Pan-cancer analyses of the PI3K/Akt/mTOR pathway revealed a series of low frequency mutations in the genes encoding AKT2, mTOR, and its associated subunits RAPTOR and RICTOR, *PIK3R1*, *PPP2R1A*, *AKT1*, *TSC1*, and *STK11* [8]. Loss of *PTEN*, a negative regulator of the PI3K pathway, has been implicated in 30% HNSCC cases. In the pan-cancer analyses, PTEN was found to be altered in 9% of cancers with deletion or a two-hit loss in 7% of cases [8]. The complexity of alterations in genes/proteins that comprise the PI3K signaling pathway underscores the importance of identifying gain-of-function or "driver" alterations that confer response to specific agents that target this pathway. Overall, *PIK3CA* is the most commonly altered gene in the PI3K pathway in HNSCC where it is amplified or mutated in about 35% of all HNSCCs. PTEN loss can also activate PI3K signaling and is found in approximately 30% of HNSCC cases.

Targeting the PI3K/Akt/mTOR pathway

Despite promising results in preclinical HNSCC models targeting the PI3K/Akt/mTOR pathway, an overarching approach to use these agents in the clinic is lacking [9]. Activation of the PI3K/Akt/mTOR pathway has been shown to confer resistance to standard treatments for HNSCC, including EGFR inhibitors [10]. Studies performed in orthotopic mouse models of HNSCC suggest that EGFR inhibitor resistance can be overcome by combining these agents with mTOR inhibitors. The combination of cetuximab and temsirolimus, an mTOR inhibitor, was demonstrated to have a synergistic anti-tumor effect in HNSCC, with strong inhibition of the MAPK + PI3K/-mTOR signaling [10].

HNSCC patient-derived xenografts (PDXs) with canonical or noncanonical *PIK3CA* mutations were found to be sensitive to a dual PI3K/mTOR inhibitor, whereas those with wild-type *PIK3CA* PDXs were refractory to treatment [11]. Therefore, *PIK3CA* mutations may predict sensitivity to PI3K/Akt/mTOR pathway inhibitors and serve as biomarkers for treatment selection.

However, in vitro and in vivo studies investigating the clinical utility of molecular biomarkers in HNSCC have yielded inconsistent results [12, 13]. Inhibitors of mTORC in HNSCC cell lines have a greater biochemical impact when combined with cetuximab compared with single agent therapy [13]. However, the effect of this treatment regimen in vivo appears to be dependent on HNSCC model selection [13]. Although improved tumor control was noted in all models tested, the extent of the effect was heterogeneous [13]. Another study found that sensitivity to PI3K inhibition differs among *PIK3CA* mutant tumors [12]. Further studies are needed to determine the impact of individual *PIK3CA* mutations on treatment response.

Pan-PI3K inhibitors

Buparlisib (BKM120)

Buparlisib is an orally-bioavailable pan-Class I PI3K inhibitor, and acts to inhibit PI3K in an ATP-competitive manner [14]. Preclinical studies demonstrated that buparlisib prevents activation of the PI3K/Akt/mTOR pathway and induces proliferation arrest. In a panel of 353 cancer cell lines, the inhibitory effects of this compound were found to be more pronounced in cell lines with mutant *PIK3CA* than those harboring wild-type *PIK3CA* [15]. One study evaluated the anti-proliferative effects of buparlisib and cetuximab with and without radiation therapy in CAL33 (HNSCC cell line containing an endogenous *PIK3CA* H1047R canonical mutation) and CAL27 (HNSCC cell line containing endogenous wild-type *PIK3CA*) [16]. A synergistic effect on inhibition of cellular proliferation was seen after sequential administration of cetuximab and buparlisib [16]. In CAL27 cells, a greater anti-proliferative effect was achieved when radiation therapy was administered in conjunction with cetuximab and buparlisib treatment [16]. In vivo studies using HNSCC xenografts also support the anti-tumor activity of buparlisib. In xenograft tumors derived from human HNSCC cell lines harboring a *PIK3CA* mutation, buparlisib significantly enhanced the anti-tumor potential of cetuximab, and this combination potentiated the anti-tumor effects of radiation therapy as well [17].

In a randomized, double-blind, placebo-controlled phase II trial of buparlisib and paclitaxel in patients with platinum-pretreated recurrent or metastatic squamous cell carcinoma of the head and neck, the median progression-free survival in the buparlisib group and placebo group was 4.6 months and 3.5 months, respectively [18]. In addition, 82% of patients in the buparlisib group and 72% of patients in the placebo group experienced a grade 3–4 adverse event, which included hyperglycemia, anemia, and neutropenia [18]. An on-going, randomized, open-label phase I/II trial is investigating the use of buparlisib and cetuximab in recurrent or metastatic head and neck cancer (NCT01816984). In addition, an ongoing phase I trial is combining buparlisib, cisplatin, and radiation treatment in high-risk patients with locally advanced HNSCC (NCT02113878) (Table 1).

TABLE 1 PI3K inhibitors in clinical trials

Drug	Trial number	Phase	Patient group	Status
Pan-PI3K inhibitors				
Buparlisib (NVP-BKM120)	NCT01816984	1	Recurrent or metastatic head and neck	Active, not recruiting
		2	Recurrent or metastatic head and neck	Active, not recruiting
	NCT02113878	1	Carcinoma, HNSCC HPV Positive Oropharyngeal Squamous Cell Carcinoma Hypopharyngeal Cancer Early Invasive Cervical Squamous Cell Carcinoma Carcinoma of Larynx Cancer of Nasopharynx	Active, not recruiting
Copanlisib (BAY 80-66946)	NCT02822482	1	Patients With Recurrent and/or Metastatic HNSCC Harboring a PI3KCA Mutation/ Amplification and/or a PTEN Loss	Recruiting
		2	Recurrent and/or Metastatic HNSCC Harboring a PI3KCA Mutation/ Amplification and/or a PTEN Loss	Recruiting
SF1126	NCT02644122	2	Metastatic Squamous Neck Cancer with Occult Primary Squamous Cell Carcinoma	Terminated—Slow enrollment No statistical analysis provided for ORR

Continued

TABLE 1 PI3K inhibitors in clinical trials—Cont'd

Drug	Trial number	Phase	Patient group	Status
Isoform-specific PI3K inhibitor				
Alpelisib (BYL-719)	NCT02145312	2	Recurrent or Metastatic HNSCC	Recruiting
	NCT01602315	1	Recurrent or Metastatic HNSCC	Early termination—Sponsor decision, slow recruitment
		2	Recurrent and Metastatic HNSCC	PFS, OS, and Kaplan-Meier medians were determined
	NCT02051751	1	Neoplasms, Breast Neoplasms, Head and Neck Neoplasms	Completed—Results not available

DLT, *dose-limiting toxicity;* HNSCC, *head and neck squamous cell carcinoma;* OS, *overall survival;* PFS, *progression free survival.*

Copanlisib (BAY 80-6946)

Copanlisib is a potent pan-PI3K inhibitor [14]. Copanlisib has been shown to inhibit cellular proliferation in various cancers including multiple myeloma, relapsed or refractory, indolent or aggressive lymphoma, and melanoma [19–21]. However, there are no studies investigating the use of copanlisib in HNSCC preclinical models. An open-label phase Ib/II trial is studying copanlisib in association with cetuximab in recurrent and/or metastatic HNSCC harboring *PIK3CA* mutation or amplification and/or PTEN loss (NCT02822482, Table 1).

Sonolisib (PX-866)

Sonolisib is a Class I pan-isoform inhibitor of PI3K [22] that irreversibly binds a lysine residue in the ATP catalytic site of the PI3K enzyme [23]. In HNSCC preclinical studies, sonolisib has been shown to enhance response to EGFR inhibitors [11]. In a phase I study, sonolisib plus cetuximab was a tolerable regimen in patients with incurable colorectal cancer or HNSCC [24]. Another phase I study confirmed the safe administration of sonolisib with docetaxel in patients with advanced solid tumors, including six patients with HNSCC [25]. In a phase II trial of cetuximab with or without sonolisib in relapsed or metastatic head and neck squamous cell carcinoma, the median progression-free survival was 80 days in both treatment groups, thus failing to demonstrate a benefit with the addition of sonolisib [26].

SF1126

SF1126 is a covalent conjugate of LY294002, a reversible PI3K inhibitor, that contains an Arg-Gly-Asp peptide modification and inhibits all PI3K isoforms [27]. The Arg-Gly-Asp (or RGD) motif binds to integrins within the tumor allowing for more rapid delivery of the active compound to the tumor site [27]. SF1126 is well-tolerated in murine systems, demonstrates favorable pharmacokinetics, and reduces phosphorylated Akt levels in vivo [27]. There are no published studies directly investigating the effects of SF1126 in HNSCC.

A randomized open-label phase II trial is evaluating the biweekly intravenous administration of SF1126 (1110 mg/m^2) and once weekly administration of SF1126 (1110 mg/m^2) in patients with recurrent or progressive HNSCC and in patients with mutations in *PIK3CA* and/or PI3K pathway genes (NCT02644122, Table 1).

Isoform-specific PI3K inhibitors

Alpelisib (BYL719)

Alpelisib is a p110 alpha isoform-selective PI3K inhibitor [28]. It inhibits both wild-type PI3K-alpha and PI3K-alpha containing canonical mutations, and is less potent against either PI3K-sigma or -gamma. In vitro studies demonstrated growth inhibition and chemo-sensitizing effects in HNSCC cell lines treated with alpelisib [29] In HNSCC cell lines containing endogenous *PIK3CA* mutations (SNU-1076 and Detroit562), alpelisib inhibited cellular proliferation. Alpelisib also demonstrated a synergistic effect with dacomitinib, a selective and irreversible EGFR inhibitor, in inhibiting cell proliferation [29]. In in vivo HNSCC models, p-selectin-mediated delivery of alpelisib using nanoparticles inhibited tumor growth and promoted radio-sensitization [28]. Compared with oral administration of alpelisib, nanoparticle encapsulated alpelisib inhibited tumor growth at much lower concentrations and led to prolonged inhibition of the PI3K-Akt-mTOR pathway with concomitant radiation therapy [28]. Phase I and phase II clinical trials are investigating the use of alpelisib in recurrent and metastatic HNSCC (NCT02145312, NCT01602315, and NCT02051751, Table 1).

Akt inhibitors

MK-2206

MK-2206 is an allosteric Akt inhibitor with affinity in the nanomolar range [30]. In vitro studies have shown that MK-2206 inhibits chemotaxis and migration, and reduces phosphorylated Akt levels under both normoxic and hypoxic conditions in HNSCC cell lines [31]. In vivo, MK-2206 has been shown to reduce tumor size [31]. Fluorescent imaging of tumor cell markers also revealed reduced metastasis to cervical lymph nodes in nude mice with human orthotopic tongue tumors treated with MK-2206 [31]. A phase II clinical trial is investigating the use of MK-2206 in recurrent and metastatic HNSCC (NCT01349933, Table 2).

TABLE 2 Akt inhibitors in clinical trials.

Drug	Trial number/DOI	Phase	Patient group	Status
MK-2206	NCT00848718	1	Locally Advanced, Metastatic Solid Tumors	Completed—Results not available
	NCT01349933	2	Recurrent and Metastatic HNSCC	Completed

HNSCC, *head and neck squamous cell carcinoma.*

Perifosine

Perifosine is an orally bioavailable alkyl-phospholipid that inhibits Akt in a dose-dependent manner. In vitro, perifosine has been shown to induce expression of the death receptors DR4 and DR5, increase c-Jun NH-terminal kinase (JNK) and c-Jun levels, and promote cell cycle arrest via induction of p21 [32]. Combination of perifosine and tumor necrosis factor-related apoptosis-inducing ligand (TRAIL) has also been shown to be more effective in inducing apoptosis and preventing tumor growth than either agent alone [32]. However, clinical trials of perifosine in HNSCC have had mixed results in terms of efficacy. In a phase II study, the median progression-free survival time was 1.7 months, and the median overall survival time was 5.5 months [33]. The most notable toxicities were gastrointestinal and fatigue. Despite the small sample size, high expression levels of p38 and Akt were found to be significantly correlated with better survival and baseline tumor size [33]. There are no active trials investigating perifosine in HNSCC.

mTOR inhibitors

Rapamycin (sirolimus)

Rapamycin is commercially known as sirolimus. In vitro studies investigating the effects of rapamycin in HNSCC cell lines have shown that it decreases the expression of phosphorylated mTOR (Ser 2448) and inhibits cell growth in a dose-dependent manner [34]. HNSCC cell lines treated with rapamycin exhibit reduced levels of mTOR downstream targets and significant inhibition of cell viability [35].

In vivo studies have demonstrated that rapamycin has potent anti-tumor activity. It inhibits DNA synthesis, induces apoptosis, and promotes tumor regression [36]. Sun et al. demonstrated that rapamycin significantly delayed the onset of the HNSCC tumors and prolonged the survival of Tgbr1 and PTEN conditional knockout mice [37]. In another study, HNSCC xenograft models treated with rapamycin showed significant inhibition of tumor growth in conjunction with decreased expression of phosphorylated-S6 expression [36].

An open-label phase I/II clinical trial (NCT01195922) investigated the therapeutic utility of rapamycin in 16 patients with advanced-stage HNSCC before they underwent either surgery or chemoradiation [38]. Rapamycin treatment was associated with reduced tumor size and standardized FDG uptake in non-nodal target lesions and lymph nodes. The most frequently cited adverse events were complications of wound healing, infections due to immune compromise and late radiation toxicities [38].

Everolimus (RAD001)

Everolimus is an orally bioavailable hydroxyethyl ether derivative of rapamycin. It binds to FKB12, complexes, and inhibits mTOR. It has a shorter half-life (18–35 h) than rapamycin (57–63 h), allowing for more immediate steady state achievement, and it exhibits both immunosuppressive and anti-proliferative effects [39, 40]. In murine HNSCC models, low doses of everolimus have been shown to inhibit both local and distant spread of disease [41]. In an orthotropic HNSCC model, everolimus was found to significantly reduce pS6-positive cells in primary tumors and metastatic lymph nodes [42]. At the end of the observation period, residual tumors in everolimus-treated mice also showed squamous differentiation and fibrosis as opposed to areas of active cellular growth and proliferation [42]. This was also reflected in the overall increased survival, reduced levels of metastases, and inhibition of tumor growth in the everolimus treated mice [42].

Everolimus has also demonstrated synergistic antitumor growth effects in combination with other pharmaceutical agents. When combined with LY2835219, a CDK4/6 inhibitor, more significant tumor growth inhibition was achieved after dual-therapy than with either agent alone [43]. In patients with locally advanced stage IV HNSCC undergoing induction chemotherapy (docetaxel, cisplatin, and fluorouracil), everolimus improved the therapeutic effects of docetaxel [44].

A phase II clinical trial of 35 patients with histologically or cytologically confirmed platinum-resistant HNSCC examined the effect of daily oral everolimus and erlotinib, an EGFR TKI [45]. The results demonstrated that everolimus combined with erlotinib achieved a partial response in three patients at 4 weeks and one patient at 12 weeks. At 4 weeks, 27 patients achieved disease stabilization and at 12 weeks, 11 were confirmed. The median progression-free survival (PFS) was 11.9 weeks, and the median overall survival (OS) was 10.25 months. The most frequently cited adverse events greater than or equal to grade 3 were diarrhea, infections, mucositis, and head and neck edema.

In another phase II clinical trial, nine patients with recurrent or metastatic HNSCC were treated with continuous 28-day cycles of everolimus 10 mg by mouth daily. Of the nine patients treated, only seven were evaluable in the first stage [46]. The proportion of patients that achieved either a complete response, a partial response, or stable disease was 28%. The median PFS and OS were 1.5 and 4.5 months, respectively. However, everolimus treatment was discontinued in three patients due to toxicity. The remaining six patients experienced toxicity that was grade 3 or higher.

A number of phase I and II clinical trials investigating the use of everolimus in head and neck cancer are ongoing (NCT01051791, NCT01133678, NCT01333085, NCT00942734, NCT01283334, NCT01313390, NCT01111058, NCT01009346, and NCT01637194, Table 1).

Everolimus at 1.0 mg/kg twice a day and 0.5 mg/kg twice a day showed statistically significant inhibition of tumor growth when compared with the cyclosporine group and the untreated animals (P // .002).

Temsirolimus (CCI-779)

Temsirolimus is a water-soluble ester that undergoes hydrolysis to sirolimus following intravenous administration. Temsirolimus has been shown to have cytotoxic and cytostatic activities in HNSCC cell lines [47]. In peripheral blood, mononuclear cells (PBMCs) of patients with newly diagnosed advanced-stage HNSCC, temsirolimus significantly decreased p-Akt, pS6 and p4E-BP1 [47].

In vitro, temsirolimus demonstrates anti-proliferative synergistic effects with cetuximab. When administered in conjunction with cetuximab in HNSCC Cal33 cell lines, temsirolimus was shown to downregulate the PI3K pathway.

In vivo, temsirolimus improved the anti-tumor effects of cetuximab monotherapy as well as cetuximab-cisplatin/fluorouracil combination therapy [10]. Compared to the mice in the temsirolimus plus cetuximab-cisplatin/fluorouracil treatment group, those in the cetuximab monotherapy treatment group did not experience a significant weight change [10]. Therefore, the cetuximab-temsirolimus treatment combination was better tolerated [10]. Cisplatin has been extensively studied for its radio-sensitizing effects, and it has demonstrated improved outcomes in patients with locally advanced head and neck cancer [48]. Exploring new therapies that offer a similar treatment benefit with fewer toxicities may be advantageous for patients. In addition, the temsirolimus and cetuximab drug combination inhibited the PI3K/Akt/mTOR pathway more significantly than either temsirolimus or cetuximab alone [10].

A single-institution phase II clinical trial consisting of 36 patients with recurrent and/or metastatic HNSCC examined the effect of the combination of temsirolimus, carboplatin, and paclitaxel [49]. The PFS was 5.9 months, and the OS was 12.8 months. In addition, the most frequently cited adverse events were hematologic toxicities [49]. Another phase II clinical trial consisting of 12 patients with recurrent and/or metastatic, platinum-refractory HNSCC and Eastern Cooperative Oncology Group Performance Status (ECOG-PS) 0–2 investigated the combination of erlotinib and temsirolimus [50]. The PFS and OS was 1.9 months and 4.0 months, respectively. An open-label multicenter phase II trial of temsirolimus in patients with relapsed or recurrent HNSCC (NCT01172769) recruited 42-patients and no results have been made available to date.

OSI-027 (ASP487)

OSI-027 is an orally bioavailable TORC1/TORC2 inhibitor that competitively binds to the ATP-binding domain of mTOR [51]. In a panel of HNSCC cell lines, OSI-027 was shown to inhibit TORC1 and TORC2 in a dose-dependent manner. It also demonstrated anti-proliferative synergistic effects with erlotinib, leading to the abrogation of phosphorylation of several downstream targets of the PI3K/Akt/mTOR pathway, including Akt (see Fig. 1) [51]. OSI-027 has also been shown to have anti-tumor efficacy. In mice bearing HNSCC xenografts, treatment with OSI-027 plus cetuximab led to more significant tumor reduction than administration of either agent alone [51] (Table 3).

Dual PI3K/mTOR inhibitors

PF-04691502

PF-04691502 is an orally-active, ATP-competitive, dual inhibitor of PI3K and mTOR that reduces levels of phosphorylated Akt and inhibits cellular proliferation [52, 53]. In preclinical studies, PF-04691502 has demonstrated a synergistic effect with radiation in UT-SCC-14 cell line derived from primary cancers [53]. There are currently no clinical studies investigating its effects in HNSCC.

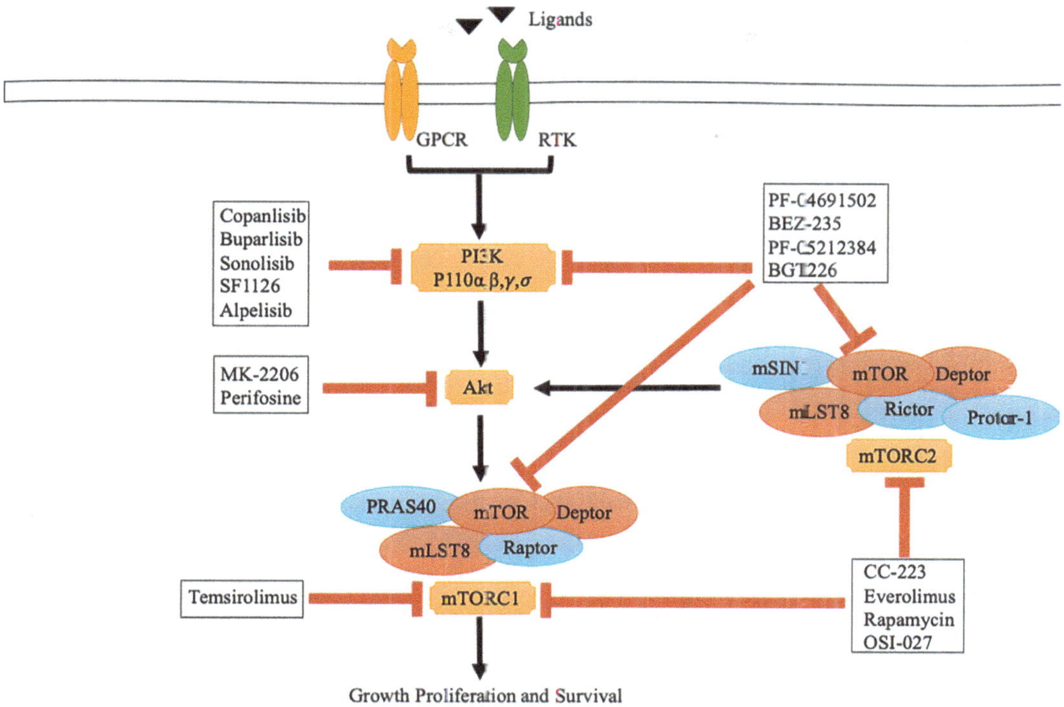

FIG. 1 Therapeutics under clinical investigation for inhibition of the PI3K/Akt/mTOR pathway in HNSCC. *Orange lines* indicate inhibition and *black arrows* indicate activation. *Akt*, protein kinase B; *Deptor*, DEP-domain-containing mTOR-interacting protein; *GPCR*, G-protein coupled receptor; *mLST8*, mammalian lethal with Sec13 protein 8; *mSIN1*, mammalian stress-activated protein kinase interacting protein; *mTORC*, mammalian target of rapamycin complex; *PI3K*, phosphoinositide 3-kinase; *PRAS40*, proline-rich Akt substrate 40 kDa; *Protor-1*, protein observed with Rictor-1; *Raptor*, regulatory associated protein of mTOR; *Rictor*, rapamycin-insensitive companion of mTOR; *RTK*, receptor tyrosine kinase.

BEZ235

BEZ-235 is a combined PI3K/mTOR inhibitor. HNSCC cell lines with *PIK3CA* canonical mutations were more sensitive to BEZ235 treatment compared to those with wild-type *PIK3CA* [7, 54]. Mice with CAL-33 xenografts were also more sensitive to BEZ235 treatment than controls with low baseline phosphorylated Akt levels and controls with WT *PIK3CA* [7].

PF-05212384 (PF-384)

PF-0212384 is an ATP-competitive, dual PI3K/mTOR inhibitor that demonstrates variable activity in HNSCC cell lines with *PIK3CA* mutations [52, 55]. In in vivo models, PF-05212384 demonstrated pro-apoptotic, anti-proliferative, and anti-angiogenic activity [55]. It inhibited the direct targets of PI3K/mTOR; however, it did not completely diminish the levels of co-activated ERK and failed to completely attenuate the growth of UMSCC-1 xenografts

TABLE 3 mTOR inhibitors in clinical trials

Drug	Trial number	Phase	Patient group	Status
Everolimus (RAD001)	NCT01051791	2	SCCHN	$n=7$, CBR: 28
	NCT01133678	1	Head and Neck Cancer	Active, not recruiting
		2	Head and Neck Cancer	Active, not recruiting
	NCT01333085	1	Head and Neck Cancer	Completed—Results not available
		2	Head and Neck Cancer	
	NCT00942734	2	Head and Neck Cancer	$n=35$, PFS: 48.57
	NCT01283334	1	Head and Neck Neoplasms Cancer of the Head and Neck	$n=6$, 2 participants $n=6$, 1 participant
		2	Head and Neck Neoplasms Cancer of the Head and Neck	
	NCT01313390	1	Head and Neck Cancer	Terminated—Lack of recruitment
		2	Head and Neck Cancer	Terminated—Lack of recruitment
	NCT01111058	2	Head and Neck Cancer	Active—Not Recruiting
	NCT01009346	1	Head and Neck Cancer	Terminated (toxicity) $n=7$, MTD: 10
		2	Head and Neck Cancer	Terminated (toxicity) $n=5$, PFS: 2.8 (1.6–13.9)
	NCT01637194	1	Metastatic or Recurrent Colon Cancer or Head and Neck Cancer	Completed—No results posted
Rapamycin	NCT01195922	1	Mouth Neoplasms Head and Neck Neoplasms Tongue Neoplasms Carcinoma, Squamous Cell	
		2	Mouth Neoplasms Head and Neck Neoplasms Tongue Neoplasms Carcinoma, Squamous Cell	See above
Temsirolimus	NCT01172769	2	HNSCC	Completed—Results not available
	NCT01015664	1	SCCHN	Terminated—Results not available
		2	SCCHN	Terminated—Results not available
	NCT01009203	2	Squamous Cell Carcinoma	Terminated—High patient withdrawal rate
	NCT01016769	1	Squamous Cell Cancer Head and Neck Cancer	Completed—Results not available
		2	Squamous Cell Cancer Head and Neck Cancer	Completed—Results not available
	NCT01256385	2	Recurrent and/or Metastatic Head and Neck Cancer Who Did Not Respond to Previous Therapy (MAESTRO HN)	Arm A ($n=40$) PFS: 105 (70–136) Arm B ($n=40$) PFS: 105 (77–147) P-value for PFS: .73

CR, *complete response rate;* CT/RT, *chemo-radiation;* HNSCC, *head and neck squamous cell carcinoma;* MTD, *maximum tolerated dose;* ORR, *overall response rate;* PFS, *progression free survival;* PR, *partial response;* RDI, *median relative dose intensities;* SCCHN, *squamous cell carcinoma of the head and neck;* SD, *stable disease.*

in vivo [55]. In nude mice bearing UMSCC-1 xenografts, intravenously administered PF-05212384 and fractionated radiation delayed xenograft regrowth [52]. PF-05212384-induced radiosensitization in in vitro and in vivo models, further supporting the potential use of PF-05212384 as a radiation modifier in exponentially-growing and/or plateau-phase HNSCC cells.

BGT226

BGT226 is a dual PI3K/mTOR inhibitor that has been shown to have antiproliferative activity against OECM1, KB, and UMSCC1 head and neck cancer cell lines [56]. There are no clinical studies investigating its therapeutic utility in HNSCC populations.

Radiation therapy and pathway inhibitors

Further investigation is needed to elucidate how radiation therapy alone impacts HNSCC models with distinct genetic alterations of the PI3K/Akt/mTOR pathway. Radiation therapy serves as an important treatment option for HNSCC patients. However, radio-resistance has been reported to arise from a variety of different mechanisms, including signaling through the PI3K/Akt/mTOR pathway [57].

Studies investigating the use of PI3K/Akt/mTOR pathway inhibitors in conjunction with radiation therapy have demonstrated advantageous effects. Radiosensitivity was achieved through the use of pathway inhibitors that target different nodes within the pathway. Radiation therapy used in conjunction with Ly294002 led to radio-sensitization in astrocytic cells. Similar effects were also seen when targeting either Akt or mTOR. Primary cultures from 120 patients with advanced HNSCC who received either primary or adjuvant radiotherapy demonstrated increased p-Akt levels when treated with radiation plus LY294002 suggesting that p-Akt levels may reflect radiosensitivity [58]. Understanding how these different nodes within the pathway impact radiation therapy is important in identifying which patients will benefit from such treatment. More studies are needed to examine the effect of PI3K or Akt inhibitors in combination with radiotherapy and to understand whether targeting this pathway may enhance treatment efficacy in genetically defined subsets of HNSCC.

Opportunities to improve treatment

The effect of specific genetic alterations in the PI3K signaling pathway such as *PIK3CA* mutation or PTEN deletion on HNSCC response to radiotherapy may help identify which patients should receive agents that target this pathway in conjunction with (or following) radiation. Given the heterogeneous mutational profile of HNSCC tumors, personalized therapeutics will require some tailoring to the individual patient's tumor biology. Genomically-stratified radiation therapies can then be developed that integrate individual biological differences into treatment. A "basket trial" approach can be implemented to study molecular targeted agents in HNSCC patients.

Acknowledgments

This work was supported in part by NIH grants R35CA231998 and R01DE023685 and by the American Cancer Society.

References

[1] Engelman JA. Targeting PI3K signalling in cancer: opportunities, challenges and limitations. Nat Rev Cancer 2009;9(8):550–62.

[2] Liu P, Cheng H, Roberts TM, Zhao JJ. Targeting the phosphoinositide 3-kinase pathway in cancer. Nat Rev Drug Discov 2009;8(8):627–44.

[3] Polivka J, Janku F. Molecular targets for cancer therapy in the PI3K/AKT/mTOR pathway. Pharmacol Ther 2014;142(2):164–75.

[4] Yap TA, Bjerke L, Clarke PA, Workman P. Drugging PI3K in cancer: refining targets and therapeutic strategies. Curr Opin Pharmacol 2015;23:98–107.

[5] Thorpe LM, Yuzugullu H, Zhao JJ. PI3K in cancer: divergent roles of isoforms, modes of activation and therapeutic targeting. Nat Rev Cancer 2015;15(1):7–24.

[6] Cancer Genome Atlas Network. Comprehensive genomic characterization of head and neck squamous cell carcinomas. Nature 2015;517(7536):576–82.

[7] Lui VWY, Hedberg ML, Li H, Vangara BS, Pendleton K, Zeng Y, Lu Y, Zhang Q, Du Y, Gilbert BR, Freilino M, Sauerwein S, Peyser ND, Xiao D, Diergaarde B, Wang L, Chiosea S, Seethala R, Johnson JT, Kim S, Duvvuri U, Ferris RL, Romkes M, Nukui T, Kwok-Shing Ng P, Garraway LA, Hammerman PS, Mills GB, Grandis JR. Frequent mutation of the PI3K pathway in head and neck cancer defines predictive biomarkers. Cancer Discov 2013;3(7):761–9.

[8] Zhang Y, Kwok-Shing Ng P, Kucherlapati M, Chen F, Liu Y, Tsang YH, de Velasco G, Jeong KJ, Akbani R, Hadjipanayis A, Pantazi A, Bristow CA, Lee E, Mahadeshwar HS, Tang J, Zhang J, Yang L, Seth S, Lee S, Ren X, Song X, Sun H, Seidman J, Luquette LJ, Xi R, Chin L, Protopopov A, Westbrook TF, Shelley CS, Choueiri TK, Ittmann M, Van Waes C, Weinstein JN, Liang H, Henske EP, Godwin AK, Park PJ, Kucherlapati R, Scott KL, Mills GB, Kwiatkowski DJ, Creighton CJ. A pan-cancer proteogenomic atlas of PI3K/AKT/mTOR pathway alterations. Cancer Cell 2017;31(6):832.e3.

[9] Isaacsson Velho PH, Castro G, Chung CH. Targeting the PI3K pathway in head and neck squamous cell carcinoma. Am Soc Clin Oncol Educ Book 2015;123–8.

[10] Bozec A, Ebran N, Radosevic-Robin N, Sudaka A, Monteverde M, Toussan N, Etienne-Grimaldi M-C, Nigro CL, Merlano M, Penault-Llorca F, Milano G. Combination of mTOR and EGFR targeting in an orthotopic xenograft model of head and neck cancer. Laryngoscope 2016;126(4):156.

[11] Keysar SB, Astling DP, Anderson RT, Vogler BW, Bowles DW, Morton JJ, Paylor JJ, Glogowska MJ, Le PN, Eagles-Soukup JR, Kako SL, Takimoto SM, Sehrt DB, Umpierrez A, Pittman MA, Macfadden SM, Helber RM, Peterson S, Hausman DF, Said S, Leem TH, Goddard JA, Arcaroli JJ, Messersmith WA, Robinson WA, Hirsch FR, Varella-Garcia M, Raben D, Wang X-J, Song JI, Tan A-C, Jimeno A. A patient tumor transplant model of squamous cell cancer identifies PI3K inhibitors as candidate therapeutics in defined molecular bins. Mol Oncol 2013;7(4):776–90.

[12] Ruicci KM, Meens J, Sun RX, Rizzo G, Pinto N, Yoo J, Fung K, MacNeil D, Mymryk JS, Barrett JW, Boutros PC, Ailles L, Nichols AC. A controlled trial of HNSCC patient-derived xenografts reveals broad efficacy of PI3Kα inhibition in controlling tumor growth. Int J Cancer 2018;143:1–7.

[13] Swick AD, Prabakaran PJ, Miller MC, Javaid AM, Fisher MM, Sampene E, Ong IM, Hu R, Iida M, Nickel KP, Bruce JY, Wheeler DL, Kimple RJ. Cotargeting mTORC and EGFR signaling as a therapeutic strategy in HNSCC. Mol Cancer Ther 2017;16(7):1257–68.

[14] Liu N, Rowley BR, Bull CO, Schneider C, Haegebarth A, Schatz CA, Fracasso PR, Wilkie DP, Hentemann M, Wilhelm SM, Scott WJ, Mumberg D, Ziegelbauer K. BAY 80-6946 is a highly selective intravenous PI3K inhibitor with potent p110α and p110δ activities in tumor cell lines and xenograft models. Mol Cancer Ther 2013;12(11):2319–30.

[15] Maira S-M, Pecchi S, Huang A, Burger M, Knapp M, Sterker D, Schnell C, Guthy D, Nagel T, Wiesmann M, Brachmann S, Fritsch C, Dorsch M, Chène P, Shoemaker K, De Pover A, Menezes D, Martiny-Baron G, Fabbro D, Wilson CJ, Schlegel R, Hofmann F, García-Echeverría C, Sellers WR, Voliva CF. Identification and

characterization of NVP-BKM120, an orally available pan-class I PI3-kinase inhibitor. Mol Cancer Ther 2012;11(2):317–28.

[16] Lattanzio L, Tonissi F, Monteverde M, Vivenza D, Russi E, Milano G, Merlano M, Lo NC. Treatment effect of buparlisib, cetuximab and irradiation in wild-type or PI3KCA-mutated head and neck cancer cell lines. Investig New Drugs 2015;33(2):310–20.

[17] Bozec A, Ebran N, Radosevic-Robin N, Chamorey E, Yahia HB, Marcie S, Gautier M, Penault-Llorca F, Milano G. Combination of phosphotidylinositol-3-kinase targeting with cetuximab and irradiation: a preclinical study on an orthotopic xenograft model of head and neck cancer. Head Neck 2017;39(1):151–9.

[18] Soulières D, Faivre S, Mesía R, Remenár É, Li S-H, Karpenko A, Dechaphunkul A, Ochsenreither S, Kiss LA, Lin J-C, Nagarkar R, Tamás L, Kim S-B, Erfán J, Alyasova A, Kasper S, Barone C, Turri S, Chakravartty A, Chol M, Aimone P, Hirawat S, Licitra L. Buparlisib and paclitaxel in patients with platinum-pretreated recurrent or metastatic squamous cell carcinoma of the head and neck (BERIL-1): a randomised, double-blind, placebo-controlled phase 2 trial. Lancet Oncol 2017;18(3):323–35.

[19] Dreyling M, Morschhauser F, Bouabdallah K, Bron D, Cunningham D, Assouline SE, Verhoef G, Linton K, Thieblemont C, Vitolo U, Hiemeyer F, Giurescu M, Garcia-Vargas J, Gorbatchevsky I, Liu L, Koechert K, Peña C, Neves M, Childs BH, Zinzani PL. Phase II study of copanlisib, a PI3K inhibitor, in relapsed or refractory, indolent or aggressive lymphoma. Ann Oncol 2017;28(9):2169–78.

[20] Okabe S, Tanaka Y, Tauchi T, Ohyashiki K. Copanlisib, a novel phosphoinositide 3-kinase inhibitor, combined with carfilzomib inhibits multiple myeloma cell proliferation. Ann Hematol 2018;98(3):723–33.

[21] Schneider P, Schön M, Pletz N, Seitz CS, Liu N, Ziegelbauer K, Zachmann K, Emmert S, Schön MP. The novel PI3 kinase inhibitor, BAY 80-6946, impairs melanoma growth in vivo and in vitro. Exp Dermatol 2014;23(8):579–84.

[22] Ihle NT, Williams R, Chow S, Chew W, Berggren MI, Paine-Murrieta G, Minion DJ, Halter RJ, Wipf P, Abraham R, Kirkpatrick L, Powis G. Molecular pharmacology and antitumor activity of PX-866, a novel inhibitor of phosphoinositide-3-kinase signaling. Mol Cancer Ther 2004;3(7):763–72.

[23] Wipf P, Minion DJ, Halter RJ, Berggren MI, Ho CB, Chiang GG, Kirkpatrick L, Abraham R, Powis G. Synthesis and biological evaluation of synthetic viridins derived from C(20)-heteroalkylation of the steroidal PI-3-kinase inhibitor wortmannin. Org Biomol Chem 2004;2(13):1911–20.

[24] Bowles DW, Senzer N, Hausman D, Peterson S, Vo A, Walker L, Cohen RB, Jimeno A. A multicenter phase 1 study of PX-866 and cetuximab in patients with metastatic colorectal carcinoma or recurrent/metastatic squamous cell carcinoma of the head and neck. Investig New Drugs 2014;32(6):1197–203.

[25] Bowles DW, Ma WW, Senzer N, Brahmer JR, Adjei AA, Davies M, Lazar AJ, Vo A, Peterson S, Walker L, Hausman D, Rudin CM, Jimeno A. A multicenter phase 1 study of PX-866 in combination with docetaxel in patients with advanced solid tumours. Br J Cancer 2013;109(5):1085–92.

[26] Jimeno A, Shirai K, Choi M, Laskin J, Kochenderfer M, Spira A, Cline-Burkhardt V, Winquist E, Hausman D, Walker L, Cohen RB. A randomized, phase II trial of cetuximab with or without PX-866, an irreversible oral phosphatidylinositol 3-kinase inhibitor, in patients with relapsed or metastatic head and neck squamous cell cancer. Ann Oncol 2015;26(3):556–61.

[27] Garlich JR, De P, Dey N, Su JD, Peng X, Miller A, Murali R, Lu Y, Mills GB, Kundra V, Shu HK, Feng Q, Durden DL. A vascular targeted pan phosphoinositide 3-kinase inhibitor prodrug, SF1126, with antitumor and antiangiogenic activity. Cancer Res 2008;68(1):206–15.

[28] Mizrachi A, Shamay Y, Shah J, Brook S, Soong J, Rajasekhar VK, Humm JL, Healey JH, Powell SN, Baselga J, Heller DA, Haimovitz-Friedman A, Scaltriti M. Tumour-specific PI3K inhibition via nanoparticle-targeted delivery in head and neck squamous cell carcinoma. Nat Commun 2017;8:14292.

[29] Keam B, Kim S, Ahn Y-O, Kim TM, Lee S-H, Kim D-W, Heo DS. In vitro anticancer activity of PI3K alpha selective inhibitor BYL719 in head and neck cancer. Anticancer Res 2015;35(1):175–82.

[30] Brolih S, Parks SK, Vial V, Durivault J, Mostosi L, Pouysségur J, Pagès G, Picco V. AKT1 restricts the invasive capacity of head and neck carcinoma cells harboring a constitutively active PI3 kinase activity. BMC Cancer 2018;18(1):249.

[31] Knowles JA, Golden B, Yan L, Carroll WR, Helman EE, Rosenthal EL. Disruption of the AKT pathway inhibits metastasis in an orthotopic model of head and neck squamous cell carcinoma. Laryngoscope 2011;121(11):2359–65.

[32] Fu L, Lin Y-D, Elrod HA, Yue P, Oh Y, Li B, Tao H, Chen GZ, Shin DM, Khuri FR, Sun S-Y. c-Jun NH2-terminal kinase-dependent upregulation of DR5 mediates cooperative induction of apoptosis by perifosine and TRAIL. Mol Cancer 2010;9:315.

[33] Argiris A, Cohen E, Karrison T, Esparaz B, Mauer A, Ansari R, Wong S, Lu Y, Pins M, Dancey J, Vokes E. A phase II trial of perifosine, an oral alkylphospholipid, in recurrent or metastatic head and neck cancer. Cancer Biol Ther 2006;5(7):766–70.
[34] Aissat N, Le Tourneau C, Ghoul A, Serova M, Bieche I, Lokiec F, Raymond E, Faivre S. Antiproliferative effects of rapamycin as a single agent and in combination with carboplatin and paclitaxel in head and neck cancer cell lines. Cancer Chemother Pharmacol 2008;62(2):305–13.
[35] Singh T, Gupta NA, Xu S, Prasad R, Velu SE, Katiyar SK. Honokiol inhibits the growth of head and neck squamous cell carcinoma by targeting epidermal growth factor receptor. Oncotarget 2015;6(25):21268–82.
[36] Hu M, Ekshyyan O, Herman Ferdinandez L, Rong X, Caldito G, Nathan C-AO. Efficacy and comparative effectiveness of sirolimus as an anticancer drug. Laryngoscope 2011;121(5):978–82.
[37] Sun Z-J, Zhang L, Hall B, Bian Y, Gutkind JS, Kulkarni AB. Chemopreventive and chemotherapeutic actions of mTOR inhibitor in genetically defined head and neck squamous cell carcinoma mouse model. Clin Cancer Res 2012;18(19):5304–13.
[38] Shirai K, Day TA, Szabo E, Van Waes C, O'Brien PE, Matheus MG, Godwin K, Sood AJ, Vick JA, Martin D, Vitale-Cross L, Molinolo A, Gutkind JS. A pilot, single arm, prospective trial using neoadjuvant rapamycin prior to definitive therapy in head and neck squamous cell carcinoma. J Clin Oncol 2015;33(15_suppl):6071.
[39] Raimondo L, D'Amato V, Servetto A, Rosa R, Marciano R, Formisano L, Di Mauro C, Orsini RC, Cascetta P, Ciciola P, De Maio AP, Di Renzo MF, Cosconati S, Bruno A, Randazzo A, Napolitano F, Montuori N, Veneziani BM, De Placido S, Bianco R. Everolimus induces Met inactivation by disrupting the FKBP12/Met complex. Oncotarget 2016;7(26):40073–84.
[40] Kirchner GI, Meier-Wiedenbach I, Manns MP. Clinical pharmacokinetics of everolimus. Cham: Adis International; 2004. p. 83–95.
[41] Khariwala SS, Kjaergaard J, Lorenz R, Van Lente F, Shu S, Strome M. Everolimus (RAD) inhibits in vivo growth of murine squamous cell carcinoma (SCC VII). Laryngoscope 2006;116(5):814–20.
[42] Patel V, Marsh CA, Dorsam RT, Mikelis CM, Masedunskas A, Amornphimoltham P, Nathan C-AO, Nathan CA, Singh B, Weigert R, Molinolo AA, Gutkind JS. Decreased lymphangiogenesis and lymph node metastasis by mTOR inhibition in head and neck cancer. Cancer Res 2011;71(22):7103–12.
[43] Ku BM, Yi SY, Koh J, Bae Y-H, Sun J-M, Lee S-H, Ahn JS, Park K, Ahn M-J. The CDK4/6 inhibitor LY2835219 has potent activity in combination with mTOR inhibitor in head and neck squamous cell carcinoma. Oncotarget 2016;7(12):14803–13.
[44] Li S-H, Lin W-C, Huang T-L, Chen C-H, Chiu T-J, Fang F-M, Huang W-T, Hsu C-M, Luo S-D, Lai C-C, Su Y-Y, Chuang H-C, Chien C-Y. Significance of mammalian target of rapamycin in patients with locally advanced stage IV head and neck squamous cell carcinoma receiving induction chemotherapy with docetaxel, cisplatin, and fluorouracil. Head Neck 2016;38(Suppl 1):844.
[45] Massarelli E, Lin H, Ginsberg LE, Tran HT, Lee JJ, Canales JR, Williams MD, Blumenschein GR, Lu C, Heymach JV, Kies MS, Papadimitrakopoulou V. Phase II trial of everolimus and erlotinib in patients with platinum-resistant recurrent and/or metastatic head and neck squamous cell carcinoma. Ann Oncol 2015;26(7):1476–80.
[46] Geiger JL, Bauman JE, Gibson MK, Gooding WE, Varadarajan P, Kotsakis A, Martin D, Gutkind JS, Hedberg ML, Grandis JR, Argiris A. Phase II trial of everolimus in patients with previously treated recurrent or metastatic head and neck squamous cell carcinoma. Head Neck 2016;38(12):1759–64.
[47] Ekshyyan O, Mills GM, Lian T, Amirghahari N, Rong X, Lowery-Nordberg M, Abreo F, Veillon DM, Caldito G, Speicher L, Glass J, Nathan C-AO. Pharmacodynamic evaluation of temsirolimus in patients with newly diagnosed advanced-stage head and neck squamous cell carcinoma. Head Neck 2010;32(12):1619–28.
[48] Bernier J, Domenge C, Ozsahin M, Matuszewska K, Lefèbvre J-L, Greiner RH, Giralt J, Maingon P, Rolland F, Bolla M, Cognetti F, Bourhis J, Kirkpatrick A, van Glabbeke M. Postoperative irradiation with or without concomitant chemotherapy for locally advanced head and neck cancer. N Engl J Med 2004;350(19):1945–52.
[49] Dunn LA, Fury MG, Xiao H, Baxi SS, Sherman EJ, Korte S, Pfister C, Haque S, Katabi N, Ho AL, Pfister DG. A phase II study of temsirolimus added to low-dose weekly carboplatin and paclitaxel for patients with recurrent and/or metastatic (R/M) head and neck squamous cell carcinoma (HNSCC). Ann Oncol 2017;28(10):2533–8.
[50] Bauman JE, Arias-Pulido H, Lee S-J, Fekrazad MH, Ozawa H, Fertig E, Howard J, Bishop J, Wang H, Olson GT, Spafford MJ, Jones DV, Chung CH. A phase II study of temsirolimus and erlotinib in patients with recurrent and/or metastatic, platinum-refractory head and neck squamous cell carcinoma. Oral Oncol 2013;49(5):461–7.

[51] Cassell A, Freilino ML, Lee J, Barr S, Wang L, Panahandeh MC, Thomas SM, Grandis JR. Targeting TORC1/2 enhances sensitivity to EGFR inhibitors in head and neck cancer preclinical models. Neoplasia 2012;14(11):1005–14.

[52] Leiker AJ, DeGraff W, Choudhuri R, Sowers AL, Thetford A, Cook JA, Van Waes C, Mitchell JB. Radiation enhancement of head and neck squamous cell carcinoma by the dual PI3K/mTOR inhibitor PF-05212384. Clin Cancer Res 2015;21(12):2792–801.

[53] Tonlaar N, Galoforo S, Thibodeau BJ, Ahmed S, Wilson TG, Yumpo Cardenas P, Marples B, Wilson GD. Antitumor activity of the dual PI3K/MTOR inhibitor, PF-04691502, in combination with radiation in head and neck cancer. Radiother Oncol 2017;124(3):504–12.

[54] Wirtz ED, Hoshino D, Maldonado AT, Tyson DR, Weaver AM. Response of head and neck squamous cell carcinoma cells carrying PIK3CA mutations to selected targeted therapies. JAMA Otolaryngol Head Neck Surg 2015;141(6):543–9.

[55] Mohan S, Vander Broek R, Shah S, Eytan DF, Pierce ML, Carlson SG, Coupar JF, Zhang J, Cheng H, Chen Z, Van Waes C. MEK inhibitor PD-0325901 overcomes resistance to PI3K/mTOR inhibitor PF-5212384 and potentiates antitumor effects in human head and neck squamous cell carcinoma. Clin Cancer Res 2015;21(17):3946–56.

[56] Chang K-Y, Tsai S-Y, Wu C-M, Yen C-J, Chuang B-F, Chang J-Y. Novel phosphoinositide 3-kinase/mTOR dual inhibitor, NVP-BGT226, displays potent growth-inhibitory activity against human head and neck cancer cells in vitro and in vivo. Clin Cancer Res 2011;17(22):7116–26.

[57] Bussink J, van der Kogel AJ, Kaanders JHAM. Activation of the PI3-K/AKT pathway and implications for radioresistance mechanisms in head and neck cancer. Lancet Oncol 2008;9(3):288–96.

[58] Freudlsperger C, Horn D, Weißfuß S, Weichert W, Weber K-J, Saure D, Sharma S, Dyckhoff G, Grabe N, Plinkert P, Hoffmann J, Freier K, Hess J. Phosphorylation of AKT(Ser473) serves as an independent prognostic marker for radiosensitivity in advanced head and neck squamous cell carcinoma. Int J Cancer 2015;136(12):2775–85.

CHAPTER

6

Improving responses to immunotherapy in head and neck squamous cell carcinoma

Wenda Ye[a], Nicole C. Schmitt[a,b], Robert L. Ferris[c], Clint T. Allen[a,b]

[a]National Institute on Deafness and other Communication Disorders, National Institutes of Health, Bethesda, MD, United States [b]Department of Otolaryngology-Head and Neck Surgery, Johns Hopkins School of Medicine, Baltimore, MD, United States [c]University of Pittsburgh Medical Center and Cancer Institute, Pittsburgh, PA, United States

Abstract

Although many forms of T cell and natural killer cell-based immunotherapies are under clinical investigation, immune checkpoint inhibitors (ICI) targeting the programmed death (PD) pathway represent a widely available and easy to use FDA-approved immunotherapy available for head and neck squamous cell carcinoma. Patients that benefit from ICI demonstrate durable responses, but overall response rates are low. Rational, mechanism-based combination treatments aimed at enhancing response rates to ICI may improve survival in patients with advanced or recurrent head and neck cancer. A thorough understanding of how anti-tumor immunity should develop, why it may be ineffective, how ICIs work and how altering the immune microenvironment may enhance T cell immunity are needed to design rational combinations to study pre-clinically and in trials. Here, we review these concepts and summarize current studies aimed at enhancing response rates to ICI in patients with advanced or relapsed head and neck cancer.

Abbreviations

ADCC	antibody-dependent cell-mediate cytotoxicity
APC	antigen processing cell
APM	antigen processing machinery
Arg1	arginase-1
ATRA	all-trans retinoic acid
CAR	chimeric antigen receptor
CCR	C-C chemokine receptor
CTL	cytotoxic T-lymphocyte

https://doi.org/10.1016/B978-0-12-817868-3.00006-8

CTLA-4	cytotoxic T-lymphocyte antigen-4
CXCR	C-X-C motif chemokine receptor
DAMP	damage associated molecular pattern
DC	dendritic cell
EGFR	epidermal growth factor receptor
G-CSF	granulocyte colony stimulating factor
GM-CSF	granulocyte-macrophage colony stimulating factor
HNSCC	head and neck squamous cell carcinoma
ICD	immunogenic cell death
ICI	immune checkpoint inhibitor
IDO	indoleamine 2,3-dioxygenase
IFN	interferon
IR	ionizing radiation
MDSC	myeloid derived suppressor cell
MHC	major histocompatibility complex
MΦ	macrophage
NK	natural killer
NO	nitric oxide
PD-1	programmed death receptor-1
PDE5	phosphodiesterase-5
PD-L1	programmed death-ligand 1
STAT	signal transducer and activator of transcription
STING	stimulator of interferon genes
TAA	tumor associated antigens
TAM	tumor associated macrophages
TCGA	the cancer genome atlas
TCR	T cell receptor
TGFβ	transforming growth factor-β
TIL	tumor infiltrating lymphocytes
TLR	toll-like receptor
T_{reg}	regulatory T lymphocyte

Conflict of interest

No potential conflicts of interest were disclosed.

Introduction

Advanced understanding of the role of the immune system in the development and progression of malignancies has directed an increased focus on immunotherapy to treat cancer. Recent analyses of the head and neck squamous cell carcinoma (HNSCC) cohort from The Cancer Genome Atlas (TCGA) has characterized frequent genomic alterations [1]. Although efforts have been made to target these alterations and associated downstream signaling pathways therapeutically, single-target small molecule inhibition is often unable to maintain long-lasting anti-tumor effects due to tumor heterogeneity and adaptive resistance after treatment [2]. Instead, enhancing the immune system to treat HNSCC may be independent of tumor-specific genomic mutations and is more likely to generate durable anti-tumor responses [3, 4]. Multiple FDA-approved immunotherapies, namely immune checkpoint inhibitors (ICIs), are now widely used for recurrent/metastatic HNSCC [3, 4]. Although patients that respond to these immunotherapies tend to have durable responses, overall response rates are low (<20%). Many current and planned pre-clinical and clinical studies aim to enhance response rates to these existing immunotherapies [5]. A fundamental understanding of how anti-tumor immunity should function and how this is disrupted in the tumor microenvironment is required to understand how immunotherapy responses can be improved with rational, mechanism-based combination treatments. In this chapter, we summarize current paradigms of anti-tumor immunity, ways in which it is suppressed to allow immune escape, and current pre-clinical and clinical studies designed to augment and improve responses to ICI immunotherapy in patients with HNSCC.

How does the adaptive immune system recognize and eliminate cancer cells?

In order for a normal cell to transform into cancer, it must overcome many obstacles designed to prevent uncontrolled proliferation and survival. In response to perturbations causing DNA damage, cellular stress, or uncontrolled growth, normal cellular signaling mechanisms are activated that eventually lead to programmed cell death [6, 7]. In addition to these cell-intrinsic pathways, the immune system serves as an extrinsic mechanism of limiting the survival of abnormal cells, recognizing stress-induced danger signals and eliminating cells deemed as "non-self" [8]. The immune system's protective role against cancer was recognized in early studies demonstrating increased risk of cancer development in immunocompromised transplant patients [9, 10]. Since then, the mechanisms by which the immune system detects and eliminates cancer cells have become clearer.

The human immune system is classically divided into two branches: a rapid, nonspecific innate immune response and a highly specific, durable adaptive immune response. Innate immune cells are comprised of pro-inflammatory dendritic cells (DCs), macrophages (MΦ), and natural killer cells (NKs) that are responsible for both immediate cytotoxicity of abnormal cells and priming of adaptive immunity. The adaptive immune response is mediated by B and T cells that generate long-lasting memory responses [11].

The immune cascade begins with activation of innate immunity. In some circumstances, innate myeloid cells such as macrophages and NK cells may play a direct role in elimination of tumor cells [12]. The conditions under which these cells are able to mediate direct anti-tumor immunity are currently being studied. Both macrophages and dendritic cells play a clearer role in antigen presentation to T cells [13, 14]. As tumor cells die, either due to immune killing or physical factors such as lack of blood supply or nutrients, their debris is phagocytosed by myeloid cells. Proteins from this debris are then processed through proteasomes (garbage disposal of the cell). Some of the small peptides that are generated from proteasomal degradation are loaded onto major histocompatibility complex (MHC) molecules. These MHCs subsequently move to the surface of the cell where they "present" peptides to T cells If a peptide presented on MHC is recognized by a T cell receptor (TCR) leading to T cell activation, that peptide is called a T cell antigen [15]. Dendritic cells and macrophages are termed "professional" antigen presenting cells (APCs) as they express all the machinery required to present antigens to both CD4 and CD8 T cells. Type I interferon (IFNα or IFNβ) is absolutely required to initiate activation of APCs [16, 17]. Without type I interferon, APC do not present antigen to T cells and adaptive T cell immunity does not develop.

In the setting of a pathogenic infection, specific molecules from the pathogen will activate innate immune receptors expressed on myeloid and other innate immune cells leading to the production of required type I IFN [18]. The classic example of this is LPS from gram negative bacteria binding toll-like receptor (TLR) 4, leading to type I IFN production. It was a mystery for many years how type I IFN production and innate immune activation occurs in cancer where there is, at least theoretically, a "sterile" environment that lacks pathogens. We now understand two clear mechanisms of innate immune activation in cancer. First, when some tumor cells die, they release stress molecules (also called danger-associated molecular patterns, or DAMPs) such as calreticulin, ATP and HMGB1 that activate different innate immune receptors leading to production of type I interferon [19]. This process is called immunogenic cell death (ICD). Second, dying tumor cells release DNA that can be phagocytosed by myeloid cells. Once it enters myeloid cells, cytoplasmic DNA results in the activation of an intracellular innate immune receptor called STING that leads to type I interferon production [20, 21]. Recognition of these processes by which dying tumor cells can initiate activation of innate immunity represent a recent major advance in tumor immunology.

Multiple signaling and cytokine pathways other than type I interferon are required for effective activation of T cells by APCs. Each T cells harbors a unique TCR that can bind a peptide presented on MHC [15]. Peptides that are 8–10 AA in length can load into MHC class I to be presented to TCRs on $CD8^+$ T cells. Longer peptides that are 15–20 AA in length are presented by MHC class II to $CD4^+$ T cells. When a professional APC breaks down a peptide, it produces peptides of variable lengths and "cross presents" peptides on MHC class I and II to $CD8^+$ and $CD4^+$ T cells, respectively. As APCs cross present antigen, they produce cytokines that dictate how T cells are primed [22]. Release of IL-12 by APCs induces naïve $CD4^+$ T cells to polarize into a Th1 helper phenotype and secrete IFNγ, TNFα and IL-2. This Th1 profile of cytokines is required for naïve $CD8^+$ T cells to mature into cytotoxic T lymphocytes (CTL) [23]. The ultimate goal of this whole process is the generation of primed $CD8^+$ CTL that are ready to clonally expand and kill upon encountering a target cell presenting antigen in MHC class I for which their TCR is specific.

Binding of the TCR to the antigen:MHC complex for which it is specific is one of several important signals required for T cell activation [15]. Another important signal is binding of CD28 present on T cells to either CD80 or CD86 present on the APC or target cell. A third signal is best described as a summation of many different factors including the cytokine milieu present within the microenvironment as well as expression of proteins that can activate or suppress T cell function. One class of these proteins, called immune checkpoints, includes cytotoxic T-lymphocyte atigen-4 (CTLA-4) and programmed death receptor-1 (PD-1) expressed on T cells [24]. When these checkpoints bind their ligands, they can suppress T cell function. There are many immune checkpoints that can activate or inhibit T cells, but CTLA-4 and PD-1 are the most extensively studied [25]. Current FDA-approved ICIs are monoclonal antibodies that bind and block the function of either CTLA-4 or PD-1 or the primary ligand for PD-1, programmed death-ligand 1 (PD-L1) [26].

It is important to understand the origin of peptides that can serve as T cell antigens. Through VDJ genetic recombination and elimination of self-reactive T cells in the thymus, T cells expressing a large repertoire of TCRs are able to detect foreign or abnormal antigens [27]. Virally-driven cancer cells can process and present strong viral antigens. Proteins that are overexpressed in tumor cells compared to normal cells can also serve as T cell antigens, but these tumor-associated antigens (TAAs) are not specific for tumor cells [28]. Recent work has revealed the existence of tumor cell-specific neoantigens, or peptide products generated from tumor-specific gene mutations that express an abnormal protein [29]. Since underlying genomic alterations are somatic mutations that are not present in normal cells and are unique between tumors, neoantigens are tumor-specific. Ultimately, the development of an adaptive T cell response against tumor-specific neoantigens may be required for immune-mediated elimination of a cancer. This likely happens very efficiently and frequently in healthy individuals. Only transformed cells that have the ability to evade this detection and elimination by the immune system can progress into a clinically relevant cancer.

What are ways in which this is derailed?

Examination of HNSCC tumor specimens reveals that most tumors exhibit CD8^{+} tumor-infiltrating lymphocytes (TILs) in the tumor microenvironment, suggesting the presence of an underlying adaptive immune response [30, 31]. Genomic analysis of HNSCC specimens also demonstrates the presence of many genomic alterations that are predicted to yield abnormal proteins that could serve as neoantigens [1]. If HNSCC have many neoantigens and evidence of an underlying immune response with tumor infiltration of CD8^{+} TIL, how does a clinically-relevant cancer form in the face of anti-tumor immunity? The underlying mechanisms of immune escape present within HNSCCs can generally be classified into two arms: dysfunctional development of effective anti-tumor immunity and suppression of an existing anti-tumor response [5].

Problems with development of effective anti-tumor immunity

Some cancer cells with low mutational burden may not express strong T cell antigens such that effective adaptive immunity does not develop [32]. Although some variability in the number of genomic alterations exists, most HNSCC harbor tumor cells with many mutations, making this unlikely [1]. Tumors cells more likely express multiple T cell antigens [33]. Antigens that induce the strongest and most rapid T cell responses are termed immunodominant [34]. Early in tumor development, strong T cell responses against immunodominant antigens dampers T cell responses against immunorecessive antigens [35, 36]. As tumor cells expressing immunodominant antigens are eliminated by T cells, tumor cells expressing immunorecessive antigens are "selected" [37]. Strong pre-clinical evidence suggests that even with loss of tumor cells expressing the immunodominant antigen, T cells specific for the immunodominant antigen (and not the immunorecessive antigen) remain [38]. Thus, tumor infiltrating T cells may be specific for an antigen that has been immunoedited and is no longer present.

An alternative mechanism for selection of tumor cells that escape T cell immunity has to do with altered tumor cell immunogenicity. Some tumor cells, either through genomic alterations or epigenetic regulation of expression, do not express one or more components required for antigen processing or presentation. Proteins important for processing and loading antigens onto MHC molecules, collectively termed antigen processing machinery (APM), can be downregulated in HNSCC [39]. APM expression can be rescued with exposure to interferon (IFN), suggesting that low or absent APM expression is not directly due to mutations in the genes encoding these proteins [40, 41]. Since the expression of these proteins is

IFN-dependent, alterations in IFN-responses and downstream signaling components can mimic APM loss. Mutation or loss of expression in IFN receptors, JAK and STAT proteins have all been demonstrated in cancers [42, 43]. Similar to altered antigenicity, cancer cells most resistant to immune detection and elimination through altered immunogenicity are similarly selected for during the process of immunoediting [44]. Cancer cells most susceptible to immune killing for whatever reason are eliminated, leaving more resistance cells to populate a progressing tumor.

Some related but mechanistically distinct scenarios exist whereby tumor cells themselves may be sufficiently antigenic and immunogenic, but meaningful antigen-specific T cell responses fail to develop. One example is the development of T cell tolerance to antigens to which they are chronically exposed. Chronic exposure to antigens in the premalignant phase of tumor establishment may lead to T-lymphocyte anergy [45–47]. Allergists take advantage of this principle with immunotherapy (chronic exposure to low-dose antigens) for patients with severe food and environmental allergies. This suggests that T cells may be blind to some of the antigens being presented on tumor cells that are fully capable of antigen processing and presentation.

Another separate but equally important issue may be the absence of a robust innate immune response. Without type I IFN production downstream of innate immune receptor activation, myeloid cells (mainly DCs) fail to mature and cross present antigen to $CD4^+$ and $CD8^+$ T cells [48]. Understanding the underlying genomic or epigenetic alterations that lead to resistance to innate immune receptor signaling and type I IFN production is being extensively studied.

Problems with suppression of existing immune response

Growing evidence demonstrates that anti-tumor immunity does in fact develop within most progressing cancers, but that it is suppressed in the local tumor microenvironment [30, 31]. Clinically relevant tumor cells can directly exert an inhibitory effect on the existing anti-tumor immune response [49]. This tumor-induced immune suppression is mediated through multiple mechanisms, including the release of cytokines and chemokines that are directly immunosuppressive. Tumor cell production of transforming growth factor β (TGF-β) and IL-10 can directly inhibit T-cell proliferation and effector immune cell killing [50]. Several forms of local immunosuppression within the tumor microenvironment are summarized in Fig. 1.

As described above, expression of immune checkpoint ligands and receptors can suppress activated T cell function. Several important concepts explain how PD-1 and PD-L1 are expressed on cells within the tumor microenvironment. Expression of PD-1 on the surface of T cells consistently occurs following T cell activation, so PD-1 expression on T cells is expected [51]. This is a normal response, and under physiologic conditions serves to limit uncontrolled T cell responses after activation. Tumor cell PD-L1 expression can sometimes be driven by upstream genomic alterations, but this is uncommon [52]. More commonly, tumor cell and infiltrating myeloid cell PD-L1 expression, which is highly IFN-responsive, is driven by the presence of IFN as part of the underlying immune response. This paracrine driven PD-L1 expression in response to IFN from effector immune cells is termed adaptive

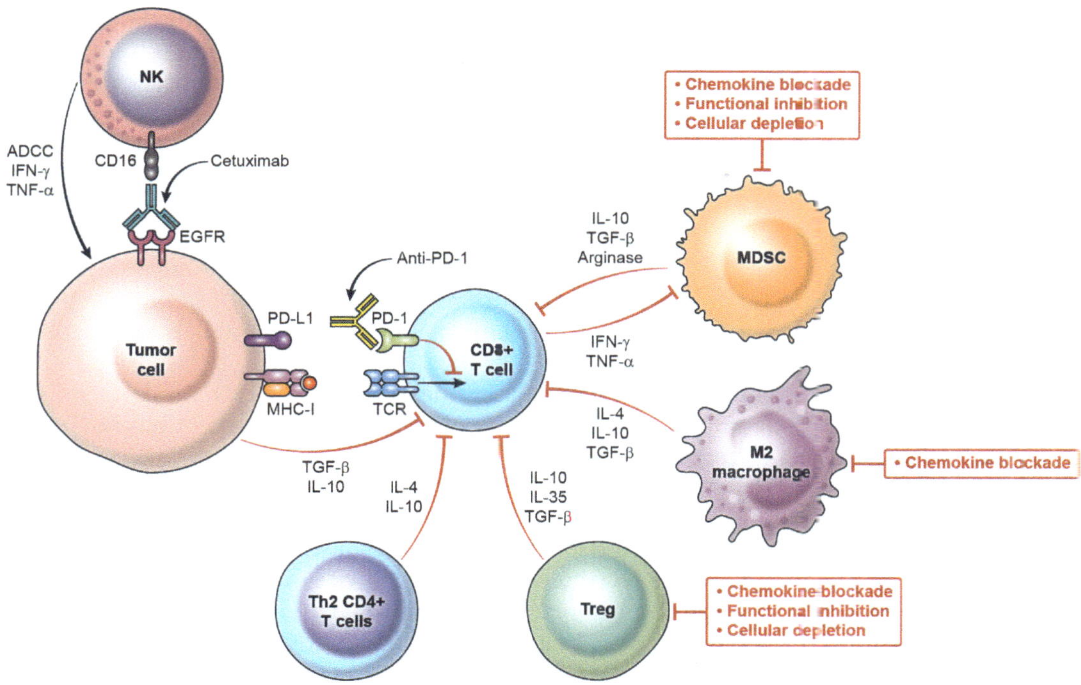

FIG. 1 Illustration of known interactions between tumor infiltrating immune cells and tumor cells within the tumor microenvironment. Described mechanisms of how immunosuppressive immune and tumor cells can suppress CD8$^+$ T cell function are displayed. Inset in *red boxes* are mechanisms of potential intervention to limit the function or recruitment of immunosuppressive immune cells into tumors. *ADCC*, antibody-dependent cell-mediated cytotoxicity; *EGFR*, epidermal growth factor receptor; *IFN*, interferon; *IL*, interleukin; *MDSC*, myeloid derived suppressor cell; *MHC*, major histocompatibility; *NK*, natural killer cell; *PD-1*, programmed death receptor-1; *PD-L1*, programmed death-ligand 1; *TCR*, T cell receptor; *TGF*, transforming growth factor; *TNF*, tumor necrosis factor; T_{reg}, regulatory CD4$^+$ T cell.

immune resistance [53]. Thus, expression of PD-1 and PD-L1 within a site of chronic T cell activation is a normal physiologic process that serves to ultimately suppress T cell responses in sites of chronic inflammation and antigen exposure—like the tumor microenvironment.

Tumor cells can also exert indirect immunosuppressive effects via recruitment of myeloid derived suppressor cells (MDSCs) and other immune inhibitory hematopoietic cells via expression of chemokines such as CCLs, CXCLs, and VEGF [54]. Specific chemo-attractants such as granulocyte-macrophage colony stimulating factor (GM-CSF), CXCL1, and IL-8 have also been implicated in the recruitment of CXCR2$^+$ and/or CCR2$^+$ MDSCs [55–57]. MDSCs are a heterogeneous group of immature myeloid cells that have potent immunosuppressive effects in many cancers [58]. MDSCs have been extensively studied in both mouse and human models and can be divided into two subpopulations: monocytic MDSC (M-MDSC) and neutrophilic MDSC (PMN-MDSC) [54]. Normal neutrophils and neutrophilic MDSC can only be differentiated via functional assays—by definition, if a granulocytic cell can suppress T cell function, it is a PMN-MDSC [59]. A more immature (early) human MDSC has been proposed as well [60–62].

MDSC immunosuppression can be conceptually divided into two arms: expansion and activation [62]. Once recruited into the tumor by chemokines, MDSC terminal differentiation (and loss of suppressive capacity) is blocked by growth factors such as granulocyte colony stimulating factor (G-CSF), GM-CSF, and VEGF [54, 60]. MDSCs in the periphery do have T cell suppressive capacity, but MDSC within the tumor microenvironment appear to have enhanced suppressive capacity [54, 60, 62]. MDSCs mechanistically exert immunosuppressive effects through nitric oxide (NO) and arginase-1 (ARG-1) production, PD-L1 expression, indoleamine 2,3-dioxygenase (IDO) upregulation, $CD4^+$ regulatory T cell (T_{reg}) differentiation/expansion, CD40 expression, and TGF-β and IL-10 secretion [63–72]. Activated MDSCs are also able to generate a tumor-supportive environment independent of immunosuppression through expression of VEGF, matrix metalloproteinases (MMP) and basic fibroblast growth factor [62, 73, 74].

Other hematopoietic cells that support tumor immune escape include Th2-polarized $CD4^+$ T cells, M2 macrophages (comprising the majority of macrophages seen in HNSCC), and $CD4^+FoxP3^+CD25^+CTLA\text{-}4^+$ regulatory T cells (T_{reg}s). MΦs polarized into the M2 phenotype secrete immunosuppressive cytokines such as IL-4, IL-10 and TGF-β to also support protumor Th2 responses [75]. $CD4^+$ T_{reg}s, characterized by $FoxP3^+$ expression, are another subset of immunosuppressive cells that infiltrate the tumor microenvironment [76]. Although there is some debate on whether T_{reg}s are recruited via cytokines vs functionally converted into T_{reg}s from naïve $CD4^+$ T cells by the tumor microenvironment, T_{reg}s accumulate early in tumor progression and potently suppress effector immunity in HNSCCs [76–78].

A lack of type I interferon production, immunosuppressive cytokine production by tumor cells, and chemokine driven tumor infiltration of immunosuppressive cells are all potentially modifiable mechanisms of tumor cell immune escape. The mechanism-based concepts for the rationale for many combination immunotherapies currently under pre-clinical or clinical study.

What are current immunotherapies?

Multiple immunotherapies are already FDA-approved for the treatment of HNSCC. Risks associated with these treatments are quite low, but unfortunately so are response rates. Current FDA-approved immunotherapies for HNSCC include.

Cetuximab

Epidermal growth factor receptor (EGFR) is a cell surface receptor that is correlated with poor prognosis when overexpressed in malignancies [79]. Intracellular signaling downstream of EGFR through the MAPK, PI3K/AKT, and JAK/STAT pathways promotes cell proliferation and survival [80]. Cetuximab, an anti-EGFR monoclonal antibody, is FDA approved for the treatment of HNSCC based upon enhanced locoregional control of advanced HNSCC when combined with radiotherapy compared to radiotherapy alone. Given that cetuximab binds EGFR, assumptions about the mechanism of action being related to decreased EGFR signaling were rational [81]. However, further studies revealed that levels of tumor cell EGFR expression and decreased signaling downstream of EGFR were not correlated to treatment response. Rather, the mechanism of action appeared to be related to the ability of cetuximab, an IgG1 mAb, to mediate antibody-dependent cell-mediated cytotoxicity (ADCC) [82]. Specifically, antibody-coated target cells (in this case tumor cells expressing EGFR covered by cetuximab) are recognized and killed by Fc-receptor (CD16) bearing NK cells [83]. This process also augments the adaptive immune response through the release of antibody-coated tumor fragments containing antigen that are phagocytosed and processed by APCs [84, 85]. This ability to elicit ADCC is seemingly antibody isotype restricted, with panitumumab (IgG2 anti-EGFR antibody) unable to recapitulate the same results [86]. Although cetuximab improves outcomes in HNSCC, only a subset of patients respond [87]. Biological differences in an individual's ADCC activity and CD16 receptor alleles predict effective clinical response to cetuximab or other IgG1-based antibody therapies [88, 89]. Cetuximab has also been found to exert different levels of effect on T_{reg}s and MDSCs in patients who respond to therapy vs those who do not [90, 91].

Immune checkpoint inhibitors

Monoclonal antibodies that bind and block the function of either PD-1 or PD-L1 reverse adaptive immune resistance and negate one of the major signals designed to limit T cell function [92]. PD-blocking ICIs have been the most extensively studied in numerous cancer types including HNSCC. Two landmark clinical trials have led to the FDA-approval of two different PD-1 mAb ICIs for second-line treatment of recurrent/metastatic HNSCC, pembrolizumab and nivolumab [3, 4]. Both are IgG4 PD-1 monoclonal antibodies that work by blocking the interaction between PD-1 and PD-L1—they are not capable of mediating ADCC. The CheckMate 141 phase III clinical trial compared nivolumab with the physician's choice of second line single agents for the second line treatment of recurrent/metastatic HNSCC (docetaxel, methotrexate or cetuximab) and demonstrated higher median overall survival (7.5 vs 5.1 months), lower risk of death (HR: 0.70; 95% CI 0.51–0.96), and greater 6-month progression free survival (19.7% vs 9.9%) [3]. Furthermore, nivolumab demonstrated significant improvements in treatment safety measures compared to physician's choice (13.1% vs 35.1% in grade 3 or 4 adverse events). In the 2-year follow-up of CheckMate 141, nivolumab continued to maintain the initially observed increased overall survival (16.9% vs 6%) and decreased adverse safety events (15.3% vs 36.9%) as compared to the physician's choice regimen [93]. Pembrolizumab was studied in phase Ib trial KEYNOTE-012 as second line treatment for recurrent/metastatic HNSCC, demonstrating a response rate of 18%, median overall survival of 8 months, and a favorable side effect profile [4]. Following this, phase III trial KEYNOTE-040 compared pembrolizumab with standard therapies for the second line treatment of recurrent/metastatic HNSCC [94]. This study demonstrated a median overall survival improvement of 8.4 vs 7.1 months in pembrolizumab compared to standard therapy. In follow-up analyses, pembrolizumab demonstrated statistically significant improved overall survival of 8.4 vs 6.9 months as compared to standard of care (HR: 0.80; 95% CI: 0.65–0.98) [95].

The use of PD-blocking ICIs does not come without risk. Immune related adverse events that lead to a patient's immune system attacking off-tumor tissues can occur with non-specific activation of T cells. As such, individuals with a prior history of autoimmunity have been found to be more susceptible to these adverse outcomes [96]. However, in a case series of 114 patients with metastatic HNSCC treated with anti-PD-1 therapy, individuals who experienced immune related adverse events had improved median overall survival (12.5 vs 6.8 months) and progression free survival (6.9 vs 2.1 months) compared to those without [92]. This suggests that immune related adverse events may be an external manifestation of effective restoration of anti-tumor immunity. A similar phenomenon is observed with cetuximab whereby patients that develop a rash are more likely to have an objective tumor response.

What therapies are being investigated to enhance responses to immunotherapy?

Many forms of immunotherapy other than ICI are being studied in HNSCC both preclinically and clinically. These include therapeutic vaccines and cellular therapies using either T cells engineered to recognize specific antigen:MHC class I complexes or T cells engineered to express a chimeric antigen receptor (CAR) that recognizes a surface antigen. These all have tremendous potential but are studied by few specialized groups of physicians and scientists and are very cost and labor intensive in their current forms. Because of this, the vast majority of immunotherapy research is aimed at improving durable response rates to ICIs. Given the low risk of serious adverse events, and the durable anti-tumor responses observed, much effort is being devoted to understanding mechanisms of resistance to PD-1/L1 mAbs and how the use of these agents can be optimized to increase the number of patients that respond. Table 1 summarizes many current immunotherapy trials designed to enhance the efficacy of ICIs in HNSCC.

TABLE 1 Clinical trials investigating how to enhance the efficacy of immune checkpoint inhibitors

Immunotherapy regimen	Trial/phase	Description	Setting
Altered timing of ICI	Phase III KEYNOTE-048 (NCT02358031)	Pembrolizumab Pembrolizumab + cisplatin/carboplatin + 5-fluorouracil Cetuximab + cisplatin/carboplatin + 5-fluorouracil	First line treatment for R/M HNSCC
	Phase III KEYNOTE-412 (NCT03040999)	Pembrolizumab + chemoradiation Placebo + chemoradiation	Newly diagnosed, treatment-naïve locally advanced HNSCC
	Phase II PATHWay Study (NCT02841748)	Adjuvant pembrolizumab Placebo	HNSCC at high-risk for recurrence after curative intent therapy
	Phase II (NCT02764593)	Nivolumab + standard cetuximab-radiotherapy	Newly diagnosed, intermediate/high risk locally advanced HNSCC

TABLE 1 Clinical trials investigating how to enhance the efficacy of immune checkpoint inhibitors—Cont'd

Immunotherapy regimen	Trial/phase	Description	Setting
	Phase II (NCT02641093)	Pembrolizumab + surgery + radiotherapy +/− cisplatin	Locally advanced HNSCC eligible for surgical resection
	Phase II (NCT02296684)	Neoadjuvant pembrolizumab	Locally advanced HNSCC treated with current standard of care surgical approaches
	Phase II (NCT03174275)	Induction chemotherapy with carboplatin, nab-paclitaxel, and durvalumab	Previously untreated stage III and IV HNSCC amenable to surgical resection
	Phase II OPTIMA-II (NCT03107182)	Induction chemotherapy with carboplatin, nab-paclitaxel, and nivolumab Followed by TransOral Robotic Surgery + radiotherapy or radiotherapy alone	Locoregionally advanced HPV-related oropharyngeal cancer
	Phase II (NCT02827838)	Neoadjuvant durvalumab	HNSCC of oral cavity or oropharynx eligible for surgical resection
	Phase I/II CheckMate-358 (NCT02488759)	Nivolumab Neoadjuvant nivolumab Nivolumab + ipilimumab Nivolumab + anti-LAG-3 antibody relatlimab Nivolumab + anti-CD38 daratumumab	Virus-positive and virus-negative solid tumors
	Phase I/II NIRT-HNC (NCT03247712)	Neoadjuvant nivolumab + radiotherapy followed by additional nivolumab after surgery	HNSCC eligible for surgical resection
	Phase I (NCT02812524)	Neoadjuvant intratumoral ipilimumab injection	HNSCC eligible for surgical resection
	Phase I (NCT02775812)	Adjuvant pembrolizumab + cisplatin + intensity modulated radiotherapy	High risk, stage III–IV HNSCC with completed gross total resection
ICI + ICI combination	Phase III CheckMate-651 (NCT02741570)	Nivolumab + ipilimumab Standard of care (Extreme study regimen)	First line treatment for R/M HNSCC
	Phase III EAGLE (NCT02369874)	Durvalumab Durvalumab – tremelimumab Standard of care in R/M HNSCC	Second line treatment for R/M HNSCC

Continued

TABLE 1 Clinical trials investigating how to enhance the efficacy of immune checkpoint inhibitors—Cont'd

Immunotherapy regimen	Trial/phase	Description	Setting
	Phase III KESTREL (NCT02551159)	Durvalumab Durvalumab + tremelimumab Standard of care chemotherapy in treatment-naïve R/M HNSCC	First line treatment for R/M HNSCC
	Phase II CONDOR (NCT02207530)	Durvalumab Durvalumab + tremelimumab Tremelimumab	Second line treatment for R/M HNSCC
	Phase II CheckMate-714 (NCT02823574)	Nivolumab + ipilimumab Nivolumab + placebo	Second line treatment for R/M HNSCC
	Phase II (NCT03082534)	Pembrolizumab + cetuximab	Second line treatment for R/M HNSCC
ICI + CRT	Phase III (NCT03349710)	Nivolumab + radiotherapy Cetuximab + radiotherapy Nivolumab + cisplatin + radiotherapy Cisplatin + radiotherapy	Locally advanced HNSCC
	Phase III REACH (NCT02999087)	Avelumab + cetuximab + radiotherapy Standard of care	Locally advanced HNSCC
	Phase III JAVELIN (NCT02952586)	Avelumab + standard of care chemoradiation therapy Placebo + standard of care chemoradiation therapy	Locally advanced HNSCC
	Phase II/III (NCT03258554)	Durvalumab + radiotherapy Cetuximab + radiotherapy	Stage III-IVB HNSCC with contraindication to cisplatin
	Phase II PembroRad (NCT02707588)	Pembrolizumab + radiotherapy Cetuximab + radiotherapy	Locally advanced HNSCC
	Phase II (NCT02609503)	Pembrolizumab + intensity modulated radiotherapy	Locally advanced HNSCC with contraindication to cisplatin
	Phase II (NCT02777385)	Pembrolizumab + cisplatin + intensity modulated radiotherapy simultaneously Pembrolizumab after cisplatin + intensity modulated radiotherapy	Intermediate or high risk, previously untreated, locally advanced HNSCC
	Phase I (NCT02586207)	Pembrolizumab + cisplatin + radiotherapy	Locally advanced HNSCC

TABLE 1 Clinical trials investigating how to enhance the efficacy of immune checkpoint inhibitors—Cont'd

Immunotherapy regimen	Trial/phase	Description	Setting
ICI + innate immune activator	Phase I/III MASTERKEY-232 (NCT02697591)	Oncolytic immunotherapy talimogene laherparepvec + pembrolizumab	Second line treatment for R/M HNSCC
	Phase I/II (NCT01714739)	Anti-KIR (killer Ig-like receptors) antibody lirilumab + nivolumab	Advanced solid tumors
	Phase I/II KEYNOTE-184 (NCT02521870)	TLR9 agonist SD-101 + pembrolizumab	Metastatic melanoma or R/M HNSCC
	Phase I (NCT02675439)	Ipilimumab + STING agonist MIW815 (ADU-S100) Pembrolizumab	Advanced/metastatic solid tumors or lymphomas
ICI + MDSC inhibitor	Phase III (NCT03358472)	Pembrolizumab Pembrolizumab + IDO inhibitor epacadostat Standard of care (EXTREME study regimen)	First line treatment for R/M HNSCC
	Phase II KEO (NCT03325465)	Pembrolizumab Pembrolizumab + IDO inhibitor epacadostat	Neoadjuvant therapy prior to curative surgical care for HNSCC
	Phase I/II (NCT02499328)	Durvalumab + STAT3 inhibitor AZD9150 Durvalumab + CXCR2 inhibitor AZD5069	Advanced solid tumors and relapsed metastatic HNSCC
	Phase I (NCT03343613)	IDO inhibitor LY3381916 LY3381916 + anti-PD-L1 antibody LY3300054	Advanced/metastatic solid tumors
ICI + macrophage inhibitor	Phase I/II (NCT02452424)	CSF1-R inhibitor PLX3397 + pembrolizumab	Advanced melanoma and other solid tumors
	Phase I (NCT02526017)	CSF1-R inhibitor cabiralizumab Cabiralizumab + nivolumab	Selected advanced cancers
ICI + T_{reg} inhibitor	Phase II JAVELIN Medley (NCT02554812)	Avelumab + OX40 agonist PF-04518600	Locally advanced or metastatic solid tumors
	Phase I (NCT02740270)	Anti-GITR antibody GWN323 GWN323 + anti-PD-1 antibody PDR001	Advanced malignancies and lymphomas

HNSCC, *head and neck squamous cell carcinoma;* ICI, *immune checkpoint inhibitor;* R/M, *recurrent/metastatic.*

Altered timing and combinations of ICIs

The most basic altered approach with ICIs involves altering the setting of administration of these agents. PD-1 ICIs are FDA-approved for the second-line treatment of recurrent/metastatic HNSCC. However, published response rates of <20% are seen in this patient population [3, 4]. Most of these patients are heavily pretreated, often with multiple rounds of systemic chemotherapy, and are arguably very immunosuppressed. One strategy to enhance response rates may be in the use of PD-1/L1 mAb ICIs in less immunosuppressed patients. The Keynote-048 trial studies the performance of pembrolizumab in the first-line treatment of recurrent/metastatic HNSCC, where patients will not have been treated with multiple rounds of systemic chemotherapy [97]. Preliminary results are promising with improved overall survival in the pembrolizumab arm compared to the standard chemotherapy regimen arm. Based on these results, it is very likely that pembrolizumab will be FDA-approved as a first line treatment for recurrent/metastatic HNSCC in the near future.

Using the same reasoning, treatment of locoregionally advanced, previously untreated HNSCC in the upfront setting is also being investigated. Ongoing trials such as the phase III KEYNOTE-412 are investigating pembrolizumab in combination with standard of care non-surgical, upfront treatment of locoregionally advanced HNSCC (NCT03040999). Nivolumab (phase I trial NCT02764593) is also being studied in intermediate and high risk locoregionally advanced HNSCC in the upfront treatment setting [98]. Multiple studies are now underway that also explore the effects of neoadjuvant ICI before definitive surgical resection [99, 100]. The use of all of these approaches may enhance anti-tumor immunity and response rates. The use of PD-based ICIs before definitive standard-of-care treatment may allow maximum activation of antigen-specific T cell immunity in the presence of high antigen load followed by cytoreduction and reversal of immunosuppression by surgical removal or radiotherapy ablation of the tumor. Upfront immunotherapy may reduce the unacceptably high rates of both locoregional and distant failure observed following definitive treatment of advanced HNSCC.

Although the PD-1/L1 signaling axis has been the most extensively studied immune checkpoint, other immune checkpoints likely also limit the function of activated T cells. Similar to activation of redundant growth and survival intracellular signaling pathways observed with use of single small molecule inhibitors, activation or increased expression of other immune checkpoints may be a mechanism of resistance to PD-based ICI monotherapy. One ICI that has been extensively studies in other tumor types such as melanoma but less so in HNSCC in CTLA-4. Ongoing clinical trials focusing on combination therapies such as pembrolizumab + cetuximab (NCT03082534) and nivolumab + ipilimumab vs nivolumab alone (CheckMate 714) are exploring synergistic interactions between multiple ICIs to increase treatment efficacy and response [98].

Combining ICIs with treatments that induce anti-tumor immunity

As detailed above, there is no evidence that PD-1/L1 mAb ICI induces de novo anti-tumor immunity. It reverses blockade of pre-existing anti-tumor immunity being restrained by the expression of PD-1/L1, most likely through adaptive immune resistance [53]. Thus, tumors that do not have some degree of anti-tumor immunity at baseline are less likely to respond to PD-1/L1 mAb treatment. Other treatments, including standard anti-cancer therapies such as chemotherapy and ionizing radiation (IR), or other targeted therapies may actually induce anti-tumor immunity. In this situation, treatment induced expression of PD-L1 in the tumor microenvironment may limit effective anti-tumor immunity, and the combination of PD-1/L1 mAb ICI with these immunity-inducing treatments could produce addictive or synergistic effects.

Combining ICIs with standard of care chemotherapy and/or ionizing radiation

Although cytotoxic chemotherapy is well known to have immunosuppressive effects by inducing bone marrow toxicity and lymphopenia, preclinical studies suggest that chemotherapy can enhance some aspects of anti-tumor immunity. In addition to promoting direct release of tumor antigens in the tumor microenvironment by direct toxicity to tumor cells, some chemotherapy drugs may induce ICD. Cisplatin, the most commonly used chemotherapeutic drug for HNSCC, is thought to be a suboptimal inducer of ICD compared with the newer-generation drug oxaliplatin, based on prior preclinical studies suggesting that oxaliplatin induces a more robust translocation of calreticulin to the tumor cell surface [101–103]. In addition, mice vaccinated with tumor cells killed by cisplatin, when rechallenged with live tumor cells, are less likely to reject tumor formation than mice vaccinated with cells killed by oxaliplatin or other chemotherapy drugs [102]. However, many of these prior studies compared supraphysiologic doses of cisplatin and oxaliplatin. More recent studies using modest, clinically-relevant doses demonstrate that cisplatin can induce calreticulin exposure, HMGB1 release, and other hallmarks of ICD [104]. Further, combination of cisplatin chemotherapy with other agents that enhance p53 or ER stress may lead to robust ICD [101, 102, 105]. Given that radiation is a known ICD inducer, it is plausible that cisplatin enhances ICD in the tumor microenvironment of HNSCC patients treated with chemoradiation.

In addition to ICD, cisplatin and other chemotherapy drugs may enhance other aspects of anti-tumor immunity. Multiple studies have shown that cisplatin can induce upregulation of MHC class I and other components of the antigen processing machinery, leading to enhanced antigen presentation and antigen-specific T cell responses [106–108]. Multiple clinical trials combining platinum chemotherapy with anti-PD-1/PD-L1 antibodies are currently underway [109]. While survival data for these chemoimmunotherapy trials are not yet available for head and neck cancer, similar combinations have been established as the new standard of care for lung cancer based on large, Phase III clinical trials showing improved survival [110, 111].

Similar to platinum chemotherapy, ionizing radiation (IR) may induce ICD and immune activation in the local microenvironment [112]. Pre-clinical data supports that this immune activation leads to the activation of immune checkpoints, and that combining IR with ICI leads to sometimes dramatic tumor control [113]. Emerging evidence suggests that how IR is administered can determine whether it has a combinatorial effect with ICI.

For example, daily fractionated IR consisting of 2 Gy treatments over 30–35 days may be very immunosuppressive, whereas single or few higher doses of IR (8–12 Gy) may be very immunostimulatory [114]. Accordingly, pre-clinical evidence demonstrates that combining daily fractionated IR with ICI has little effect, whereas combining higher dose hypofractionated IR with ICI produces dramatic tumor control [115, 116]. These concepts are being explored in multiple clinical trials combining different IR schemas with ICI in patients with HNSCC [117].

Combining ICIs with innate immune receptor agonists

Innate immune receptor activation is required for type I IFN production and innate immune priming. Therapeutic agonists of innate immune receptors may lead to production of type I IFN and promote the development of innate and adaptive immunity. Stimulator of Interferon Genes (STING) agonists, such as synthetic cyclic dinucleotides, induce type I IFN production. This not only can directly induce maturation of and cross presentation of antigen by DCs; it also reverses M2 tumor-associated macrophage (TAM) polarization [118, 119]. Activation of the STING pathway has been shown to enhance cetuximab mediated NK cell activation and DC maturation, and in another study elicited regression of established tumors in mouse models treated in combination with checkpoint inhibitors [120, 121]. These findings suggest that STING agonists have great potential to serve as adjuvant immune therapy in HNSCC.

Combining ICIs with MDSC inhibitors

It is clear that immunosuppressive myeloid cells play a pivotal role in determining the immunosuppressive or immunopermissive status of the tumor immune microenvironment. Based on approaches described in extensive pre-clinical studies, multiple approaches to inhibiting MDSCs are currently under investigation. These include forced differentiation, small molecule inhibitors and chemotaxis blockade [58].

As MDSCs are immature myeloid cells, inducing terminal differentiation is a logical approach to inhibiting their function. All-trans retinoic acid (ATRA) has been shown to drive differentiation of MDSCs into macrophages and DCs in murine and human samples through blockade of retinoic acid signal transduction [122]. In clinical trials for metastatic renal cell carcinoma and late stage small cell lung cancer, ATRA has demonstrated efficacy in reducing MDSC frequencies and improving patient survival [123, 124]. Another agent that has demonstrated promise in driving MDSC differentiation is 1,25-dihydroxyvitamin D3. Treatment with 1,25-dihydroxyvitamin D3 pushed $CD34^+$ MDSCs isolated from the peripheral blood of HNSCC patients to differentiate into DCs [125]. Other studies have demonstrated reductions in MDSC accumulation, increased IL-12 and IFN-γ concentrations, and increased mature dendritic cells, $CD4^+$ T cells and $CD8^+$ T cells after treatment [126, 127]. A clinical trial of combination treatment 1,25-dihydroxyvitamin D3 and COX2 inhibitor celecoxib in surgically resectable oral cavity HNSCC was recently completed with pending results (NCT00953849) [58].

Other mechanisms for MDSC-induced immunosuppression include Arg-1 and nitric oxide synthase (iNOS) mediated depletion of L-arginine, which directly inhibits T-cells [58]. Although small molecule inhibitors against these proteins have demonstrated effective downregulation of murine and human MDSCs, they are limited by their toxicity in humans [58, 128]. Another class of small molecule inhibitors, IDO antagonists, are also being studied, as MDSC-related IDO expression has been implicated in local tryptophan depletion, production of immunosuppressive metabolites, and subsequent T-cell suppression [62]. There is currently one active phase III clinical trial comparing pembrolizumab + epacadostat (IDO inhibitor) vs pembrolizumab monotherapy in recurrent/metastatic HNSCC (NCT03358472). Two other clinical trials (phase I and I/II) involving IDO inhibitors in HNSCC are in the patient recruitment phase (NCT03343613, NCT03085914), and one other phase II trial is not yet recruiting (NCT03325465). In the same pathway, signal transducer and activator of transcription 3 (STAT3) is a transcriptional regulator of Arg-1 and IDO expression in MDSCs and thus is another promising target for therapy [58, 62]. AZD9150, an antisense oligonucleotide inhibitor of STAT3, has already demonstrated positive results in lymphoma and lung cancer and is being studied in a phase I/II trial in combination with durvalumab (ICI) in relapsed metastatic HNSCC (NCT02499328). Other phase I and II clinical trials exploring oral STAT3 inhibition and Janus kinase (JAK) inhibition are ongoing (NCT03195699, NCT03153982). Phosphoinositide 3-kinase (PI3K) is another important signaling pathway in cellular growth and survival that is frequently upregulated in HNSCC [58]. In preclinical studies, low dose inhibition of PI3Kδ and PI3Kγ isoforms elicited functional inhibition of MDSC and enhanced response to PD-L1 blockade [129]. Numerous ongoing clinical trials are exploring PI3K inhibitor monotherapy and combination immune or standard therapy in HNSCC.

In addition to differentiation-inducing agents and small molecule inhibitors designed to block suppressive function, therapies blocking MDSC recruitment are another rational approach to inhibiting MDSC immunosuppression. Chemokine receptors and ligands involved in MDSC recruitment include C-X-C motif chemokine receptor (CXCR)2 and its ligand, CXCL1, which is significantly expressed in the majority of HNSCCs [58, 62]. Monoclonal antibody or small molecule blockade of CXCR2 results in decreased MDSC recruitment as well as tumor growth delay, and combination therapy with checkpoint inhibitors results in enhanced checkpoint inhibitor treatment response rates in various animal models [55, 130]. A phase I/II clinical trial studying durvalumab + STAT3 inhibitor vs durvalumab + CXCR2 inhibitor in recurrent/metastatic HNSCC is ongoing (NCT02499328).

Phosphodiesterase 5 (PDE5) inhibitors, often thought of in their use to treat erectile dysfunction and pulmonary hypertension, are another drug class able to exert an inhibitory effect on MDSCs due to their suppression of Arg-1 and iNOS [131]. Preclinical models have demonstrated that sildenafil (PDE5 inhibitor) prolongs survival, reduces MDSC levels and activity, and restores $CD8^+$ T cell infiltration and function [132]. Clinical trials in HNSCC and metastatic melanoma have shown tadalafil (PDE5 inhibitor) decreases peripheral and tumor-infiltrating MDSCs, leading to improved clinical outcome and anti-tumor response [133]. There are currently two clinical trials (phase I and I/II) exploring tadalafil's immune activating role in HNSCC (NCT03238365, NCT02544880).

Combining ICIs with macrophage inhibitors

TAMs represent another myeloid cell type that may be immunosuppressive. TAMs may be polarized into either anti-tumor M1 or pro-tumor M2 subtypes, with primarily immunosuppressive M2 macrophages present in the HNSCC TME [134]. M2 macrophage immunosuppression may be mediated by expression of Arg-1 and iNOS, surface PD-L1, checkpoint ligands CD80/86, and immunosuppressive cytokines IL-1β, IL-6, IL-10, and TGFβ. They, similar to tumor cells themselves, also play a role in T_{reg} recruitment through expression of chemokines CCL22, CCL20, and CCL5 [58, 135, 136]. Strategies to inhibit TAMs revolve around blocking chemotaxis and reversing macrophage polarization. Chemotactic factors such as colony stimulating factor (CSF)1 and its receptor CSF1R have been implicated in both M-MDSC and TAM recruitment. Subsequent small molecule inhibition of CSF1/CSF1R signaling has been shown to deplete TAMs from the TME and drive M2 macrophages toward the M1 phenotype [137]. In addition to this, combination treatment of CSF1/CSF1R blockade with checkpoint inhibitors in preclinical models resulted in tumor regression [137]. Two clinical trials (phase I and phase I/II) of CSF1R blockade with PD-1 inhibitors in advanced cancers are currently ongoing (NCT02526017, NCT02452424). C-C motif chemokine ligand (CCL) 2 and its receptors C-C chemokine receptor (CCR)2 and 4 have also been shown to play roles in M-MDSC and TAM recruitment [57, 138]. As a result, inhibiting CCL2/CCR2 signaling is another way to target TAM immunosuppression. In clinical trials, orally administered CCR2 inhibitor CCX872-b improved overall survival in patients with advanced/metastatic pancreatic cancer when used in combination with FOLFIRINOX (fluorouracil [5-FU], leucovorin, irinotecan, oxaliplatin) vs FOLFIRINOX alone [139–141]. There are currently no ongoing clinical trials with CCR2 inhibitors in HNSCC.

Combos with T_{reg} inhibitors

Regulatory $CD4^+$ T-cells are another class of immunosuppressive agents that inhibit effector immune function in the TME. They exert their function by secreting cytokines IL-10 and TGF-β, killing NK and $CD8^+$ T-cells via granzyme-B and perforin, limiting CD28 binding/co-stimulation of naïve T-cells, and catalyzing the degradation of ATP into immunosuppressive adenosine [76, 142–144]. Strategies to inhibit T_{reg} function have revolved around surface markers CD25, CCR4, CTLA4, OX40 and GITR [58].

CD25 is the α-chain of the IL-2 receptor and is expressed on the surface of many T_{reg}s. Although preclinical data with anti-CD25 treatment showed promise, there have been conflicting results in clinical trials. One study on anti-CD25 treatment in combination with dendritic cell vaccination showed that activated effector T-cells expressing CD25 were collaterally depleted along with T_{reg}s while another trial on anti-CD25 treatment in combination with peptide vaccination demonstrated improved peptide-specific $CD4^+$/$CD8^+$ T-cell response with concurrent T_{reg} depletion [145, 146]. A phase I study is currently underway with ADCT-301, a novel antibody drug conjugate targeting CD25 in HNSCC and other advanced solid tumors (NCT03621982).

CCR4 is a cell surface marker that is preferentially expressed on the surface of activated T_{reg}s. Preclinical work using anti-CCR4 treatment depleted T_{reg} levels in peripheral blood mononuclear cells (PBMCs) of healthy individuals, and also reduced both T_{reg}s and adult

T-cell leukemia-lymphoma (ATL) cells in PBMCs of patients with ATL [147]. There is one ongoing phase III clinical trial of KW-0761 (anti-CCR4) vs vorinostat in relapsed/refractory cutaneous T-cell lymphoma (NCT01728805) and no current trials in HNSCC.

CTLA-4 is an immune checkpoint that acts to downregulate the immune response, and as a result, is a rational target for immune therapy in cancer. Ipilimumab, an anti-CTLA-4 checkpoint inhibitor, is currently FDA-approved for metastatic melanoma and in advanced renal cell carcinoma when used in combination with nivolumab. Although anti-CTLA-4 checkpoint inhibitors have been classically thought to elicit therapeutic response through blockade of negative B7-CTLA-4 interactions, recent work has suggested that it may instead be mediated by selective T_{reg} depletion in the TME [148]. There are numerous ongoing clinical trials using CTLA-4 inhibitors alone or in combination with other immune and/or standard therapies in HNSCC.

OX40 and GITR are two commonly expressed costimulatory molecules of the TNF receptor family. OX40 has been shown to generate a positive costimulatory signal when found on effector T-cells and APCs, and an inhibitory signal when found on infiltrating T_{reg}s, leading to an overall immune-activating phenotype [58, 149]. Preclinical studies have demonstrated improved survival and anti-tumor immunity in mouse models treated with OX40 agonist antibodies, with evidence that this effect is contingent on OX40 agonism on both T_{reg}s and effector T-cells [149, 150]. There are currently two phase I clinical trials on anti-OX40 antibodies as monotherapy in HNSCC (NCT02274155, NCT03336606) and another phase I study studying anti-OX40 alone vs combination anti-OX40 + 4-1BB agonist in various locally advanced/ metastatic neoplasms (NCT02315066). An ongoing phase II clinical trial is also studying anti-OX40 in combination with avelumab (anti-PD-L1 antibody) in advanced malignancies (NCT02554812). GITR is another costimulatory molecule found on the surface of APCs and $CD4^+$ T-cells that exhibits an inhibitory effect on T_{reg} immunosuppression when directly targeted by an agonist antibody. Preclinical work on GITR agonist antibodies have shown direct inhibition of naïve T_{reg}-mediated immunosuppression in vitro and decreased FoxP3 expression and T_{reg} IL-10 production in vivo [58, 151, 152]. Numerous clinical trials with anti-GITR antibodies alone or in combination with other therapies are underway in advanced malignancies (NCT01239134, NCT03295942, NCT02583165, NCT02740270, NCT02697591).

Concluding remarks

Recognition of the role of PD-1/L1 signaling in the tumor microenvironment has led to the clinical development of monoclonal antibody therapies that are easy to administer and widely available and appear to have an acceptable safety profile. Although response rates to PD-1/L1 ICI in the recurrent/metastatic setting are quite low, responses tend to be durable. Emerging evidence suggests that PD-1/L1 ICI as a first-line treatment for disease relapse or in the upfront setting in previously untreated patients may provide significant benefit. For these reasons, enhancing the number of patients that respond to ICI represents a major research focus. Through enhanced understanding of how effective anti-tumor immunity is abrogated and the development of novel agents targeting these immunosuppressive mechanisms, rational combinations of treatments can be designed to try and enhance response rates to ICIs. Many trials studying such combinations are underway. While single agent ICI used in the recurrent/metastatic setting has a more favorable toxicity profile compared to standard-of-care systemic chemotherapy regimens, a major challenge lies in determining how to best activate anti-tumor immunity with combination therapies while continuing to minimize risk of immune-related adverse events. Systemic ICIs activate the immune system systemically, with the hope of an anti-tumor effect as a result of this systemic activation. The use of therapeutic vaccines or cellular therapies that induce tumor-specific immunity may allow more targeted activation of the immune system in the future, but issues surrounding cost and technical expertise required for such treatments may remain major issues. In the meantime, carefully executed clinical trials studying rationally designed combinations of agents that activate anti-tumor immunity with ICI, and the incorporation of biomarkers that predict likelihood of response to these agents may get us closer to personalized immunotherapy for HNSCC.

References

[1] Cancer Genome Atlas Network. Comprehensive genomic characterization of head and neck squamous cell carcinomas. Nature 2015;517(7536):576–82.

[2] Massard C, Michiels S, Ferté C, Le Deley M-C, Lacroix L, Hollebecque A, Verlingue L, Ileana E, Rosellini S, Ammari S, Ngo-Camus M, Bahleda R, Gazzah A, Varga A, Postel-Vinay S, Loriot Y, Even C, Breuskin I, Auger N, Job B, De Baere T, Deschamps F, Vielh P, Scoazec J-Y, Lazar V, Richon C, Ribrag V, Deutsch E, Angevin E, Vassal G, Eggermont A, André F, Soria J-C. High-throughput genomics and clinical outcome in hard-to-treat advanced cancers: results of the MOSCATO 01 trial. Cancer Discov 2017;7(6):586–95.

[3] Ferris RL, Blumenschein G, Fayette J, Guigay J, Colevas AD, Licitra L, Harrington K, Kasper S, Vokes EE, Even C, Worden F, Saba NF, Iglesias Docampo LC, Haddad R, Rordorf T, Kiyota N, Tahara M, Monga M, Lynch M, Geese WJ, Kopit J, Shaw JW, Gillison ML. Nivolumab for recurrent squamous-cell carcinoma of the head and neck. N Engl J Med 2016;375(19):1856–67.

[4] Seiwert TY, Burtness B, Mehra R, Weiss J, Berger R, Eder JP, Heath K, McClanahan T, Lunceford J, Gause C, Cheng JD, Chow LQ. Safety and clinical activity of pembrolizumab for treatment of recurrent or metastatic squamous cell carcinoma of the head and neck (KEYNOTE-012): an open-label, multicentre, phase 1b trial. Lancet Oncol 2016;17(7):956–65.

[5] Allen CT, Clavijo PE, Van Waes C, Chen Z. Anti-tumor immunity in head and neck cancer: understanding the evidence, how tumors escape and immunotherapeutic approaches. Cancers (Basel) 2015;7(4):2397–414.

[6] Hanahan D, Weinberg RA. The hallmarks of cancer. Cell 2000;100(1):57–70.

[7] Molinolo AA, Amornphimoltham P, Squarize CH, Castilho RM, Patel V, Gutkind JS. Dysregulated molecular networks in head and neck carcinogenesis. Oral Oncol 2009;45(4–5):324–34.

[8] Kono H, Rock KL. How dying cells alert the immune system to danger. Nat Rev Immunol 2008;8(4):279–89.

[9] Birkeland SA, Storm HH, Lamm LU, Barlow L, Blohmé I, Forsberg B, Eklund B, Fjeldborg O, Friedberg M, Frödin L. Cancer risk after renal transplantation in the Nordic countries, 1964–1986. Int J Cancer 1995;60(2):183–9.

[10] Pham SM, Kormos RL, Landreneau RJ, Kawai A, Gonzalez-Cancel I, Hardesty RL, Hattler BG, Griffith BP. Solid tumors after heart transplantation: lethality of lung cancer. Ann Thorac Surg 1995;60(6):1623–6.

[11] Murphy KM, Weaver C. Janeway's immunobiology. 9th ed. New York, NY: W. W. Norton & Company; 2016.

[12] O'Sullivan T, Saddawi-Konefka R, Vermi W, Koebel CM, Arthur C, White JM, Uppaluri R, Andrews DM, Ngiow SF, Teng MWL, Smyth MJ, Schreiber RD, Bui JD. Cancer immunoediting by the innate immune system in the absence of adaptive immunity. J Exp Med 2012;209(10):1869–82.

[13] Banchereau J, Steinman RM. Dendritic cells and the control of immunity. Nature 1998;392(6673):245–52.

[14] Blum JS, Wearsch PA, Cresswell P. Pathways of antigen processing. Annu Rev Immunol 2013;31:443–73.

[15] Krogsgaard M, Davis MM. How T cells 'see' antigen. Nat Immunol 2005;6(3):239–45.

[16] Diamond MS, Kinder M, Matsushita H, Mashayekhi M, Dunn GP, Archambault JM, Lee H, Arthur CD, White JM, Kalinke U, Murphy KM, Schreiber RD. Type I interferon is selectively required by dendritic cells for immune rejection of tumors. J Exp Med 2011;208(10):1989–2003.

[17] Fuertes MB, Kacha AK, Kline J, Woo S-R, Kranz DM, Murphy KM, Gajewski TF. Host type I IFN signals are required for antitumor CD8+ T cell responses through CD8{alpha}+ dendritic cells. J Exp Med 2011;208(10):2005–16.

[18] Kumar H, Kawai T, Akira S. Pathogen recognition by the innate immune system. Int Rev Immunol 2011;30(1):16–34.

[19] Kepp O, Senovilla L, Vitale I, Vacchelli E, Adjemian S, Agostinis P, Apetoh L, Aranda F, Barnaba V, Bloy N, Bracci L, Breckpot K, Brough D, Buqué A, Castro MG, Cirone M, Colombo MI, Cremer I, Demaria S, Dini L, Eliopoulos AG, Faggioni A, Formenti SC, Fučíková J, Gabriele L, Gaipl US, Galon J, Garg A, Ghiringhelli F, Giese NA, Guo ZS, Hemminki A, Herrmann M, Hodge JW, Holdenrieder S, Honeychurch J, Hu H-M, Huang X, Illidge TM, Kono K, Korbelik M, Krysko DV, Loi S, Lowenstein PR, Lugli E, Ma Y, Madeo F, Manfredi AA, Martins I, Mavilio D, Menger L, Merendino N, Michaud M, Mignot G, Mossman KL, Multhoff G, Oehler R, Palombo F, Panaretakis T, Pol J, Proietti E, Ricci J-E, Riganti C, Rovere-Querini P, Rubartelli A, Sistigu A, Smyth MJ, Sonnemann J, Spisek R, Stagg J, Sukkurwala AQ, Tartour E, Thorburn A, Thorne SH, Vandenabeele P, Velotti F, Workenhe ST, Yang H, Zong W-X, Zitvogel L, Kroemer G, Galluzzi L. Consensus guidelines for the detection of immunogenic cell death. Oncoimmunology 2014;3(9): e955691.

[20] Corrales L, Glickman LH, McWhirter SM, Kanne DB, Sivick KE, Katibah GE, Woo S-R, Lemmens E, Banda T, Leong JJ, Metchette K, Dubensky TW, Gajewski TF. Direct activation of STING in the tumor microenvironment leads to potent and systemic tumor regression and immunity. Cell Rep 2015;11(7):1018–30.

[21] Woo S-R, Fuertes MB, Corrales L, Spranger S, Furdyna MJ, Leung MYK, Duggan R, Wang Y, Barber GN, Fitzgerald KA, Alegre M-L, Gajewski TF. STING-dependent cytosolic DNA sensing mediates innate immune recognition of immunogenic tumors. Immunity 2014;41(5):830–42.

[22] Embgenbroich M, Burgdorf S. Current concepts of antigen cross-presentation. Front Immunol 2018;9:1643.

[23] Romagnani S. T-cell subsets (Th1 versus Th2). Ann Allergy Asthma Immunol 2000;85(1):9–18. quiz 18, 21.

[24] Riley JL. PD-1 signaling in primary T cells. Immunol Rev 2009;229(1):114–25.

[25] Buchbinder EI, Desai A. CTLA-4 and PD-1 pathways. Am J Clin Oncol 2016;39(1):98–106.

[26] Hargadon KM, Johnson CE, Williams CJ. Immune checkpoint blockade therapy for cancer: an overview of FDA-approved immune checkpoint inhibitors. Int Immunopharmacol 2018;62:29–39.

[27] Davis MM, Bjorkman PJ. T-cell antigen receptor genes and T-cell recognition. Nature 1988;334(6181): 395–402.

[28] Finn OJ. Human tumor antigens yesterday, today, and tomorrow. Cancer Immunol Res 2017;5(5):347–54.

[29] Schumacher TN, Schreiber RD. Neoantigens in cancer immunotherapy. Science 2015;348(6230):69–74.

[30] Keck MK, Zuo Z, Khattri A, Stricker TP, Brown CD, Imanguli M, Rieke D, Endhardt K, Fang P, Brägelmann J, DeBoer R, El-Dinali M, Aktolga S, Lei Z, Tan P, Rozen SG, Salgia R, Weichselbaum RR, Lingen MW, Story MD, Ang KK, Cohen EEW, White KP, Vokes EE, Seiwert TY. Integrative analysis of head and neck cancer identifies two biologically distinct HPV and three non-HPV subtypes. Clin Cancer Res 2015;21(4):870–81.

[31] Wolf GT, Chepeha DB, Bellile E, Nguyen A, Thomas D, McHugh J, University of Michigan Head and Neck SPORE Program, Neck SP. Tumor infiltrating lymphocytes (TIL) and prognosis in oral cavity squamous carcinoma: a preliminary study. Oral Oncol 2015;51(1):90–5.

[32] Goodman AM, Kato S, Bazhenova L, Patel SP, Frampton GM, Miller V, Stephens PJ, Daniels GA, Kurzrock R. Tumor mutational burden as an independent predictor of response to immunotherapy in diverse cancers. Mol Cancer Ther 2017;16(11):2598–608.

[33] Wortzel RD, Philipps C, Schreiber H. Multiple tumour-specific antigens expressed on a single tumour cell. Nature 1983;304(5922):165–7.

[34] Sandberg JK, Grufman P, Wolpert EZ, Franksson L, Chambers BJ, Kärre K. Superdominance among immunodominant H-2Kb-restricted epitopes and reversal by dendritic cell-mediated antigen delivery. J Immunol 1998;160(7):3163–9.

[35] Rowley DA, Stach RM. A first or dominant immunization. I. Suppression of simultaneous cytolytic T cell responses to unrelated alloantigens. J Exp Med 1993;178(3):835–40.

[36] Van Waes C, Monach PA, Urban JL, Wortzel RD, Schreiber H. Immunodominance deters the response to other tumor antigens thereby favoring escape: prevention by vaccination with tumor variants selected with cloned cytolytic T cells in vitro. Tissue Antigens 1996;47(5):399–407.

[37] Urban JL, Kripke ML, Schreiber H. Stepwise immunologic selection of antigenic variants during tumor growth. J Immunol 1986;137(9):3036–41.

[38] Seung S, Urban JL, Schreiber H. A tumor escape variant that has lost one major histocompatibility complex class I restriction element induces specific CD8+ T cells to an antigen that no longer serves as a target. J Exp Med 1993;178(3):933–40.
[39] Ferris RL, Whiteside TL, Ferrone S. Immune escape associated with functional defects in antigen-processing machinery in head and neck cancer. Clin Cancer Res 2006;12(13):3890–5.
[40] Martini M, Testi MG, Pasetto M, Picchio MC, Innamorati G, Mazzocco M, Ugel S, Cingarlini S, Bronte V, Zanovello P, Krampera M, Mosna F, Cestari T, Riviera AP, Brutti N, Barbieri O, Matera L, Tridente G, Colombatti M, Sartoris S. IFN-gamma-mediated upmodulation of MHC class I expression activates tumor-specific immune response in a mouse model of prostate cancer. Vaccine 2010;28(20):3548–57.
[41] Respa A, Bukur J, Ferrone S, Pawelec G, Zhao Y, Wang E, Marincola FM, Seliger B. Association of IFN-γ signal transduction defects with impaired HLA class I antigen processing in melanoma cell lines. Clin Cancer Res 2011;17(9):2668–78.
[42] Duncan TJ, Rolland P, Deen S, Scott IV, Liu DTY, Spendlove I, Durrant LG. Loss of IFN gamma receptor is an independent prognostic factor in ovarian cancer. Clin Cancer Res 2007;13(14):4139–45.
[43] Thomas SJ, Snowden JA, Zeidler MP, Danson SJ. The role of JAK/STAT signalling in the pathogenesis, prognosis and treatment of solid tumours. Br J Cancer 2015;113(3):365–71.
[44] Dunn GP, Old LJ, Schreiber RD. The three Es of cancer immunoediting. Annu Rev Immunol 2004;22:329–60.
[45] Janicki CN, Jenkinson SR, Williams NA, Morgan DJ. Loss of CTL function among high-avidity tumor-specific CD8+ T cells following tumor infiltration. Cancer Res 2008;68(8):2993–3000.
[46] Pradeu T, Carosella ED. On the definition of a criterion of immunogenicity. Proc Natl Acad Sci U S A 2006;103(47):17858–61.
[47] Willimsky G, Czéh M, Loddenkemper C, Gellermann J, Schmidt K, Wust P, Stein H, Blankenstein T. Immunogenicity of premalignant lesions is the primary cause of general cytotoxic T lymphocyte unresponsiveness. J Exp Med 2008;205(7):1687–700.
[48] Pan J, Zhang M, Wang J, Wang Q, Xia D, Sun W, Zhang L, Yu H, Liu Y, Cao X. Interferon-gamma is an autocrine mediator for dendritic cell maturation. Immunol Lett 2004;94(1–2):141–51.
[49] Zitvogel L, Tesniere A, Kroemer G. Cancer despite immunosurveillance: immunoselection and immunosubversion. Nat Rev Immunol 2006;6(10):715–27.
[50] Bedi A, Chang X, Noonan K, Pham V, Bedi R, Fertig EJ, Considine M, Califano JA, Borrello I, Chung CH, Sidransky D, Ravi R. Inhibition of TGF-β enhances the in vivo antitumor efficacy of EGF receptor-targeted therapy. Mol Cancer Ther 2012;11(11):2429–39.
[51] Simon S, Labarriere N. PD-1 expression on tumor-specific T cells: friend or foe for immunotherapy? Oncoimmunology 2017;7(1).
[52] Heineman TE, Widman A, Kuan EC, St John M. The genetic landscape of programmed death ligand-1 (PD-L1) alterations in head and neck cancer. Laryngoscope Investig Otolaryngol 2017;2(3):99–103.
[53] Ribas A. Adaptive immune resistance: how cancer protects from immune attack. Cancer Discov 2015;5(9):915–9.
[54] Gabrilovich DI, Nagaraj S. Myeloid-derived suppressor cells as regulators of the immune system. Nat Rev Immunol 2009;9(3):162–74.
[55] Highfill SL, Cui Y, Giles AJ, Smith JP, Zhang H, Morse E, Kaplan RN, Mackall CL. Disruption of CXCR2-mediated MDSC tumor trafficking enhances anti-PD1 efficacy. Sci Transl Med 2014;6(237):237ra67.
[56] Katoh H, Wang D, Daikoku T, Sun H, Dey SK, Dubois RN. CXCR2-expressing myeloid-derived suppressor cells are essential to promote colitis-associated tumorigenesis. Cancer Cell 2013;24(5):631–44.
[57] Lesokhin AM, Hohl TM, Kitano S, Cortez C, Hirschhorn-Cymerman D, Avogadri F, Rizzuto GA, Lazarus JJ, Pamer EG, Houghton AN, Merghoub T, Wolchok JD. Monocytic CCR2(+) myeloid-derived suppressor cells promote immune escape by limiting activated CD8 T-cell infiltration into the tumor microenvironment. Cancer Res 2012;72(4):876–86.
[58] Davis RJ, Van Waes C, Allen CT. Overcoming barriers to effective immunotherapy: MDSCs, TAMs, and Tregs as mediators of the immunosuppressive microenvironment in head and neck cancer. Oral Oncol 2016;58:59–70.
[59] Gabrilovich DI. Myeloid-derived suppressor cells. Cancer Immunol Res 2017;5(1):3–8.
[60] Bronte V, Brandau S, Chen S-H, Colombo MP, Frey AB, Greten TF, Mandruzzato S, Murray PJ, Ochoa A, Ostrand-Rosenberg S, Rodriguez PC, Sica A, Umansky V, Vonderheide RH, Gabrilovich DI. Recommendations for myeloid-derived suppressor cell nomenclature and characterization standards. Nat Commun 2016;7:12150.

[61] Solito S, Marigo I, Pinton L, Damuzzo V, Mandruzzato S, Bronte V. Myeloid-derived suppressor cell heterogeneity in human cancers. Ann N Y Acad Sci 2014;1319:47–65.
[62] Weber R, Fleming V, Hu X, Nagibin V, Groth C, Altevogt P, Utikal J, Umansky V. Myeloid-derived suppressor cells hinder the anti-cancer activity of immune checkpoint inhibitors. Front Immunol 2018;9:1310.
[63] Bogdan C. Nitric oxide and the immune response. Nat Immunol 2001;2(10):907–16.
[64] Munn DH, Sharma MD, Baban B, Harding HP, Zhang Y, Ron D, Mellor AL. GCN2 kinase in T cells mediates proliferative arrest and anergy induction in response to indoleamine 2,3-dioxygenase. Immunity 2005;22(5): 633–42.
[65] Nagaraj S, Gupta K, Pisarev V, Kinarsky L, Sherman S, Kang L, Herber DL, Schneck J, Gabrilovich DI. Altered recognition of antigen is a mechanism of CD8+ T cell tolerance in cancer. Nat Med 2007;13(7):828–35.
[66] Noman MZ, Desantis G, Janji B, Hasmim M, Karray S, Dessen P, Bronte V, Chouaib S. PD-L1 is a novel direct target of HIF-1α, and its blockade under hypoxia enhanced MDSC-mediated T cell activation. J Exp Med 2014;211(5):781–90.
[67] Ostrand-Rosenberg S, Horn LA, Haile ST. The programmed death-1 immune-suppressive pathway: barrier to antitumor immunity. J Immunol 2014;193(8):3835–41.
[68] Pan P-Y, Ma G, Weber KJ, Ozao-Choy J, Wang G, Yin B, Divino CM, Chen S-H. Immune stimulatory receptor CD40 is required for T-cell suppression and T regulatory cell activation mediated by myeloid-derived suppressor cells in cancer. Cancer Res 2010;70(1):99–108.
[69] Rodriguez PC, Quiceno DG, Zabaleta J, Ortiz B, Zea AH, Piazuelo MB, Delgado A, Correa P, Brayer J, Sotomayor EM, Antonia S, Ochoa JB, Ochoa AC. Arginase I production in the tumor microenvironment by mature myeloid cells inhibits T-cell receptor expression and antigen-specific T-cell responses. Cancer Res 2004;64(16):5839–49.
[70] Rodriguez PC, Quiceno DG, Ochoa AC. L-arginine availability regulates T-lymphocyte cell-cycle progression. Blood 2007;109(4):1568–73.
[71] Sinha P, Clements VK, Bunt SK, Albelda SM, Ostrand-Rosenberg S. Cross-talk between myeloid-derived suppressor cells and macrophages subverts tumor immunity toward a type 2 response. J Immunol 2007;179(2): 977–83.
[72] Yu J, Du W, Yan F, Wang Y, Li H, Cao S, Yu W, Shen C, Liu J, Ren X. Myeloid-derived suppressor cells suppress antitumor immune responses through IDO expression and correlate with lymph node metastasis in patients with breast cancer. J Immunol 2013;190(7):3783–97.
[73] Baniyash M. Myeloid-derived suppressor cells as intruders and targets: clinical implications in cancer therapy. Cancer Immunol Immunother 2016;65(7):857–67.
[74] Schmid MC, Varner JA. Myeloid cells in tumor inflammation. Vasc Cell 2012;4(1):14.
[75] Allen CT, Judd NP, Bui JD, Uppaluri R. The clinical implications of antitumor immunity in head and neck cancer. Laryngoscope 2012;122(1):144–57.
[76] Strauss L, Bergmann C, Szczepanski M, Gooding W, Johnson JT, Whiteside TL. A unique subset of CD4+CD25highFoxp3+ T cells secreting interleukin-10 and transforming growth factor-beta1 mediates suppression in the tumor microenvironment. Clin Cancer Res 2007;13(15 Pt 1):4345–54.
[77] Lau KM, Cheng SH, Lo KW, SAKW L, JKS W, van Hasselt CA, Lee SP, Rickinson AB, Ng MHL. Increase in circulating Foxp3+CD4+CD25(high) regulatory T cells in nasopharyngeal carcinoma patients. Br J Cancer 2007;96(4):617–22.
[78] Schaefer C, Kim GG, Albers A, Hoermann K, Myers EN, Whiteside TL. Characteristics of CD4−CD25+ regulatory T cells in the peripheral circulation of patients with head and neck cancer. Br J Cancer 2005;92(5):913–20.
[79] Nicholson RI, Gee JM, Harper ME. EGFR and cancer prognosis. Eur J Cancer 2001;37(Suppl 4):S9–15.
[80] Tian Y, Lin J, Tian Y, Zhang G, Zeng X, Zheng R, Zhang W, Yuan Y. Efficacy and safety of anti-EGFR agents administered concurrently with standard therapies for patients with head and neck squamous cell carcinoma: a systematic review and meta-analysis of randomized controlled trials. Int J Cancer 2018;142(11):2198–206.
[81] Bonner JA, Harari PM, Giralt J, Azarnia N, Shin DM, Cohen RB, Jones CU, Sur R, Raben D, Jassem J, Ove R, Kies MS, Baselga J, Youssoufian H, Amellal N, Rowinsky EK, Ang KK. Radiotherapy plus cetuximab for squamous-cell carcinoma of the head and neck. N Engl J Med 2006;354(6):567–78.
[82] Ferris RL, Jaffee EM, Ferrone S. Tumor antigen-targeted, monoclonal antibody-based immunotherapy: clinical response, cellular immunity, and immunoescape. J Clin Oncol 2010;28(28):4390–9.

[83] Gómez Román VR, Murray JC, Weiner LM. Chapter 1—antibody-dependent cellular cytotoxicity (ADCC). In: Ackerman ME, Nimmerjahn F, editors. Antibody Fc. Boston: Academic Press; 2014. p. 1–27.
[84] Lee SC, Srivastava RM, López-Albaitero A, Ferrone S, Ferris RL. Natural killer (NK): dendritic cell (DC) cross talk induced by therapeutic monoclonal antibody triggers tumor antigen-specific T cell immunity. Immunol Res 2011;50(2–3):248–54.
[85] Srivastava RM, Lee SC, Andrade Filho PA, Lord CA, Jie H-B, Davidson HC, López-Albaitero A, Gibson SP, Gooding WE, Ferrone S, Ferris RL. Cetuximab-activated natural killer and dendritic cells collaborate to trigger tumor antigen-specific T-cell immunity in head and neck cancer patients. Clin Cancer Res 2013;19(7):1858–72.
[86] Trivedi S, Srivastava RM, Concha-Benavente F, Ferrone S, Garcia-Bates TM, Li J, Ferris RL. Anti-EGFR targeted monoclonal antibody isotype influences antitumor cellular immunity in head and neck cancer patients. Clin Cancer Res 2016;22(21):5229–37.
[87] Vermorken JB, Trigo J, Hitt R, Koralewski P, Diaz-Rubio E, Rolland F, Knecht R, Amellal N, Schueler A, Baselga J. Open-label, uncontrolled, multicenter phase II study to evaluate the efficacy and toxicity of cetuximab as a single agent in patients with recurrent and/or metastatic squamous cell carcinoma of the head and neck who failed to respond to platinum-based therapy. J Clin Oncol 2007;25(16):2171–7.
[88] Ferris RL, Lenz H-J, Trotta AM, García-Foncillas J, Schulten J, Audhuy F, Merlano M, Milano G. Rationale for combination of therapeutic antibodies targeting tumor cells and immune checkpoint receptors: harnessing innate and adaptive immunity through IgG1 isotype immune effector stimulation. Cancer Treat Rev 2018;63:48–60.
[89] Lo Nigro C, Ricci V, Vivenza D, Monteverde M, Strola G, Lucio F, Tonissi F, Miraglio E, Granetto C, Fortunato M, Merlano MC. Evaluation of antibody-dependent cell-mediated cytotoxicity activity and cetuximab response in KRAS wild-type metastatic colorectal cancer patients. World J Gastrointest Oncol 2016;8(2):222–30.
[90] Jie H-B, Schuler PJ, Lee SC, Srivastava RM, Argiris A, Ferrone S, Whiteside TL, Ferris RL. CTLA-4^{+} regulatory T cells increased in cetuximab-treated head and neck cancer patients suppress NK cell cytotoxicity and correlate with poor prognosis. Cancer Res 2015;75(11):2200–10.
[91] Li J, Srivastava RM, Ettyreddy A, Ferris RL. Cetuximab ameliorates suppressive phenotypes of myeloid antigen presenting cells in head and neck cancer patients. J Immunother Cancer 2015;3:54.
[92] Forster MD, Devlin M-J. Immune checkpoint inhibition in head and neck cancer. Front Oncol 2018;8:310.
[93] Ferris RL, Blumenschein GR, Fayette J, Guigay J, Colevas AD, Licitra L, Harrington KJ, Kasper S, Vokes EE, Even C, Worden F, Saba NF, Docampo LCI, Haddad R, Rordorf T, Kiyota N, Tahara M, Lynch M, Jayaprakash V, Li L, Gillison ML. Nivolumab vs investigator's choice in recurrent or metastatic squamous cell carcinoma of the head and neck: 2-year long-term survival update of CheckMate 141 with analyses by tumor PD-L1 expression. Oral Oncol 2018;81:45–51.
[94] Cohen EE, Harrington KJ, Le Tourneau C, Dinis J, Licitra L, Ahn MJ, Soria A, Machiels JP, Mach N, Mehra R, Burtness B, Wang Y, Tuozzo AJ, Swaby R, Soulieres D. LBA45_PRPembrolizumab (pembro) vs standard of care (SOC) for recurrent or metastatic head and neck squamous cell carcinoma (R/M HNSCC): phase 3 KEYNOTE-040 trial. Ann Oncol 2017;28(suppl_5):628.
[95] Soulieres D, Cohen E, Tourneau CL, Dinis J, Licitra L, Ahn M-J, Soria A, Machiels J-P, Mach N, Mehra R, Burtness B, Zhang P, Cheng J, Swaby R, Harrington KJ. Abstract CT115: updated survival results of the KEYNOTE-040 study of pembrolizumab vs standard-of-care chemotherapy for recurrent or metastatic head and neck squamous cell carcinoma. Cancer Res 2018;78(13 Supplement).
[96] Postow MA, Sidlow R, Hellmann MD. Immune-related adverse events associated with immune checkpoint blockade. N Engl J Med 2018;378(2):158–68.
[97] Burtness B, Harrington KJ, Greil R, Soulières D, Tahara M, De Castro Jr G, Psyrri A, Rotllan NB, Neupane PC, Bratland Å, Fuereder T, Hughes BGM, Mesia R, Ngamphaiboon N, Rordorf T, Ishak WZW, Roy A, Cheng J, Jin F, Rischin D. Rischin KEYNOTE-048: phase 3 study of first-line pembrolizumab (P) for recurrent/metastatic head and neck squamous cell carcinoma (R/M HNSCC). Ann Oncol 2018;29(Suppl 8):729.
[98] Samra B, Tam E, Baseri B, Shapira I. Checkpoint inhibitors in head and neck cancer: current knowledge and perspectives. J Investig Med 2018;66(7):1023–30.
[99] Ferris RL, Gonçalves A, Baxi SS, Martens UM, Gauthier H, Langenberg M, Spanos WC, Leidner RS, Kang H, Russell J, Chiosea S, Soumaoro I, Rao S, Cao ZA, Topalian S. LBA46An open-label, multicohort, phase 1/2 study in patients with virus-associated cancers (CheckMate 358): safety and efficacy of

neoadjuvant nivolumab in squamous cell carcinoma of the head and neck (SCCHN). Ann Oncol 2017;28(suppl_5):628–9.

[100] Uppaluri R, Zolkind P, Lin T, Nussenbaum B, Jackson RS, Rich J, Pipkorn P, Paniello R, Thorstad W, Michel L, Mudianto T, Oppelt PJ, Wildes TM, Dunn GP, Piccirillo JF, Kallogjeri D, Rodig S, Hagemann IS, Chernock R, Adkins D. Neoadjuvant pembrolizumab in surgically resectable, locally advanced HPV negative head and neck squamous cell carcinoma (HNSCC). J Clin Oncol 2017;35(15_suppl):6012.

[101] Bezu L, Gomes-de-Silva LC, Dewitte H, Breckpot K, Fucikova J, Spisek R, Galluzzi L, Kepp O, Kroemer G. Combinatorial strategies for the induction of immunogenic cell death. Front Immunol 2015;6.

[102] Martins I, Kepp O, Schlemmer F, Adjemian S, Tailler M, Shen S, Michaud M, Menger L, Gdoura A, Tajeddine N, Tesniere A, Zitvogel L, Kroemer G. Restoration of the immunogenicity of cisplatin-induced cancer cell death by endoplasmic reticulum stress. Oncogene 2011;30(10):1147–58.

[103] Terenzi A, Pirker C, Keppler BK, Berger W. Anticancer metal drugs and immunogenic cell death. J Inorg Biochem 2016;165:71–9.

[104] Beyranvand Nejad E, van der Sluis TC, van Duikeren S, Yagita H, Janssen GM, van Veelen PA, Melief CJM, van der Burg SH, Arens R. Tumor eradication by cisplatin is sustained by CD80/86-mediated costimulation of CD8+ T cells. Cancer Res 2016;76(20):6017–29.

[105] Cirone M, Garufi A, Di Renzo L, Granato M, Faggioni A, D'Orazi G. Zinc supplementation is required for the cytotoxic and immunogenic effects of chemotherapy in chemoresistant p53-functionally deficient cells. Oncoimmunology 2013;2(9).

[106] de Biasi AR, Villena-Vargas J, Adusumilli PS. Cisplatin-induced antitumor immunomodulation: a review of preclinical and clinical evidence. Clin Cancer Res 2014;20(21):5384–91.

[107] Hato SV, Khong A, IJMd V, Lesterhuis WJ. Molecular pathways: the immunogenic effects of platinum-based chemotherapeutics. Clin Cancer Res 2014;20(11):2831–7.

[108] Tran L, Allen CT, Xiao R, Moore E, Davis R, Park S-J, Spielbauer K, Van Waes C, Schmitt NC. Cisplatin alters antitumor immunity and synergizes with PD-1/PD-L1 inhibition in head and neck squamous cell carcinoma. Cancer Immunol Res 2017;5(12):1141–51.

[109] Spielbauer K, Cunningham L, Schmitt N. PD-1 inhibition minimally affects cisplatin-induced toxicities in a murine model. Otolaryngol Head Neck Surg 2018;159(2):343–6.

[110] Gandhi L, Rodríguez-Abreu D, Gadgeel S, Esteban E, Felip E, Angelis FD, Domine M, Clingan P, Hochmair MJ, Powell SF, Cheng SYS, Bischoff HG, Peled N, Grossi F, Jennens RR, Reck M, Hui R, Garon EB, Boyer M, Rubio-Viqueira B, Novello S, Kurata T, Gray JE, Vida J, Wei Z, Yang J, Raftopoulos H, Pietanza MC, Garassino MC. Pembrolizumab plus chemotherapy in metastatic non–small-cell lung cancer. N Engl J Med 2018;378:2078–92.

[111] Langer CJ, Gadgeel SM, Borghaei H, Papadimitrakopoulou VA, Patnaik A, Powell SF, Gentzler RD, Martins RG, Stevenson JP, Jalal SI, Panwalkar A, Yang JC-H, Gubens M, Sequist LV, Awad MM, Fiore J, Ge Y, Raftopoulos H, Gandhi L, KEYNOTE-021 investigators. Carboplatin and pemetrexed with or without pembrolizumab for advanced, non-squamous non-small-cell lung cancer: a randomised, phase 2 cohort of the open-label KEYNOTE-021 study. Lancet Oncol 2016;17(11):1497–508.

[112] Golden EB, Pellicciotta I, Demaria S, Barcellos-Hoff MH, Formenti SC. The convergence of radiation and immunogenic cell death signaling pathways. Front Oncol 2012;2.

[113] Derer A, Frey B, Fietkau R, Gaipl US. Immune-modulating properties of ionizing radiation: rationale for the treatment of cancer by combination radiotherapy and immune checkpoint inhibitors. Cancer Immunol Immunother 2016;65(7):779–86.

[114] Kaur P, Asea A. Radiation-induced effects and the immune system in cancer. Front Oncol 2012;2.

[115] Demaria S, Kawashima N, Yang AM, Devitt ML, Babb JS, Allison JP, Formenti SC. Immune-mediated inhibition of metastases after treatment with local radiation and CTLA-4 blockade in a mouse model of breast cancer. Clin Cancer Res 2005;11(2 Pt 1):728–34.

[116] Tang C, Wang X, Soh H, Seyedin S, Cortez MA, Krishnan S, Massarelli E, Hong D, Naing A, Diab A, Gomez D, Ye H, Heymach J, Komaki R, Allison JP, Sharma P, Welsh JW. Combining radiation and immunotherapy: a new systemic therapy for solid tumors? Cancer Immunol Res 2014;2(9):831–8.

[117] Morisada M, Chamberlin M, Allen C. Exploring the rationale for combining ionizing radiation and immune checkpoint blockade in head and neck cancer. Head Neck 2018;40(6):1321–34.

[118] Baird JR, Friedman D, Cottam B, Dubensky TW, Kanne DB, Bambina S, Bahjat K, Crittenden MR, Gough MJ. Radiotherapy combined with novel STING-targeting oligonucleotides results in regression of established tumors. Cancer Res 2016;76(1):50–61.

[119] Downey CM, Aghaei M, Schwendener RA, Jirik FR. DMXAA causes tumor site-specific vascular disruption in murine non-small cell lung cancer, and like the endogenous non-canonical cyclic dinucleotide STING agonist, 2′3′-cGAMP, induces M2 macrophage repolarization. PLoS ONE 2014;9(6).

[120] Gadkaree SK, Fu J, Sen R, Korrer MJ, Allen C, Kim YJ. Induction of tumor regression by intratumoral STING agonists combined with anti-programmed death-L1 blocking antibody in a preclinical squamous cell carcinoma model. Head Neck 2017;39(6):1086–94.

[121] Lu S, Concha-Benavente F, Shayan G, Srivastava RM, Gibson SP, Wang L, Gooding WE, Ferris RL. STING activation enhances cetuximab-mediated NK cell activation and DC maturation and correlates with HPV+ status in head and neck cancer. Oral Oncol 2018;78:186–93.

[122] Nefedova Y, Fishman M, Sherman S, Wang X, Beg AA, Gabrilovich DI. Mechanism of all-trans retinoic acid effect on tumor-associated myeloid-derived suppressor cells. Cancer Res 2007;67(22):11021–8.

[123] Iclozan C, Antonia S, Chiappori A, Chen D-T, Gabrilovich D. Therapeutic regulation of myeloid-derived suppressor cells and immune response to cancer vaccine in patients with extensive stage small cell lung cancer. Cancer Immunol Immunother 2013;62(5):909–18.

[124] Mirza N, Fishman M, Fricke I, Dunn M, Neuger AM, Frost TJ, Lush RM, Antonia S, Gabrilovich DI. All-trans-retinoic acid improves differentiation of myeloid cells and immune response in cancer patients. Cancer Res 2006;66(18):9299–307.

[125] Garrity T, Pandit R, Wright MA, Benefield J, Keni S, Young MR. Increased presence of CD34+ cells in the peripheral blood of head and neck cancer patients and their differentiation into dendritic cells. Int J Cancer 1997;73(5):663–9.

[126] Lathers DMR, Clark JI, Achille NJ, Young MRI. Phase 1B study to improve immune responses in head and neck cancer patients using escalating doses of 25-hydroxyvitamin D3. Cancer Immunol Immunother 2004;53(5):422–30.

[127] Walsh JE, Clark A-M, Day TA, Gillespie MB, Young MRI. Use of alpha,25-dihydroxyvitamin D3 treatment to stimulate immune infiltration into head and neck squamous cell carcinoma. Hum Immunol 2010;71(7):659–65.

[128] Serafini P, Borrello I, Bronte V. Myeloid suppressor cells in cancer: recruitment, phenotype, properties, and mechanisms of immune suppression. Semin Cancer Biol 2006;16(1):53–65.

[129] Davis RJ, Moore EC, Clavijo PE, Friedman J, Cash H, Chen Z, Silvin C, Van Waes C, Allen C. Anti-PD-L1 efficacy can be enhanced by inhibition of myeloid-derived suppressor cells with a selective inhibitor of PI3Kδ/γ. Cancer Res 2017;77(10):2607–19.

[130] Keane MP, Belperio JA, Xue YY, Burdick MD, Strieter RM. Depletion of CXCR2 inhibits tumor growth and angiogenesis in a murine model of lung cancer. J Immunol 2004;172(5):2853–60.

[131] Serafini P, Meckel K, Kelso M, Noonan K, Califano J, Koch W, Dolcetti L, Bronte V, Borrello I. Phosphodiesterase-5 inhibition augments endogenous antitumor immunity by reducing myeloid-derived suppressor cell function. J Exp Med 2006;203(12):2691–702.

[132] Meyer C, Sevko A, Ramacher M, Bazhin AV, Falk CS, Osen W, Borrello I, Kato M, Schadendorf D, Baniyash M, Umansky V. Chronic inflammation promotes myeloid-derived suppressor cell activation blocking antitumor immunity in transgenic mouse melanoma model. Proc Natl Acad Sci U S A 2011;108(41):17111–6.

[133] Califano JA, Khan Z, Noonan KA, Rudraraju L, Zhang Z, Wang H, Goodman S, Gourin CG, Ha PK, Fakhry C, Saunders J, Levine M, Tang M, Neuner G, Richmon JD, Blanco R, Agrawal N, Koch WM, Marur S, Weed DT, Serafini P, Borrello I. Tadalafil augments tumor specific immunity in patients with head and neck squamous cell carcinoma. Clin Cancer Res 2015;21(1):30–8.

[134] Costa NL, Valadares MC, Souza PPC, Mendonça EF, Oliveira JC, Silva TA, Batista AC. Tumor-associated macrophages and the profile of inflammatory cytokines in oral squamous cell carcinoma. Oral Oncol 2013;49(3):216–23.

[135] Chang L-Y, Lin Y-C, Mahalingam J, Huang C-T, Chen T-W, Kang C-W, Peng H-M, Chu Y-Y, Chiang J-M, Dutta A, Day Y-J, Chen T-C, Yeh C-T, Lin C-Y. Tumor-derived chemokine CCL5 enhances TGF-β-mediated killing of CD8(+) T cells in colon cancer by T-regulatory cells. Cancer Res 2012;72(5):1092–102.

[136] Curiel TJ, Coukos G, Zou L, Alvarez X, Cheng P, Mottram P, Evdemon-Hogan M, Conejo-Garcia JR, Zhang L, Burow M, Zhu Y, Wei S, Kryczek I, Daniel B, Gordon A, Myers L, Lackner A, Disis ML, Knutson KL, Chen L, Zou W. Specific recruitment of regulatory T cells in ovarian carcinoma fosters immune privilege and predicts reduced survival. Nat Med 2004;10(9):942–9.

[137] Zhu Y, Knolhoff BL, Meyer MA, Nywening TM, West BL, Luo J, Wang-Gillam A, Goedegebuure SP, Linehan DC, DeNardo DG. CSF1/CSF1R blockade reprograms tumor-infiltrating macrophages and improves response to T-cell checkpoint immunotherapy in pancreatic cancer models. Cancer Res 2014;74(18):5057–69.

[138] Zhang J, Patel L, Pienta KJ. CC chemokine ligand 2 (CCL2) promotes prostate cancer tumorigenesis and metastasis. Cytokine Growth Factor Rev 2010;21(1):41–8.

[139] Linehan D, Noel MS, Hezel AF, Wang-Gillam A, Eskens F, Sleijfer S, Desar IME, Erdkamp F, Wilmink J, Diehl J, Potarca A, Zhao N, Miao S, Deng J, Hillson J, Bekker P, Schall TJ, Singh R. Overall survival in a trial of orally administered CCR2 inhibitor CCX872 in locally advanced/metastatic pancreatic cancer: correlation with blood monocyte counts. J Clin Oncol 2018;36(5_suppl):92.

[140] Noel MS, Hezel AF, Linehan D, Wang-Gillam A, Eskens F, Sleijfer S, Desar I, Erdkamp F, Wilmink J, Diehl J, Potarca A, Zhao N, Deng J, Lohr L, Miao S, Charo I, Singh R, Schall TJ, Bekker P. Orally administered CCR2 selective inhibitor CCX872-b clinical trial in pancreatic cancer. J Clin Oncol 2017;35(4_suppl):276.

[141] Nywening TM, Wang-Gillam A, Sanford DE, Belt BA, Panni RZ, Cusworth BM, Toriola AT, Nieman RK, Worley LA, Yano M, Fowler KJ, Lockhart AC, Suresh R, Tan BR, Lim K-H, Fields RC, Strasberg SM, Hawkins WG, DeNardo DG, Goedegebuure SP, Linehan DC. Targeting tumour-associated macrophages with CCR2 inhibition in combination with FOLFIRINOX in patients with borderline resectable and locally advanced pancreatic cancer: a single-centre, open-label, dose-finding, non-randomised, phase 1b trial. Lancet Oncol 2016;17(5):651–62.

[142] Bergmann C, Strauss L, Wang Y, Szczepanski MJ, Lang S, Johnson JT, Whiteside TL. T regulatory type 1 cells in squamous cell carcinoma of the head and neck: mechanisms of suppression and expansion in advanced disease. Clin Cancer Res 2008;14(12):3706–15.

[143] Cao X, Cai SF, Fehniger TA, Song J, Collins LI, Piwnica-Worms DR, Ley TJ. Granzyme B and perforin are important for regulatory T cell-mediated suppression of tumor clearance. Immunity 2007;27(4):635–46.

[144] Deaglio S, Dwyer KM, Gao W, Friedman D, Usheva A, Erat A, Chen J-F, Enjyoji K, Linden J, Oukka M, Kuchroo VK, Strom TB, Robson SC. Adenosine generation catalyzed by CD39 and CD73 expressed on regulatory T cells mediates immune suppression. J Exp Med 2007;204(6):1257–65.

[145] Jacobs JFM, Punt CJA, Lesterhuis WJ, Sutmuller RPM, Brouwer HM-LH, Scharenborg NM, Klasen IS, Hilbrands LB, Figdor CG, de Vries IJM, Adema GJ. Dendritic cell vaccination in combination with anti-CD25 monoclonal antibody treatment: a phase I/II study in metastatic melanoma patients. Clin Cancer Res 2010;16(20):5067–78.

[146] Rech AJ, Mick R, Martin S, Recio A, Aqui NA, Powell DJ, Colligon TA, Trosko JA, Leinbach LI, Pletcher CH, Tweed CK, DeMichele A, Fox KR, Domchek SM, Riley JL, Vonderheide RH. CD25 blockade depletes and selectively reprograms regulatory T cells in concert with immunotherapy in cancer patients. Sci Transl Med 2012;4(134):134ra62.

[147] Sugiyama D, Nishikawa H, Maeda Y, Nishioka M, Tanemura A, Katayama I, Ezoe S, Kanakura Y, Sato E, Fukumori Y, Karbach J, Jäger E, Sakaguchi S. Anti-CCR4 mAb selectively depletes effector-type FoxP3+CD4+ regulatory T cells, evoking antitumor immune responses in humans. Proc Natl Acad Sci U S A 2013;110(44):17945–50.

[148] Selby MJ, Engelhardt JJ, Quigley M, Henning KA, Chen T, Srinivasan M, Korman AJ. Anti-CTLA-4 antibodies of IgG2a isotype enhance antitumor activity through reduction of intratumoral regulatory T cells. Cancer Immunol Res 2013;1(1):32–42.

[149] Weinberg AD, Rivera MM, Prell R, Morris A, Ramstad T, Vetto JT, Urba WJ, Alvord G, Bunce C, Shields J. Engagement of the OX-40 receptor in vivo enhances antitumor immunity. J Immunol 2000;164(4):2160–9.

[150] Piconese S, Valzasina B, Colombo MP. OX40 triggering blocks suppression by regulatory T cells and facilitates tumor rejection. J Exp Med 2008;205(4):825–39.

[151] Ko K, Yamazaki S, Nakamura K, Nishioka T, Hirota K, Yamaguchi T, Shimizu J, Nomura T, Chiba T, Sakaguchi S. Treatment of advanced tumors with agonistic anti-GITR mAb and its effects on tumor-infiltrating Foxp3+CD25+CD4+ regulatory T cells. J Exp Med 2005;202(7):885–91.

[152] Shimizu J, Yamazaki S, Takahashi T, Ishida Y, Sakaguchi S. Stimulation of CD25(+)CD4(+) regulatory T cells through GITR breaks immunological self-tolerance. Nat Immunol 2002;3(2):135–42.

CHAPTER

7

Overcoming head and neck cancer stem cells

Alexandra E. Oklejas[a], *Jacques E. Nör*[a,b,c,d]

[a]Department of Cariology, Restorative Sciences, Endodontics, University of Michigan School of Dentistry, Ann Arbor, MI, United States [b]Department of Otolaryngology, University of Michigan School of Medicine, Ann Arbor, MI, United States [c]University of Michigan Comprehensive Cancer Center, Ann Arbor, MI, United States [d]Department of Biomedical Engineering, University of Michigan College of Engineering, Ann Arbor, MI, United States

Abstract

Head and neck cancer is a frequent and deadly malignancy. Despite significant advances in the understanding of the pathobiology of head and neck cancer, patient morbidity associated with invasive treatment and the high frequency of tumor recurrence/metastasis typically result in an unacceptably short patient survival and poor quality of life. The cancer stem cell hypothesis attempts to explain the observed heterogeneity of cancer cells within a tumor, with many malignant features of a cancer cell deriving from a shift towards stem-like features. Cancer stem cells function as drivers of tumor initiation, mediate therapeutic evasion, and promote recurrence and metastases of head and neck cancer. Notably, cancer stem cells are highly resistant to conventional chemotherapy and radiotherapy. Therefore, cancer stem cells represent an attractive target for developmental therapeutics. In this chapter, we discuss cancer stem cell-mediated resistance to treatment and approaches that might sensitize patients to therapy by targeting these cells.

Abbreviations

ACC	adenoid cystic carcinoma
ALDH	aldehyde-dehydrogenase
CAF	cancer-associated fibroblast
CSC	cancer stem cell
EV	extracellular vesicle
FACS	fluorescence-activated cell sorting
HNSCC	head and neck squamous cell carcinoma
IGF	insulin-like growth factor

Improving the Therapeutic Ratio in Head and Neck Cancer
https://doi.org/10.1016/B978-0-12-817868-3.00007-X

IGFBP7	insulin-like growth factor binding protein 7
IL-6	interleukin-6
LDA	limiting dilution assay
MAST1	microtubule-associated serine/threonine kinase 1
MFG-E8	milk-fat globule-epidermal growth factor-VIII
NCCN	national comprehensive cancer network
PD-L1	programmed death-ligand 1
PDX	patient-derived xenograft
PNI	perineural invasion
PRC1	polycomb repressive complex 1
PTK7	protein tyrosine kinase 7
SDF-1	stromal cell-derived factor-1
STAT3	signal transducer and activator of transcription 3
TEC	tumor endothelial cell

Conflict of interest

No potential conflicts of interest were disclosed.

Introduction

Key malignant features of a cancer cell have been shown to derive from the acquisition of stem-like features [1]. Indeed, we will discuss here the scientific evidence that suggests that cancer stem cells (CSC) define the pathobiology and the response to treatment of head and neck squamous cell carcinomas (HNSCC) [2–4]. Cancer stem cells represent a self-renewing, multipotent, and highly tumorigenic subpopulation of cancer cells generally ranging from 1 to 10% of the overall tumor cell population in HNSCC. These cells have been shown to engage in a cellular crosstalk within the dynamic tumor microenvironment, which plays a major role in governing their maintenance and cell fate. The preferred chemotherapeutic treatment for HNSCC remains platinum-based agents, Cisplatin being the most commonly used. Though their survival benefit is well recognized, Cisplatin-based treatment regimens are associated with substantial patient morbidity [5] and disease relapse, with locoregional recurrence accounting for 20–40% of HNSCC mortality of patients at 5years [6]. Cancer stem cells have been shown to be highly resistant to conventional chemotherapy [7] and radiation [8], which may account for the fact that many head and neck cancer patients experience tumor recurrence. Advancements in the knowledge of cancer stem cell biology have provided a scientific rationale for the targeted elimination of the cancer stem cell pool to benefit the survival and quality of life of patients with head and neck malignancies. Here, we will review the evidence on the biology of head and neck cancer stem cells and will discuss their therapeutic targeting as a means to overcome treatment resistance and disease progression.

The cancer stem cell hypothesis

Over the past decade, the idea of stem-like cells within tumors has received significant attention. It has now been accepted that not all tumor cells are equal, and that a tumor is vastly heterogenous. The CSC hypothesis postulates that among the bulk of neoplastic cells, cancer stem cells make up a small fraction of cells endowed with characteristics of self-renewal, multipotency, and uniquely high tumorigenic potential [1]. These characteristics are demonstrated by (1) the unique ability of CSCs to initiate and propagate tumors in vivo, (2) the observation that the resulting tumors consist of a heterogeneous population that recapitulates the CSC's original tumor, and (3) the capability of these tumors to be serially transplanted [9]. On the other hand, the bulk of the tumor is composed of rapidly proliferating transit amplifying cells and postmitotic differentiated cells, both of which are not capable of initiating tumors. These cells lack longevity and have only limited proliferative potential, as opposed to the cancer stem cell population [1].

A principal feature of stem cells is their ability to recapitulate a heterogeneous organ from a single cell. Rather than proposing a tumor arises from normal stem cells, the CSC hypothesis attempts to explain the observation that a tumor may be hierarchically organized in much the same manner as physiological tissues [10]. Similar to their physiological counterparts, cancer stem cells share the attributes of self-renewal and the capacity to differentiate into committed progenitor cells. However, a distinguishing feature of cancer stem cells is their uniquely high tumorigenic potential, making these cells a crucial part of tumor initiation and disease progression to metastasis and/or recurrence [1]. In parallel to the functional role of stem cells in physiological tissues, the relatively small number of cancer stem cells fuels tumor growth and gives rise to the bulk tumor cells. The bulk tumor cells are not able to self-renew and do not contribute to the long-term viability of the tumor, therefore becoming less relevant in the grand scheme of disease progression and treatment [11]. Unfortunately, the line between what is a cancer stem cell and what is a bulk tumor cell is blurred. Cancer stem cells maintain plasticity to transition between epithelial- and mesenchymal cell stages, with molecular and environmental cues shifting the equilibrium of stem-like versus nonstem-like characteristics [12].

The CSC hypothesis is especially appealing due to the explanation it provides for several poorly understood clinical phenomena. A patient may present a robust response to chemo- and radiation therapy in the form of tumor shrinkage and disease remission. However, a small fraction of cancer stem cells survives these treatment attempts. Thus, the inevitable outcome of seemingly effective treatments is oftentimes local recurrence or metastatic spread. Cancer stem cells are resilient cells that can maintain their population through self-renewal, and their quiescent nature allows them to evade the majority of chemotherapeutic approaches. Cancer stem cells are highly migratory to distant sites within the body, where they can initiate a new tumor and lead to metastasis, even long after the patient was initially treated. In clinical assessments, the CSC content within tumors has been shown to be a predictive factor [13] and to define the tumor's heterogenous phenotypical and molecular traits [14]. Therefore, cancer stem cells present an enticing avenue for research into more effective therapeutic developments, as the goal is to target the "warhead" of the tumor instead of simply eradicating the, by comparison, harmless tumor bulk.

Identifying head and neck cancer stem cells

The identification of cancer stem cells is an area of ongoing research in many different cancer types. Cancer stem cell identification relies predominantly on the expression of cell surface molecules or intracellular enzymes, which can vary depending on the cancer. Flow cytometry/fluorescence-activated cell sorting (FACS) remains the most commonly used technique to identify and isolate cancer stem cells based on cell-specific markers.

In HNSCC, cancer stem cells were first identified as $CD44^+$ expressing cells by Prince and colleagues [2]. CD44 is a cell surface adhesion receptor involved in cell-to-cell interaction, adhesion, and migration, which can bind to hyaluronic acid (among other extracellular matrix ligands). The small fraction of $CD44^+$ cells flow-sorted from primary HNSCC tumor samples was shown to be highly tumorigenic when implanted into immunodeficient mice, as compared to the $CD44^-$ cancer cells [2]. Interestingly, the $CD44^+$ cancer cells co-stained with epithelial basal cell marker Cytokeratin 5/14, which is highly expressed in mitotically active basal stem cells and downregulated as these cells differentiate [15]. The tumors generated from $CD44^+$ cells reproduced a heterogenous tumor expressing both $CD44^+$ and $CD44^-$ cells, which demonstrated the multipotency of these cells. Gene expression analysis also revealed differential expression of the self-renewal marker Bmi-1 in $CD44^+$ versus $CD44^-$ cells. This work was among the first to unveil a subpopulation of cancer cells in HNSCC that follow the principles put forth by the CSC hypothesis. It has since then been shown that CD44 is highly expressed in several stem cells and many cancers, in which it helps promote the migration and invasion processes involved in metastasis [16]. However, though CD44 expression has been shown to select for a subset of cells with stem-like characteristics, it is not sufficient to distinguish between normal, benign, or malignant epithelia of the head and neck [17].

Another widely used marker for cancer stem cell identification is aldehyde dehydrogenase (ALDH), which has emerged as a rather universal CSC marker. The aldehyde dehydrogenase family of enzymes is crucial for maintenance, differentiation, and normal development of stem cells, and functions in cellular detoxification, retinoic acid metabolism and signaling, and protection from reactive oxygen species [18]. Increased ALDH activity was first shown to isolate cancer stem cells in leukemia, where $ALDH^{high}$ cells demonstrated superior tumorigenic potential as compared to $ALDH^{low}$ cells, and were co-expressed with the leukemia stem cell marker CD34 [19]. Concurrently in the breast cancer field, the Wicha laboratory identified high ALDH activity as a marker for a subpopulation of tumor cells characterized by multipotency, self-renewal, and enhanced tumorigenicity [20]. It was a few years later that the role of ALDH was first determined in the context of head and neck cancer stem cells, where Chen and colleagues demonstrated $ALDH^{high}$ HNSCC cells displayed evidence of epithelial-mesenchymal transition (EMT) shifting, radioresistance, and high tumorigenicity as compared to $ALDH^{low}$ cells in a xenotransplanted tumorigenicity study [21]. Further, it was shown that the $ALDH^{high}$ and $CD44^+$ cells purified an even more tumorigenic population, with 1000 $ALDH^{high}CD44^+$ cells capable of initiating more tumors in vivo (i.e., in 13 mice out of 15), as compared to 10,000 $ALDH^{low}$ $CD44^-$ cells (i.e., in 2 mice out of 15) [22]. In another study, tumors generated from $ALDH^{high}$ cells were able to be serially passaged in vivo while reproducing the original tumor heterogeneity, thus confirming ALDH as a highly selective marker for cancer stem cells in HNSCC [23]. Currently, ALDH activity is being successfully used as a cancer stem cell marker in many cancers including melanoma [24], colon [25], lung [26], prostate [27], liver [28], pancreatic [29], and salivary gland cancers [30] among many others.

The increasing evidence for the use of both CD44 and ALDH as cancer stem cell markers in HNSCC led to studies in which the combined expression of both markers further enhanced the ability to isolate cancer stem cells. $ALDH^{high}CD44^+$ cells were confirmed to be highly tumorigenic, capable of self-renewing as orospheres, and were able to be serially passaged in vitro and in vivo while reproducing the original tumor histology and heterogeneity. These studies also demonstrated that the $ALDH^{high}CD44^+$ subpopulation of cells localized in close proximity to blood vessels, suggesting the existence of a perivascular cancer stem cell niche in HNSCC [22]. In head and neck cancer patients, CD44 and ALDH expression/activity in tumors have been correlated with worse prognosis, disease grade, and overall survival [31–33]. The prognostic significance of ALDH and CD44 suggests that they could be perhaps considered as potential diagnostic tools for patients [34].

In addition to CD44 and ALDH, a large cohort of other potential cancer stem cell markers have been identified in HNSCC. One of these includes the transmembrane glycoprotein CD133, which has been implicated as a potential marker for cancer stem cells in HNSCC. $CD133^+$ cells showed a marked capacity for self-renewal and multilineage differentiation in vitro [35], as well as correlated with worse patient survival when co-expressed with Oct-4 and Nanog [36]. In Cisplatin-resistant HNSCC cells, increased $CD10^+$ expression was shown to enhance sphere formation in vitro and tumorigenicity in vivo, and may be associated with therapeutic resistance and CSC-like properties of these tumors [37]. In yet another study, $CD166^+$ stem-like HNSCC cells were shown to be significantly related to poorer clinical outcome, as compared to the correlation of CD44 [38]. There have been a plethora of other efforts to identify CSC-specific biomarkers in HNSCC, including more functional approaches such as dye-dependent isolation [39] or measuring intracellular reactive oxygen species [40] to enrich for cancer stem cells.

CSC functional assays

The perfect CSC-specific marker, i.e., only expressed in cancer stem cells, has not been identified. This complicates the attempt to isolate absolutely pure populations of cancer stem cells and to perform CSC-targeted developmental therapeutics studies, as the majority of cancer stem cell markers are also expressed in physiological stem cells [41]. Efforts to develop CSC-specific biomarkers could also provide insight into whether these markers are only useful for identifying this subpopulation or whether they also have a functional role in modulating the cancer stem cell phenotype. Other methods that allow enrichment and investigation of cancer stem cells that are independent of CSC biomarkers include functional cancer stem cell assays, which utilize the abilities of cancer stem cells to self-renew, differentiate into other cell types, and generate new tumors.

For example, to study the stem-like phenotype in vitro, the orosphere assay is a method which exploits the fact that stem cells are capable of surviving and proliferating in low/no serum and anchorage-independent conditions, unlike differentiated cells [42]. Culturing cancer cells in serum-free ultra-low attachment conditions allows only cancer stem cells to survive in the form of spheres, and their passaging is an indication of self-renewal [43]. Head and neck cancer cells cultured as orospheres exhibited high expression/activity of ALDH and CD44 [42], were more radio- and chemo-resistant, and demonstrated enhanced tumorigenic potential in vivo [44]. The stem-like phenotype can also be investigated in vivo, where FACS-sorted CSCs and non-CSCs can be transplanted into immunodeficient mice by a limiting dilution assay (LDA) and serially passaged [45]. This in vivo LDA is a common approach for cancer stem cell research and is used to study differences in tumorigenic potential, self-renewal, and differentiation capacity of cancer stem cells in vivo, with orosphere assay complementing this method in vitro. Most recently, cancer stem cells have been studied in 3D tumor organoids, which can recapitulate the in vivo architecture and functionality of the original tumor microenvironment and have been shown to enable stable culture of the cancer stem cell population [46]. This method can provide a more biologically representative in vitro assay to study metabolic effects of a variety of anticancer therapies [47].

The cancer stem cell niche

Although there are significant efforts in place to discover CSC-specific biomarkers, the ones identified so far cannot be therapeutically targeted. However, this prompts other therapeutic approaches to be attractive, such as targeting signaling pathways that are active in cancer stem cells specifically. Because physiological stem cells and cancer stem cells share many attributes, it seems reasonable that many pathways that regulate normal stem cell development and function may be hijacked by cancer cells. Similarly, this suggests that cancer stem cells depend on their immediate microenvironment for cues that regulate their self-renewal and differentiation just as normal stem cells do [48].

Stem cells are typically found in perivascular niches in close proximity to blood vessels, which enable ease of access to signaling cues and provide a dynamic and protective microenvironment [22,49]. In this niche, stem cells can engage in a crosstalk with stromal cells to maintain their self-renewal and regulate their differentiation. As with many similarities between normal stem cells and CSCs, their choice in residence bares no exception (Fig. 1). The exploration of a CSC niche was ignited in brain cancers, as existing knowledge suggested that endothelial-cell secreted factors maintain neural stem cells [50]. In fact, some of the pioneering research about the CSC niche was performed in brain tumors, where cancer stem cells in glioblastoma multiforme were shown to reside in close proximity to endothelial cells [51].

Head and neck cancer stem cells reside in the invasive fronts within the tumor in perivascular niches, where they engage in crosstalks with stromal cells, endothelial cells, cancer-associated fibroblasts, immune cells, and others [22, 52]. Other important constituents of the CSC niche are comprised by signaling molecules secreted from a variety of cells and the extracellular matrix [53]. Cancer stem cell survival depends on the tumor microenvironment and their location within perivascular niches, which also provides a protective environment [3, 22].

Cell-cell crosstalk

Tumors are often referred to as "wounds that do not heal". It is reasonable to believe then, that similar cell types involved in the response to injury and wound healing may contribute to the pathobiology of a tumor. There is increasing evidence that growth and survival of cancer stem cells are highly influenced by the tumor microenvironment and molecular signaling crosstalk. An important component of solid tumors are nonepithelial stromal cells, which

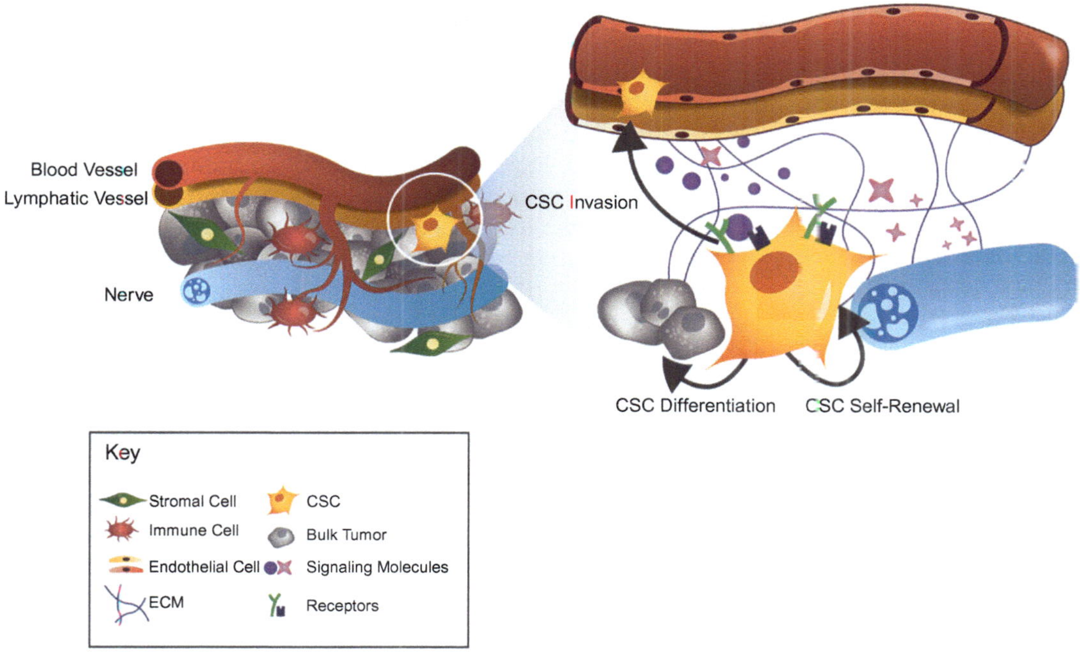

FIG. 1 The cancer stem cell niche. Cancer stem cells reside in a perivascular niche that serves as a protective environment and consists of multiple cell types, including endothelial cells, fibroblasts, immune cells, and nerve cells. Cellular crosstalk between cancer stem cells and stromal cells regulate cancer stem cell self-renewal, differentiation, and invasion. These processes are supported by the cancer stem cell niche and enable dissemination of cancer stem cells, ultimately leading to metastasis. Interfering with environmental cues that sustain the cancer stem cell niche may allow targeted elimination of cancer stem cells.

include vascular endothelial cells, fibroblasts, and inflammatory cells that together play a key role in tumor growth and pathobiology through either cell-cell contact or secreted signaling effectors [54]. The stromal cell component in HNSCC tumors has also been shown to contribute to invasion and metastasis [55].

Endothelial cells

The close proximity between cancer stem cells and endothelial cells of blood vessels in the perivascular niche enables their active crosstalk. It has been demonstrated that the crosstalk between CSCs and endothelial cells through endothelial cell-secreted factors enhances cancer stem cell self-renewal and survival [22, 56, 57]. One of the most potent CSC-inducing endothelial cell-secreted factors is interleukin-6 (IL-6), an important inflammatory cytokine. IL-6 specifically enhances the tumorigenic potential and self-renewal of cancer stem cells via JAK/STAT3 signaling [58, 59]. It has also been shown that the $ALDH^+CD44^+$ head and neck cancer cells express higher levels of both IL-6 co-receptors (IL-6R and gp130) than non-CSCs [59].

Other endothelial cell-secreted candidate targets have been shown to play a role in cancer stem cell maintenance [60]. Endothelial-cell secreted EGF was shown to induce epithelial to mesenchymal transition (EMT) and promote the stem-like phenotype in HNSCC cancer cells [58]. Orosphere formation and cell motility was enhanced by EGF in vitro, and silencing of EGF expression in co-transplanted endothelial cells decreased the fraction of cancer stem cells in tumor xenografts engineered in vivo. Furthermore, conditioned media from primary human endothelial cells protected head and neck cancer stem cells from anoikis [57]. Collectively, these findings suggest that endothelial-cell secreted factors within the perivascular niche maintain the cancer stem cell population and induce a more stem-like phenotype in HNSCC cells.

Cancer-associated fibroblasts

Along with other cell types of the CSC niche, cancer-associated fibroblasts (CAFs) activate pathways that play major roles in cancer cell stemness and secrete growth factors, chemokines, and extracellular matrix components involved in angiogenic recruitment of other stromal cells [61]. Given the physiological roles of fibroblasts, CAFs function as builders of the tumor microenvironment and regulators of cell differentiation, inflammation, and tumor progression. Stromal myofibroblasts were shown to activate the canonical Wnt signaling in colorectal cancer stem cells. The cancer stem cells with high Wnt activity were located in close proximity to these fibroblasts. Myofibroblast-secreted factors were also capable of restoring the stem-like phenotype in more differentiated cancer cells [62]. In lung cancer stem cells, IGF1R signaling activation by CAF-secreted IGF-II was shown to be necessary to sustain cancer cell stemness [63]. In HNSCC, fibroblasts recruit cancer stem cells to supportive niches through secretion of stromal cell-derived factor-1 (SDF-1) and stimulate podia formation in cancer cells [64]. The SDF-1/CXCR4 signaling axis has been shown to play an important role in tumor metastasis in various cancers [65, 66]. Periostin secreted by CAFs has also been linked to metastasis, and in HNSCC was shown to promote the stem-like phenotype via the protein tyrosine kinase 7 (PTK7)-Wnt/β-catenin signaling, as well as induce tumor initiation and invasion in vivo [67]. These findings signify the role of CAFs within the tumor microenvironment in supporting the CSC niche.

Immune cells

One of the hallmarks of cancer is the ability of tumor cells to evade destruction by the immune system [68]. The immune system interacts with cancer cells before and after clinical detection of a tumor and engages in a process called tumor immunoediting, which consists of three phases: elimination, equilibrium, and escape [69]. During elimination, most of the cancer cells are recognized and destroyed by the immune system. However, some cancer cells manage to escape this immunosurveillance, leading to a phase of equilibrium between tumor growth and immunological elimination. Cancer stem cells have been shown to have advantages in evading immune detection and elimination, thereby augmenting their tumorigenic ability, ultimately leading to the escape phase of tumor immunoediting [70]. In HNSCC, the

T-cell inhibitory molecule PD-L1 has been shown to be preferentially expressed on CD44− tumor-initiating cells, which suppressed T-cell mediated immunity and supported an EMT phenotype of the cancer stem cells [71, 72]. Likewise, head and neck cancer stem cells downregulate various immunogenic components and can inhibit proliferation of T-cells, regulatory T-cells, and myeloid-derived suppressor cells [73]. Immune cells can also help sustain the stem-like phenotype, as MFG-E8 and IL-6 secreted by tumor-associated macrophages have been shown to enhance cancer stem cell tumorigenicity and drug resistance [74]. Although the immune system does not primarily play a CSC-supporting role in the tumor microenvironment, it plays a crucial role in allowing cancer stem cells to overcome immunosurveillance of tumor growth. This presents another fascinating crosstalk to be investigated for sensitization towards eliminating cancer stem cells through targeted activation of host immune defenses.

Neural cells

Lastly, the nervous system has been shown to have direct effects on proliferation patterns of epithelial stem cells and plays a role in various contexts of tissue development and regeneration [75]. The stimulatory role of nerves is not restricted to physiological tissues, as denervation of primary tumors has been shown to suppress tumor growth and metastasis, suggesting a function for nerves in cancer progression [76, 77]. This denervation was linked with suppression of stem cell expansion and tumorigenesis, as well as inhibition of Wnt signaling [76]. In HNSCC, a high incidence of perineural invasion (PNI) is associated with tumor recurrence and poor survival [78], as nerves present a significant route of tumor metastasis [79]. As the nerve-tumor distance decreases in human biopsy specimens, the estimated relative death rate increases [80]. Nerves in HNSCC tumors have been shown to secrete a variety of neurotrophic factors, adhesion molecules, and chemokines that function in crosstalk with cancer cells, which also attract nerve fibers [81]. A recent study has shown that human cancer stem cells from gastric and colorectal carcinoma can differentiate into functional neurons that are involved in tumor neurogenesis and progression in tumor xenograft models [82].

We can gather from these findings investigating the CSC niche that the tumor microenvironment is highly complex, which may contribute to the functional diversity of cancer cells (Fig. 1). Although there are many players in these interactions between cancer stem cells and stromal cells, an underlying common ground can be found in the signaling pathways that are employed by the various cell types.

CSC signaling pathways

In addition to studying the cellular crosstalks between cancer stem cells and other cells in the tumor microenvironment, a pathway-oriented perspective can be taken to understand cancer stem cell biology. Elucidating basic molecular mechanisms can provide insight into what governs phenotypic changes within the cancer cell population that lead to the acquisition of this stem-like phenotype.

In epidemiological studies, high levels of pretreatment serum IL-6 and tumor IL-6 receptor (IL-6R) expression correlate with poor survival of HNSCC patients and a higher rate of tumor recurrence, which propounds the use of IL-6 as a marker to predict survival and risk of recurrence in HNSCC [41, 83]. IL-6 was found to regulate head and neck cancer stem cell function and induces their self-renewal, as shown by an increase in expression of the self-renewal marker Bmi-1 [7]. There is a direct correlation between levels of IL-6 secreted by tumor-associated endothelial cells and the tumorigenicity of cancer stem cells in vivo [59]. Endothelial cell-secreted IL-6 enhances orosphere formation, STAT3 activation, and self-renewal of cancer stem cells in vitro. Activation of STAT3 plays an important role in the self-renewal and stemness of a variety of normal stem cells and cancer stem cells [84, 85]. STAT3 is constitutively phosphorylated in head and neck cancer stem cells ($ALDH^{high}CD44^{+}$), as compared to the $ALDH^{low}CD44^{-}$ population [59]. This STAT3 expression is enhanced by endothelial cell-secreted IL-6 and EGF, and its inhibition reduces the sphere-forming ability of cancer stem cells [57, 59]. In a study investigating cancer stem cell interactions with T-cell-mediated immunity, constitutive phosphorylation of STAT3 was observed in $CD44^{+}$ cells that selectively expressed PD-L1 [86]. Activation of STAT3 in head and neck cancer overall has been implicated in treatment resistance and immune escape [87].

The protein Bmi-1 is an important transcriptional repressor and component of the polycomb repressive complex 1 (PRC1). It is highly expressed in self-renewing normal stem cells [88, 89], in which it inhibits cell cycle regulators $p16^{Ink4a}$ and $p19^{Arf}$ to regulate various important stem cell functions [90, 91]. In HNSCC, Bmi-1 has been shown to be highly expressed specifically in cancer stem cells, and that this effect is enhanced by endothelial cell-secreted factors [22]. The importance of Bmi-1 in the maintenance of cancer stem cells has also been implicated [2], as $Bmi\text{-}1^{+}$ head and neck cancer cells express the known cancer stem cell markers ALDH and CD44 [92]. Currently no mechanistic studies have determined the regulation of Bmi-1 in head and neck cancer stem cells, but a body of literature has demonstrated IL-6/STAT3 signaling to orchestrate epithelial-mesenchymal transition (EMT) in breast cancer through activation of Bmi-1 by transcription factor Twist [93, 94], thus conferring tumors with migratory and self-renewal abilities.

A variety of other signaling pathways have been implicated in cancer stem cell biology in head and neck cancers specifically. For example, Wnt signaling serves a vital role in normal stem cell regulation, especially during embryonic development, but also in adult tissue self-renewal and cancer [95]. In HNSCC, Wnt/β-catenin signaling was shown to be crucial in the maintenance of cancer stem cell self-renewal and tumorigenicity [96]. In these studies, β-catenin was co-expressed with ALDH and CD44 staining, suggesting that this signaling axis is primarily restricted to cancer stem cells. They demonstrated that β-catenin positively regulated sphere formation and expression of the stem cells markers Oct4 and Sox2, as well as chemoresistance to Cisplatin. Other signaling pathways that have been shown to play significant roles in head and neck cancer stem cells include the Notch and Hedgehog pathways [97, 98].

Many of these pathways regulate common downstream embryonic stem cell transcription factors that are relevant in cancer stem cell function. Octamer-binding transcription factor 4 (Oct4) regulates many genes in embryonic stem cell development and cancer, and has been linked to a less differentiated phenotype, increased tumor aggressiveness, and resistance to radiation therapy in HNSCC [99]. Ectopic expression of Oct4 promotes the stem-like phenotype, sphere formation, invasion capacity, and xenograft tumorigenicity in head and neck cancer [100]. Nanog is another transcription factor required for the maintenance of embryonic stem cells [101], and has been recently implicated as a prognostic cancer stem cell marker in HNSCC [102]. Another major downstream signaling effector is the protein p53. One of the most frequent genomic mutations in HNSCC is within the *TP53* gene, which has been correlated with poor patient survival [103, 104]. Normally, p53 functions to repair DNA damage by either transiently arresting the cell cycle or irreversibly triggering apoptosis. In both normal and cancer stem cells, p53 is involved in mechanisms of self-renewal and differentiation [105, 106], specifically affecting stem cell expansion [107]. MDM2 functions as a major negative regulator of p53 by targeting it for proteasomal degradation and has also been shown to independently drive stemness characteristics [108, 109].

Epithelial-mesenchymal transition and cell plasticity

When discussing the acquisition of stem-like features, the physiological phenomenon of epithelial-mesenchymal transition has been strongly associated with cancer progression, endowing cancer cells with enhanced migratory characteristics and invasive ability [110]. Induction of EMT in normal epithelial cells has been shown to transform these cells into cancer stem cells capable of forming spheres in vitro and tumors in preclinical models of breast cancer in vivo [111]. In HNSCC, EGF and Snail were shown to induce EMT and maintain the cancer stem cell phenotype [57, 112]. Endothelial cell-derived IL-6 has also been shown to induce EMT, motility, and invasive capacity of head and neck cancer stem cells by creating a chemotactic gradient towards blood vessels [41]. EMT induced by IL-6 converted nonmetastatic tumors into metastatic ones in vivo, also through STAT3/Snail pathway activation [113].

These studies incite the idea of cancer cell plasticity and the conversion of bulk cancer cells to cancer stem cells, or vice versa. This leads to consideration of how these signaling pathways in cancer stem cell biology might influence fate decisions during cell division. Drawing another parallel to physiological tissues, the regulation of cell fate in cancer stem cells is a homeostatic balance of self-renewal and cell differentiation. This becomes apparent through one of many examples in which a pure population of sorted cancer stem cells will revert to the same percentage of CSCs versus non-CSCs prior to the sort [23]. The decision that cancer stem cells face between self-renewal and differentiation into bulk tumor cells comes down to a single cycle of mitosis. Cancer stem cell divisions can be symmetrical, creating two self-same daughter cells, or asymmetrical, creating one cancer stem cell and one differentiated cell [114]. Cues from the tumor microenvironment can affect these decision processes [85, 115]. The signaling pathways implicated in connecting cancer cell plasticity to EMT transitions have been reviewed elsewhere [12, 116]. Collectively, these findings support the roles of various signaling pathways in head and neck cancer stem cells. Many pathways share common downstream effectors, which may provide opportunities for investigating potential therapeutic targets.

Therapeutic resistance

Head and neck squamous cell carcinoma (HNSCC) is the sixth most-common solid tumor [117]. Although multiple treatment modalities exist, including surgical resection, radiation therapy, and/or chemotherapy, they frequently are associated with unacceptably poor patient survival and low quality of life [118]. Unfortunately, the improved control of tumor-growth achieved with platinum-based drugs correlates with an increased incidence of evasive resistance and recurrence of HNSCC [119]. Although platinum-based drugs have improved control of local disease [120], increases in incidence of resistance and metastasis have essentially nullified their benefits [121].

Evasive resistance is a process by which cancer cells acquire an invasive phenotype to escape the unfriendly tumor microenvironment generated by chemotherapy [119]. Clinical observations suggest that some tumor cells are resistant to chemotherapy and have the capability to migrate and initiate a new tumor (Fig. 2). Indeed, traditional platinum-based chemotherapy induces phenotypic changes in cancer cells, suggesting that conventional therapy causes a shift in the tumor cell population towards self-renewal [122–124]. This primes residual HNSCC cells for a more aggressive phenotype and leads to therapeutic resistance and disease progression [125]. Cancer stem cells have been shown to drive tumor initiation, therapeutic evasion and resistance, metastasis, and recurrence in HNSCC [1, 2, 57, 126]. Due to their defining characteristics of self-renewal, multipotency, and tumorigenic potential, it is not surprising that cancer stem cells are capable of reforming an entire heterogenous tumor

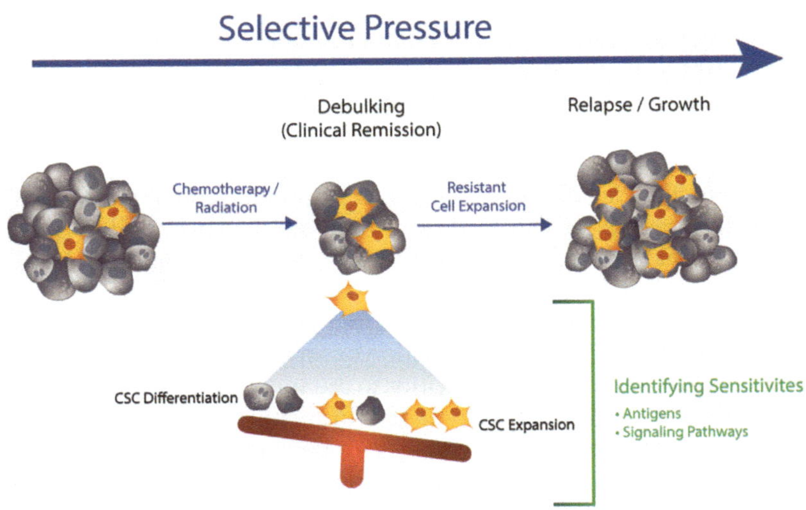

FIG. 2 Conventional anticancer therapies successfully debulk tumors, leading to clinical remission. However, cancer stem cells are resistant, leading to their expansion within the tumor and acquisition of a more aggressive disease phenotype and therapeutic evasion. Cancer stem cell proliferation can occur symmetrically or asymmetrically, leading to their expansion, maintenance, or differentiation. These cell division events are regulated by signaling pathways that can be targeted therapeutically to shift the scale towards cancer stem cell differentiation, which supports their sensitization to conventional anticancer therapies.

posttreatment. Not surprisingly, many of the same pathways that sustain the cancer stem cell niche also play a role in CSC-mediated therapeutic resistance.

Cisplatin is considered the standard of care in the treatment of HNSCC patients (NCCN Practice Guidelines in Oncology-Head Neck Cancer). Similar to other cytotoxic chemotherapies and platinum-based agents, Cisplatin forms DNA crosslinks to trigger apoptosis in dividing cells. Both Cisplatin and radiotherapy target actively dividing, rapidly-proliferating cells. Cancer stem cells have been shown to be slow-cycling cells, which may explain their resistance to conventional chemotherapeutics and radiation therapy, as both target actively dividing cells [127, 128]. Cisplatin causes a shift in the head and neck cancer cell population towards more stem-like features [7, 122, 129], but there is no well-defined mechanistic evidence demonstrating how this cancer stem cell increase occurs. Cisplatin has been shown to increase the cancer stem cell fraction and induce stem cell self-renewal in a dose-dependent manner, as determined by Bmi-1 expression [7]. When comparing Cisplatin-resistant with Cisplatin-sensitive head and neck cancer cells, the resistant cells show a marked increase in expression of stemness markers, particularly in Oct4, Nanog, and Bmi-1 [7, 122].

These findings suggest that the cancer stem cell increase postchemotherapy is due to preferential killing of non-CSCs while simultaneously supporting symmetric division of surviving cancer stem cells (Fig. 2). Both a (1) numerator and (2) denominator effect can be responsible for this shift in balance towards cancer stem cell expansion. In (1), conventional chemotherapy may kill the bulk tumor cells, but simultaneously induce self-renewal mechanisms that support cell plasticity or symmetric division of cancer stem cells. In (2), conventional chemotherapy may kill the bulk tumor cells and decrease the total tumor volume, whereas the cancer stem cells survive and then compose a greater fraction of total cells. The evidence provided in the literature supports both of these hypotheses, with the biologically observed effect likely being due to a combination of these factors.

The crosstalk between endothelial cells and cancer stem cells has been shown to play a crucial role in HNSCC resistance to conventional chemotherapy [130–132]. The IL-6 pathway is activated upon Cisplatin exposure [130] and potentiates the Cisplatin-induced increase in the fraction of $ALDH^{high}CD44^{+}$ and Bmi-1 expression [7]. In Cisplatin-resistant versions of HNSCC parental cell lines, IL-6 was shown to enhance their resistance to Cisplatin in vitro. The resistant cell lines express higher levels of Bmi-1 and pSTAT3, and have higher sphere-forming ability compared to parental cells [7]. IL-6 and MFG-E8 secreted in context of tumor-associated inflammation has also been shown to amplify anticancer drug resistance through activating both STAT3 and Hedgehog signaling [133]. A large body of literature has demonstrated that endothelial cells regulate cancer cell aggressiveness. Recently, Cao and colleagues described a mechanism by which tumor-associated endothelial cells (TECs) confer stem-like activity to indolent cancer cells [134]. They show that naïve tumor cells are converted to chemoresistant cancer stem cells through a complex feedback loop between TECs and cancer cells. Normally, insulin growth factor (IGF) binding protein-7 (IGFBP7) expressed by TECs blocks IGF1 signaling to cancer cells, thereby inhibiting their stem-like traits. However, conventional chemotherapy deregulates this balance, inhibiting IGFBP7 and elevating IGF1 in TECs, which activates a positive angiocrine cascade that promotes CSC-mediated invasiveness and progression.

In a recent publication, Jin and colleagues defined MAST1 as the main driver of Cisplatin-resistance across various human cancers [135]. They demonstrated that Cisplatin, but no other DNA-damaging agent, inhibits the MAPK pathway by dissociating cRaf from MEK1. MAST1 replaces cRaf and was shown to contribute to platinum resistance and worse clinical outcome both before and after Cisplatin induction. Additionally, the authors showed that inhibiting MAST1 with the targeted agent Lestaurtinib restores Cisplatin sensitivity. These findings suggest that Cisplatin-resistance may be explained by the fact that platinum-based therapies directly affect major signaling pathways that are also important in cancer stem cell biology, as inactivation of MAPK has been shown to contribute to stem-like properties [136]. Other novel findings have described the wide-ranging effects of cytotoxic chemotherapy in eliciting release of tumor-derived extracellular vesicles (EVs), which have been shown to promote seeding and growth of metastatic cancer cells in distant organ sites in vivo [137]. These chemotherapy-induced EVs were shown to promote NF-kB-dependent activation of endothelial cells, one of the key players in the maintenance of the cancer stem cell niche.

Although therapeutic reduction in tumor size is successful at first, long-term clinical improvement is often not attained. This can be accounted for by chemoresistance of the cancer stem cell subpopulation. Growing knowledge about CSC-associated resistance mechanisms in HNSCC has demonstrated a need for treatment strategies that take into account CSC-driven tumor biology and target these cells specifically. Thus, cancer stem cells constitute an attractive target for developmental therapeutics studies in head and neck cancer.

Targeting cancer stem cells

Targeting cancer stem cells is difficult, as they share many characteristics and markers with normal stem cells and exhibit high drug efflux through ATP-binding cassette transporters [138]. Currently, no HNSCC treatment strategies for patients exist that directly target cancer stem cells despite their important pathological properties and identifiable population. Notably, the goal of CSC-targeted therapy is not necessarily to "shrink tumors." Rather, the aim is to eliminate the fraction of tumor cells ($\sim$5%), which are therapy-resistant cancer stem cells that drive tumor recurrence and HNSCC progression. Therefore, CSC-driven tumor biology requires a combination therapy approach to expose the effects of CSC-targeted treatment on the HNSCC cell population. The increased knowledge about cancer stem cells has sparked an avid interest in developing novel therapeutic strategies that incorporate targeted ablation of cancer stem cells in anticancer treatments. Here, we highlight a select number of preclinical and clinical trials for targeting cancer stem cells in head and neck cancer (Table 1). As mentioned by Saygin and colleagues in a cross-cancer review of CSC-targeted clinical trials [139], the interpretation of these trials is limited by the fact that most of these studies lack a concrete stem cell readout to prove the efficacy of these therapies in eradicating cancer stem cells specifically. However, we will focus here on clinical trials with mechanistic approaches that have been validated in head and neck cancer stem cell biology.

The importance of cellular crosstalk within the perivascular niche, particularly on endothelial cell-secreted IL-6, has been discussed above. Therefore, the IL-6 pathway presents an appealing target within CSC-specific therapies. It has been shown that IL-6 defines the tumorigenic potential of head and neck cancer stem cells [59], and that inhibition of IL-6R signaling with Tocilizumab (Genentech) decreases the cancer stem cell fraction in preclinical models of different head and neck cancers in vitro and in vivo [58, 140]. The clinically relevant agent

TABLE 1 Clinical trials targeting cancer stem cells in head and neck cancer

Drug class/ mechanism	Agent name	Title	Suggested patient population	Phase	Enrollment	Status	Identifier
EGFR inhibitor/ Hedgehog inhibitor	Cetuximab/IPI-926	Pilot Study of Cetuximab and the Hedgehog Inhibitor IPI-926 in Recurrent Head and Neck Cancer	Recurrent head and neck cancer	I	9	Completed	NCT01255800
Tyrosine kinase inhibitor	Sunitinib/ cetuximab	Sunitinib, Cetuximab, and Radiation Therapy in Treating Patients With Locally Advanced or Recurrent Squamous Cell Carcinoma of the Head and Neck	Head and neck squamous cell carcinoma	I	36	Terminated	NCT00906360
MDM2 Inhibitor	APG-115 +/− Carboplatin	A Phase I/II Trial of APG-115 in Patients With Salivary Gland Carcinoma	Malignant salivary gland cancer	I/II	62	Not yet recruiting	NCT03781986
Cancer stem cell vaccine	N/A	The Immunotherapy of Nasopharyngeal Cancer Using Cancer Stem Cells Vaccine	Nasopharyngeal cancer	I/II	40	Completed	NCT02115958
Histone deacetylase inhibitor/Cytotoxic chemotherapy	Vorinostat/ Capecitabine	Capecitabine and Vorinostat in Treating Patients With Recurrent and/or Metastatic Head and Neck Cancer	Recurrent and/or metastatic head and neck cancer	II	25	Terminated	NCT01267240
mTor inhibitor	Temsirolimus	Efficacy Study of Temsirolimus to Treat Head and Neck Cancer	Head and neck squamous cell carcinoma	II	42	Completed	NCT01172769
Tyrosine kinase inhibitor	Dasatinib	Dasatinib in Treating Patients With Recurrent or Metastatic Malignant Salivary Gland Tumors	Salivary gland tumors	II	55	Completed	NCT00859937
FGFR inhibitor	BGJ398	BGJ398 in Treating Patients With FGFR Positive Recurrent Head and Neck Cancer	Recurrent head and neck cancer	II	20	Recruiting	NCT02706691

Tocilizumab (Genentech) is a humanized, monoclonal anti-IL-6R antibody and was FDA-approved for rheumatoid arthritis in 2010. An already approved agent such as Tocilizumab presents an attractive adjunct therapy, as developing a new FDA-approved drug proves highly difficult and costly. In one study, combination therapy with Tocilizumab enhanced the antitumor effects of Cisplatin and Paclitaxel chemotherapy, and decreased VEGF expression and tumor microvessel density [140]. Inhibition of IL-6 signaling did not affect cancer cell viability, but robustly inhibited STAT3 signaling. Within the endothelial-cancer cell crosstalk, the clinical results of antiangiogenic therapies have been attractive. However, issues arise in the fact that a decrease in tumor microvasculature leads to hypoxia, which creates a hypoxic environment that cancer stem cells thrive in [141]. This suggests that the observed decrease in the cancer stem cell fraction and enhanced antitumor effect may either be due to preferential killing of only the small cancer stem cell population, or due to blocking essential mechanisms that support cancer stem cell self-renewal, forcing these cells to lose their stem-like traits instead.

Many other signaling pathways that are implicated in head and neck cancer stem cell self-renewal have been targeted in preclinical studies. In nasopharyngeal carcinoma, both a Wnt inhibitor (Wnt-C59) and an mTor inhibitor (Rapamycin) were shown to downregulate cancer stem cell properties [142, 143]. The findings from inhibiting mTor signaling corroborated the successfully completed phase II clinical trial studying the efficacy of an mTor inhibitor (Temsirolimus) in head and neck cancer (Table 1). Inhibiting Bmi-1 (PCT-209) was also shown to robustly abrogate cancer stem cell function in HNSCC and colorectal cancer, and sensitized cancer stem cells to Cisplatin therapy [92, 144]. Approaches such as these are riveting, as they suggest a method for sensitizing otherwise resistant cancer stem cells to conventional chemotherapy by targeting the cancer stem cell population in addition to having a direct effect on the more differentiated cells. Epigenetic regulations have also been shown to influence tumor behavior. The effects of a histone deacetylase inhibitor (Vorinostat) was studied in a phase II clinical trial, but was terminated due to the targeted agent showing no clinical activity (Table 1). This may be explained by the observation that although HDAC inhibition decreased the fraction of $ALDH^{high}CD44^{+}$ cells, it paradoxically induced epithelial to mesenchymal transition and expression of Bmi-1 [145]. This suggests that histone acetylation status may bivalently regulate different mechanisms involved in cancer stem cell biology.

Cancer stem cells play an important role in tumor progression, particularly in locoregional tumor recurrence [146]. A possible explanation for tumor recurrence might be that conventional platinum-based therapies and radiotherapy target fast-proliferating cells, but spare the quiescent cancer stem cells that can re-establish the tumor. Based on the cancer stem cell hypothesis, eliminating cancer stem cells should prevent tumor recurrence [147]. In head and neck cancer, MDM2 inhibition has been shown to activate endogenous p53 in tumors in vivo [148] and sensitize HNSCC tumors to chemotherapy [149]. Preclinical recurrence studies mimicking a clinical trial using adenoid cystic carcinoma (ACC) patient-derived xenograft (PDX) models have been conducted in vivo, where therapeutic inhibition of the MDM2-p53 interaction decreased the cancer stem cell fraction and sensitized adenoid cystic carcinoma PDX tumors to Cisplatin therapy [150]. Neoadjuvant administration of the targeted therapy followed by surgical resection of the primary tumors completely eliminated tumor recurrence with a postsurgical follow-up of over 300 days, as compared to 62.5% tumor recurrence in the control group. Ablation of cancer stem cells through MDM2-p53 inhibition was also demonstrated in mucoepidermoid carcinoma [151]. MDM2 inhibitors have been in clinical trials for the treatment of several cancers [76, 152], and now are on the cusp of beginning a Phase I/II clinical trial in salivary gland cancers (Table 1).

In context of immunotherapy in head and neck cancer, most targeted therapies have focused on modulating T-cell interactions with bulk tumor cells. Although these strategies produce promising results, the same long-term problems may arise through CSC-mediated cancer progression and may require targeting cancer stem cells specifically through immunologic approaches. ALDH1A1, the most specific cancer stem cell marker in HNSCC, has been investigated as a potential CD8+ T-cell-defined tumor antigen in head and neck cancer [153, 154]. The development of an anticancer vaccine has been a promising, ongoing area of research. Antitumor immunity was achieved by vaccinating immunocompetent mice with sera from cancer stem cells of syngeneic murine tumors [155]. In a recent phase 1 clinical trial (Table 1), the safety and efficacy of a nasopharyngeal cancer stem cell vaccine was preliminarily proven, where the CSC-specific patient response was significantly enhanced with few side effects [156].

Novel treatment strategies involving targeting cancer stem cells are highly complex. However, it is very apparent that in comparison to other cancers, HNSCC is one of the last to follow with new paradigms. This is directly reflected in the number of past and current clinical trials of CSC-targeted therapies in head and neck tumors (Table 1), which are few and far between CSC-targeting clinical trials in other cancers [139]. Collectively, these studies present at least two strategies for incorporating CSC-targeted therapies into novel treatment paradigms for head and neck cancer: (A) an approach targeting crucial crosstalks between cancer stem cells and the tumor microenvironment (i.e., cutting off the supportive supply), or (B) an approach targeting vital signaling pathways in cancer stem cell maintenance (i.e., using intrinsic cancer stem cell mechanisms to either eliminate or differentiate these cells). Preclinical evidence supports the efficacy of both strategies in suppressing CSC-mediated tumor progression, but more clinical trials are direly necessary to determine their impact, especially in head and neck cancer.

Conclusion

The existence of heterogeneity within the cancer cell population of a tumor has become indisputable with recent advancements in tumor biology. Head and neck tumors have been demonstrated to be composed of cancer cells that exhibit significant plasticity in phenotypic characteristics, directed by cues within the tumor microenvironment. In this chapter, we have explored the biology and clinical relevance of the cancer stem cell subpopulation, which can be therapeutically targeted.

Evidence suggests that while conventional chemotherapy is effective in debulking the tumor, it does not eradicate the uniquely resistant cancer stem cells. Conversely, while using a CSC-targeted approach alone would eliminate the small fraction of cancer stem cells within the tumor, it would leave behind the bulk tumor cells that still have growth potential and are capable of repopulating the cancer stem cell niche through cell plasticity. In attempt to achieve optimal therapeutic outcomes, both cancer stem cells and tumor bulk cells need to be targeted with a combination therapy approach, which could effectively inhibit the cancer stem cell population in HNSCC and prevent disease progression. Therefore, elucidating critical processes that define the fate of head and neck cancer stem cells will inform mechanism-based therapies to prevent CSC-driven tumor growth and recurrence. Targeting pathways that play a significant role in the maintenance of cancer stem cells could sensitize these cells to conventional therapies and result in better treatment outcomes. Strategic and innovative approaches to understand CSC-driven tumor biology will fundamentally advance and provide rationale for novel treatment paradigms, which aim to improve the survival and quality of life of patients with head and neck cancer.

Acknowledgments

We are thankful to the support received by grants R01-DE021139 and R01-DE023220 from the National Institutes of Health/National Institutes of Dental and Craniofacial Research (NIH/NIDCR).

References

[1] Reya T, Morrison SJ, Clarke MF, Weissman IL. Stem cells, cancer, and cancer stem cells. Nature 2001;414(6859):105–11.
[2] Prince ME, Sivanandan R, Kaczorowski A, Wolf GT, Kaplan MJ, Dalerba P, Weissman IL, Clarke MF, Ailles LE. Identification of a subpopulation of cells with cancer stem cell properties in head and neck squamous cell carcinoma. Proc Natl Acad Sci U S A 2007;104(3):973–8.
[3] Krishnamurthy S, Nör JE. Head and neck cancer stem cells. J Dent Res 2012;91(4):334–40.
[4] Steinbichler TB, Dudás J, Skvortsov S, Ganswindt U, Riechelmann H, Skvortsova II. Therapy resistance mediated by cancer stem cells. Semin Cancer Biol 2018;53:156–67.
[5] Vermorken JB, Mesia R, Rivera F, Remenar E, Kawecki A, Rottey S, Erfan J, Zabolotnyy D, Kienzer HR, Cupissol D, Peyrade F, Benasso M, Vynnychenko I, De Raucourt D, Bokemeyer C, Schueler A, Amellal N, Hitt R. Platinum-based chemotherapy plus cetuximab in head and neck cancer. N Engl J Med 2008;359(11):1116–27.
[6] Carvalho AL, Nishimoto IN, Califano JA, Kowalski LP. Trends in incidence and prognosis for head and neck cancer in the United States: a site-specific analysis of the SEER database. Int J Cancer 2005;114(5):806–16.
[7] Nör C, Zhang Z, Warner KA, Bernardi L, Visioli F, Helman JI, Roesler R, Nör JE. Cisplatin induces Bmi-1 and enhances the stem cell fraction in head and neck cancer. Neoplasia 2014;16(2):137–46.
[8] Reid PA, Wilson P, Li Y, Marcu LG, Bezak E. Current understanding of cancer stem cells: review of their radiobiology and role in head and neck cancers. Head Neck 2017;39(9):1920–32.
[9] Prince ME, Ailles LE. Cancer stem cells in head and neck squamous cell cancer. J Clin Oncol 2008;26(17):2871–5.
[10] Shackleton M, Quintana E, Fearon ER, Morrison SJ. Heterogeneity in cancer: cancer stem cells versus clonal evolution. Cell 2009;138(5):822–9.
[11] Clevers H. The cancer stem cell: premises, promises and challenges. Nat Med 2011;17(3):313–9.
[12] Zhu Y, Luo M, Brooks M, Clouthier SG, Wicha MS. Biological and clinical significance of cancer stem cell plasticity. Clin Transl Med 2014;3(1):32.
[13] Tao Y, Li H, Huang R, Mo D, Zeng T, Fang M, Li M. Clinicopathological and prognostic significance of cancer stem cell markers in ovarian cancer patients: evidence from 52 studies. Cell Physiol Biochem 2018;46(4):1716–26.
[14] Pece S, Tosoni D, Confalonieri S, Mazzarol G, Vecchi M, Ronzoni S, Bernard L, Viale G, Pelicci PG, Di Fiore PP. Biological and molecular heterogeneity of breast cancers correlates with their cancer stem cell content. Cell 2010;140(1):62–73.
[15] Alam H, Sehgal L, Kundu ST, Dalal SN, Vaidya MM. Novel function of keratins 5 and 14 in proliferation and differentiation of stratified epithelial cells. Mol Biol Cell 2011;22(21):4068–78.
[16] Senbanjo LT, Chellaiah MA. CD44: a multifunctional cell surface adhesion receptor is a regulator of progression and metastasis of cancer cells. Front Cell Dev Biol 2017;5:18.
[17] Mack B, Gires O. CD44s and CD44v6 expression in head and neck epithelia. PLoS One 2008;3(10).

[18] Clark DW, Palle K. Aldehyde dehydrogenases in cancer stem cells: potential as therapeutic targets. Ann Transl Med 2016;4(24):518.
[19] Cheung AM, Wan TS, Leung JC, Chan LY, Huang H, Kwong YL, Liang R, Leung AY. Aldehyde dehydrogenase activity in leukemic blasts defines a subgroup of acute myeloid leukemia with adverse prognosis and superior NOD/SCID engrafting potential. Leukemia 2007;21(7):1423–30.
[20] Ginestier C, Hur MH, Charafe-Jauffret E, et al. ALDH1 is a marker of normal and malignant human mammary stem cells and a predictor of poor clinical outcome. Cell Stem Cell 2007;1(5):555–67.
[21] Chen YC, Chen YW, Hsu HS, Tseng LM, Huang PI, Lu KH, Chen DT, Tai LK, Yung MC, Chang SC, Ku HH, Chiou SH, Lo WL. Aldehyde dehydrogenase 1 is a putative marker for cancer stem cells in head and neck squamous cancer. Biochem Biophys Res Commun 2009;385(3):307–13.
[22] Krishnamurthy S, Dong Z, Vodopyanov D, Imai A, Helman JI, Prince ME, Wicha MS, Nör JE. Endothelial cell-initiated signaling promotes the survival and self-renewal of cancer stem cells. Cancer Res 2010;70(23):9969–78.
[23] Clay MR, Tabor M, Owen JH, Carey TE, Bradford CR, Wolf GT, Wicha MS, Prince ME. Single-marker identification of head and neck squamous cell carcinoma cancer stem cells with aldehyde dehydrogenase. Head Neck 2010;32(9):1195–201.
[24] Boonyaratanakornkit JB, Yue L, Strachan LR, et al. Selection of tumorigenic melanoma cells using ALDH. J Invest Dermatol 2010;130(12):2799–808.
[25] Carpentino JE, Hynes MJ, Appelman HD, et al. Aldehyde dehydrogenase-expressing colon stem cells contribute to tumorigenesis in the transition from colitis to cancer. Cancer Res 2009;69(20):8208–15.
[26] Jiang F, Qiu Q, Khanna A, et al. Aldehyde dehydrogenase 1 is a tumor stem cell-associated marker in lung cancer. Mol Cancer Res 2009;7(3):330–8.
[27] Li T, Su Y, Mei Y, et al. ALDH1A1 is a marker for malignant prostate stem cells and predictor of prostate cancer patients' outcome. Lab Investig 2009;90(2):234–44.
[28] Ma S, Chan KW, Lee TK, Tang KH, Wo JY, Zheng BJ, Guan XY. Aldehyde dehydrogenase discriminates the CD133 liver cancer stem cell populations. Mol Cancer Res 2008;6(7):1146–53.
[29] Rasheed ZA, Yang J, Wang Q, et al. Prognostic significance of tumorigenic cells with mesenchymal features in pancreatic adenocarcinoma. J Natl Cancer Inst 2010;102(5):340–51.
[30] Adams A, Warner K, Pearson AT, et al. ALDH/CD44 identifies uniquely tumorigenic cancer stem cells in salivary gland mucoepidermoid carcinomas. Oncotarget 2015;6(29):26633–50.
[31] Chen J, Zhou J, Lu J, Xiong H, Shi X, Gong L. Significance of CD44 expression in head and neck cancer: a systemic review and meta-analysis. BMC Cancer 2014;14:15.
[32] Kokko LL, Hurme S, Maula SM, Alanen K, Grenman R, Kinnunen I, Ventela S. Significance of site-specific prognosis of cancer stem cell marker CD44 in head and neck squamous-cell carcinoma. Oral Oncol 2011;47(6):510–6.
[33] Dong Y, Ochsenreither S, Cai C, Kaufmann AM, Albers AE, Qian X. Aldehyde dehydrogenase 1 isoenzyme expression as a marker of cancer stem cells correlates to histopathological features in head and neck cancer: a meta-analysis. PLoS One 2017;12(11).
[34] Qian X, Wagner S, Ma C, Coordes A, Gekeler J, Klussman JP, Hummel M, Kaufmann AM, Albers AE. Prognostic significanc of ALDH1A1-positive cancer stem cells in patients with locally advanced, metastasized head and neck squamous cell carcinoma. J Cancer Res Clin Oncol 2014;140(7):1151–8.
[35] Zhou L, Wei X, Cheng L, Tian J, Jiang JJ. CD133, one of the markers of cancer stem cells in Hep-2 cell line. Laryngoscope 2007;117(3):455–60.
[36] Chiou SH, Yu CC, Huang CY, Lin SC, Liu CJ, Tsai TH, Chou SH, Chien CS, Ku HH, Lo JF. Positive correlations of Oct-4 and Nanog in oral cancer stem-like cells and high-grade oral squamous cell carcinoma. Clin Cancer Res 2008;14(13):4085–95.
[37] Fukusumi T, Ishii H, Konno M, et al. CD10 as a novel marker of therapeutic resistance and cancer stem cells in head and neck squamous cell carcinoma. Br J Cancer 2014;111(3):506–14.
[38] Yan M, Yang X, Wang L, et al. Plasma membrane proteomics of tumor spheres identify CD166 as a novel marker for cancer stem-like cells in head and neck squamous cell carcinoma. Mol Cell Proteomics 2013;12(11):3271–84.
[39] Song J, Chang I, Chen Z, Kang M, Wang CY. Characterization of side populations in HNSCC: highly invasive, chemoresistant and abnormal Wnt signaling. PLoS One 2010;5(7).
[40] Chang CW, Chen YS, Chou SH, Han CL, Chen YJ, Yang CC, Huang CY, Lo JF. Distinct subpopulations of head and neck cancer cells with different levels of intracellular reactive oxygen species exhibit diverse stemness, proliferation, and chemosensitivity. Cancer Res 2014;74(21):6291–305.

[41] Kim WT, Ryu CJ. Cancer stem cell surface markers on normal stem cells. BMB Rep 2017;50(6):285–98.
[42] Krishnamurthy S, Nör JE. Orosphere assay: a method for propagation of head and neck cancer stem cells. Head Neck 2013;35(7):1015–21.
[43] Dontu G, Abdallah WM, Foley JM, Jackson KW, Clarke MF, Kawamura MJ, Wicha MS. In vitro propagation and transcriptional profiling of human mammary stem/progenitor cells. Genes Dev 2003;17:1253–70.
[44] Chen SF, Chang YC, Nieh S, Liu CL, Yang CY, Lin YS. Nonadhesive culture system as a model of rapid sphere formation with cancer stem cell properties. PLoS One 2012;7(2).
[45] O'Brien CA, Kreso A, Jamieson CH. Cancer stem cells and self-renewal. Clin Cancer Res 2010;16(12):3113–20.
[46] Dutta D, Heo I, Clevers H. Disease modeling in stem cell-derived 3D organoid systems. Trends Mol Med 2017;23(5):393–410.
[47] Shah AT, Heaster TM, Skala MC. Metabolic imaging of head and neck cancer organoids. PLoS One 2017;12(1).
[48] Fuchs E, Tumbar T, Guasch G. Socializing with the neighbors: stem cells and their niche. Cell 2004;116 (6):769–78.
[49] Oh M, Nör JE. The perivascular niche and self-renewal of stem cells. Front Physiol 2015;6:367.
[50] Jin K, Mao XO, Sun Y, Xie L, Greenberg DA. Stem cell factor stimulates neurogenesis in vitro and in vivo. J Clin Invest 2002;110(3):311–9.
[51] Calabrese C, Poppleton H, Kocak M, Hogg TL, Fuller C, Hamner B, Oh EY, Gaber MW, Finklestein D, Allen M, Frank A, Bayazitov IT, Zakharenko SS, Gajjar A, Davidoff A, Gilbertson RJ. A perivascular niche for brain tumor stem cells. Cancer Cell 2007;11(1):69–82.
[52] Hildebrand LC, Carvalho AL, Lauxen IS, Nör JE, Cerski CT, Sant'Ana M. Spatial distribution of cancer stem cells in head and neck squamous cell carcinomas. J Oral Pathol Med 2014;43(7):499–506.
[53] Korkaya H, Liu S, Wicha MS. Breast cancer stem cells, cytokine networks, and the tumor microenvironment. J Clin Invest 2011;121(10):3804–9.
[54] Kenny PA, Lee GY, Bissell MJ. Targeting the tumor microenvironment. Front Biosci 2007;12:3468–74.
[55] Markwell SM, Weed SA. Tumor and stromal-based contributions to head and neck squamous cell carcinoma invasion. Cancers (Basel) 2015;7(1):382–406.
[56] Campos M, Neiva K, Meyers K, Krishnamurthy S, Nör JE. Endothelial derived factors inhibit anoikis of head and neck cancer stem cells. Oral Oncol 2012;48(1):26–32.
[57] Zhang Z, Dong Z, Lauxen I, Sant'Anna M, Nör JE. Endothelial cell-secreted EGF induces epithelial to mesenchymal transition and endows head and neck cancer cells with stem-like phenotype. Cancer Res 2014;74(10):2869–81.
[58] Kim HS, Chen YC, Nör F, Warner KA, Andrews A, Wagner VP, Zhang Z, Zhang Z, Martins MD, Pearson AT, Yoon E, Nör JE. Endothelial-derived interleukin-6 induces cancer stem cell motility by generating a chemotactic gradient towards blood vessels. Oncotarget 2017;8(59):100339–52.
[59] Krishnamurthy S, Warner KA, Dong Z, Imai A, Nör C, Ward BB, Helman JI, Taichman RS, Bellile EL, McCauley LK, Polverini PJ, Prince ME, Wicha MS, Nör JE. Endothelial interleukin-6 defines the tumorigenic potential of primary human cancer stem cells. Stem Cells 2014;32(11):2845–57.
[60] Neiva KG, Zhang Z, Miyazawa M, Warner KA, Karl E, Nör JE. Crosstalk initiated by endothelial cells enhances migration and inhibits anoikis of squamous cell carcinoma cells through STAT3/Akt/ERK signaling. Neoplasia 2009;11:583–93.
[61] Kalluri R, Zeisberg M. Fibroblasts in cancer. Nat Rev Cancer 2006;6(5):392–401.
[62] Vermeulen L, E Melo F DS, van der Heijden M, Cameron K, de Jong JH, Borovski T, Tuynman JB, Todaro M, Merz C, Rodermond H, Sprick MR, Kemper K, Richel DJ, Stassi G, Medema JP. Wnt activity defines colon cancer stem cells and is regulated by the microenvironment. Nat Cell Biol 2010;12(5):468–76.
[63] Chen WJ, Ho CC, Chang YL, Chen HY, Lin CA, Ling TY, Yu SL, Yuan SS, Chen YJ, Lin CY, Pan SH, Chou HY, Chen YJ, Chang GC, Chu WC, Lee YM, Lee JY, Lee PJ, Li KC, Chen HW, Yang PC. Cancer-associated fibroblasts regulate the plasticity of lung cancer stemness via paracrine signaling. Nat Commun 2014;5:3472.
[64] Faber A, Aderhold C, Goessler UR, Hoermann K, Schultz JD, Umbreit C, Walliczek U, Stern-Straeter J. Interaction of a CD44+ head and neck squamous cell carcinoma cell line with a stromal cell-derived factor-1-expressing supportive niche: an in vitro model. Oncol Lett 2014;7(1):82–6.
[65] Taichman RS, Cooper C, Keller ET, Pienta KJ, Taichman NS, McCauley LK. Use of the stromal cell-derived factor-1/CXCR4 pathway in prostate cancer metastasis to bone. Cancer Res 2002;62(6):1832–7.
[66] Geminder H, Sagi-Assif O, Goldberg L, Meshel T, Rechavi G, Witz IP, Ben-Baruch A. A possible role for CXCR4 and its ligand, the CXC chemokine stromal cell-derived factor-1, in the development of bone marrow metastases in neuroblastoma. J Immunol 2001;167(8):4747–57.

[67] Yu B, Wu K, Wang X, Zhang J, Wang L, Jiang Y, Zhu X, Chen W, Yan M. Periostin secreted by cancer-associated fibroblasts promotes cancer stemness in head and neck cancer by activating protein tyrosine kinase 7. Cell Death Dis 2018;9(11):1082.

[68] Hanahan D, Weinberg RA. Hallmarks of cancer: the next generation. Cell 2011;144(5):646–74.

[69] Dunn GP, Old LJ, Schreiber RD. The three Es of cancer immunoediting. Annu Rev Immunol 2004;22:329–60.

[70] Sultan M, Coyle KM, Vidovic D, Thomas ML, Gujar S, Marcato P. Hide-and-seek: the interplay between cancer stem cells and the immune system. Carcinogenesis 2017;38(2):107–18.

[71] Lee Y, Sunwoo J. PD-L1 is preferentially expressed on CD44+ tumor-initiating cells in head and neck squamous cell carcinoma. J Immunother Cancer 2014;2(Suppl. 3):P270.

[72] Lee Y, Shin JH, Longmire M, Wang H, Kohrt HE, Chang HY, Sunwoo JB. CD44+ cells in head and neck squamous cell carcinoma suppress T-cell-mediated immunity by selective constitutive and inducible expression of PD-L1. Clin Cancer Res 2016;22(14):3571–81.

[73] Chikamatsu K, Takahashi G, Sakakura K, Ferrone S, Masuyama K. Immunoregulatory properties of CD44+ cancer stem-like cells in squamous cell carcinoma of the head and neck. Head Neck 2011;33(2):208–15.

[74] Jinushi M, Chiba S, Yoshiyama H, Masutomi K, Kinoshita I, Dosaka-Akita H, Yagita H, Takaoka A, Tahara H. Tumor-associated macrophages regulate tumorigenicity and anticancer drug responses of cancer stem/initiating cells. Proc Natl Acad Sci U S A 2011;108(30):12425–30.

[75] Davis EA, Zhou W, Dailey MJ. Evidence for a direct effect of the autonomic nervous system on intestinal epithelial stem cell proliferation. Phys Rep 2018;6(12).

[76] Zhao Y, Aguilar A, Bernard D, Wang S. Small-molecule inhibitors of the MDM2-p53 protein–protein interaction (MDM2 inhibitors) in clinical trials for cancer treatment. J Med Chem 2015;58(3):1038–52.

[77] Magnon C, Hall SJ, Lin J, Xue X, Gerber L, Freedland SJ, Frenette PS. Autonomic nerve development contributes to prostate cancer progression. Science 2013;341(6142):1236361.

[78] Kurtz KA, Hoffman HT, Zimmerman MB, Robinson RA. Perineural and vascular invasion in oral cavity squamous carcinoma: increased incidence on re-review of slides and by using immunohistochemical enhancement. Arch Pathol Lab Med 2005;129:354–9.

[79] Johnston M, Yu E, Kim J. Perineural invasion and spread in head and neck cancer. Expert Rev Anticancer Ther 2012;12:359–71.

[80] Schmitd LB, Beesley LJ, Russo N, Bellile EL, Inglehart RC, Liu M, Romanowicz G, Wolf GT, Taylor JMG, D'Silva NJ. Redefining perineural invasion: integration of biology with clinical outcome. Neoplasia 2018;20(7):657–67.

[81] Boilly B, Faulkner S, Jobling P, Hondermarck H. Nerve dependence: from regeneration to cancer. Cancer Cell 2017;31(3):342–54.

[82] Lu R, Fan C, Shangguan W, Liu Y, Li Y, Shang Y, Yin D, Zhang S, Huang Q, Li X, Meng W, Xu H, Zhou Z, Hu J, Li W, Liu L, Mo X. Neurons generated from carcinoma stem cells support cancer progression. Signal Transduct Target Ther 2017;2:16036.

[83] Duffy SA, Taylor JM, Terrell JE, Islam M, Li Y, Fowler KE, Wolf GT, Teknos TN. Interleukin-6 predicts recurrence and survival among head and neck cancer patients. Cancer 2008;113:750–7.

[84] Marotta LL, Almendro V, Marusyk A, et al. The JAK2/STAT3 signaling pathway is required for growth of CD44+CD24- stem cell-like breast cancer cells in human tumors. J Clin Invest 2011;121:2723–35.

[85] Matthews JR, Sansom OJ, Clarke AR. Absolute requirement for STAT3 function in small-intestine crypt stem cell survival. Cell Death Differ 2011;18:1934–43.

[86] Lee G, Hall 3rd RR, Ahmed AU. Cancer stem cells: cellular plasticity, niche, and its clinical relevance. J Stem Cell Res Ther 2016;6(10):363.

[87] Geiger JL, Grandis JR, Bauman JE. The STAT3 pathway as a therapeutic target in head and neck cancer: barriers and innovations. Oral Oncol 2016;56:84–92.

[88] Molofsky AV, Pardal R, Iwashita T, Park IK, Clarke MF, Morrison SJ. Bmi-1 dependence distinguishes neural stem cell self-renewal from progenitor proliferation. Nature 2003;425(6961):962–7.

[89] Park IK, Qian D, Kiel M, Becker MW, Pihalja M, Weissman IL, Morrison SJ, Clarke MF. Bmi-1 is required for maintenance of adult self-renewing haematopoietic stem cells. Nature 2003;423(6937):302–5.

[90] Jacobs JJ, Kieboom K, Marino S, DePinho RA, van Lohuizen M. The oncogene and polycomb-group gene bmi-1 regulates cell proliferation and senescence through the ink4a locus. Nature 1999;397(6715):164–8.

[91] Molofsky AV, He S, Bydon M, Morrison SJ, Pardal R. Bmi-1 promotes neural stem cell self-renewal and neural development but not mouse growth and survival by repressing the p16Ink4a and p19Arf senescence pathways. Genes Dev 2005;19(12):1432–7.

[92] Chen D, Wu M, Li Y, Chang I, Yuan Q, Ekimyan-Salvo M, Deng P, Yu B, Yu Y, Dong J, Szymanski JM, Ramadoss S, Li J, Wang CY. Targeting BMI1+ cancer stem cells overcomes chemoresistance and inhibits metastases in squamous cell carcinoma. Cell Stem Cell 2017;20(5):621–634.e6.
[93] Sullivan NJ, Sasser AK, Axel AE, Vesuna F, Raman V, Ramirez N, Oberyszyn TM, Hall BM. Interleukin-6 induces an epithelial-mesenchymal transition phenotype in human breast cancer cells. Oncogene 2009;28(33):2940–7.
[94] Cheng GZ, Zhang WZ, Sun M, Wang Q, Coppola D, Mansour M, Xu LM, Costanzo C, Cheng JQ, Wang LH. Twist is transcriptionally induced by activation of STAT3 and mediates STAT3 oncogenic function. J Biol Chem 2008;283(21):14665–73.
[95] Clevers H. Wnt/β-catenin signaling in development and disease. Cell 2006;127:469–80.
[96] Lee SH, Koo BS, Kim JM, Huang S, Rho YS, Bae WJ, Kang HJ, Kim YS, Moon JH, Lim YC. Wnt/β- catenin signaling maintains self-renewal and tumourigenicity of head and neck squamous cell carcinoma stem-like cells by activating Oct4. J Pathol 2014;234(1):99–107.
[97] Mao L, Zhao ZL, Yu GT, Wu L, Deng WW, Li YC, Liu JF, Bu LL, Liu B, Kulkarni AB, Zhang WF, Zhang L, Sun ZJ. γ-Secretase inhibitor reduces immunosuppressive cells and enhances tumour immunity in head and neck squamous cell carcinoma. Int J Cancer 2018;142(5):999–1009. https://doi.org/10.1002/ijc.31115.
[98] Gan GN, Eagles J, Keysar SB, Wang G, Glogowska MJ, Altunbas C, Anderson RT, Le PN, Morton JJ, Frederick B, Raben D, Wang XJ, Jimeno A. Hedgehog signaling drives radioresistance and stroma-driven tumor repopulation in head and neck squamous cancers. Cancer Res 2014;74(23):7024–36.
[99] Ventelä S, Sittig E, Mannermaa L, Mäkelä JA, Kulmala J, Löyttyniemi E, Strauss L, Cárpen O, Toppari J, Grénman R, Westermarck J. CIP2A is an Oct4 target gene involved in head and neck squamous cell cancer oncogenicity and radioresistance. Oncotarget 2015;6(1):144–58.
[100] Koo BS, Lee SH, Kim JM, Huang S, Kim SH, Rho YS, Bae WJ, Kang HJ, Kim YS, Moon JH, Lim YC. Oct4 is a critical regulator of stemness in head and neck squamous carcinoma cells. Oncogene 2015;34(18):2317–24.
[101] Mitsui K, Tokuzawa Y, Itoh H, Segawa K, Murakami M, Takahashi K, Maruyama M, Maeda M, Yamanaka S. The homeoprotein Nanog is required for maintenance of pluripotency in mouse epiblast and ES cells. Cell 2003;113(5):631–42.
[102] Fan Z, Li M, Chen X, Wang J, Liang X, Wang H, Wang Z, Cheng B, Xia J. Prognostic value of cancer stem cell markers in head and neck squamous cell carcinoma: a meta-analysis. Sci Rep 2017;7:43008.
[103] Cancer Genome Atlas Network. Comprehensive genomic characterization of head and neck squamous cell carcinomas. Nature 2015;517(7536):576–82.
[104] Poeta ML, Manola J, Goldwasser MA, Forastiere A, Benoit N, Califano JA, Ridge JA, Goodwin J, Kenady D, Saunders J, Westra W, Sidransky D, Koch WM. TP53 mutations and survival in squamous-cell carcinoma of the head and neck. N Engl J Med 2007;357(25):2552–61.
[105] Stiewe T. The p53 family in differentiation and tumorigenesis. Nat Rev Cancer 2007;7(3):165–8.
[106] Verga Falzacappa MV, Ronchini C, Reavie LB, Pelicci PG. Regulation of self-renewal in normal and cancer stem cells. FEBS J 2012;279(19):3559–72.
[107] Cicalese A, Bonizzi G, Pasi CE, Faretta M, Ronzoni S, Giulini B, Brisken C, Minucci S, Di Fiore PP, Pelicci PG. The tumor suppressor p53 regulates polarity of self-renewing divisions in mammary stem cells. Cell 2009;138(6):1083–95.
[108] Kubbutat MH, Jones SN, Vousden KH. Regulation of p53 stability by MDM2. Nature 1997;387(6630):299–303.
[109] Wienken M, Dickmanns A, Nemajerova A, Kramer D, Najafova Z, Weiss M, Karpiuk O, Kassem M, Zhang Y, Lozano G, Johnsen SA, Moll UM, Zhang X, Dobbelstein M. MDM2 associates with polycomb repressor complex 2 and enhances stemness promoting chromatin modifications independent of p53. Mol Cell 2016;61(1):68–83.
[110] De Craene B, Berx G. Regulatory networks defining EMT during cancer initiation and progression. Nat Rev Cancer 2013;13(2):97–110.
[111] Mani SA, Guo W, Liao MJ, Eaton EN, Ayyanan A, Zhou AY, Brooks M, Reinhard F, Zhang CC, Shipitsin M, Campbell LL, Polyak K, Brisken C, Yang J, Weinberg RA. The epithelial-mesenchymal transition generates cells with properties of stem cells. Cell 2008;133(4):704–15.
[112] Ota I, Masui T, Kurihara M, Yook JI, Mikami S, Kimura T, Shimada K, Konishi N, Yane K, Yamanaka T, Kitahara T. Snail-induced EMT promotes cancer stem cell-like properties in head and neck cancer cells. Oncol Rep 2016;35(1):261–6.
[113] Yadav A, Kumar B, Datta J, Teknos TN, Kumar P. IL-6 promotes head and neck tumor metastasis by inducing epithelial-mesenchymal transition via the JAK-STAT3-SNAIL signaling pathway. Mol Cancer Res 2011;9(12):1658–67.

[114] Pearson AT, Ingram P, Bai S, O'Hayer P, Chung J, Yoon E, Jackson T, Buckanovich RJ. Sampling from single-cell observations to predict tumor cell growth *in-vitro* and *in-vivo*. Oncotarget 2017;8(67):111176–89.

[115] Heddleston JM, Li Z, McLendon RE, Hjelmeland AB, Rich JN. The hypoxic microenvironment maintains glioblastoma stem cells and promotes reprogramming towards a cancer stem cell phenotype. Cell Cycle 2009;8(20):3274–84.

[116] Morrison SJ, Kimble J. Asymmetric and symmetric stem-cell divisions in development and cancer. Nature 2006;441(7097):1068–74.

[117] Siegel R, Ma J, Zou Z, Jemal A. Cancer statistics, 2014. CA Cancer J Clin 2014;64(1):9–29.

[118] Jemal A, Bray F, Center MM, Ferlay J, Ward E, Forman D. Global cancer statistics. CA Cancer J Clin 2011;61:69–90.

[119] Bergers G, Hanahan D. Modes of resistance to anti-angiogenic therapy. Nat Rev Cancer 2008;8(8):592–603.

[120] Forastiere AA. Chemotherapy in the treatment of locally advanced head and neck cancer. J Surg Oncol 2008;97:701–7.

[121] Seiwert TY, Salama JK, Vokes EE. The chemoradiation paradigm in head and neck cancer. Nat Clin Pract Oncol 2007;4(3):156–71.

[122] Tsai LL, Yu CC, Chang YC, Yu CH, Chou MY. Markedly increased Oct4 and Nanog expression correlates with cisplatin resistance in oral squamous cell carcinoma. J Oral Pathol Med 2011;40(8):621–8.

[123] Goldman A, Majumder B, Dhawan A, Ravi S, Goldman D, Kohandel M, Majumder PK, Sengupta S. Temporally sequenced anticancer drugs overcome adaptive resistance by targeting a vulnerable chemotherapy-induced phenotypic transition. Nat Commun 2015;6:6139.

[124] Sharma SV, Lee DY, Li B, Quinlan MP, Takahashi F, Maheswaran S, McDermott U, Azizian N, Zou L, Fischbach MA, Wong KK, Brandstetter K, Wittner B, Ramaswamy S, Classon M, Settleman J. A chromatin-mediated reversible drug-tolerant state in cancer cell subpopulations. Cell 2010;141(1):69–80.

[125] Davis SJ, Divi V, Owen JH, Bradford CR, Carey TE, Papagerakis S, Prince ME. Metastatic potential of cancer stem cells in head and neck squamous cell carcinoma. Arch Otolaryngol Head Neck Surg 2010;136(12):1260–6.

[126] Chinn SB, Darr OA, Owen JH, Bellile E, McHugh JB, Spector ME, Papagerakis SM, Chepeha DB, Bradford CR, Carey TE, Prince ME. Cancer stem cells: mediators of tumorigenesis and metastasis in head and neck squamous cell carcinoma. Head Neck 2015;37(3):317–26.

[127] Chang JC. Cancer stem cells: role in tumor growth, recurrence, metastasis, and treatment resistance. Medicine (Baltimore) 2016;95(1 Suppl. 1):S20–5.

[128] Chaffer CL, Weinberg RA. A perspective on cancer cell metastasis. Science 2011;331(6024):1559–64.

[129] McDermott SC, Rodriguez-Ramirez C, McDermott SP, Wicha MS, Nör JE. FGFR signaling regulates resistance of head and neck cancer stem cells to cisplatin. Oncotarget 2018;9(38):25148–65.

[130] Eyler CE, Rich JN. Survival of the fittest: cancer stem cells in therapeutic resistance and angiogenesis. J Clin Oncol 2008;26:2839–45.

[131] Li Z, Bao S, Wu Q, Wang H, Eyler C, Sathornsumetee S, Shi Q, Cao Y, Lathia J, McLendon RE, Hjelmeland AB, Rich JN. Hypoxia-inducible factors regulate tumorigenic capacity of glioma stem cells. Cancer Cell 2009;15:501–13.

[132] Gilbert LA, Hemann MT. DNA damage-mediated induction of a chemoresistant niche. Cell 2010;143(3):355–66.

[133] Jinushi M, Chiba S, Yoshiyama H, Masutomi K, Kinoshita I, Dosaka-Akita H, Yagita H, Takaoka A, Tahara H. Tumor-associated macrophages regulate tumorigenicity and anticancer drug responses of cancer stem/initiating cells. Proc Natl Acad Sci U S A 2011;108(30):12425–30.

[134] Cao Z, Scandura JM, Inghirami GG, Shido K, Ding BS, Rafii S. Molecular checkpoint decisions made by subverted vascular niche transform indolent tumor cells into chemoresistant cancer stem cells. Cancer Cell 2017;31(1):110–26.

[135] Jin L, Chun J, Pan C, Li D, Lin R, Alesi GN, Wang X, Kang HB, Song L, Wang D, Zhang G, Fan J, Boggon TJ, Zhou L, Kowalski J, Qu CK, Steuer CE, Chen GZ, Saba NF, Boise LH, Owonikoko TK, Khuri FR, Magliocca KR, Shin DM, Lonial S, Kang S. MAST1 drives cisplatin resistance in human cancers by rewiring cRaf-independent MEK activation. Cancer Cell 2018;34(2):315–330.e7.

[136] Fang Y, Wang J, Wang G, Zhou C, Wang P, Zhao S, Zhao S, Huang S, Su W, Jiang P, Chang A, Xiang R, Sun P. Inactivation of p38 MAPK contributes to stem cell-like properties of non-small cell lung cancer. Oncotarget 2017;8(16):26702–17.

[137] Keklikoglou I, Cianciaruso C, Güç E, Squadrito ML, Spring LM, Tazzyman S, Lambein L, Poissonnier A, Ferraro GB, Baer C, Cassará A, Guichard A, Iruela-Arispe ML, Lewis CE, Coussens LM, Bardia A, Jain RK,

Pollard JW, De Palma M. Chemotherapy elicits pro-metastatic extracellular vesicles in breast cancer models. Nat Cell Biol 2019;21:190–202.
[138] Zhou S, Schuetz JD, Bunting KD, Colapietro AM, Sampath J, Morris JJ, Lagutina I, Grosveld GC, Osawa M, Nakauchi H, Sorrentino BP. The ABC transporter Bcrp1/ABCG2 is expressed in a wide variety of stem cells and is a molecular determinant of the side-population phenotype. Nat Med 2001;7:1028–34.
[139] Saygin C, Matei D, Majeti R, Reizes O, Lathia JD. Targeting cancer stemness in the clinic: from hype to hope. Cell Stem Cell 2019;24(1):25–40.
[140] Mochizuki D, Adams A, Warner KA, Zhang Z, Pearson AT, Misawa K, McLean SA, Wolf GT, Nör JE. Anti-tumor effect of inhibition of IL-6 signaling in mucoepidermoid carcinoma. Oncotarget 2015;6(26):22822–35.
[141] Schöning JP, Monteiro M, Gu W. Drug resistance and cancer stem cells: the shared but distinct roles of hypoxia-inducible factors HIF1α and HIF2α. Clin Exp Pharmacol Physiol 2017;44(2):153–61.
[142] Cheng Y, Phoon YP, Jin X, Chong SY, Ip JC, Wong BW, Lung ML. Wnt-C59 arrests stemness and suppresses growth of nasopharyngeal carcinoma in mice by inhibiting the Wnt pathway in the tumor microenvironment. Oncotarget 2015;6(16):14428–39.
[143] Yang C, Zhang Y, Zhang Y, Zhang Z, Peng J, Li Z, Han L, You Q, Chen X, Rao X, Zhu Y, Liao Z. Downregulation of cancer stem cell properties via mTOR signaling pathway inhibition by rapamycin in nasopharyngeal carcinoma. Int J Oncol 2015;47(3):909–17.
[144] Kreso A, van Galen P, Pedley NM, Lima-Fernandes E, Frelin C, Davis T, Cao L, Baiazitov R, Du W, Sydorenko N, Moon YC, Gibson L, Wang Y, Leung C, Iscove NN, Arrowsmith CH, Szentgyorgyi E, Gallinger S, Dick JE, O'Brien CA. Self-renewal as a therapeutic target in human colorectal cancer. Nat Med 2014;20(1):29–36.
[145] Giudice FS, Pinto Jr DS, Nör JE, Squarize CH, Castilho RM. Inhibition of histone deacetylase impacts cancer stem cells and induces epithelialmesenchyme transition of head and neck cancer. PLoS One 2013;8(3).
[146] Islam F, Gopalan V, Smith RA, Lam AK. Translational potential of cancer stem cells: a review of the detection of cancer stem cells and their roles in cancer recurrence and cancer treatment. Exp Cell Res 2015;335(1):135–47.
[147] Schatton T, Frank NY, Frank MH. Identification and targeting of cancer stem cells. Bioessays 2009;31(10):1038–49.
[148] Warner KA, Nör F, Acasigua GA, Martins MD, Zhang Z, McLean SA, Spector ME, Chepeha DB, Helman J, Wick MJ, Moskaluk CA, Castilho RM, Pearson AT, Wang S, Nör JE. Targeting MDM2 for treatment of adenoid cystic carcinoma. Clin Cancer Res 2016;22(14):3550–9.
[149] Roh JL, Kang SK, Minn I, Califano JA, Sidransky D, Koch WM. p53-Reactivating small molecules induce apoptosis and enhance chemotherapeutic cytotoxicity in head and neck squamous cell carcinoma. Oral Oncol 2011;47(1):8–15.
[150] Nör F, Warner KA, Zhang Z, Acasigua GA, Pearson AT, Kerk SA, Helman JI, Sant'Ana Filho M, Wang S, Nör JE. Therapeutic inhibition of the MDM2-p53 interaction prevents recurrence of adenoid cystic carcinomas. Clin Cancer Res 2017;23(4):1036–48.
[151] Andrews A, Warner K, Rodriguez-Ramirez C, Pearson AT, Nör F, Zhang Z, Kerk S, Kulkarni A, Helman JI, Brenner JC, Wicha MS, Wang S, Nör JE. Ablation of cancer stem cells by therapeutic inhibition of the MDM2-p53 interaction in mucoepidermoid carcinoma. Clin Cancer Res 2019;25(5):1588–600.
[152] de Jonge M, de Weger VA, Dickson MA, Langenberg M, Le Cesne A, Wagner AJ, Hsu K, Zheng W, Mace S, Tuffal G, Thomas K, Schellens JH. A phase I study of SAR405838 a novel human double minute 2 (HDM2) antagonist in patients with solid tumours. Eur J Cancer 2017;76:144–51.
[153] Visus C, Ito D, Amoscato A, Maciejewska-Franczak M, Abdelsalem A, Dhir R, Shin DM, Donnenberg VS, Whiteside TL, DeLeo AB. Identification of human aldehyde dehydrogenase 1 family member A1 as a novel CD8+ T-cell-defined tumor antigen in squamous cell carcinoma of the head and neck. Cancer Res 2007;67(21):10538–45.
[154] Visus C, Wang Y, Lozano-Leon A, Ferris RL, Silver S, Szczepanski MJ, Brand RE, Ferrone CR, Whiteside TL, Ferrone S, DeLeo AB, Wang X. Targeting ALDHbright human carcinoma-initiating cells with ALDH1A1-specific CD8+ T cells. Clin Cancer Res 2011;17(19):6174–84.
[155] Ning N, Pan Q, Zheng F, Teitz-Tennenbaum S, Egenti M, Yet J, Li M, Ginestier C, Wicha MS, Moyer JS, Prince ME, Xu Y, Zhang XL, Huang S, Chang AE, Li Q. Cancer stem cell vaccination confers significant antitumor immunity. Cancer Res 2012;72(7):1853–64.
[156] Lin M, Yuan Y, Xu J, Cai X, Liu S, Niu L, Chen J, Li Q, Xu K. Safety and efficacy study of nasopharyngeal cancer stem cell vaccine. Immunol Lett 2015;165(1):26–31.

CHAPTER

8

TAM family proteins and therapy resistance

Nellie K. McDaniel, Samantha R. Fischbach, Olivia J. Ondracek, Noah B. Welke, Mari Iida, Deric L. Wheeler

University of Wisconsin School of Medicine and Public Health, University of Wisconsin, Madison, WI, United States

Abstract

The TAM family of receptor tyrosine kinases have been identified in HNSCC as novel molecular targets. This chapter summarizes the current literature concerning the relationship of the TAM family with cancer and therapeutic resistance. Through basic science characterization, preclinical modeling, and clinical trial testing, significant associations have been identified correlating TAM expression and signaling with cancer prognosis and resistance to therapy. These discoveries emphasize the need to study TAM family proteins and target these receptors in cancer patients.

Abbreviations

AML	acute myeloid leukemia
CML	chronic myeloid leukemia
CNS	central nervous system
DC	dendritic cell
EGFR	epidermal growth factor
EMT	epithelial-mesenchymal transition
ER	estrogen receptor
HNSCC	head and neck squamous cell carcinoma
HPV	human papillomavirus
IFN	interferon
mAb	monoclonal antibody
NK	natural killer
NSCLC	non-small cell lung cancer
PDX	patient-derived xenograft

Improving the Therapeutic Ratio in Head and Neck Cancer
https://doi.org/10.1016/B978-0-12-817868-3.00008-1

PI3K	phosphatidyl inositol 3 kinase
PLC	phospholipase C
RCC	renal cell carcinoma
RTK	receptor tyrosine kinase
TAM	Tyro, Axl, MerTK
TNBC	triple negative breast cancer
VSMC	vascular smooth muscle cell

Conflict of interest

No potential conflicts of interest were disclosed.

Introduction

This chapter focuses on the TAM family proteins: Tyro, Axl, and MerTK. This is a family of receptor tyrosine kinases (RTKs) that has recently increased in popularity after they were discovered to be indicators of cancer prognosis and therapeutic response. In head and neck squamous cell carcinoma (HNSCC), all three members have been observed with a particular focus on the family member Axl. Axl expression has been shown to correlate with worse disease in HNSCC which warranted further examination of its signaling [1]. More studies revealed Axl's role in mediating resistance to several treatment modalities including radiation, chemotherapy, and molecular-targeted therapy [1, 2]. This discovery of Axl's role in HNSCC has led to further investigations regarding the involvement of the entire TAM family of receptors in this disease. As more of their normal and abnormal biology are elucidated, targeting strategies have emerged and have advanced rapidly to the clinic.

Because of the rapidly developing interest in the TAM family proteins, this chapter has been compiled to provide a general summary of what is currently known about the TAMs and highlight areas where additional research is needed. This chapter (1) introduces a basic science overview of the TAM family of RTKs (2) discusses the family's involvement in cancer (3) explores roles in therapeutic resistance (4) reviews targeting approaches and (5) considers clinical developments of these targeting approaches. Together this information reinforces the importance of studying the TAM family proteins and how advances in this area can provide benefit to cancer patients.

TAM family basic science

The TAM family receptors were originally discovered by multiple groups in the early 1990s and as such were known by a variety of names [3–6]. Since their early discovery, these proteins have been designated as Tyro3, Axl, and MerTK and are grouped into their own distinct family of RTKs based on structural similarities [7, 8] (see Fig. 1).

RTKs contain extracellular, transmembrane, and intracellular domains; however, each RTK family has a distinctive and unique make-up of these domains. For the TAM receptors, the ectodomain that mediates ligand binding is comprised of two immunoglobulin-related domains followed by two fibronectin type III domains. There is a single-pass transmembrane domain that connects the ectodomain to the intracellular tyrosine kinase domain. The kinase contains a conserved sequence of amino acids KW(I/L)A(I/L)ES that is unique to the TAMs

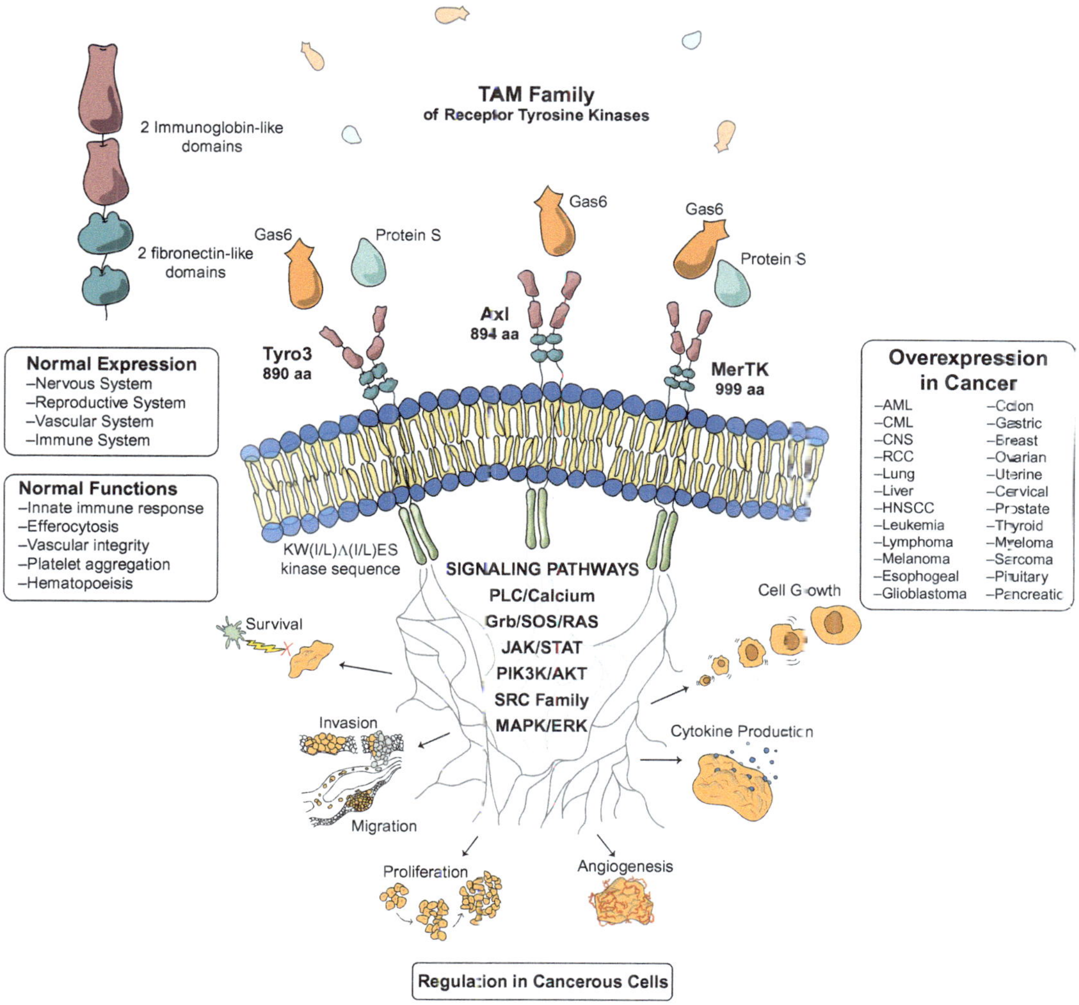

FIG. 1 Overview. The TAM family of receptor tyrosine kinases consists of Tyro3, Axl, and MerTK. These receptors are activated by ligands Gas6 and/or Protein S and normally signal through an assortment of pathways depending on the tissue and function. Abnormally, the TAM family is overexpressed in a variety of cancers where they regulate several different features of the disease.

[3, 9, 10]. The two ligands that have been identified to interact with the TAM receptors are the vitamin K-dependent protein Gas6 and the vitamin K-dependent anticoagulation factor Protein S, which share 42% amino acid identity [11–14]. Both of these ligands contain gamma-carboxylated glutamic acid (Gla) residues which enable them to bind to phosphatidylserine [15]. Gas6 is known to bind all three of the receptors—with higher affinity for Tyro3 and Axl than MerTK—while Protein S can activate Tyro3 and MerTK but not Axl [11, 16, 17]. The exclusive structure of these receptors and their interactions with specific ligands determine the functions that the TAMs perform and regulate in cells including proliferation, migration, and adhesion [3, 9, 10, 18–21].

The genomic and amino acid sequences of the human TAM receptors are also very similar to each other in composition. Tyro3 and Axl have 20 exons while MerTK encodes 19 [22–26]. Alternative splicing of Tyro3 creates isoforms with exons 2A, 2B, or 2C [27] while alternative splicing at exon 10 creates two Axl variants [3, 24, 26, 28]. No splice variants of MerTK have been fully characterized yet [26, 29]. Proteolytic cleavage in the extracellular domains of Axl and MerTK by a metalloproteinase produces soluble forms of the receptors that bind Gas6 [30–33]. The similarities in TAM genes translate to similarities in protein, with the receptors sharing over 50% similar amino acids in the extracellular region and 75% similarity in the intracellular domains. The total lengths of the human receptors are Tyro3 with 890 amino acids, Axl with 894, and MerTK with 999. The actual molecular weights are around 140 kDa for Axl and Tyro3 and 170 kDa for MerTK [3, 23, 26, 29].

The TAM receptors are expressed broadly throughout the mouse and human adult tissues including nervous, reproductive, vascular, and immune systems [34]. These receptors have overlapping and cooperative roles in signaling and function. First attempts at defining the biological roles of the TAM receptors were slow since mouse loss-of-function mutants were viable and fertile [23, 35, 36]. This is considered unusual since RTKs typically play a critical role in embryonic development. However, degenerative phenotypes of the TAM mutants did develop in adult knockout mice, indicating essential roles of the receptors in various signaling activities [23, 35–40].

The TAM receptors have been shown to normally function in a variety of homeostatic processes including the innate immune response, phagocytosis of apoptotic cells, vascular integrity, platelet aggregation, and hematopoiesis. Research is still ongoing as to the many functions of these receptors in normal cellular activities in order to gain a better understanding of the roles they play in disease [26, 34, 36].

Development of TAM receptor knockout mice have indicated the important role of these receptors in the immune system since the knockouts always develop autoimmune diseases such as rheumatoid arthritis and lupus [41, 42]. This is because TAM receptors and their ligands are upregulated by inflammatory signaling and act as a brake to inhibit inflammation. Toll-like receptor engagement and type I interferon (IFN) signaling upregulate TAM receptors in dendritic cells (DCs). The TAM receptors then upregulate suppressor cytokine signaling (i.e., SOCS1 and SOCS2) via STAT1 to stop inflammation. Additionally, the ligand Protein S is produced by activated T cells and can stimulate TAM signaling to restrain DC activation [20, 43, 44].

In addition to signaling that halts the inflammatory response, the TAM receptors have been shown to mediate phagocytosis of apoptotic cells, a process also known as efferocytosis [45]. Macrophages and dendritic cells clear neutrophils that have accumulated as a result of inflammation [46]. MerTK then mediates signaling changes in cell shape required for engulfment of the apoptotic cells [38, 41, 47–49]. Specifically, MerTK mediates macrophage clearance as compared to dendritic cell clearance of apoptotic cells which involves Tyro3 or Axl [50, 51].

TAM receptors also play an important role in normal platelet aggregation and the integrity and permeability of blood vessels. Gas6 knockout mice demonstrated defective platelet aggregation, and knockouts of the TAM receptors revealed reduced platelet stabilization as well [52–54]. Injured blood vessels upregulate Axl and Gas6 [55], and TAM signaling is shown to participate in the formation of new blood vessels [56]. Because Gas6 and Axl signaling have been shown to prevent apoptosis in vascular smooth muscle cells (VSMCs) [57], the TAM receptors' role is currently being investigated in vascular diseases such as atherosclerosis.

While the three receptors are very similar in many aspects, each receptor does exhibit some specific characteristics that are important in understanding the family better.

Axl

The Axl receptor (also known as Ark, Tyro7, and Ufo) is the most studied of the TAM receptors. Structurally, there are three conserved tyrosines (698, 702, 703) which correspond to matching residues in the other family members (MerTK and Tyro3). These have been shown to be phosphorylated, but autophosphorylation of Axl has not been reported at these three sites. Rather, tyrosines 779, 821, and 866 are three residues in the c-terminal domain that act as autophosphorylation sites [58, 59]. It has been shown that these sites mediate the interaction of Axl with other signaling molecules such as phospholipase C (PLC), phosphatidyl inositol 3 kinase (PI3K), and Grb2 [58, 60, 61].

Axl is ubiquitously expressed in cell types, and as such there are many different consequences of Axl over-activation or down-regulation in cells and tissues. The earliest studies revealed Axl activation of Grb2, Ras, Raf1, MEK-1, and ERK1/2 via the expression of a chimeric EGFR/Axl receptor [60]. Additional studies have demonstrated Axl's mediation of cell proliferation through the MAPK/ERK pathway, cell survival through the PI3K/AKT pathway, and inflammation control through the JAK/STAT signaling [20, 43, 58, 60, 62–69].

MerTK

The MerTK receptor (also known as Eyk, Mer, Nyk, and Tyro12) is expressed in many hematopoietic lineages including monocytes/macrophages, dendritic cells, natural killer (NK) cells, NK T cells, megakaryocytes, and platelets. Additionally, MerTK is found in retina, prostate, testis, ovary, lung, brain, kidney, heart, and skeletal muscle [10, 29, 40, 51, 52, 70–72]. The primary autophosphorylation sites for MerTK are 749, 753, and 754, which are conserved residues in Axl and Tyro3 [59].

Original understanding of MerTK signaling arose from work with the FMS-MerTK receptor chimera developed by Ling and Kung in 1995 [73]. This indicates that MerTK signals through similar pathways to that of Axl including PI3K/AKT, PLC, and MAPK/ERK [47, 49, 59, 72, 74, 75]. MerTK is also considered to be one of the cell surface markers that discerns a macrophage from a dendritic cell or a monocyte [76]. Because of its presence in immune cells, studies of MerTK describe its role in stimulating both pro-inflammatory and anti-inflammatory signaling [35, 77].

Tyro3

The Tyro3 receptor (also known as Brt, Dtk, Rse, Sky, Tif, Etk-2, Rek) is the least studied and least characterized of the TAM receptors. Tyro3 is abundantly expressed in the nervous system as well as the ovary, testis, breast, lung, kidney, osteoclasts, retina, monocytes/macrophages, and platelets [9, 37, 40, 52, 78, 79]. Tyrosines 681, 685, and 686 correspond to the conserved residues in Axl and MerTK [80], but autophosphorylation sites of murine Tyro3 have been mapped to tyrosines 723 and 756 while K540 is required for kinase activity [81]. Coimmunoprecipitation of the receptor indicates interactions with the phosphorylated Src family kinase (SFK), Ran-binding protein M (RanBPM), protein phosphatase 1 (PP1), and the p85 subunit of PI3K. Signaling through an EGFR/Tyro3 chimera suggests activation of PI3K and AKT [82–84].

TAM family and cancer

The TAM receptors and ligands have gained much interest in recent years because of their involvement in cancer. Over-expression or up-regulation of Tyro3, Axl, MerTK, and/or Gas6 have been observed in a variety of solid and hematological tumors. Aberrant expression has even been associated with worse disease in several malignancies indicating the potential of these receptors and ligands as biomarkers of disease (see Fig. 1).

The oncogenic potential of these receptors has been demonstrated in their ability to transform NIH3T3 fibroblasts and BaF3 lymphocytes in culture [72, 73, 84]. Beyond associations and correlations, though, there has been no definitive relationship of causality between TAMs and cancer [36]. Nonetheless, abnormal expression of one or more TAM receptors is associated with cancer progression, metastasis, and resistance to therapy [85–88].

Proto-oncogenes can be activated through mechanisms of gene amplification, gene mutations, and altered protein expression. Most of the TAM receptor gene mutations reported involve MerTK and retinal degeneration [25, 89–92]. No activating TAM receptor mutations have been associated with the development of cancer. The oncogenic potential of these receptors is related to the abnormal regulation of signaling that leads to migration, invasion, angiogenesis, cell survival and tumor growth [36].

Clinically, the over-expression of Axl and MerTK in solid cancers is associated with more aggressive and invasive malignant behaviors as well as poorer patient survival outcomes [26, 93–99]. High expression of MerTK and Tyro3 in colorectal tumor tissue has also been linked to shorten patient survival [100]. In addition to the receptors being over-expressed, the TAM family ligand Gas6 is over-expressed in human cancers, implying that TAMs participate in autocrine circuits by over-expressing both receptors and ligands [94, 96, 101, 102].

Among the TAM family receptors, there are some unique differences in the expression and signaling in various cancers.

Axl

Altered Axl expression has been reported in many malignancies including AML and CML [5, 28, 103–105], erythroid leukemia [103], megakaryocytic leukemia [103], melanoma [106–108], head and neck squamous cell carcinoma [1, 109], esophageal carcinoma [110, 111], thyroid cancer [112–114], lung cancer [99, 115], breast cancer [116–118], ovarian cancer [119, 120], uterine endometrial cancer [121], prostate cancer [122–124], gastric cancer [98, 125], colon cancer [126], pancreatic cancer [127, 128], liver cancer [129], renal cell carcinoma [130], astrocytoma [67], glioblastoma [101], and osteosarcoma [131].

Axl has also been attributed to an adverse prognosis in acute myeloid leukemia [105], glioblastoma [132, 133], pancreatic cancer [127], esophageal adenocarcinoma [111, 134], non-small cell lung cancer [99, 135], breast cancer [136], head and neck squamous cell carcinoma [1, 109], renal cell carcinoma [137], and ovarian cancer [138]. The expression of Axl and MZF1 (myeloid zinc finger protein 1) was highly correlated with the clinical stage of cervical cancer and HPV16/18 infection [139].

Axl expression is also associated with increased metastasis [128, 136] and has been shown to be an essential factor for metastatic disease in ovarian cancer [140]. Furthermore, inhibition of Axl has been shown to reduce metastasis [141]. Axl over-expression activates the canonical EMT-inducing transcriptional factors TWIST, SNAIL, and SLUG, which can enhance cell motility, adhesion, and invasion. TWIST and SNAIL have also been shown to stimulate the *Axl* gene expression which reinforces the EMT process [142–145]. The Axl correlation with metastasis is further strengthened by its ability to enhance expression of matrix metalloproteinase 9 (MMP9) which promotes tissue remodeling and invasion [146].

MerTK

Aberrant expression of MerTK has been linked to acute lymphoblastic leukemias [10, 97, 147], mantle cell lymphoma [148], melanoma [106–108], rhabdomyosarcoma [149], pituitary adenoma [150], gastric cancer [98, 125], prostate cancer [95, 122–124], and breast cancer [116–118, 151].

In tumor cells, MerTK signals through the NF-kB pathway to activate anti-apoptotic and survival mechanisms. This is opposite of its normal inhibition of NF-kB to suppress the inflammatory response [35, 72, 75, 77, 144, 146, 152, 153]. MerTK also promotes survival by activating activated CDC42 kinase 1 (ACK1) in prostate cancer cells. This results in ubiquitylation and degradation of the multifunctional pro-apoptotic tumor suppressor WW domain-containing oxidoreductase (WWOX) [74].

Tyro3

Tyro3 is found primarily in the CNS [7], but abnormal expression has been associated with AML and CML [5, 28, 103–105], multiple myeloma [154], hepatocellular carcinoma [155], colon cancer [100, 156], thyroid cancer [157], and melanoma [158].

Similar to MerTK, Tyro3 uses the AKT/NF-kB signaling pathways to promote cell survival [104]. A genomic screen by Zhu S et al. identified TYRO3 as a MITF regulator in malignant melanoma indicating its potential as a therapeutic target [159]. Additionally, a study by Chien CW et al. revealed that Tyro3 over-expression in colon cancer increased metastatic potential though EMT signatures. The inhibition of Tyro3 increased sensitivity to anticancer drugs and induced cell apoptosis [156]. Genetic silencing of Tyro3 also suppressed proliferation in luminal-type breast cancer [160] and ovarian cancer cell lines [161].

TAM family and therapeutic resistance

A major clinical concern in the treatment of cancer patients is therapy resistance. Resistance can be present at the very beginning—never responding to treatment—or it can develop after a period of time of previously responding to treatment. These types of resistance—intrinsic and acquired—are difficult to predict and cause significant setbacks in patient care. There are several strategies of treating cancer, but resistance can be observed for each method.

This challenging problem creates a critical need to identify ways that cancer cells promote resistance in the hopes of overcoming these processes. Identification of proteins that drive resistance opens the opportunity to resensitize cancer cells to therapy. The TAM family of receptors has been associated with resistance to several treatment modalities including chemotherapy, radiation therapy, small molecule inhibitors, antibody therapeutics, and immunotherapy (see Fig. 2). As more is understood regarding the role of the TAM family in therapeutic resistance, greater is the possibility of inhibiting these proteins and increasing the efficacy of treatment for cancer patients.

Chemotherapy

Chemotherapy is a common treatment modality for many cancers. As one of the oldest therapies available for cancer treatment, these drugs do not identify targets but poison all

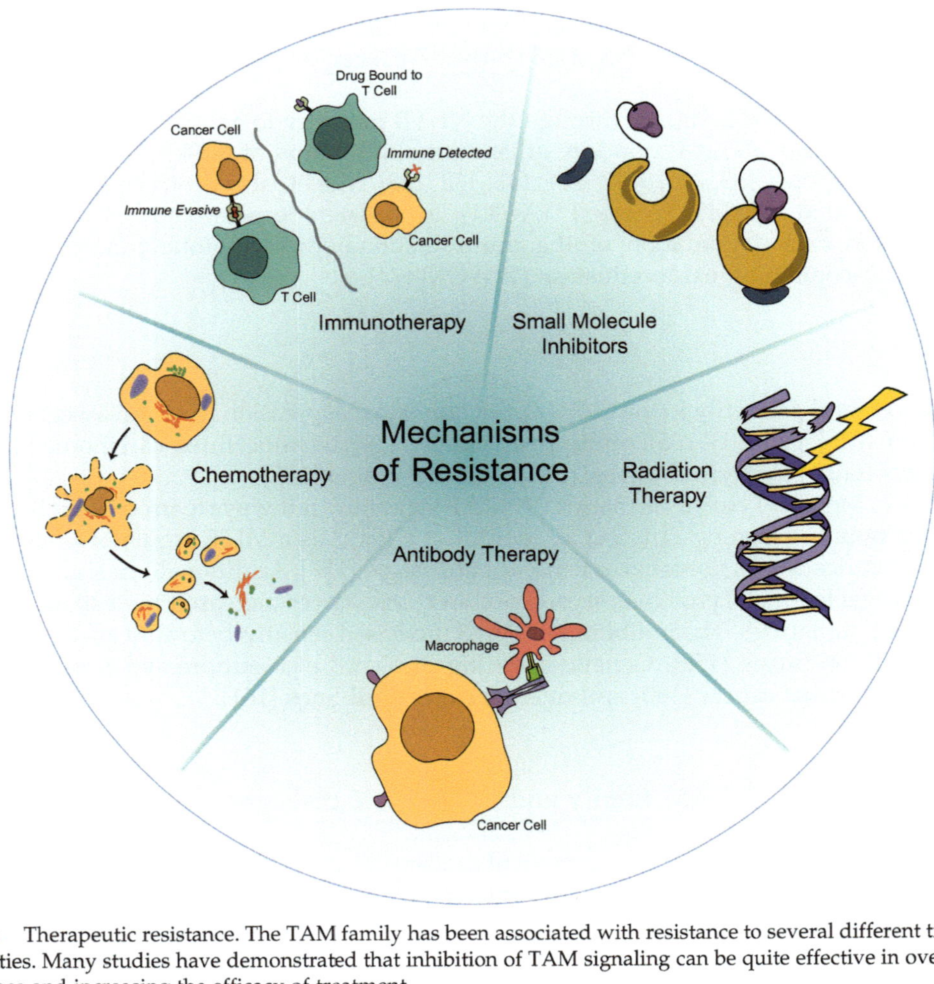

FIG. 2 Therapeutic resistance. The TAM family has been associated with resistance to several different treatment modalities. Many studies have demonstrated that inhibition of TAM signaling can be quite effective in overcoming resistance and increasing the efficacy of treatment.

rapidly dividing cells in the body. Because of this lack of specificity, they are known for their toxic effects. Therefore, their prolonged use can be dangerous for patients and resistance to this therapy is quite problematic.

The TAM receptors have been observed in various cancers to mediate resistance to different chemotherapeutics. In AML, Axl has been shown to promote resistance to the chemotherapy drugs doxorubicin, VP16, and cisplatin. This was demonstrated through higher expression of Axl in intrinsically resistant cell lines and the induction of Axl expression after exposure to chemotherapy [142]. In addition, refraction to cisplatin coincides with induction of Axl expression in NSCLC [162], and Axl inhibition resensitizes breast cancer stem cells to chemotherapy [144]. Axl inhibition has also been shown to increase the sensitivity of HNSCC cells to chemotherapies cisplatin and carboplatin [1]. Targeting Axl also overcomes resistance to docetaxel therapy in advanced prostate cancer [163].

MerTK and Tyro3 have also been shown to increase resistance to chemotherapies. MerTK over-expression in high-grade glioblastomas increases resistance to etoposide [164] while MerTK inhibition by mAb treatment increased sensitivity to carboplatin in NSCLC [165]. Tyro3 promotes chemoresistance in breast cancer [160], and treatment with Tyro3 siRNA was shown to reduce the proliferative capacity of ovarian cancer cells in the presence of paclitaxel treatment [161]. Treatment with an anti-Tyro3 antibody has been shown to increase sensitivity of primary colon cancer cells to paclitaxel, oxaliplactin, and 5FU by inducing cell apoptosis [156].

Studies of the family of receptors have shown that inhibition of Axl and MerTK leads to increased chemosensitivity to temozolomide, carboplatin, and vincristine in astrocytoma [67] and to cisplatin and vincristine in neuroblastoma [166]. Downregulation of Axl and Tyro3 has also been observed to reverse paclitaxel resistance in ovarian cancer cells [161, 167, 168]. In vitro inhibition of Tyro3 and MerTK reduced cell proliferation and sensitized colorectal cancer cells to 5FU therapy [100].

Radiation therapy

Radiation therapy is another common method of treating cancer. This involves the directed use of beams of intense energy at malignant cells. Many radiation oncologists consider radiotherapy to be a kind of "sledge hammer" that initiates damages in the tumor DNA followed by a cascade of changes in the microenvironment that augments the killing effect.

A few studies have associated the TAM family members Axl and MerTK with radiation therapy. Integrative analysis of HNSCC tumors indicated a significant association of Axl and radiation resistance [169]. Axl is over-expressed and activated in radiation-resistant HNSCC patient derived xenografts (PDXs) and targeting of Axl was shown to enhance the efficacy of radiation therapy in this cancer [1]. Over-expression of Axl has also been observed in syngeneic breast cancer tumors in mice unresponsive to radiation therapy. Knockout of Axl in these tumors increased the sensitivity of tumors to radiation [170]. Loss of MerTK combined with radiation therapy was observed to be sufficient for tumor regression in immunogenic tumors. In tumors with poor immune responses, the addition of TGFβ inhibitors was necessary to see similar results [171]. MerTK inhibition with the small molecule inhibitor UNC2025 combined with radiotherapy was shown to have a strong therapeutic effect in a syngeneic orthotopic glioblastoma mouse model [172]. No studies of Tyro3 association and radiation have been reported at this time.

Small molecule inhibitors

In recent decades, cancer research has begun focusing on personalizing treatment strategies to provide more effective care with minimalized toxicity. This shift has involved strategies to first discover the different ways that cancer survives, grows, and spreads, followed by researching how to target those specific characteristics. Rather than treat all patients with cytotoxic chemotherapies, this strategy seeks to treat the cancer more specifically based on certain genes or proteins that the cancer uses to advance. This often involves the use of small molecule inhibitors: organic compounds of less than 500 Da in weight. The majority of pharmaceuticals are considered small molecule inhibitors, and these inhibitors can be designed to interact with single or multiple targets. As the use of these inhibitors increases, so does our knowledge of resistance mechanisms that must be addressed [173].

Extensive work has been completed on studying Axl and resistance to small molecule inhibitors. In CML, increased Axl expression was correlated with resistance to the small molecule inhibitor imatinib [174]. Treatment with the Axl inhibitor BGB324 sensitized resistant CML cells independent of the BCR-Abl mutational status [175]. Axl has also been shown to promote resistance to ALK inhibitors in neuroblastoma through EMT induction [85, 176]. Inhibition of Axl induced reversal of EMT and sensitized cells to inhibition of the DNA repair protein PARP1 in TNBC, HNSCC, and NSCLC cells [177]. Targeting of Axl with R428 [178] and also along with the receptor tyrosine kinase Met using the drug cabozantinib [179] has been demonstrated to overcome resistance to sunitinib therapy in RCC.

Much of the work on the Axl's role with resistance to small molecule inhibitors has identified Axl as a key player in resistance to EGFR inhibitors. Axl has been observed to mediate resistance alone and also in cooperation with other RTKs [86]. In lung cancer, Axl has been associated with resistance to erlotinib and gefitinib [180, 181] and PI3K inhibitors [182]. In HNSCC resistant to PI3K inhibitors, dimerization of Axl with EGFR can activate the PKC/mTOR pathway to sustain tumor progression [183]. In TNBC, Axl predicted a lack of response to lapatinib and erlotinib [86], and Axl inhibition in HER2-positive, ER-positive breast cancer cells resensitized to lapatinib and trastuzumab therapies [184]. Axl was also identified as a key marker of resistance to drugs targeting EGFR and PDGFR in melanoma [185]. A study in glioblastoma showed that the TNF-JNK-Axl-ERK signaling axis can also mediate primary resistance to EGFR inhibition [186].

A few studies have indicated the MerTK's role in resistance to small molecule inhibitors as well. MerTK has been shown to be essential in NSCLC for increased resistance to erlotinib [187]. MerTK has also been observed to promote resistance to irreversible EGFR TKIs (Co-1686 and osimertinib) in NSCLC expressing the wild-type *EGFR* [188]. In addition, a novel MerTK-FLT3 inhibitor has been developed to overcome resistance-conferring FLT3 mutations in AML [189]. In melanoma resistant to BRAF inhibitors, MerTK was found to be upregulated and co-targeting of MerTK and BRAFV600 significantly delayed tumor growth in mice [190].

Investigation into family associations shows that Axl and MerTK over-expression leads to increased resistance to imatinib in gastrointestinal tumors [191] and NSCLC [192]. However, there is a significant lack of studies of small molecule resistance and Tyro3 association. More research is needed that focuses on examining the role of Tyro3 in resistance to small molecule inhibitors.

Antibody therapeutics

There are several ways that naked monoclonal antibodies (mAbs) can treat cancer. These antibodies can boost the immune response by flagging a cancer cell for the immune system to destroy. They can also boost the immune response by targeting the immune system checkpoints and work to remove these inhibitors of the T-cell responses. mAbs can also trigger cell-membrane destruction, prevent blood vessel growth, and attach to and block specific antigens on cancer cells that help cancer grow and spread. mAbs can be designed to target almost any extracellular or cell surface target [193].

More specifically, mAbs can be additionally modified to be considered bispecific or conjugated to improve their specificity. Bispecific antibodies are able to attach to two different proteins at the same time [194]. This can bring the cancer cells and the immune cells together to encourage immune-mediated elimination of the cancer. Antibody drug conjugates are another kind of antibody that circulates in the bloodstream and after binding to its specific antigen, it delivers a toxic substance or drug to the cell to which it is bound. Additionally, conjugated monoclonal antibodies can be radiolabeled to deliver radioactivity more specifically to cells or conjugated with chemotherapies to precisely direct these drugs to cancer cells [195].

Only three examples of TAM family-mediating resistance to monoclonal antibody therapy could be identified for this work. Each of these examples have associated Axl with resistance; no reports of MerTK or Tyro3 being associated with antibody therapy resistance could be found.

First, in NSCLC and HNSCC, Axl has been shown to be sufficient to mediate resistance to cetuximab, a monoclonal antibody against EGFR. Further investigation revealed a positive feedback loop in which EGFR regulated the expression of Axl mRNA to maintain EGFR activation in the presence of cetuximab therapy [2]. Another mechanism of cetuximab resistance discovered by Li et al. involves the translocation of EGFR to the nucleus in NSCLC [196]. Axl was identified as being critical for mediating this translocation as genetic loss of Axl reduced the accumulation of nuclear EGFR [197].

Second, trastuzumab is a monoclonal antibody for the HER2 receptor tyrosine kinase and this antibody is used in the treatment of HER2 positive breast cancer. Increased Axl expression was identified in breast cancer clones resistant to lapatinib and trastuzumab therapies. Inhibition of Axl with a multi-kinase inhibitor targeting Axl, Met, and VEGFR restored sensitivity in these cells [184].

Third, treatment with the immune checkpoint inhibitor nivolumab revealed an increased presence of intratumoral macrophages and activation of Axl. Inhibition of Axl with the small molecule inhibitor BGB324 resensitized to nivolumab treatment, and the combination synergistically prolonged the survival of mice bearing glioblastoma tumors [198].

Immunotherapy

A rapidly developing area of therapy involves targeting and manipulation of the immune system. Cancer cells have the ability to suppress the immune system from responding and destroying malignant cells. Therapy can target these checkpoints and silence their activity, encouraging the immune system to begin attacking the cancer cells. Other immunotherapy approaches involve stimulation of immune cells to boost their ability to respond and eliminate the cancer cells. The impact of immunotherapy in certain cancers has been remarkable, even achieving cures in some situations. However, these results are not shared by all patients and resistance to immunotherapy—both intrinsic and acquired—is a serious problem that must be investigated.

Axl expression is correlated with resistance to immune checkpoint based therapies in mouse mammary tumor models and cancer patients [170, 199]. Aguilera et al. showed that Axl suppresses antigen presentation through MHCI and enhances cytokine secretion. This promotes a suppressive microenvironment which sets the stage for use of checkpoint inhibitors such as anti-PD1 and anti-CTLA4 [170]. Analysis of somatic mutanomes and transcriptomes discovered that Axl is a key gene expressed in melanoma tumors from patients innately resistant to anti-PD1 immunotherapy [199]. Both of these studies indicate the potential benefit of combined Axl inhibition and checkpoint inhibition. Wnuk-Lipinska et al. confirms this hypothesis by combining the Axl inhibitor BGB324 with anti-CTLA4 and anti-PD1 in mouse models of pancreatic carcinoma. Results show that this combination enhances tumor clearance, survival, and tumor infiltration of cytotoxic T cells [200]. An additional study in syngeneic mouse models of ovarian (ID8) and breast (4T1) tumors that combined treatment of Axl inhibition and PD1 blockade (mAb SGI-7079) elicited a potent anti-tumor response [201]. Collectively, these results strongly suggest that Axl targeting could provide a beneficial approach to overcoming resistance to immune checkpoint based therapy.

In addition to Axl, MerTK has also been shown to be associated with immunotherapy resistance. MerTK has been shown to induce PD-L1 expression in breast cancer cells [94]. As this study suggests, efferocytosis driven by MerTK may be utilized by cancer cells to advance tumor progression. This was confirmed in an additional study by Kasikara et al. Here they showed that Axl and MerTK are able to upregulate the immune checkpoint molecule PD-L1 as well as promote phosphatidylserine-dependent efferocytosis and AKT-mediated chemoresistance in breast tumor cells [202]. In another study, MerTK signaling has been demonstrated to promote breast tumor growth via reduction of inflammatory cytokines and induction of wound-healing cytokines [203]. The identification of a phosphatidylserine-TAM-PD-L1 signaling axis in breast cancer and the stimulation of anti-tumor immune responses following MerTK inhibition in tumor leukocytes, provides rationale for targeting TAMs in the tumor microenvironment.

Though there are few studies that have been published regarding the TAM family and resistance to immunotherapy, it is widely acknowledged that targeting of the TAMs provides an opportunity to increase anti-tumor responses. Because the TAMs are key players in the regulation of the innate immune system in the tumor microenvironment, targeted inhibition of this family of receptors could reverse the immunosuppressive microenvironment [204]. This would only enhance the activity of immune checkpoint inhibitors in activating the immune signaling response. More studies are needed that use TAM inhibitors in settings where tumors have demonstrated resistance to immunotherapy. As immunotherapy gains popularity as a therapeutic approach, such resistant modeling will become increasingly necessary.

TAM family targeting approaches

Because of their prominent role in cancer, targeted treatment strategies have been rapidly developing for TAM receptors and ligands. The most common approaches include (1) small molecule kinase inhibitors that compete with ATP to inhibit receptor signaling, (2) monoclonal antibodies (mAbs) that disrupt receptor signaling through ligand binding, receptor endocytosis, or receptor degradation, and (3) monoclonal antibodies or fusion proteins that can act as ligand "sinks" to prevent ligand-dependent receptor activation [205]. Many of these therapeutics are directed towards one receptor (i.e., Axl or MerTK), but development of TAM family inhibitors has been increasing recently given the ability of the receptors to compensate in activity for each other [206]. Initial studies show promise of TAM-targeted therapeutics in reducing tumor progression.

Small molecule inhibitors

There are many small molecule inhibitors that have been developed over the years to target TAM receptors individually, as a family, or have been observed to have an impact on TAM family signaling. Among the most advanced inhibitors for Axl is a compound called BGB324—also known as R428—which was originally studied in breast cancer [143]. Inhibitors for MerTK include several compounds developed at the University of North Carolina including UNC569 [207], UNC1666 [208], UNC2025 [209], UNC2371A also known as MRX-2843 [189]. Often because of the similarities in structures of RTKs, small molecule inhibitors designed to target one RTK have the ability to inhibit other RTKs like the TAM family. An example of this would be the Met inhibitor called BMS777607 which has been shown to have activity against Tyro3, MerTK, and Axl [210]. This inhibitor though designed to target Met was used to target Axl in glioblastoma xenografts resulting in significant reduction in tumor volume [211].

Monoclonal antibodies (mAbs)

A few mAbs have been designed to target Axl. These include YW327.6S2 an Axl mAb that attenuates xenograft tumor growth in NSCLC and breast cancer models [212]. D9 and E8 mAbs were developed by Leconet et al. to internalize Axl and inhibit tumor growth and cell migration in pancreatic ductal adenocarcinoma [213] and triple negative breast cancer [214]. Mab173 is a mAb that induces receptor degradation and reduced tumor growth in Kaposi sarcoma in vivo xenograft studies [215]. AXL-107 MMAE is an antibody drug conjugate that was generated by Boshuizen et al. to target Axl specifically with the microtubule-disrupting agent MMAE. This agent was shown to cause tumor regression or tumor growth inhibition in patient-derived xenograft models of pancreas, esophageal, lung, thyroid, ovarian, cervical, melanoma, and sarcoma [216].

Ligand sinks

Another form of RTK inhibition is through the sequestering of ligands. This can be via a mAb for ligands, soluble receptors, or engineered decoys. The production of Axl and MerTK soluble receptors that lack tyrosine kinase domains and therefore can limit receptor activity by isolating ligands has been reported in a few studies [31–33, 217]. An Axl decoy receptor has also been designed to silence the Gas6-Axl signaling axis which can inhibit metastasis and disease progression in vivo [218].

Other strategies

Two additional methods of targeting TAM receptors that have been reported include natural compounds and aptamers. First, the use of a natural compound called MTE—water extract of the herb *Marsdenia tenacissima*—is able to overcome Axl and Met mediated erlotinib and gefitinib resistance in NSCLC cells [219]. Second, nucleic acid aptamers have been designed to bind to Axl and inhibit its activity leading to tumor growth inhibition in ovarian cancer [141] and NSCLC mouse xenograft models [220].

TAM family clinical developments

As the role of the TAM family in cancer becomes more established and targeting strategies for these receptors improve, translational studies have emphasized the need to begin investigating the potential of TAM family inhibition in the clinical realm. Thus, there are a multitude of promising clinical trials that are currently focused on the regulation of the TAM family receptors (see Table 1 for listing in order of appearance in text). These clinical trials include drug, genetic, and biologic interventions for all three of the TAM family receptors, of which Axl is the most prominent target.

BGB324, also known as bemcentinib or R428, is a small molecule inhibitor of Axl currently undergoing 6 different clinical trials. BerGenBio ASA is currently sponsoring a Phase Ib/II trial looking at BGB324 as a monotherapy in patients with acute myeloid lymphoma (AML) or myelodysplastic syndrome (MDS), followed by further review of BGB324 in combination with cytarabine or decitabine in AML patients. The study consists of a dose escalation phase followed by cohort expansion (ClinicalTrials.gov Identifier: NCT02488408). BerGenBio is further sponsoring a phase I/II open label study of BGB324 combined with erlotinib in stage IIIb and stage IV non-small cell lung cancer (NSCLC) (NCT02424617). In addition, there are two combined studies looking at BGB324 in combination with pembrolizumab in metastatic TNBC or TN-IBC who have been previously treated, is locally advanced, or unresectable (NCT03184558), as well as patients with previously treated advanced carcinoma of the lung (NCT03184571). The University of Texas Southwestern Medical Center is also sponsoring a phase I/II trial in collaboration with BerGenBio ASA to determine the complete response rate of BGB324 plus chemotherapy in patients with metastatic pancreatic adenocarcinoma (NCT03649321). Lastly, the University of Leichester is collaborating with BerGenBio ASA and the British Lung Foundation to complete a multi-arm phase II trial in patients with relapsed mesothelioma, of which one arm includes combined treatment with BGB324 and pembrolizumab (NCT03654833).

Another promising Axl kinase inhibitor with multiple clinical trials is TP-0903. In vitro TP-0903 has been shown to reverse the mesenchymal phenotype as well as resensitize cells to

TABLE 1 Clinical developments

#	TAM	ID	Title	Type	Phase	Condition	Intervention	Status	Est completion
1	Axl	NCT02488408	A Phase Ib/II Multicenter Open-label Study of **BGB324** in Patients With AML or MDS	Interventional	Phase 1, Phase 2	Acute Myeloid Leukemia, Myelodysplastic Syndromes	Drug: **BGB324**, Drug: Cytarabine, Drug: Decitabine	Recruiting	Aug-2018
2	Axl	NCT02424617	A Study of **BGB324** in Combination With Erlotinib in Patients With Non-Small Cell Lung Cancer	Interventional	Phase 1, Phase 2	Non-Small Cell Lung Cancer	Drug: **BGB324**, Drug: Erlotinib	Recruiting	Jul-2018
3	Axl	NCT03184558	Bemcentinib (**BGB324**) in Combination With Pembrolizumab in Patients With TNBC	Interventional	Phase 2	Triple Negative Breast Cancer, Inflammatory Breast Cancer Stage IV	Drug: **BGB324**, Drug: Pembrolizumab	Recruiting	Dec-2019
4	Axl	NCT03184571	Bemcentinib (**BGB324**) in Combination With Pembrolizumab in Patients With Advanced NSCLC	Interventional	Phase 2	Lung Cancer Metastatic, NSCLC Stage IV, Adenocarcinoma of Lung	Drug: **BGB324**, Drug: Pembrolizumab	Recruiting	Dec-2019
5	Axl	NCT03649321	Chemotherapy With Nab-paclitaxel/ Germcitabine/ Cisplatin +/− **BGB324** With Pancreatic Cancer	Interventional	Phase 1, Phase 2	Pancreatic Cancer	Drug: **BGB324**, Drug: Nab-paclitaxel, Drug: Gemcitabine, Drug: Cisplatin	Not yet recruiting	May-2022
6	Axl	NCT03654833	Mesothelioma Stratified Therapy (MiST): A Multi-drug Phase II Trial in Malignant Mesothelioma	Interventional	Phase 2	Malignant Mesothelioma	Drug: Rucaparib, Drug: Abemaciclib, Drug: **BGB324** & Pembrolizumab, Drug: Atezolizumab & Bevacizumab	Not yet recruiting	Nov-2021

Continued

TABLE 1 Clinical developments—Cont'd

#	TAM	ID	Title	Type	Phase	Condition	Intervention	Status	Est completion
7	Axl	NCT02729298	First-in-human Study of Oral **TP-0903** (a Novel Inhibitor of AXL Kinase) in Patients With Advanced Solid Tumors	Interventional	Phase 1	Advanced Solid Tumors, EGFR Positive NSCLC, Colorectal Carcinoma, Recurrent Ovarian Carcinoma, BRAF-Mutated Melanoma	Drug: **TP-0903**	Recruiting	Sep-2019
8	Axl	NCT03572634	Phase 1/2 Study of **TP-0903** (an Inhibitor of AXL Kinase) in Patients With Previously Treated CLL	Interventional	Phase 1, Phase 2	Chronic Lymphocytic Leukemia, Small Lymphocytic Lymphoma	Drug: **TP-0903**, Drug: Ibrutinib	Not yet recruiting	Apr-2021
9	Axl	NCT01639508	**Cabozantinib** in Patients With RET Fusion-Positive Advanced Non-Small Cell Lung Cancer and Those With Other Genotypes: ROS1 or NTRK Fusions or Increased MET or AXL Activity	Interventional	Phase 2	Non-Small Cell Lung Cancer	Drug: **Cabozantinib**	Recruiting	Jul-2019
10	Axl	NCT03425201	Niraparib in Combination With **Cabozantinib** (XL184) in Patients With Advanced Urothelial Cancer (NICARAGUA)	Interventional	Phase 1, Phase 2	Urothelial Cancer	Drug: Niraparib, Drug: **Cabozantinib**	Not yet recruiting	Dec-2019

11	Axl	NCT02034981	Phase 2 Study Assessing Efficacy and Safety of **Crizotinib** in Patients Harboring an Alteration on ALK, MET or ROS1	Interventional	Phase 2	Hematologic Cancers, Solid Tumors, Metastatic Cancer	Drug: **Crizotinib**	Active, not recruiting	Jul-2022
12	Axl	NCT00697632	Safety Study of Oral **MGCD265** Administered Without Interruption to Subjects With Advanced Malignancies	Interventional	Phase 1	Advanced Cancer	Drug: **MGCD265**	Active, not recruiting	Dec-2018
13	Axl	NCT02478866	Safety and Pharmacokinetics of **BPI-9016M** in Patients With Advanced Solid Tumors	Interventional	Phase 1	Solid Tumors	Drug: **BPI-9016M**	Recruiting	Dec-2017
14	Axl	NCT03536208	Biological Effect of **Warfarin** on Pancreatic Cancer	Interventional	Early Phase 1	Pancreatic Cancer	Drug: **Warfarin**	Not yet recruiting	Jun-2021
15	Axl	NCT03607955	Paclitaxel + Carboplatin With **AVB-S6-500** in Women With Stage III or IV Epithelial Ovarian, Primary Peritoneal, or Fallopian Tube Cancer Receiving Neoadjuvant Chemotherapy	Interventional	Phase 1	Epithelial Ovarian Cancer, Primary Peritoneal Carcinoma, Fallopian Tube Cancer	Biological: **AVB-S6-500**, Drug: Paclitaxel, Drug: Carboplatin	Not yet recruiting	Mar-2026

Continued

TABLE 1 Clinical developments—Cont'd

#	TAM	ID	Title	Type	Phase	Condition	Intervention	Status	Est completion
16	Axl	NCT03425279	**CAB-AXL-ADC** Safety and Efficacy Study in Patients With Solid Tumors	Interventional	Phase 1, Phase 2	Solid Tumor, Non Small Cell Lung Cancer, Castration-resistant Prostate Cancer, Pancreatic Cancer	Biological: **CAB-AXL-ADC**	Recruiting	Jan-2022
17	Axl	NCT02988817	**HuMax-AXL-ADC** Safety Study in Patients With Solid Tumors	Interventional	Phase 1, Phase 2	Ovarian Cancer, Cervical Cancer, Endometrial Cancer, Non Small Cell Lung Cancer, Thyroid Cancer, Melanoma, Sarcoma	Biological: **HuMax-AXL-ADC**	Recruiting	Jun-2021
18	Mer	NCT03510104	Pharmacokinetic and Safety Study of **MRX-2843** in Adults With Relapsed/Refractory Advanced and/or Metastatic Solid Tumors	Interventional	Phase 1	Advanced Cancer, Metastatic Cancer, Neoplasms, Neoplasm Metastasis, Neoplastic Processes, Pathologic Processes	Drug: **MRX-2843**	Recruiting	Nov-2019
19	Mer	NCT01482195	Trial of Ocular Subretinal Injection of a Recombinant Adeno-Associated Virus (rAAV2-VMD2-hMERTK) Gene Vector to Patients With Retinal Disease Due to MERTK Mutations	Interventional	Phase 1	Retinal Disease	Genetic: Recombinant Adeno-Associated Virus	Recruiting	Aug-2023

20	Mer + Axl	NCT02219711	Phase 1/1b Study of **MGCD516** in Patients With Advanced Cancer	Interventional	Phase 1	Advanced Cancer	Drug: **MGCD516**	Recruiting	Mar-2019
21	Mer + Axl	NCT03176277	A Study of **ONO-7475** in Patients With Acute Leukemias	Interventional	Phase 1	Acute Leukemia	Drug: **ONO-7475**	Recruiting	Nov-2019
22	TAM	NCT03522142	A Study Exploring the Safety and Tolerability of **INCB081776** in Participants With Advanced Solid Tumors	Interventional	Phase 1	Advanced Solid Tumors	Drug: **INCB081776**, Drug: Nivolumab	Recruiting	Nov-2019
23	TAM	NCT03454243	Oral Immunomodulatory Tyrosine Kinase Inhibitor in Patients With Locally Advanced or Metastatic Solid Tumors	Interventional	Early Phase 1	Advanced or Metastatic Solid Tumors	Drug: **RXDX-106**	Terminated	Jun-2018
24	Mer +Axl	NCT01139333	Blood and Tumor Tissue Samples From Young Patients With Acute Myeloid Leukemia	Observational	N/A	Leukemia	Genetic: gene expression analysis, Genetic: western blotting, Other: flow cytometry, Other: laboratory biomarker analysis	Completed	May-2016

There are a variety of clinical trials currently investigating the potential of TAM-targeted therapies in patients. These trials include drugs and biologicals directed at Axl, MerTK, and the entire TAM family.

treatment with other targeted agents such as cetuximab and erlotinib [221]. Tolero Pharmaceuticals is currently sponsoring two clinical trials with TP-0903, the first being a phase Ia study to determine the maximum tolerated dose of orally administered TP-0903 in patients with advanced solid tumors, then moving into a phase Ib study for data collection and to explore potential biomarkers and to further confirm safety (NCT02729298). Secondly, Tolero is sponsoring a phase I/II study of TP-0903 in patients with chronic lymphocytic leukemia and small lymphocytic leukemia in combination with ibrutinib (NCT03572634).

Cabozantinib is another drug-based Axl therapy currently being investigated in clinical trials as well as being approved by the FDA for treatment of progressive, metastatic medullary thyroid cancer. Cabozantinib is an oral, small molecule tyrosine kinase inhibitor which affects multiple targets, primarily Met, VEGFR1-3, RET, Axl, FLT3, and KIT [222]. Memorial Sloan Kettering Cancer Center is currently conducting a phase II monotherapy trial looking at cabozantinib in RET fusion positive NSCLC as well as NSCLC with increased Met or Axl activity or ROS1 and NTRK fusions (NCT01639508). The Fundacion CRIS de Investigación para Vencer el Cáncer is also sponsoring a phase I/II study for urothelial cancer looking at cabozantinib in combination with niraparib, a PARP inhibitor (NCT03425201).

There are many other Axl-targeting drugs currently being investigated as well. Crizotinib has been shown to decrease Axl phosphorylation in vitro and is currently undergoing a phase II clinical trial sponsored by UNICANCER for patients suffering from hematologic cancers, solid tumors, and metastatic cancers focusing primarily on inflammatory myofibroblastic tumor (IMT), glioblastoma, and rhabdomyosarcoma (NCT02034981). Mirati Therapeutics Inc. is sponsoring MGCD265, a Met/Axl tyrosine kinase inhibitor shown to slow tumor growth. This phase I study is going to focus on orally administered MGCD265 in advanced malignancies to evaluate dose escalation as well as pharmacokinetics and pharmacodynamics (NCT00697632). Betta Pharmaceuticals Co., Ltd. is sponsoring a phase I trial looking at patients with locally advanced or metastatic solid tumors and the potent small molecule inhibitor BPI-9016 M, a Met/Axl kinase inhibitor which has been shown to have anti-tumor activity both in vitro and in vivo (NCT02478866). Warfarin is a vitamin K antagonist that has been used in the clinic for prevention of thrombosis for decades. A recent study by Kirane et al. has shown that inhibition of Gas6-dependent Axl signaling via warfarin can block pancreatic tumor cell migration, invasiveness, and proliferation as well as resensitize cells to chemotherapy [223]. Because of this preclinical evidence, the University of Texas Southwestern Medical Center is sponsoring a clinical trial planned to investigate the effects of warfarin on Axl signaling in patients with pancreatic cancer (NCT03536208).

Beyond small molecule inhibitors of Axl, there are multiple biological-focused studies of Axl inhibition. The FDA has fast tracked a phase I study of ABV-S6-500, a soluble Fc-fusion protein with binding affinity for Gas6, in patients with ovarian cancer. The study is focusing on combining ABV-S6-500 with carboplatin and paclitaxel (NCT03607955). A phase I/II clinical study is also currently ongoing, looking at a conditionally active biologic Axl antibody drug conjugate (CAB-AXL-ADC). This method utilizes antibodies that bind to Axl in the context of the tumor microenvironment but do not bind to healthy tissues. The current clinical trial sponsored by BioAlta, LLC is focusing on the treatment of solid tumors in NSCLC, castration-resistant prostate cancer, and pancreatic cancer (NCT03425279). Genmab is further sponsoring a phase I/II clinical study looking at HuMax-Axl-ADC in ovarian and cervical cancers as well as NSCLC, thyroid cancer, and endometrial cancer (NCT02988817).

HuMax-Axl-ADC is another antibody drug conjugate consisting of an Axl antibody conjugated to the cytotoxic agent monomethyl auristatin E (MMAE). Once the monoclonal Axl antibody moiety of the conjugate binds to Axl, MMAE is released into the cancer cell and tubulin polymerization is inhibited.

Although Axl is by far the most prominent of the TAM family members currently being researched in clinical trials, there are translational studies and clinical trials focusing on MerTK inhibition in cancer and other diseases. Meryx is currently sponsoring a phase I study of MRX-2843, a MerTK/FLT3 small molecule inhibitor, to look at safety and a dose escalation in solid tumors (NCT03510104). As previously mentioned, mutations of MerTK have been observed to play a role in retinal disease. To combat this defect, a recombinant viral gene vector has been developed and used for gene replacement therapy in rats with retinitis pigmentosa [224]. Because of the preclinical data indicating considerable correction and preservation of the retinal function in rats, a clinical trial has started recruiting in order to treat patients with this gene therapy for mutations of MerTK in retinal disease (NCT01482195).

Although these monotherapeutic approaches to TAM family inhibition are promising, many efforts are being focused on dual inhibitors of Axl/MerTK or inhibitors of the entire TAM family. Mirati Therapeutics Inc. is sponsoring a phase I study of MGCD516, an RTK inhibitor shown to inhibit Met, Axl, and MerTK along with a variety of related family members. The study will focus on multiple cancer types with various genetic alterations including, but not limited to, Met, Axl, RET, TRK, DDR2, KDR, PDGFRA, KIT, or CBL (NCT02219711). Additionally, ONO-7475, sponsored by Ono Pharmaceuticals Co. Ltd., is a selective inhibitor of Axl and MerTK, acting via blocking Axl and MerTK signal transduction pathways (NCT03176277). The phase I study is focusing on dose escalation cohorts in acute leukemia patients. Another phase I clinical trial utilizing dual Axl/MerTK inhibition is INCB081776, which along with activity against Axl and MerTK, has also been shown to be selective against Tyro3 as well [225]. Sponsored by Incyte Corporation, the phase Ia/Ib trial will focus on monotherapeutics with INCB081776 before moving on to dual therapy with nivolumab (NCT03522142). Lastly, a novel TAM kinase inhibitor called RXDX-106 has been shown to have significant potential as a therapeutic agent in gastric cancer [226] as well as syngeneic breast cancer and colon cancer models [227]. Further tumor growth inhibition was observed in combination with anti-CTLA4 and anti-PD1 antibodies [227]. An early phase I clinical trial of RXDX was planned for advanced or metastatic solid tumors; however, this study was recently terminated by Ignyta, Inc. who was acquired by Roche (NCT03454243).

Conclusion

As this chapter has discussed, the TAM family of receptor tyrosine kinases plays an important role in cancer signaling and targeting of these receptors offers significant potential as a cancer treatment. Because of the role the receptors play in therapy resistance, inhibition of these proteins could enhance a variety of treatment strategies. Specific to head and neck cancer, this combination treatment of Axl and/or MerTK could improve radiation therapy, chemotherapy, and cetuximab therapy.

Further investigation is needed, though, regarding the specific biologic mechanisms that lead to therapy resistance. There have been a few indicators of different ways that the TAM family achieves resistance. Signaling through AKT for survival and MAPK for proliferation has been determined as one way of superseding therapy [2, 192, 228, 229]. Other mechanisms utilizing transcription factors to upregulate the TAM receptors has also been observed [2, 230]. Evasive signaling involved in upregulating immune checkpoints has been detected but not fully delineated [94, 170, 199, 202]. Compensatory signaling between family members—such as Axl and MerTK—has also been demonstrated as a way of circumventing treatment [206]. Better definition of these resistant mechanisms is needed to equip clinicians to address intrinsic resistance prior to treatment and combat acquired resistance as therapy continues.

Another area where additional preclinical data is critical is in the long-term inhibition of TAM proteins and how that could affect patients receiving this type of therapy for months, even years. Because of the early studies that revealed the susceptibility of TAM knockout mice to autoimmune disease, additional studies involving immunotherapy approaches should be completed to observe the lasting effects on the host immune system. These studies should be done with TAM monotherapeutics as well as TAM family inhibitors. Perhaps there will be no observed consequences of long-term inhibition of the TAMs, but confirmation of this is essential considering the number of patients already beginning to undergo TAM inhibition therapies.

The past decade has seen remarkable advances of our knowledge of the TAM family and a consequent increase in the number of inhibitors for these receptors. As studies of the TAM family in cancer and resistance continue to emerge, collecting human data will be imperative to translating these preclinical results to cancer patients.

Acknowledgments

None.

References

[1] Brand TM, Iida M, Stein AP, Corrigan KL, Braverman CM, Coan JP, Pearson HE, Bahrar H, Fowler TL, Bednarz BP, Saha S, Yang D, Gill PS, Lingen MW, Saloura V, Villaflor VM, Salgia R, Kimple RJ, Wheeler DL. AXL is a logical molecular target in head and neck squamous cell carcinoma. Clin Cancer Res 2015;21(11):2601–12.

[2] Brand TM, Iida M, Stein AP, Corrigan KL, Braverman CM, Luthar N, Toulany M, Gill PS, Salgia R, Kimple RJ, Wheeler DL. AXL mediates resistance to cetuximab therapy. Cancer Res 2014;74(18):5152–64.

[3] O'Bryan JP, Frye RA, Cogswell PC, Neubauer A, Kitch B, Prokop C, Espinosa 3rd R, Le Beau MM, Earp HS, Liu ET. Axl, a transforming gene isolated from primary human myeloid leukemia cells, encodes a novel receptor tyrosine kinase. Mol Cell Biol 1991;11(10):5016–31.

[4] Janssen JW, Schulz AS, Steenvoorden AC, Schmidberger M, Strehl S, Ambros PF, Bartram CR. A novel putative tyrosine kinase receptor with oncogenic potential. Oncogene 1991;6(11):2113–20.

[5] Liu E, Hjelle B, Bishop JM. Transforming genes in chronic myelogenous leukemia. Proc Natl Acad Sci U S A 1988;85(6):1952–6.

[6] McCloskey P, Pierce J, Koski RA, Varnum B, Liu ET. Activation of the Axl receptor tyrosine kinase induces mitogenesis and transformation in 32D cells. Cell Growth Differ 1994;5(10):1105–17.

[7] Lai C, Lemke G. An extended family of protein-tyrosine kinase genes differentially expressed in the vertebrate nervous system. Neuron 1991;6(5):691–704.
[8] Manning G, Whyte DB, Martinez R, Hunter T, Sudarsanam S. The protein kinase complement of the human genome. Science 2002;298(5600):1912–34.
[9] Lai C, Lemke G. Structure and expression of the Tyro 10 receptor tyrosine kinase. Oncogene 1994;9(3):377–83.
[10] Graham DK, Dawson TL, Mullaney DL, Snodgrass HR, Earp HS. Cloning and mRNA expression analysis of a novel human protooncogene, c-mer. Cell Growth Differ 1994;5(6):647–57.
[11] Stitt TN, Conn G, Gore M, Lai C, Bruno J, Radziejewski C, Mattsson K, Fisher J, Gies DR, Jones PF, et al. The anticoagulation factor protein S and its relative, Gas6, are ligands for the Tyro 3/Axl family of receptor tyrosine kinases. Cell 1995;80(4):661–70.
[12] Varnum BC, Young C, Elliott G, Garcia A, Bartley TD, Fridell YW, Hunt RW, Trail G, Clogston C, Toso RJ, et al. Axl receptor tyrosine kinase stimulated by the vitamin K-dependent protein encoded by growth-arrest-specific gene 6. Nature 1995;373(6515):623–6.
[13] Hafizi S, Dahlback B. Gas6 and protein S. Vitamin K-dependent ligands for the Axl receptor tyrosine kinase subfamily. FEBS J 2006;273(23):5231–44.
[14] Manfioletti G, Brancolini C, Avanzi G, Schneider C. The protein encoded by a growth arrest-specific gene (gas6) is a new member of the vitamin K-dependent proteins related to protein S, a negative coregulator in the blood coagulation cascade. Mol Cell Biol 1993;13(8):4976–85.
[15] Huang M, Rigby AC, Morelli X, Grant MA, Huang G, Furie B, Seaton B, Furie BC. Structural basis of membrane binding by Gla domains of vitamin K-dependent proteins. Nat Struct Biol 2003;10(9):751–6.
[16] Nagata K, Ohashi K, Nakano T, Arita H, Zong C, Hanafusa H, Mizuno K. Identification of the product of growth arrest-specific gene 6 as a common ligand for Axl, Sky, and Mer receptor tyrosine kinases. J Biol Chem 1996;271 (47):30022–7.
[17] Bhattacharyya S, Zagorska A, Lew ED, Shrestha B, Rothlin CV, Naughton J, Diamond MS, Lemke G, Young JA. Enveloped viruses disable innate immune responses in dendritic cells by direct activation of TAM receptors. Cell Host Microbe 2013;14(2):136–47.
[18] Heiring C, Dahlback B, Muller YA. Ligand recognition and homophilic interactions in Tyro3: structural insights into the Axl/Tyro3 receptor tyrosine kinase family. J Biol Chem 2004;279(8):6952–8.
[19] Sasaki T, Knyazev PG, Clout NJ, Cheburkin Y, Gohring W, Ullrich A, Timpl R, Hohenester E. Structural basis for Gas6-Axl signalling. EMBO J 2006;25(1):80–7.
[20] Lemke G, Rothlin CV. Immunobiology of the TAM receptors. Nat Rev Immunol 2008;8(5):327–36.
[21] Axelrod H, Pienta KJ. Axl as a mediator of cellular growth and survival. Oncotarget 2014;5(19):8818–52.
[22] Lewis PM, Crosier KE, Wood CR, Crosier PS. Analysis of the murine Dtk gene identifies conservation of genomic structure within a new receptor tyrosine kinase subfamily. Genomics 1996;31(1):13–9.
[23] Lu Q, Gore M, Zhang Q, Camenisch T, Boast S, Casagranda F, Lai C, Skinner MK, Klein R, Matsushima GK, Earp HS, Goff SP, Lemke G. Tyro-3 family receptors are essential regulators of mammalian spermatogenesis. Nature 1999;398(6729):723–8.
[24] Schulz AS, Schleithoff L, Faust M, Bartram CR, Janssen JW. The genomic structure of the human UFO receptor. Oncogene 1993;8(2):509–13.
[25] Gal A, Li Y, Thompson DA, Weir J, Orth U, Jacobson SG, Apfelstedt-Sylla E, Vollrath D. Mutations in MERTK, the human orthologue of the RCS rat retinal dystrophy gene, cause retinitis pigmentosa. Nat Genet 2000;26 (3):270–1.
[26] Linger RM, Keating AK, Earp HS, Graham DK. TAM receptor tyrosine kinases: biologic functions, signaling, and potential therapeutic targeting in human cancer. Adv Cancer Res 2008;100:35–83.
[27] Biesecker LG, Giannola DM, Emerson SG. Identification of alternative exons, including a novel exon, in the tyrosine kinase receptor gene Etk2/tyro3 that explain differences in 5′ cDNA sequences. Oncogene 1995;10 (11):2239–42.
[28] Neubauer A, Fiebeler A, Graham DK, O'Bryan JP, Schmidt CA, Barckow P, Serke S, Siegert W, Snodgrass HR, Huhn D, et al. Expression of axl, a transforming receptor tyrosine kinase, in normal and malignant hematopoiesis. Blood 1994;84(6):1931–41.
[29] Graham DK, Bowman GW, Dawson TL, Stanford WL, Earp HS, Snodgrass HR. Cloning and developmental expression analysis of the murine c-mer tyrosine kinase. Oncogene 1995;10(12):2349–59.

[30] O'Bryan JP, Fridell YW, Koski R, Varnum B, Liu ET. The transforming receptor tyrosine kinase, Axl, is post-translationally regulated by proteolytic cleavage. J Biol Chem 1995;270(2):551–7.
[31] Weinger JG, Omari KM, Marsden K, Raine CS, Shafit-Zagardo B. Up-regulation of soluble Axl and Mer receptor tyrosine kinases negatively correlates with Gas6 in established multiple sclerosis lesions. Am J Pathol 2009;175 (1):283–93.
[32] D'Arcangelo D, Ambrosino V, Giannuzzo M, Gaetano C, Capogrossi MC. Axl receptor activation mediates laminar shear stress anti-apoptotic effects in human endothelial cells. Cardiovasc Res 2006;71(4):754–63.
[33] Sather S, Kenyon KD, Lefkowitz JB, Liang X, Varnum BC, Henson PM, Graham DK. A soluble form of the Mer receptor tyrosine kinase inhibits macrophage clearance of apoptotic cells and platelet aggregation. Blood 2007;109(3):1026–33.
[34] Rothlin CV, Carrera-Silva EA, Bosurgi L, Ghosh S. TAM receptor signaling in immune homeostasis. Annu Rev Immunol 2015;33:355–91.
[35] Camenisch TD, Koller BH, Earp HS, Matsushima GK. A novel receptor tyrosine kinase, Mer, inhibits TNF-alpha production and lipopolysaccharide-induced endotoxic shock. J Immunol 1999;162(6):3498–503.
[36] Lemke G. Biology of the TAM receptors. Cold Spring Harb Perspect Biol 2013;5(11):a009076.
[37] Lu Q, Lemke G. Homeostatic regulation of the immune system by receptor tyrosine kinases of the Tyro 3 family. Science 2001;293(5528):306–11.
[38] Scott RS, McMahon EJ, Pop SM, Reap EA, Caricchio R, Cohen PL, Earp HS, Matsushima GK. Phagocytosis and clearance of apoptotic cells is mediated by MER. Nature 2001;411(6834):207–11.
[39] Duncan JL, LaVail MM, Yasumura D, Matthes MT, Yang H, Trautmann N, Chappelow AV, Feng W, Earp HS, Matsushima GK, Vollrath D. An RCS-like retinal dystrophy phenotype in mer knockout mice. Invest Ophthalmol Vis Sci 2003;44(2):826–38.
[40] Prasad D, Rothlin CV, Burrola P, Burstyn-Cohen T, Lu Q, Garcia de Frutos P, Lemke G. TAM receptor function in the retinal pigment epithelium. Mol Cell Neurosci 2006;33(1):96–108.
[41] Cohen PL, Caricchio R, Abraham V, Camenisch TD, Jennette JC, Roubey RA, Earp HS, Matsushima G, Reap EA. Delayed apoptotic cell clearance and lupus-like autoimmunity in mice lacking the c-mer membrane tyrosine kinase. J Exp Med 2002;196(1):135–40.
[42] Lemke G, Lu Q. Macrophage regulation by Tyro 3 family receptors. Curr Opin Immunol 2003;15(1):31–6.
[43] Rothlin CV, Ghosh S, Zuniga EI, Oldstone MB, Lemke G. TAM receptors are pleiotropic inhibitors of the innate immune response. Cell 2007;131(6):1124–36.
[44] Carrera Silva EA, Chan PY, Joannas L, Errasti AE, Gagliani N, Bosurgi L, Jabbour M, Perry A, Smith-Chakmakova F, Mucida D, Cheroutre H, Burstyn-Cohen T, Leighton JA, Lemke G, Ghosh S, Rothlin CV. T cell-derived protein S engages TAM receptor signaling in dendritic cells to control the magnitude of the immune response. Immunity 2013;39(1):160–70.
[45] Hochreiter-Hufford A, Ravichandran KS. Clearing the dead: apoptotic cell sensing, recognition, engulfment, and digestion. Cold Spring Harb Perspect Biol 2013;5(1):a008748.
[46] Stark MA, Huo Y, Burcin TL, Morris MA, Olson TS, Ley K. Phagocytosis of apoptotic neutrophils regulates granulopoiesis via IL-23 and IL-17. Immunity 2005;22(3):285–94.
[47] Guttridge KL, Luft JC, Dawson TL, Kozlowska E, Mahajan NP, Varnum B, Earp HS. Mer receptor tyrosine kinase signaling: prevention of apoptosis and alteration of cytoskeletal architecture without stimulation or proliferation. J Biol Chem 2002;277(27):24057–66.
[48] Hu B, Jennings JH, Sonstein J, Floros J, Todt JC, Polak T, Curtis JL. Resident murine alveolar and peritoneal macrophages differ in adhesion of apoptotic thymocytes. Am J Respir Cell Mol Biol 2004;30(5):687–93.
[49] Todt JC, Hu B, Curtis JL. The receptor tyrosine kinase MerTK activates phospholipase C gamma2 during recognition of apoptotic thymocytes by murine macrophages. J Leukoc Biol 2004;75(4):705–13.
[50] Seitz HM, Camenisch TD, Lemke G, Earp HS, Matsushima GK. Macrophages and dendritic cells use different Axl/Mertk/Tyro3 receptors in clearance of apoptotic cells. J Immunol 2007;178(9):5635–42.
[51] Behrens EM, Gadue P, Gong SY, Garrett S, Stein PL, Cohen PL. The mer receptor tyrosine kinase: expression and function suggest a role in innate immunity. Eur J Immunol 2003;33(8):2160–7.
[52] Angelillo-Scherrer A, de Frutos P, Aparicio C, Melis E, Savi P, Lupu F, Arnout J, Dewerchin M, Hoylaerts M, Herbert J, Collen D, Dahlback B, Carmeliet P. Deficiency or inhibition of Gas6 causes platelet dysfunction and protects mice against thrombosis. Nat Med 2001;7(2):215–21.
[53] Angelillo-Scherrer A, Burnier L, Flores N, Savi P, DeMol M, Schaeffer P, Herbert JM, Lemke G, Goff SP, Matsushima GK, Earp HS, Vesin C, Hoylaerts MF, Plaisance S, Collen D, Conway EM, Wehrle-Haller B,

Carmeliet P. Role of Gas6 receptors in platelet signaling during thrombus stabilization and implications for antithrombotic therapy. J Clin Invest 2005;115(2):237–46.

[54] Wang H, Chen S, Chen Y, Wang H, Wu H, Tang H, Xiong W, Ma J, Ge Y, Lu Q, Han D. The role of Tyro 3 subfamily receptors in the regulation of hemostasis and megakaryocytopoiesis. Haematologica 2007;92 (5):643–50.

[55] Melaragno MG, Wuthrich DA, Poppa V, Gill D, Lindner V, Berk BC, Corson MA. Increased expression of Axl tyrosine kinase after vascular injury and regulation by G protein-coupled receptor agonists in rats. Circ Res 1998;83(7):697–704.

[56] Burstyn-Cohen T, Heeb MJ, Lemke G. Lack of protein S in mice causes embryonic lethal coagulopathy and vascular dysgenesis. J Clin Invest 2009;119(10):2942–53.

[57] Melaragno MG, Cavet ME, Yan C, Tai LK, Jin ZG, Haendeler J, Berk BC. Gas6 inhibits apoptosis in vascular smooth muscle: role of Axl kinase and Akt. J Mol Cell Cardiol 2004;37(4):881–7.

[58] Braunger J, Schleithoff L, Schulz AS, Kessler H, Lammers R, Ullrich A, Bartram CR, Janssen JW. Intracellular signaling of the Ufo/Axl receptor tyrosine kinase is mediated mainly by a multi-substrate docking-site. Oncogene 1997;14(22):2619–31.

[59] Ling L, Templeton D, Kung HJ. Identification of the major autophosphorylation sites of Nyk/Mer, an NCAM-related receptor tyrosine kinase. J Biol Chem 1996;271(31):18355–62.

[60] Fridell YW, Jin Y, Quilliam LA, Burchert A, McCloskey P, Spizz G, Varnum B, Der C, Liu ET. Differential activation of the Ras/extracellular-signal-regulated protein kinase pathway is responsible for the biological consequences induced by the Axl receptor tyrosine kinase. Mol Cell Biol 1996;16(1):135–45.

[61] Pao-Chun L, Chan PM, Chan W, Manser E. Cytoplasmic ACK1 interaction with multiple receptor tyrosine kinases is mediated by Grb2: an analysis of ACK1 effects on Axl signaling. J Biol Chem 2009;284(50):34954–63.

[62] Goruppi S, Ruaro E, Varnum B, Schneider C. Requirement of phosphatidylinositol 3-kinase-dependent pathway and Src for Gas6-Axl mitogenic and survival activities in NIH 3T3 fibroblasts. Mol Cell Biol 1997;17 (8):4442–53.

[63] Goruppi S, Ruaro E, Varnum B, Schneider C. Gas6-mediated survival in NIH3T3 cells activates stress signalling cascade and is independent of Ras. Oncogene 1999;18(29):4224–36.

[64] Ming Cao W, Murao K, Imachi H, Sato M, Nakano T, Kodama T, Sasaguri Y, Wong NC, Takahara J, Ishida T. Phosphatidylinositol 3-OH kinase-Akt/protein kinase B pathway mediates Gas6 induction of scavenger receptor a in immortalized human vascular smooth muscle cell line. Arterioscler Thromb Vasc Biol 2001;21 (10):1592–7.

[65] Son BK, Kozaki K, Iijima K, Eto M, Nakano T, Akishita M, Ouchi Y. Gas6/Axl-PI3K/Akt pathway plays a central role in the effect of statins on inorganic phosphate-induced calcification of vascular smooth muscle cells. Eur J Pharmacol 2007;556(1–3):1–8.

[66] Weinger JG, Gohari P, Yan Y, Backer JM, Varnum B, Shafit-Zagardo B. In brain, Axl recruits Grb2 and the p85 regulatory subunit of PI3 kinase; in vitro mutagenesis defines the requisite binding sites for downstream Akt activation. J Neurochem 2008;106(1):134–46.

[67] Keating AK, Kim GK, Jones AE, Donson AM, Ware K, Mulcahy JM, Salzberg DB, Foreman NK, Liang X, Thorburn A, Graham DK. Inhibition of Mer and Axl receptor tyrosine kinases in astrocytoma cells leads to increased apoptosis and improved chemosensitivity. Mol Cancer Ther 2010;9(5):1298–307.

[68] Lijnen HR, Christiaens V, Scroyen L. Growth arrest-specific protein 6 receptor antagonism impairs adipocyte differentiation and adipose tissue development in mice. J Pharmacol Exp Ther 2011;337(2):457–64.

[69] Ou WB, Corson JM, Flynn DL, Lu WP, Wise SC, Bueno R, Sugarbaker DJ, Fletcher JA. AXL regulates mesothelioma proliferation and invasiveness. Oncogene 2011;30(14):1643–52.

[70] Paolino M, Penninger JM. The role of TAM family receptors in immune cell function: implications for cancer therapy. Cancers (Basel) 2016;8(10):97.

[71] Schoumacher M, Burbridge M. Key roles of AXL and MER receptor tyrosine kinases in resistance to multiple anticancer therapies. Curr Oncol Rep 2017;19(3):19.

[72] Georgescu MM, Kirsch KH, Shishido T, Zong C, Hanafusa H. Biological effects of c-Mer receptor tyrosine kinase in hematopoietic cells depend on the Grb2 binding site in the receptor and activation of NF-kappaB. Mol Cell Biol 1999;19(2):1171–81.

[73] Ling L, Kung HJ. Mitogenic signals and transforming potential of Nyk, a newly identified neural cell adhesion molecule-related receptor tyrosine kinase. Mol Cell Biol 1995;15(12):6582–92.

[74] Mahajan NP, Earp HS. An SH2 domain-dependent, phosphotyrosine-independent interaction between Vav1 and the Mer receptor tyrosine kinase: a mechanism for localizing guanine nucleotide-exchange factor action. J Biol Chem 2003;278(43):42596–603.
[75] Tibrewal N, Wu Y, D'Mello V, Akakura R, George TC, Varnum B, Birge RB. Autophosphorylation docking site Tyr-867 in Mer receptor tyrosine kinase allows for dissociation of multiple signaling pathways for phagocytosis of apoptotic cells and down-modulation of lipopolysaccharide-inducible NF-kappaB transcriptional activation. J Biol Chem 2008;283(6):3618–27.
[76] Xue J, Schmidt SV, Sander J, Draffehn A, Krebs W, Quester I, De Nardo D, Gohel TD, Emde M, Schmidleithner L, Ganesan H, Nino-Castro A, Mallmann MR, Labzin L, Theis H, Kraut M, Beyer M, Latz E, Freeman TC, Ulas T, Schultze JL. Transcriptome-based network analysis reveals a spectrum model of human macrophage activation. Immunity 2014;40(2):274–88.
[77] Sen P, Wallet MA, Yi Z, Huang Y, Henderson M, Mathews CE, Earp HS, Matsushima G, Baldwin Jr. AS, Tisch RM. Apoptotic cells induce Mer tyrosine kinase-dependent blockade of NF-kappaB activation in dendritic cells. Blood 2007;109(2):653–60.
[78] Katagiri M, Hakeda Y, Chikazu D, Ogasawara T, Takato T, Kumegawa M, Nakamura K, Kawaguchi H. Mechanism of stimulation of osteoclastic bone resorption through Gas6/Tyro 3, a receptor tyrosine kinase signaling, in mouse osteoclasts. J Biol Chem 2001;276(10):7376–82.
[79] Mark MR, Scadden DT, Wang Z, Gu Q, Goddard A, Godowski PJ. Rse, a novel receptor-type tyrosine kinase with homology to Axl/Ufo, is expressed at high levels in the brain. J Biol Chem 1994;269(14):10720–8.
[80] Graham DK, DeRyckere D, Davies KD, Earp HS. The TAM family: phosphatidylserine sensing receptor tyrosine kinases gone awry in cancer. Nat Rev Cancer 2014;14(12):769–85.
[81] Shao H, Wang A, Lauffenburger D, Wells A. Tyro3-mediated phosphorylation of ACTN4 at tyrosines is FAK-dependent and decreases susceptibility to cleavage by m-Calpain. Int J Biochem Cell Biol 2018;95:73–84.
[82] Toshima J, Ohashi K, Iwashita S, Mizuno K. Autophosphorylation activity and association with Src family kinase of Sky receptor tyrosine kinase. Biochem Biophys Res Commun 1995;209(2):656–63.
[83] Hafizi S, Dahlback B. Signalling and functional diversity within the Axl subfamily of receptor tyrosine kinases. Cytokine Growth Factor Rev 2006;17(4):295–304.
[84] Lan Z, Wu H, Li W, Wu S, Lu L, Xu M, Dai W. Transforming activity of receptor tyrosine kinase tyro3 is mediated, at least in part, by the PI3 kinase-signaling pathway. Blood 2000;95(2):633–8.
[85] Debruyne DN, Bhatnagar N, Sharma B, Luther W, Moore NF, Cheung NK, Gray NS, George RE. ALK inhibitor resistance in ALK(F1174L)-driven neuroblastoma is associated with AXL activation and induction of EMT. Oncogene 2016;35(28):3681–91.
[86] Meyer AS, Miller MA, Gertler FB, Lauffenburger DA. The receptor AXL diversifies EGFR signaling and limits the response to EGFR-targeted inhibitors in triple-negative breast cancer cells. Sci Signal 2013;6(287):ra66.
[87] Giles KM, Kalinowski FC, Candy PA, Epis MR, Zhang PM, Redfern AD, Stuart LM, Goodall GJ, Leedman PJ. Axl mediates acquired resistance of head and neck cancer cells to the epidermal growth factor receptor inhibitor erlotinib. Mol Cancer Ther 2013;12(11):2541–58.
[88] Akalu YT, Rothlin CV, Ghosh S. TAM receptor tyrosine kinases as emerging targets of innate immune checkpoint blockade for cancer therapy. Immunol Rev 2017;276(1):165–77.
[89] D'Cruz PM, Yasumura D, Weir J, Matthes MT, Abderrahim H, LaVail MM, Vollrath D. Mutation of the receptor tyrosine kinase gene Mertk in the retinal dystrophic RCS rat. Hum Mol Genet 2000;9(4):645–51.
[90] McHenry CL, Liu Y, Feng W, Nair AR, Feathers KL, Ding X, Gal A, Vollrath D, Sieving PA, Thompson DA. MERTK arginine-844-cysteine in a patient with severe rod-cone dystrophy: loss of mutant protein function in transfected cells. Invest Ophthalmol Vis Sci 2004;45(5):1456–63.
[91] Tada A, Wada Y, Sato H, Itabashi T, Kawamura M, Tamai M, Nishida K. Screening of the MERTK gene for mutations in Japanese patients with autosomal recessive retinitis pigmentosa. Mol Vis 2006;12:441–4.
[92] Tschernutter M, Jenkins SA, Waseem NH, Saihan Z, Holder GE, Bird AC, Bhattacharya SS, Ali RR, Webster AR. Clinical characterisation of a family with retinal dystrophy caused by mutation in the Mertk gene. Br J Ophthalmol 2006;90(6):718–23.
[93] Nguyen KQ, Tsou WI, Kotenko S, Birge RB. TAM receptors in apoptotic cell clearance, autoimmunity, and cancer. Autoimmunity 2013;46(5):294–7.
[94] Nguyen KQ, Tsou WI, Calarese DA, Kimani SG, Singh S, Hsieh S, Liu Y, Lu B, Wu Y, Garforth SJ, Almo SC, Kotenko SV, Birge RB. Overexpression of MERTK receptor tyrosine kinase in epithelial cancer cells drives efferocytosis in a gain-of-function capacity. J Biol Chem 2014;289(37):25737–49.

[95] Wu YM, Robinson DR, Kung HJ. Signal pathways in up-regulation of chemokines by tyrosine kinase MER/NYK in prostate cancer cells. Cancer Res 2004;64(20):7311–20.
[96] Verma A, Warner SL, Vankayalapati H, Bearss DJ, Sharma S. Targeting Axl and Mer kinases in cancer. Mol Cancer Ther 2011;10(10):1763–73.
[97] Graham DK, Salzberg DB, Kurtzberg J, Sather S, Matsushima GK, Keating AK, Liang X, Lovell MA, Williams SA, Dawson TL, Schell MJ, Anwar AA, Snodgrass HR, Earp HS. Ectopic expression of the proto-oncogene Mer in pediatric T-cell acute lymphoblastic leukemia. Clin Cancer Res 2006;12(9):2662–9.
[98] Wu CW, Li AF, Chi CW, Lai CH, Huang CL, Lo SS, Lui WY, Lin WC. Clinical significance of AXL kinase family in gastric cancer. Anticancer Res 2002;22(2B):1071–8.
[99] Shieh YS, Lai CY, Kao YR, Shiah SG, Chu YW, Lee HS, Wu CW. Expression of axl in lung adenocarcinoma and correlation with tumor progression. Neoplasia 2005;7(12):1058–64.
[100] Schmitz R, Valls AF, Yerbes R, von Richter S, Kahlert C, Loges S, Weitz J, Schneider M, Ruiz de Almodovar C, Ulrich A, Schmidt T. TAM receptors Tyro3 and Mer as novel targets in colorectal cancer. Oncotarget 2016;7(35):56355–70.
[101] Vajkoczy P, Knyazev P, Kunkel A, Capelle HH, Behrndt S, von Tengg-Kobligk H, Kiessling F, Eichelsbacher U, Essig M, Read TA, Erber R, Ullrich A. Dominant-negative inhibition of the Axl receptor tyrosine kinase suppresses brain tumor cell growth and invasion and prolongs survival. Proc Natl Acad Sci U S A 2006;103(15):5799–804.
[102] Holland SJ, Powell MJ, Franci C, Chan EW, Friera AM, Atchison RE, McLaughlin J, Swift SE, Pali ES, Yam G, Wong S, Lasaga J, Shen MR, Yu S, Xu W, Hitoshi Y, Bogenberger J, Nor JE, Payan DG, Lorens JB. Multiple roles for the receptor tyrosine kinase axl in tumor formation. Cancer Res 2005;65(20):9294–303.
[103] Challier C, Uphoff CC, Janssen JW, Drexler HG. Differential expression of the ufo/axl oncogene in human leukemia-lymphoma cell lines. Leukemia 1996;10(5):781–7.
[104] Crosier PS, Hall LR, Vitas MR, Lewis PM, Crosier KE. Identification of a novel receptor tyrosine kinase expressed in acute myeloid leukemic blasts. Leuk Lymphoma 1995;18(5–6):443–9.
[105] Rochlitz C, Lohri A, Bacchi M, Schmidt M, Nagel S, Fopp M, Fey MF, Herrmann R, Neubauer A. Axl expression is associated with adverse prognosis and with expression of Bcl-2 and CD34 in de novo acute myeloid leukemia (AML): results from a multicenter trial of the Swiss Group for Clinical Cancer Research (SAKK). Leukemia 1999;13(9):1352–8.
[106] Gyorffy B, Lage H. A web-based data warehouse on gene expression in human malignant melanoma. J Invest Dermatol 2007;127(2):394–9.
[107] Quong RY, Bickford ST, Ing YL, Terman B, Herlyn M, Lassam NJ. Protein kinases in normal and transformed melanocytes. Melanoma Res 1994;4(5):313–9.
[108] van Ginkel PR, Gee RL, Shearer RL, Subramanian L, Walker TM, Albert DM, Meisner LF, Varnum BC, Polans AS. Expression of the receptor tyrosine kinase Axl promotes ocular melanoma cell survival. Cancer Res 2004;64(1):128–34.
[109] Lee CH, Yen CY, Liu SY, Chen CK, Chiang CF, Shiah SG, Chen PH, Shieh YS. Axl is a prognostic marker in oral squamous cell carcinoma. Ann Surg Oncol 2012;19(Suppl. 3):S500–8.
[110] Nemoto T, Ohashi K, Akashi T, Johnson JD, Hirokawa K. Overexpression of protein tyrosine kinases in human esophageal cancer. Pathobiology 1997;65(4):195–203.
[111] Hector A, Montgomery EA, Karikari C, Canto M, Dunbar KB, Wang JS, Feldmann G, Hong SM, Haffner MC, Meeker AK, Holland SJ, Yu J, Heckrodt TJ, Zhang J, Ding P, Goff D, Singh R, Roa JC, Marimuthu A, Riggins GJ, Eshleman JR, Nelkin BD, Pandey A, Maitra A. The Axl receptor tyrosine kinase is an adverse prognostic factor and a therapeutic target in esophageal adenocarcinoma. Cancer Biol Ther 2010;10(10):1009–18.
[112] Ito T, Ito M, Naito S, Ohtsuru A, Nagayama Y, Kanematsu T, Yamashita S, Sekine I. Expression of the Axl receptor tyrosine kinase in human thyroid carcinoma. Thyroid 1999;9(6):563–7.
[113] Ito M, Nakashima M, Nakayama T, Ohtsuru A, Nagayama Y, Takamura N, Demedchik EP, Sekine I, Yamashita S. Expression of receptor-type tyrosine kinase, Axl, and its ligand, Gas6, in pediatric thyroid carcinomas around chernobyl. Thyroid 2002;12(11):971–5.
[114] Tanaka K, Nagayama Y, Nakano T, Takamura N, Namba H, Fukada S, Kuma K, Yamashita S, Niwa M. Expression profile of receptor-type protein tyrosine kinase genes in the human thyroid. Endocrinology 1998;139(3):852–8.
[115] Wimmel A, Glitz D, Kraus A, Roeder J, Schuermann M. Axl receptor tyrosine kinase expression in human lung cancer cell lines correlates with cellular adhesion. Eur J Cancer 2001;37(17):2264–74.

[116] Berclaz G, Altermatt HJ, Rohrbach V, Kieffer I, Dreher E, Andres AC. Estrogen dependent expression of the receptor tyrosine kinase axl in normal and malignant human breast. Ann Oncol 2001;12(6):819–24.
[117] Meric F, Lee WP, Sahin A, Zhang H, Kung HJ, Hung MC. Expression profile of tyrosine kinases in breast cancer. Clin Cancer Res 2002;8(2):361–7.
[118] Zantek ND, Walker-Daniels J, Stewart J, Hansen RK, Robinson D, Miao H, Wang B, Kung HJ, Bissell MJ, Kinch MS. MCF-10A-NeoST: a new cell system for studying cell-ECM and cell-cell interactions in breast cancer. Clin Cancer Res 2001;7(11):3640–8.
[119] Macleod K, Mullen P, Sewell J, Rabiasz G, Lawrie S, Miller E, Smyth JF, Langdon SP. Altered ErbB receptor signaling and gene expression in cisplatin-resistant ovarian cancer. Cancer Res 2005;65(15):6789–800.
[120] Sun W, Fujimoto J, Tamaya T. Coexpression of Gas6/Axl in human ovarian cancers. Oncology 2004;66(6):450–7.
[121] Sun WS, Fujimoto J, Tamaya T. Coexpression of growth arrest-specific gene 6 and receptor tyrosine kinases Axl and Sky in human uterine endometrial cancers. Ann Oncol 2003;14(6):898–906.
[122] Jacob AN, Kalapurakal J, Davidson WR, Kandpal G, Dunson N, Prashar Y, Kandpal RP. A receptor tyrosine kinase, UFO/Axl, and other genes isolated by a modified differential display PCR are overexpressed in metastatic prostatic carcinoma cell line DU145. Cancer Detect Prev 1999;23(4):325–32.
[123] Mahajan NP, Whang YE, Mohler JL, Earp HS. Activated tyrosine kinase Ack1 promotes prostate tumorigenesis: role of Ack1 in polyubiquitination of tumor suppressor Wwox. Cancer Res 2005;65(22):10514–23.
[124] Sainaghi PP, Castello L, Bergamasco L, Galletti M, Bellosta P, Avanzi GC. Gas6 induces proliferation in prostate carcinoma cell lines expressing the Axl receptor. J Cell Physiol 2005;204(1):36–44.
[125] Lin WC, Li AF, Chi CW, Chung WW, Huang CL, Lui WY, Kung HJ, Wu CW. Tie-1 protein tyrosine kinase: a novel independent prognostic marker for gastric cancer. Clin Cancer Res 1999;5(7):1745–51.
[126] Craven RJ, Xu LH, Weiner TM, Fridell YW, Dent GA, Srivastava S, Varnum B, Liu ET, Cance WG. Receptor tyrosine kinases expressed in metastatic colon cancer. Int J Cancer 1995;60(6):791–7.
[127] Koorstra JB, Karikari CA, Feldmann G, Bisht S, Rojas PL, Offerhaus GJ, Alvarez H, Maitra A. The Axl receptor tyrosine kinase confers an adverse prognostic influence in pancreatic cancer and represents a new therapeutic target. Cancer Biol Ther 2009;8(7):618–26.
[128] Song X, Wang H, Logsdon CD, Rashid A, Fleming JB, Abbruzzese JL, Gomez HF, Evans DB, Wang H. Overexpression of receptor tyrosine kinase Axl promotes tumor cell invasion and survival in pancreatic ductal adenocarcinoma. Cancer 2011;117(4):734–43.
[129] Tsou AP, Wu KM, Tsen TY, Chi CW, Chiu JH, Lui WY, Hu CP, Chang C, Chou CK, Tsai SF. Parallel hybridization analysis of multiple protein kinase genes: identification of gene expression patterns characteristic of human hepatocellular carcinoma. Genomics 1998;50(3):331–40.
[130] Chung BI, Malkowicz SB, Nguyen TB, Libertino JA, McGarvey TW. Expression of the proto-oncogene Axl in renal cell carcinoma. DNA Cell Biol 2003;22(8):533–40.
[131] Nakano T, Tani M, Ishibashi Y, Kimura K, Park YB, Imaizumi N, Tsuda H, Aoyagi K, Sasaki H, Ohwada S, Yokota J. Biological properties and gene expression associated with metastatic potential of human osteosarcoma. Clin Exp Metastasis 2003;20(7):665–74.
[132] Hutterer M, Knyazev P, Abate A, Reschke M, Maier H, Stefanova N, Knyazeva T, Barbieri V, Reindl M, Muigg A, Kostron H, Stockhammer G, Ullrich A. Axl and growth arrest-specific gene 6 are frequently overexpressed in human gliomas and predict poor prognosis in patients with glioblastoma multiforme. Clin Cancer Res 2008;14(1):130–8.
[133] Onken J, Vajkoczy P, Torka R, Hempt C, Patsouris V, Heppner FL, Radke J. Phospho-AXL is widely expressed in glioblastoma and associated with significant shorter overall survival. Oncotarget 2017;8(31):50403–14.
[134] Hsieh MS, Yang PW, Wong LF, Lee JM. The AXL receptor tyrosine kinase is associated with adverse prognosis and distant metastasis in esophageal squamous cell carcinoma. Oncotarget 2016;7(24):36956–70.
[135] Wu X, Ma W, Zhou Q, Yan H, Lim ZF, Huang M, Deng C, Yu X, Su H, Komo S, Yang H, Zhang X, Wen S, Zhang Z, Ma PC. AXL-GAS6 expression can predict for adverse prognosis in non-small cell lung cancer with brain metastases. J Cancer Res Clin Oncol 2017;143:1947–57.
[136] Gjerdrum C, Tiron C, Hoiby T, Stefansson I, Haugen H, Sandal T, Collett K, Li S, McCormack E, Gjertsen BT, Micklem DR, Akslen LA, Glackin C, Lorens JB. Axl is an essential epithelial-to-mesenchymal transition-induced regulator of breast cancer metastasis and patient survival. Proc Natl Acad Sci U S A 2010;107(3):1124–9.
[137] Zucca LE, Morini Matushita MA, da Silva Oliveira RJ, Scapulatempo-Neto C, de Lima MA, Ribeiro GG, Viana CR, Carcano FM, Reis RM. Expression of tyrosine kinase receptor AXL is associated with worse outcome of metastatic renal cell carcinomas treated with sunitinib. Urol Oncol 2018;36(1):11.e13–21.

[138] Rea K, Pinciroli P, Sensi M, Alciato F, Bisaro B, Lozneanu L, Raspagliesi F, Centritto F, Cabodi S, Defilippi P, Avanzi GC, Canevari S, Tomassetti A. Novel Axl-driven signaling pathway and molecular signature characterize high-grade ovarian cancer patients with poor clinical outcome. Oncotarget 2015;6(31):30859–75.
[139] Lee EH, Ji KY, Kim EM, Kim SM, Song HW, Choi HR, Chung BY, Choi HJ, Bai HW, Kang HS. Blockade of Axl signaling ameliorates HPV16E6-mediated tumorigenecity of cervical cancer. Sci Rep 2017;7(1):5759.
[140] Rankin EB, Fuh KC, Taylor TE, Krieg AJ, Musser M, Yuan J, Wei K, Kuo CJ, Longacre TA, Giaccia AJ. AXL is an essential factor and therapeutic target for metastatic ovarian cancer. Cancer Res 2010;70(19):7570–9.
[141] Kanlikilicer P, Ozpolat B, Aslan B, Bayraktar R, Gurbuz N, Rodriguez-Aguayo C, Bayraktar E, Denizli M, Gonzalez-Villasana V, Ivan C, Lokesh GLR, Amero P, Catuogno S, Haemmerle M, Wu SY, Mitra R, Gorenstein DG, Volk DE, de Franciscis V, Sood AK, Lopez-Berestein G. Therapeutic targeting of AXL receptor tyrosine kinase inhibits tumor growth and intraperitoneal metastasis in ovarian cancer models. Mol Ther Nucleic Acids 2017;9:251–62.
[142] Hong CC, Lay JD, Huang JS, Cheng AL, Tang JL, Lin MT, Lai GM, Chuang SE. Receptor tyrosine kinase AXL is induced by chemotherapy drugs and overexpression of AXL confers drug resistance in acute myeloid leukemia. Cancer Lett 2008;268(2):314–24.
[143] Holland SJ, Pan A, Franci C, Hu Y, Chang B, Li W, Duan M, Torneros A, Yu J, Heckrodt TJ, Zhang J, Ding P, Apatira A, Chua J, Brandt R, Pine P, Goff D, Singh R, Payan DG, Hitoshi Y. R428, a selective small molecule inhibitor of Axl kinase, blocks tumor spread and prolongs survival in models of metastatic breast cancer. Cancer Res 2010;70(4):1544–54.
[144] Asiedu MK, Beauchamp-Perez FD, Ingle JN, Behrens MD, Radisky DC, Knutson KL. AXL induces epithelial-to-mesenchymal transition and regulates the function of breast cancer stem cells. Oncogene 2014;33(10):1316–24.
[145] Lee HJ, Jeng YM, Chen YL, Chung L, Yuan RH. Gas6/Axl pathway promotes tumor invasion through the transcriptional activation of Slug in hepatocellular carcinoma. Carcinogenesis 2014;35(4):769–75.
[146] Tai KY, Shieh YS, Lee CS, Shiah SG, Wu CW. Axl promotes cell invasion by inducing MMP-9 activity through activation of NF-kappaB and Brg-1. Oncogene 2008;27(29):4044–55.
[147] Yeoh EJ, Ross ME, Shurtleff SA, Williams WK, Patel D, Mahfouz R, Behm FG, Raimondi SC, Relling MV, Patel A, Cheng C, Campana D, Wilkins D, Zhou X, Li J, Liu H, Pui CH, Evans WE, Naeve C, Wong L, Downing JR. Classification, subtype discovery, and prediction of outcome in pediatric acute lymphoblastic leukemia by gene expression profiling. Cancer Cell 2002;1(2):133–43.
[148] Ek S, Hogerkorp CM, Dictor M, Ehinger M, Borrebaeck CA. Mantle cell lymphomas express a distinct genetic signature affecting lymphocyte trafficking and growth regulation as compared with subpopulations of normal human B cells. Cancer Res 2002;62(15):4398–405.
[149] Khan J, Bittner ML, Saal LH, Teichmann U, Azorsa DO, Gooden GC, Pavan WJ, Trent JM, Meltzer PS. cDNA microarrays detect activation of a myogenic transcription program by the PAX3-FKHR fusion oncogene. Proc Natl Acad Sci U S A 1999;96(23):13264–9.
[150] Evans CO, Young AN, Brown MR, Brat DJ, Parks JS, Neish AS, Oyesiku NM. Novel patterns of gene expression in pituitary adenomas identified by complementary deoxyribonucleic acid microarrays and quantitative reverse transcription-polymerase chain reaction. J Clin Endocrinol Metab 2001;86(7):3097–107.
[151] Tavazoie SF, Alarcon C, Oskarsson T, Padua D, Wang Q, Bos PD, Gerald WL, Massague J. Endogenous human microRNAs that suppress breast cancer metastasis. Nature 2008;451(7175):147–52.
[152] Lay JD, Hong CC, Huang JS, Yang YY, Pao CY, Liu CH, Lai YP, Lai GM, Cheng AL, Su IJ, Chuang SE. Sulfasalazine suppresses drug resistance and invasiveness of lung adenocarcinoma cells expressing AXL. Cancer Res 2007;67(8):3878–87.
[153] Paccez JD, Vasques GJ, Correa RG, Vasconcellos JF, Duncan K, Gu X, Bhasin M, Libermann TA, Zerbini LF. The receptor tyrosine kinase Axl is an essential regulator of prostate cancer proliferation and tumor growth and represents a new therapeutic target. Oncogene 2013;32(6):689–98.
[154] De Vos J, Couderc G, Tarte K, Jourdan M, Requirand G, Delteil MC, Rossi JF, Mechti N, Klein B. Identifying intercellular signaling genes expressed in malignant plasma cells by using complementary DNA arrays. Blood 2001;98(3):771–80.
[155] Duan Y, Wong W, Chua SC, Wee HL, Lim SG, Chua BT, Ho HK. Overexpression of Tyro3 and its implications on hepatocellular carcinoma progression. Int J Oncol 2016;48(1):358–66.
[156] Chien CW, Hou PC, Wu HC, Chang YL, Lin SC, Lin SC, Lin BW, Lee JC, Chang YJ, Sun HS, Tsai SJ. Targeting TYRO3 inhibits epithelial-mesenchymal transition and increases drug sensitivity in colon cancer. Oncogene 2016;35(45):5872–81.

[157] Avilla E, Guarino V, Visciano C, Liotti F, Svelto M, Krishnamoorthy G, Franco R, Melillo RM. Activation of TYRO3/AXL tyrosine kinase receptors in thyroid cancer. Cancer Res 2011;71(5):1792–804.

[158] Demarest SJ, Gardner J, Vendel MC, Ailor E, Szak S, Huang F, Doern A, Tan X, Yang W, Grueneberg DA, Richards EJ, Endege WO, Harlow E, Koopman LA. Evaluation of Tyro3 expression, Gas6-mediated Akt phosphorylation, and the impact of anti-Tyro3 antibodies in melanoma cell lines. Biochemistry 2013;52(18):3102–18.

[159] Zhu S, Wurdak H, Wang Y, Galkin A, Tao H, Li J, Lyssiotis CA, Yan F, Tu BP, Miraglia L, Walker J, Sun F, Orth A, Schultz PG, Wu X. A genomic screen identifies TYRO3 as a MITF regulator in melanoma. Proc Natl Acad Sci U S A 2009;106(40):17025–30.

[160] Ekyalongo RC, Mukohara T, Funakoshi Y, Tomioka H, Kataoka Y, Shimono Y, Chayahara N, Toyoda M, Kiyota N, Minami H. TYRO3 as a potential therapeutic target in breast cancer. Anticancer Res 2014;34 (7):3337–45.

[161] Lee C. Overexpression of Tyro3 receptor tyrosine kinase leads to the acquisition of taxol resistance in ovarian cancer cells. Mol Med Rep 2015;12(1):1485–92.

[162] Kurokawa M, Ise N, Omi K, Goishi K, Higashiyama S. Cisplatin influences acquisition of resistance to molecular-targeted agents through epithelial-mesenchymal transition-like changes. Cancer Sci 2013;104 (7):904–11.

[163] Lin JZ, Wang ZJ, De W, Zheng M, Xu WZ, Wu HF, Armstrong A, Zhu JG. Targeting AXL overcomes resistance to docetaxel therapy in advanced prostate cancer. Oncotarget 2017;8(25):41064–77.

[164] Wang Y, Moncayo G, Morin Jr. P, Xue G, Grzmil M, Lino MM, Clement-Schatlo V, Frank S, Merlo A, Hemmings BA. Mer receptor tyrosine kinase promotes invasion and survival in glioblastoma multiforme. Oncogene 2013;32(7):872–82.

[165] Cummings CT, Linger RM, Cohen RA, Sather S, Kirkpatrick GD, Davies KD, DeRyckere D, Earp HS, Graham DK. Mer590, a novel monoclonal antibody targeting MER receptor tyrosine kinase, decreases colony formation and increases chemosensitivity in non-small cell lung cancer. Oncotarget 2014;5(21):10434–45.

[166] Li Y, Wang X, Bi S, Zhao K, Yu C. Inhibition of Mer and Axl receptor tyrosine kinases leads to increased apoptosis and improved chemosensitivity in human neuroblastoma. Biochem Biophys Res Commun 2015;457 (3):461–6.

[167] Suh YA, Jo SY, Lee HY, Lee C. Inhibition of IL-6/STAT3 axis and targeting Axl and Tyro3 receptor tyrosine kinases by apigenin circumvent taxol resistance in ovarian cancer cells. Int J Oncol 2015;46(3):1405–11.

[168] Kim NY, Lee HY, Lee C. Metformin targets Axl and Tyro3 receptor tyrosine kinases to inhibit cell proliferation and overcome chemoresistance in ovarian cancer cells. Int J Oncol 2015;47(1):353–60.

[169] Skinner HD, Giri U, Yang LP, Kumar M, Liu Y, Story MD, Pickering CR, Byers LA, Williams MD, Wang J, Shen L, Yoo SY, Fan YH, Molkentine DP, Beadle BM, Meyn RE, Myers JN, Heymach JV. Integrative analysis identifies a novel AXL-PI3 kinase-PD-L1 signaling axis associated with radiation resistance in head and neck cancer. Clin Cancer Res 2017;23(11):2713–22.

[170] Aguilera TA, Rafat M, Castellini L, Shehade H, Kariolis MS, Hui AB, Stehr H, von Eyben R, Jiang D, Ellies LG, Koong AC, Diehn M, Rankin EB, Graves EE, Giaccia AJ. Reprogramming the immunological microenvironment through radiation and targeting Axl. Nat Commun 2016;7:13898.

[171] Crittenden MR, Baird J, Friedman D, Savage T, Uhde L, Alice A, Cottam B, Young K, Newell P, Nguyen C, Bambina S, Kramer G, Akporiaye E, Malecka A, Jackson A, Gough MJ. Mertk on tumor macrophages is a therapeutic target to prevent tumor recurrence following radiation therapy. Oncotarget 2016;7(48):78653–66.

[172] Wu J, Frady LN, Bash RE, Cohen SM, Schorzman AN, Su YT, Irvin DM, Zamboni WC, Wang X, Frye SV, Ewend MG, Sulman EP, Gilbert MR, Earp HS, Miller CR. MerTK as a therapeutic target in glioblastoma. Neuro Oncol 2018;20(1):92–102.

[173] Hoelder S, Clarke PA, Workman P. Discovery of small molecule cancer drugs: successes, challenges and opportunities. Mol Oncol 2012;6(2):155–76.

[174] Dufies M, Jacquel A, Belhacene N, Robert G, Cluzeau T, Luciano F, Cassuto JP, Raynaud S, Auberger P. Mechanisms of AXL overexpression and function in Imatinib-resistant chronic myeloid leukemia cells. Oncotarget 2011;2(11):874–85.

[175] Ben-Batalla I, Erdmann R, Jorgensen H, Mitchell R, Ernst T, von Amsberg G, Schafhausen P, Velthaus JL, Rankin S, Clark RE, Koschmieder S, Schultze A, Mitra S, Vandenberghe P, Brummendorf TH, Carmeliet P, Hochhaus A, Pantel K, Bokemeyer C, Helgason GV, Holyoake TL, Loges S. Axl blockade by BGB324 inhibits BCR-ABL tyrosine kinase inhibitor-sensitive and -resistant chronic myeloid leukemia. Clin Cancer Res 2017;23 (9):2289–300.

[176] Xu F, Li H, Sun Y. Inhibition of Axl improves the targeted therapy against ALK-mutated neuroblastoma. Biochem Biophys Res Commun 2014;454(4):566–71.

[177] Balaji K, Vijayaraghavan S, Diao L, Tong P, Fan Y, Carey JP, Bui TN, Warner S, Heymach JV, Hunt KK, Wang J, Byers LA, Keyomarsi K. AXL inhibition suppresses the DNA damage response and sensitizes cells to PARP inhibition in multiple cancers. Mol Cancer Res 2017;15(1):45–58.

[178] van der Mijn JC, Broxterman HJ, Knol JC, Piersma SR, De Haas RR, Dekker H, Pham TV, Van Beusechem VW, Halmos B, Mier JW, Jimenez CR, Verheul HM. Sunitinib activates Axl signaling in renal cell cancer. Int J Cancer 2016;138(12):3002–10.

[179] Zhou L, Liu XD, Sun M, Zhang X, German P, Bai S, Ding Z, Tannir N, Wood CG, Matin SF, Karam JA, Tamboli P, Sircar K, Rao P, Rankin EB, Laird DA, Hoang AG, Walker CL, Giaccia AJ, Jonasch E. Targeting MET and AXL overcomes resistance to sunitinib therapy in renal cell carcinoma. Oncogene 2016;35(21):2687–97.

[180] Zhang Z, Lee JC, Lin L, Olivas V, Au V, LaFramboise T, Abdel-Rahman M, Wang X, Levine AD, Rho JK, Choi YJ, Choi CM, Kim SW, Jang SJ, Park YS, Kim WS, Lee DH, Lee JS, Miller VA, Arcila M, Ladanyi M, Moonsamy P, Sawyers C, Boggon TJ, Ma PC, Costa C, Taron M, Rosell R, Halmos B, Bivona TG. Activation of the AXL kinase causes resistance to EGFR-targeted therapy in lung cancer. Nat Genet 2012;44(8):852–60.

[181] Bae SY, Hong JY, Lee HJ, Park HJ, Lee SK. Targeting the degradation of AXL receptor tyrosine kinase to overcome resistance in gefitinib-resistant non-small cell lung cancer. Oncotarget 2015;6(12):10146–60.

[182] Byers LA, Diao L, Wang J, Saintigny P, Girard L, Peyton M, Shen L, Fan Y, Giri U, Tumula PK, Nilsson MB, Gudikote J, Tran H, Cardnell RJ, Bearss DJ, Warner SL, Foulks JM, Kanner SB, Gandhi V, Krett N, Rosen ST, Kim ES, Herbst RS, Blumenschein GR, Lee JJ, Lippman SM, Ang KK, Mills GB, Hong WK, Weinstein JN, Wistuba II, Coombes KR, Minna JD, Heymach JV. An epithelial-mesenchymal transition gene signature predicts resistance to EGFR and PI3K inhibitors and identifies Axl as a therapeutic target for overcoming EGFR inhibitor resistance. Clin Cancer Res 2013;19(1):279–90.

[183] Elkabets M, Pazarentzos E, Juric D, Sheng Q, Pelossof RA, Brook S, Benzaken AO, Rodon J, Morse N, Yan JJ, Liu M, Das R, Chen Y, Tam A, Wang H, Liang J, Gurski JM, Kerr DA, Rosell R, Teixido C, Huang A, Ghossein RA, Rosen N, Bivona TG, Scaltriti M, Baselga J. AXL mediates resistance to PI3Kalpha inhibition by activating the EGFR/PKC/mTOR axis in head and neck and esophageal squamous cell carcinomas. Cancer Cell 2015;27(4):533–46.

[184] Liu L, Greger J, Shi H, Liu Y, Greshock J, Annan R, Halsey W, Sathe GM, Martin AM, Gilmer TM. Novel mechanism of lapatinib resistance in HER2-positive breast tumor cells: activation of AXL. Cancer Res 2009;69(17):6871–8.

[185] Muller J, Krijgsman O, Tsoi J, Robert L, Hugo W, Song C, Kong X, Possik PA, Cornelissen-Steijger PD, Geukes Foppen MH, Kemper K, Goding CR, McDermott U, Blank C, Haanen J, Graeber TG, Ribas A, Lo RS, Peeper DS. Low MITF/AXL ratio predicts early resistance to multiple targeted drugs in melanoma. Nat Commun 2014;5:5712.

[186] Guo G, Gong K, Ali S, Ali N, Shallwani S, Hatanpaa KJ, Pan E, Mickey B, Burma S, Wang DH, Kesari S, Sarkaria JN, Zhao D, Habib AA. A TNF-JNK-Axl-ERK signaling axis mediates primary resistance to EGFR inhibition in glioblastoma. Nat Neurosci 2017;20(8):1074–84.

[187] Xie S, Li Y, Li X, Wang L, Yang N, Wang Y, Wei H. Mer receptor tyrosine kinase is frequently overexpressed in human non-small cell lung cancer, confirming resistance to erlotinib. Oncotarget 2015;6(11):9206–19.

[188] Yan D, Parker RE, Wang X, Frye SV, Earp 3rd HS, DeRyckere D, Graham DK. MERTK promotes resistance to irreversible EGFR tyrosine kinase inhibitors in non-small cell lung cancers expressing wild-type EGFR family members. Clin Cancer Res 2018; https://doi.org/10.1158/1078-0432.CCR-18-0040.

[189] Minson KA, Smith CC, DeRyckere D, Libbrecht C, Lee-Sherick AB, Huey MG, Lasater EA, Kirkpatrick GD, Stashko MA, Zhang W, Jordan CT, Kireev D, Wang X, Frye SV, Earp HS, Shah NP, Graham DK. The MERTK/FLT3 inhibitor MRX-2843 overcomes resistance-conferring FLT3 mutations in acute myeloid leukemia. JCI Insight 2016;1(3):e85630.

[190] Xue G, Kohler R, Tang F, Hynx D, Wang Y, Orso F, Pretre V, Ritschard R, Hirschmann P, Cron P, Roloff T, Dummer R, Mandala M, Bichet S, Genoud C, Meyer AG, Muraro MG, Spagnoli GC, Taverna D, Ruegg C, Merghoub T, Massi D, Tang H, Levesque MP, Dirnhofer S, Zippelius A, Hemmings BA, Wicki A. mTORC1/autophagy-regulated MerTK in mutant BRAFV600 melanoma with acquired resistance to BRAF inhibition. Oncotarget 2017;8(41):69204–18.

[191] Mahadevan D, Cooke L, Riley C, Swart R, Simons B, Della Croce K, Wisner L, Iorio M, Shakalya K, Garewal H, Nagle R, Bearss D. A novel tyrosine kinase switch is a mechanism of imatinib resistance in gastrointestinal stromal tumors. Oncogene 2007;26(27):3909–19.

[192] Linger RM, Cohen RA, Cummings CT, Sather S, Migdall-Wilson J, Middleton DH, Lu X, Baron AE, Franklin WA, Merrick DT, Jedlicka P, DeRyckere D, Heasley LE, Graham DK. Mer or Axl receptor tyrosine kinase inhibition promotes apoptosis, blocks growth and enhances chemosensitivity of human non-small cell lung cancer. Oncogene 2013;32(29):3420–31.

[193] Ayyar BV, Arora S, O'Kennedy R. Coming-of-age of antibodies in cancer therapeutics. Trends Pharmacol Sci 2016;37(12):1009–28.

[194] Lewis SM, Wu X, Pustilnik A, Sereno A, Huang F, Rick HL, Guntas G, Leaver-Fay A, Smith EM, Ho C, Hansen-Estruch C, Chamberlain AK, Truhlar SM, Conner EM, Atwell S, Kuhlman B, Demarest SJ. Generation of bispecific IgG antibodies by structure-based design of an orthogonal Fab interface. Nat Biotechnol 2014;32 (2):191–8.

[195] Redman JM, Hill EM, AlDeghaither D, Weiner LM. Mechanisms of action of therapeutic antibodies for cancer. Mol Immunol 2015;67(2 Pt. A):28–45.

[196] Li C, Iida M, Dunn EF, Ghia AJ, Wheeler DL. Nuclear EGFR contributes to acquired resistance to cetuximab. Oncogene 2009;28(43):3801–13.

[197] Brand TM, Iida M, Corrigan KL, Braverman CM, Coan JP, Flanigan BG, Stein AP, Salgia R, Rolff J, Kimple RJ, Wheeler DL. The receptor tyrosine kinase AXL mediates nuclear translocation of the epidermal growth factor receptor. Sci Signal 2017;10(460). https://doi.org/10.1126/scisignal.aag1064.

[198] Sadahiro H, Kang KD, Gibson JT, Minata M, Yu H, Shi J, Chhipa R, Chen Z, Lu S, Simoni Y, Furuta T, Sabit H, Zhang S, Bastola S, Yamaguchi S, Alsheikh H, Komarova S, Wang J, Kim SH, Hambardzumyan D, Lu X, Newell EW, DasGupta B, Nakada M, Lee LJ, Nabors B, Norian LA, Nakano I. Activation of the receptor tyrosine kinase AXL regulates the immune microenvironment in glioblastoma. Cancer Res 2018;78 (11):3002–13.

[199] Hugo W, Zaretsky JM, Sun L, Song C, Moreno BH, Hu-Lieskovan S, Berent-Maoz B, Pang J, Chmielowski B, Cherry G, Seja E, Lomeli S, Kong X, Kelley MC, Sosman JA, Johnson DB, Ribas A, Lo RS. Genomic and transcriptomic features of response to anti-PD-1 therapy in metastatic melanoma. Cell 2016;165(1):35–44.

[200] Wnuk-Lipinska K, Davidsen K, Blø M, Engelsen A, Kang J, Hodneland L, Lie M, Bougnaud S, Aguilera K, Ahmed L, Rybicka A, Nævdal EM, D'eyna P, Boniecka A, Oddbjørn S, Chouaib S, Brekken R, Gausdal G, Lorens JB. BGB324, a selective small molecule inhibitor of receptor tyrosine kinase AXL, abrogates tumor intrinsic and microenvironmental immune suppression and enhances immune checkpoint inhibitor efficacy in lung and mammary adenocarcinoma models [abstract]. In: Proceedings of the American Association for Cancer Research annual meeting 2017, April 1–5. Washington, DC; Philadelphia, PA: AACR; 2017.

[201] Guo Z, Li Y, Zhang D, Ma J. Axl inhibition induces the antitumor immune response which can be further potentiated by PD-1 blockade in the mouse cancer models. Oncotarget 2017;8(52):89761–74.

[202] Kasikara C, Kumar S, Kimani S, Tsou WI, Geng K, Davra V, Sriram G, Devoe C, Nguyen KN, Antes A, Krantz A, Rymarczyk G, Wilczynski A, Empig C, Freimark B, Gray M, Schlunegger K, Hutchins J, Kotenko SV, Birge RB. Phosphatidylserine sensing by TAM receptors regulates AKT-dependent chemoresistance and PD-L1 expression. Mol Cancer Res 2017;15(6):753–64.

[203] Cook RS, Jacobsen KM, Wofford AM, DeRyckere D, Stanford J, Prieto AL, Redente E, Sandahl M, Hunter DM, Strunk KE, Graham DK, Earp 3rd HS. MerTK inhibition in tumor leukocytes decreases tumor growth and metastasis. J Clin Invest 2013;123(8):3231–42.

[204] Vouri M, Hafizi S. TAM receptor tyrosine kinases in cancer drug resistance. Cancer Res 2017;77(11):2775–8.

[205] Linger RM, Keating AK, Earp HS, Graham DK. Taking aim at Mer and Axl receptor tyrosine kinases as novel therapeutic targets in solid tumors. Expert Opin Ther Targets 2010;14(10):1073–90.

[206] McDaniel NK, Cummings CT, Iida M, Hulse J, Pearson HE, Vasileiadi E, Parker RE, Orbuch RA, Ondracek OJ, Welke NB, Kang GH, Davies KD, Wang X, Frye SV, Earp HS, Harari PM, Kimple RJ, DeRyckere D, Graham DK, Wheeler DL. MERTK mediates intrinsic and adaptive resistance to AXL-targeting agents. Mol Cancer Ther 2018;17(11):2297–308.

[207] Christoph S, Deryckere D, Schlegel J, Frazer JK, Batchelor LA, Trakhimets AY, Sather S, Hunter DM, Cummings CT, Liu J, Yang C, Kireev D, Simpson C, Norris-Drouin J, Hull-Ryde EA, Janzen WP, Johnson GL, Wang X, Frye SV, Earp 3rd HS, Graham DK. UNC569, a novel small-molecule mer inhibitor with efficacy against acute lymphoblastic leukemia in vitro and in vivo. Mol Cancer Ther 2013;12(11):2367–77.

[208] Lee-Sherick AB, Zhang W, Menachof KK, Hill AA, Rinella S, Kirkpatrick G, Page LS, Stashko MA, Jordan CT, Wei Q, Liu J, Zhang D, DeRyckere D, Wang X, Frye S, Earp HS, Graham DK. Efficacy of a Mer and Flt3 tyrosine kinase small molecule inhibitor, UNC1666, in acute myeloid leukemia. Oncotarget 2015;6 (9):6722–36.

[209] Zhang W, DeRyckere D, Hunter D, Liu J, Stashko MA, Minson KA, Cummings CT, Lee M, Glaros TG, Newton DL, Sather S, Zhang D, Kireev D, Janzen WP, Earp HS, Graham DK, Frye SV, Wang X. UNC2025, a potent and orally bioavailable MER/FLT3 dual inhibitor. J Med Chem 2014;57(16):7031–41.

[210] Suarez RM, Chevot F, Cavagnino A, Saettel N, Radvanyi F, Piguel S, Bernard-Pierrot I, Stoven V, Legraverend M. Inhibitors of the TAM subfamily of tyrosine kinases: synthesis and biological evaluation. Eur J Med Chem 2013;61:2–25.

[211] Onken J, Torka R, Korsing S, Radke J, Krementeskaia I, Nieminen M, Bai X, Ullrich A, Heppner F, Vajkoczy P. Inhibiting receptor tyrosine kinase AXL with small molecule inhibitor BMS-777607 reduces glioblastoma growth, migration, and invasion in vitro and in vivo. Oncotarget 2016;7(9):9876–89.

[212] Ye X, Li Y, Stawicki S, Couto S, Eastham-Anderson J, Kallop D, Weimer R, Wu Y, Pei L. An anti-Axl monoclonal antibody attenuates xenograft tumor growth and enhances the effect of multiple anticancer therapies. Oncogene 2010;29(38):5254–64.

[213] Leconet W, Larbouret C, Chardes T, Thomas G, Neiveyans M, Busson M, Jarlier M, Radosevic-Robin N, Pugniere M, Bernex F, Penault-Llorca F, Pasquet JM, Pelegrin A, Robert B. Preclinical validation of AXL receptor as a target for antibody-based pancreatic cancer immunotherapy. Oncogene 2014;33(47):5405–14.

[214] Leconet W, Chentouf M, du Manoir S, Chevalier C, Sirvent A, Ait-Arsa I, Busson M, Jarlier M, Radosevic-Robin N, Theillet C, Chalbos D, Pasquet JM, Pelegrin A, Larbouret C, Robert B. Therapeutic activity of anti-AXL antibody against triple-negative breast cancer patient-derived xenografts and metastasis. Clin Cancer Res 2017;23(11):2806–16.

[215] Liu R, Gong M, Li X, Zhou Y, Gao W, Tulpule A, Chaudhary PM, Jung J, Gill PS. Induction, regulation, and biologic function of Axl receptor tyrosine kinase in Kaposi sarcoma. Blood 2010;116(2):297–305.

[216] Boshuizen J, Koopman LA, Krijgsman O, Shahrabi A, van den Heuvel EG, Ligtenberg MA, Vredevoogd DW, Kemper K, Kuilman T, Song JY, Pencheva N, Mortensen JT, Foppen MG, Rozeman EA, Blank CU, Janmaat ML, Satijn D, Breij ECW, Peeper DS, Parren P. Cooperative targeting of melanoma heterogeneity with an AXL antibody-drug conjugate and BRAF/MEK inhibitors. Nat Med 2018;24(2):203–12.

[217] Costa M, Bellosta P, Basilico C. Cleavage and release of a soluble form of the receptor tyrosine kinase ARK in vitro and in vivo. J Cell Physiol 1996;168(3):737–44.

[218] Kariolis MS, Miao YR, Jones 2nd DS, Kapur S, Mathews II, Giaccia AJ, Cochran JR. An engineered Axl 'decoy receptor' effectively silences the Gas6-Axl signaling axis. Nat Chem Biol 2014;10(11):977–83.

[219] Han SY, Zhao W, Han HB, Sun H, Xue D, Jiao YN, He XR, Jiang ST, Li PP. Marsdenia tenacissima extract overcomes Axl- and Met-mediated erlotinib and gefitinib cross-resistance in non-small cell lung cancer cells. Oncotarget 2017;8(34):56893–905.

[220] Cerchia L, Esposito CL, Camorani S, Rienzo A, Stasio L, Insabato L, Affuso A, de Franciscis V. Targeting Axl with an high-affinity inhibitory aptamer. Mol Ther 2012;20(12):2291–303.

[221] Soh KK, Kim W, Lee YS. Abstract 235: AXL inhibition leads to a reversal of a mesenchymal phenotype sensitizing cancer cells to targeted agents and immune-oncology therapies. Exp Mol Ther 2016;76(14 Suppl). https://doi.org/10.1158/1538-7445.AM2016-235.

[222] Yakes FM, Chen J, Tan J, Yamaguchi K, Shi Y, Yu P, Qian F, Chu F, Bentzien F, Cancilla B, Orf J, You A, Laird AD, Engst S, Lee L, Lesch J, Chou YC, Joly AH. Cabozantinib (XL184), a novel MET and VEGFR2 inhibitor, simultaneously suppresses metastasis, angiogenesis, and tumor growth. Mol Cancer Ther 2011;10(12):2298–308.

[223] Kirane A, Ludwig KF, Sorrelle N, Haaland G, Sandal T, Ranaweera R, Toombs JE, Wang M, Dineen SP, Micklem D, Dellinger MT, Lorens JB, Brekken RA. Warfarin blocks Gas6-mediated Axl activation required for pancreatic cancer epithelial plasticity and metastasis. Cancer Res 2015;75(18):3699–705.

[224] Tschernutter M, Schlichtenbrede FC, Howe S, Balaggan KS, Munro PM, Bainbridge JW, Thrasher AJ, Smith AJ, Ali RR. Long-term preservation of retinal function in the RCS rat model of retinitis pigmentosa following lentivirus-mediated gene therapy. Gene Ther 2005;12(8):694–701.

[225] Favata M, Lasky K, Lo Y, Feldman P, Li J, Chen Y, Stevens C, Ye M, Wang H, Liu K, Wynn R, Li Y, Harris J, Landman R, Li Y, Wang X, He C, Li Y-L, Xue CB, Yao W, Rios-Doria J, Gao Z, Covington M, Liu XM, Koblish H, Scherle P. Characterization of INCB081776, a potent and selective dual AXL/MER kinase inhibitor [abstract]. In: Proceedings of the American Association for Cancer Research annual meeting 2018, April 14–18. Chicago, IL; Philadelphia, PA: AACR; 2018.

[226] Kim JE, Kim Y, Li G, Kim ST, Kim K, Park SH, Park JO, Park YS, Lim HY, Lee H, Sohn TS, Kim KM, Kang WK, Lee J. MerTK inhibition by RXDX-106 in MerTK activated gastric cancer cell lines. Oncotarget 2017;8(62):105727–34.

[227] Yokoyama Y, Lew ED, Seelige R, Walsh C, Barrera M, Tindall E, … Li G, editors. Immuno-oncological efficacy of RXDX-106, a novel, selective and potent small molecule TAM (TYRO3, AXL, MER) inhibitor [abstract]. Proceedings of the American Association for Cancer Research annual meeting 2017, April 1–5. Washington, DC; Philadelphia, PA: AACR; 2017.
[228] Tian Y, Zhang Z, Miao L, Yang Z, Yang J, Wang Y, Qian D, Cai H, Wang Y. Anexelekto (AXL) increases resistance to EGFR-TKI and activation of AKT and ERK1/2 in non-small cell lung cancer cells. Oncol Res 2016;24 (5):295–303.
[229] Zuo Q, Liu J, Huang L, Qin Y, Hawley T, Seo C, Merlino G, Yu Y. AXL/AKT axis mediated-resistance to BRAF inhibitor depends on PTEN status in melanoma. Oncogene 2018;37(24):3275–89.
[230] Hong J, Maacha S, Belkhiri A. Transcriptional upregulation of c-MYC by AXL confers epirubicin resistance in esophageal adenocarcinoma. Mol Oncol 2018;12:2191–208.

CHAPTER

9

Eph/ephrin family proteins and therapeutic resistance

Shilpa Bhatia, Sana D. Karam

University of Colorado Denver, Anschutz Medical Campus, Aurora, CO, United States

Abstract

The clinical success of conventional and targeted therapies is compromised by the low response rates and development of therapeutic resistance. Dysregulated Eph-ephrin signaling has emerged as a critical factor in mediating resistance to therapeutics and radiation in human malignancies. Eph receptors belong to super-family of receptor protein tyrosine kinases that interact with their membrane-anchored ligands, ephrins, to regulate a variety of processes under physiological and pathological conditions including cancer. This book chapter sheds light on the functional role of Eph-ephrins in tumor development and progression and how their aberrant expression contributes toward therapeutic and radiation resistance. The potential of targeting this gene family to combat tumor resistance and enhance therapeutic sensitivity with an overview of the available agents currently in preclinical or clinical phase is also discussed.

Abbreviations

Eph	erythropoietin-producing hepatocellular carcinoma
EGF	epidermal growth factor
HNSCC	head and neck cancer
SCLC	small cell lung cancer
NSCLC	non-small cell lung cancer
FAK	focal adhesion kinase
GBM	glioblastoma multiforme
GPI	glycosylphosphatidylinositol
RT	radiation therapy
LBD	ligand binding domain
PDZ	postsynaptic density 95-discs large-Zonula occludentes-1
EMT	epithelial mesenchymal transition
scFv	single chain fragment variable
SHH	sonic hedge-hog
Treg	tregulatory
VM	vasculogenic mimicry
VEGF	vascular endothelial growth factor
VEGFR	vascular endothelial growth factor receptor

https://doi.org/10.1016/B978-0-12-817868-3.00009-3

Conflict of interest

No potential conflicts of interest were disclosed.

Eph receptors and ephrins-structure and bidirectional signaling

Eph (erythropoietin-producing hepatocellular carcinoma) is largest superfamily of receptor protein tyrosine kinases and together with their membrane-tethered ligands, ephrins, have been implicated in developmental processes, tissue homeostasis, and various pathological conditions including cancer [1, 2]. Based on their structural homologies and binding preference, both Eph receptors and ephrins have been categorized into two classes-class A and class B [1] (Fig. 1). In general, class A Eph receptors bind to glycosylphosphatidylinositol anchor (GPI)-linked ephrin A proteins and class B Eph receptors bind to transmembrane ephrin B proteins [1]. Some exceptions and promiscuous binding have been noted between receptors and ligand subtypes, in particular for EphA4 that binds to class B ephrins and EphB2 that binds to ephrin-A5 [2, 3].

The Eph receptors have the classical receptor tyrosine kinase domain organization (Fig. 2). The multidomain extracellular region consists of a globular ligand binding domain (LBD), a cysteine-rich region with a Sushi and an epidermal growth factor (EGF)-like domain, and two fibronectin type III domains (FN1 and FN2). FN2 is followed by a single transmembrane (TM) segment and an intracellular region that comprises of a juxtamembrane (JM) region with a tyrosine kinase domain, a sterile-α motif (SAM) protein-protein interaction domain, and a C-terminal Psd-95, Dlg and ZO1 domain (PDZ)-binding domain [2]. Both ephrin classes contain an extracellular receptor-binding domain, which is conserved across the animal phyla.

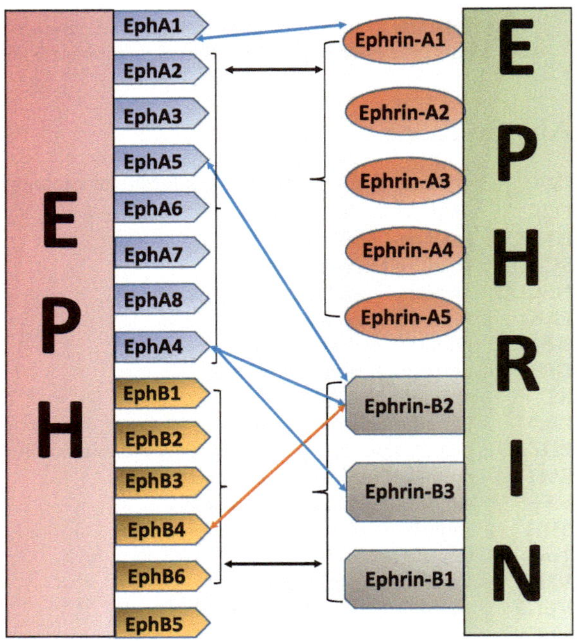

FIG. 1 Schematic representation of Eph receptors and ephrin binding affinities.

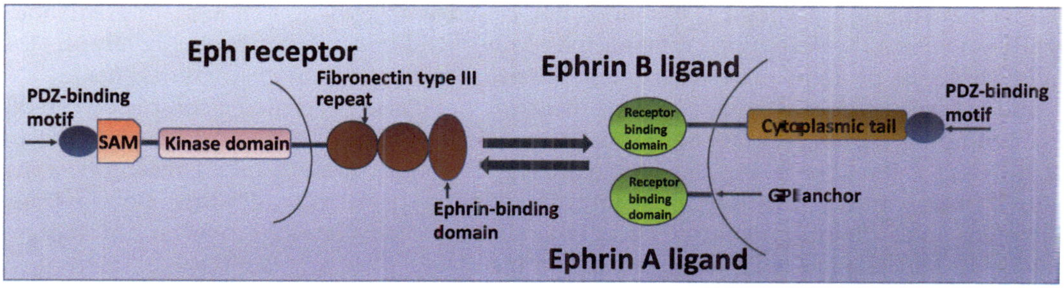

FIG. 2 Structural organization of class A and B Eph receptors and ephrin ligands. Eph receptors comprises of an extracellular region an ephrin-binding domain and two fibronectin type III repeats, and an intracellular region consisting of a tyrosine kinase domain, a SAM domain and a PDZ binding domain. Ephrin class A ligands are tethered to the cell membrane with a GPI link. Ephrin class B ligands, on the other hand, are transmembrane proteins comprising of a cytoplasmic tail and a PDZ binding domain. Interaction between membrane-bound Eph receptor with its ephrin ligand can stimulate either bidirectional signaling or can induce "forward signals" in Eph receptor-expressing cell or "reverse signals" in ephrin-expressing cell. *Reproduced with permission from Edwards CM, Mundy GR. Eph receptors and ephrin signaling pathways: a role in bone homeostasis. Int J Med Sci 2008;5(5):263–272. https://doi.org/10.7150/ijms.5.263.*

The ephrin A proteins are bound to the cell surface by a glycosylphosphatidylinositol (GPI) link whereas the ephrin B proteins possess a transmembrane segment, an intracellular domain with several conserved tyrosine residues and a C-terminal PDZ-binding motif [2] (Fig. 2).

Eph-ephrin signaling

Unlike other members of the receptor tyrosine kinase family, Eph-ephrin proteins display distinctive characteristics in terms of binding and activation. They typically mediate contact-dependent communication between cells of the same or different types to regulate cell morphology, and a variety of cellular processes including cell adhesion, migration, proliferation, survival, and differentiation [4]. Engagement between membrane-bound Eph receptor with its ligand can trigger either bidirectional signaling or can induce "forward signals" in Eph receptor-expressing cell or "reverse signals" in ephrin-expressing cell further resulting in the activation of downstream signaling pathways [1]. Reports have suggested an additional level of complexity in the Eph-ephrin signaling, which results from the fact that interactions between Eph receptors and ephrins can occur in *trans* (between two opposing cells) or in *cis* (within the same cell) [5]. While the common assumption is that *trans* interactions are activating in nature, *cis* interactions, on the other hand, can trigger inhibitory function [5].

Activation of Eph receptors by cognate ephrin ligands expressed on another cell can trigger forward signaling. It can either mediate autophosphorylation or phosphorylation of various effector proteins upon association [2]. Tyrosine phosphorylation of conserved residues allows the Eph to adopt an open conformation structure resulting in further tyrosine phosphorylation of the active kinase domain [2]. Effector proteins containing Src homology 2 (SH2), SAM, or PDZ domains can then bind further stimulating downstream signaling pathways [2].

The mechanisms underlying reverse signaling are less understood but ephrins also trigger a variety of signaling events. Class B ephrins contain intracellular tyrosine kinase residues, which upon interaction with Eph receptor are phosphorylated by Src family of protein kinases [2]. Class B ephrins also have a carboxy-terminus PDZ-binding domain, which plays a critical role in regulating physiological functions associated with ephrins [2]. Signaling emanated by GPI-anchored class A ephrins remains to be fully elucidated but they probably rely on other transmembrane proteins such as Src family kinase, Fyn to affect downstream events [2].

Interestingly, numerous reports in the literature have also suggested a complex cross-talk between Eph-ephrins and other additional molecules including the atypical receptor tyrosine kinase (RYK), the chemokine receptor (CXCR4), fibroblast growth factor receptors, extracellular matrix-interacting integrins, immunoglobulin proteins, cadherins, claudins, and connexins, and neurotransmitter receptors such as the *n*-methyl D-aspartate receptor (NMDAR) and this has been reviewed elsewhere [5–8].

Role of Eph receptors and ephrins in cancer

Research over the past decades have shed light on the involvement of Eph-ephrin signaling on different aspects of tumor development and progression including tumor angiogenesis, invasion and metastasis, vasculomimicry, and induction of tumor immune response by affecting tumor immune microenvironment [9–11].

Tumor angiogenesis

Eph/ephrin signaling has been well established to play a role in vascular development during embryogenesis. EphB4 and ephrin-B2 are expressed in venous and arterial endothelial cells respectively [12]. Bidirectional signals emanating from this receptor-ligand pair mediate vascular remodeling [13–15]. Seminal papers from Sawamiphak et al. [16], and Wang et al. [17], demonstrated that ephrin-B2, in particular, mediates internalization and signaling of vascular endothelial growth factor receptors (VEGFRs) thus affecting VEGF-induced angiogenesis under physiological and pathological conditions. Angiogenesis also marks a critical step in the transition of a small tumor mass into a large malignant tumor capable of distant metastatic spread. From the information available to date, possible mechanisms by which Eph/ephrin interactions can promote tumor angiogenesis have emerged. Ephrins present on the tumor cells may act as contact-dependent molecules to direct incoming vessel that express Eph receptor [1, 18]. In an alternate scenario, factors present in the tumor microenvironment including VEGF may induce the expression of ephrins in endothelial cells, which can engage in bidirectional signaling following interaction with Eph receptors on neighboring endothelial cells thus affecting endothelial cell sprouting, capillary tube formation, and migration [1, 18]. Interactions between the tumor cells and vascular cells may also occur allowing entry of tumor cells into the blood stream and metastatic dissemination [1, 18, 19]. Regardless of the mechanisms, it is evident that Eph-ephrins serve as key molecules facilitating tumor angiogenesis and targeted therapeutics directed against them can be beneficial to restrict tumor growth. Ephrin-A1, for instance, has been reported to affect metastasis in an in vivo mammary tumor model through an angiogenesis-dependent mechanism that is mediated by EphA receptor and vascular endothelial growth factor [20]. Ogawa et al. [21], investigated the role of Eph-ephrins in tumor neovascularization and observed that both EphA2 and its ephrin-A1 ligand were detectable throughout tumor vasculature. Another study revealed that both EphA2 and VEGF are overexpressed in squamous cell carcinoma of tongue and EphA2, VEGF, and microvessel density showed significant correlation with tumor size, clinical stage, lymph invasion, recurrence, and metastasis [22]. Ephrin-B2, a sole ligand for EphB4 and a predictor for poor prognosis [23], is also suggested to regulate the interaction between endothelial cells and pericytes or vascular smooth muscle cells [24, 25] and may play a role in stabilizing tumor vessels. Ephrin-B2 expression has been observed in the microvasculature of variety of solid tumors [13]. Thus, it is plausible that EphB4 present on tumor cells may cross-talk with endothelial ephrin-B2 to promote angiogenesis. Noren et al. [26], overexpressed a dominant negative EphB4 construct in human breast cancer cells and observed enhanced tumor growth and significant enlargement of blood vessels in a mouse xenograft model. In cultured cells transfected with cytoplasmically truncated EphB4, increased endothelial migration, proliferation, and migration was evident suggesting that EphB4-mediated stimulation of tumor growth and angiogenesis is dependent on reverse signaling by endothelial ephrin-B2 [26]. Therefore, inhibition of these interactions could be promising in targeting tumor angiogenesis and may show favorable outcomes in advanced malignancies.

Vasculogenic mimicry

Vasculogenic mimicry (VM) entails the generation of vascular-like structures independent of endothelial cells by highly aggressive tumor cells for invasion and metastatic spread [27]. VM imparts intratumor heterogeneity and survival advantage to the aggressive tumor cell phenotype and has been associated with poor prognosis in various human malignancies [28–30]. A unique feature associated with VM is the reappearance of endothelial and mesenchymal markers that are otherwise restricted to the embryonic stage. We are beginning to understand the contribution of diverse molecular mechanisms and signaling pathways involved in VM and Eph receptors have been strongly implicated to play a role. EphA2, a receptor tyrosine kinase that is activated in the presence of ephrin-A1 is reported to form vessel-like networks upon phosphorylation and the process is mediated by upregulation of PI3K and FAK signaling [31]. Hess et al. [32], further provided an evidence for the contribution of Eph proteins when they showed that specific targeting of EphA2 inhibited the formation of tubular structures in aggressive uveal melanoma cells. In another study, both EphA2 and PI3K were found to be significantly correlated with VM formation in a prostate cancer model both in vitro and in vivo [33]. A recent study reported that miR-141 inhibits vasculogenic mimicry in glioma by regulating EphA2 expression [34] thus suggesting that Eph receptors can be used as a therapeutic target to counter VM formation in cancers.

Cell migration/invasion

Eph-ephrin signaling has been implicated in regulating various aspects of cell migration and invasion in malignant conditions. In particular, Morrison et al. [35], published that EphB1 transcript levels are highly upregulated in migrating medulloblastoma cells and in medulloblastoma tumor spheres that possess high self-renewing capability. Different datasets were interrogated by multiple investigators to validate EphB1 expression in medulloblastoma patient samples [36–38]. Our group further confirmed that EphB1 is highly expressed in medulloblastoma cell lines reportedly shown to fall in the category of sonic hedge-hog (SHH) sub-type [39]. We further showed that siRNA mediated knockdown of EphB1 sensitized medulloblastoma cells toward ionizing radiation by modulating cell cycle, cell proliferation, and cell migration [39]. This is consistent with previous findings suggesting that EphB1 plays an important role in mediating cell migration [40–42]. Similar to EphB1, EphB2 was also found to be highly upregulated in medulloblastoma sub-types and its silencing sensitized medulloblastoma cells to ionizing radiation treatment in vitro and resulted in decreased tumor cell growth, viability, and invasion [43]. In agreement with our findings, Sikemma et al. [44], reported enhanced medulloblastoma cell invasion following ligand-mediated stimulation of EphB2 receptor, and this effect was inhibited following downregulation of EphB2. Findings from another study showed that EphB2 mediates a dichotomous pro-invasive and anti-proliferative role in GBM stem-like neurospheres following interaction with focal adhesion kinase (FAK) [45]. In a breast cancer model, the effect of EphB2 on migration and invasion was reported to be dependent on its kinase activity [46] suggesting that blockade of EphB2 phosphorylation could be used as an effective approach to target advanced stage metastatic breast cancer. Nakada and colleagues also investigated functional role of EphB/ephrin B signaling in imparting invasive characteristics to malignant brain tumors [47]. They used fusion proteins to force activation of ephrin-B2 ligand in GBM cell lines and observed increased migration and invasion [47]. Treatment of cells with an ephrin-B2 blocking antibody showed the opposite results [47]. EphB4 is another Eph receptor, whose expression has been associated with head and neck cancer progression from normal status through dysplasia to cancer [48]. In an over expressed state, EphB4 was shown to stimulate tumor growth, invasive potential, and tumor angiogenesis in colorectal cancer and head and neck cancer [49, 50] thus, providing cancer cells with survival advantage. EphB4 was also found to express at high levels in malignant melanoma cells that showed faster migratory capabilities [51] and positively correlated with survival in lung cancer patients [52].

Tumor microenvironment

The components of tumor microenvironment play a critical role in tumor initiation, progression, metastasis, and resistance to therapy. Multiple layers of evidence suggest that Ephs and ephrins regulate cell-cell interactions between the tumor cells and in tumor microenvironment. Tumor-derived factors and hypoxic environment is known to stimulate the expression of Eph-ephrins in vascular cells. For example, HIF-2α is suggested to play a role in vascular remodeling during tumor neovascularization and the process is mediated via activation of ephrin-A1 [53]. In another study, tumor necrosis factor-α (TNFα) was shown to upregulate ephrin-A1 levels in endothelial cells [54]. The expression of EphB4 and ephrin-B2 in tumor blood vessels is well established [13, 14, 55]. Ephrin-B2 has also been reported on pericytes and vascular smooth muscle cells (vSMCs) where it regulates the interaction between vSMCs as well as pericytes and endothelial cells [24]. Furthermore, ephrin-B2 expression on stromal cells has been shown to promote the growth and sprouting of endothelial cells and is also involved in smooth muscle cell recruitment and proliferation [56]. Multiple reports exist documenting that Eph/ephrin members modulate immune cell processes in inflammatory models [57–60]. However, our group attempted to investigate the role of Eph-ephrin proteins in context of tumor immune modulation (Bhatia et al. 2019, Cancer Research, in press). We analyzed intratumoral immune cells including Tregulatory cells (Tregs) and macrophages and found EphB4 and ephrin-B2 present on them suggesting they might be serving a pivotal role in modulating immune response as discussed in the following section.

Cancer immunity

It is known that Eph receptors can serve as sources of tumor-associated antigens (TAAs) with immunogenic potential and may play pivotal roles in modulating the tumor immune microenvironment. In a study, a CD4+ T-cell clone isolated from a melanoma patient with an exceptionally favorable clinical evolution was found to recognize EphA3 antigen and trigger selective immune response against melanoma cell lines. EphA2-derived epitopes were also reported to stimulate immunoreactivity in CD8+T cells in renal cell carcinoma and glioma samples [61]. The findings from these studies prompted to use EphA2 for vaccination strategies in murine tumor models [62–64]. Although, EphA2 peptide-pulsed dendritic cells decreased tumor burden and stimulated immune response, the non-orthotopic models used in these studies did not accurately mimic the human disease scenario and therefore was considered as a caveat. Furthermore, since these Eph/ephrins are known to be present on immune cells, we tested the hypothesis that EphB4-ephrin-B2 cross-talk plays a regulatory role in the tumor immune microenvironment by sustaining Tregs and TAMs, thus affecting the ability of CD8+ T cells to induce an anti-tumor response. Our data suggest that targeted inhibition of EphB4-ephrin-B2 signaling results in a significant tumor growth suppression and remodeling of the immune landscape in a HNSCC immunocompetent model and in a PDX model modulation (Bhatia et al. 2019, Cancer Research, In Press). Besides these studies, there are still some interesting questions that need to be addressed and warrant rigorous scientific scrutiny to fully understand the impact of Eph/ephrin on immune system. This will add to our current understanding and might help to exploit the potential of novel agents targeting these gene families to rejuvenate/induce an effective anti-tumor immune response.

Tumor suppressive functions by Eph-ephrins

Eph receptor gene family and its membrane-bound ligands represents a complex system with family members showing paradoxical roles depending on cancer type. Infact, both ligand-dependent and ligand-independent functions of the Eph receptor have emerged, which in turn regulates its dichotomous role as a tumor-promoter or tumor-suppressor. For instance, in addition to acting as a tumor promoter, EphA2 has been shown to trigger tumor-suppressing signals based on its phosphorylation status [65, 66]. This was particularly evident in a study where EphA4, following ephrin-A1 stimulation effectively blocked signaling emanating from the AKT-mTORC1 pathway that favors tumor growth in glioblastoma model [66]. EphA3, another member of Eph class A subfamily, has been reported to have a tumor-suppressive function in small cell lung cancer. It was further shown that low levels of EphA3 contributed toward multidrug resistance by PI3K/BMX/STAT3 pathway [67] and EphA3 upregulation in turn showed opposite effects and suppressed tumor growth in an in vivo model [67]. Similar to EphA3, EphA4 was demonstrated to have a protective role as its overexpression blocked cell migration and invasion [68] in a non-small cell lung cancer (NSCLC) model. The data is in congruence with a study showing positive correlation between EphA4 expression and patient survival outcomes with high EphA4 expression resulting in better survival and low EphA4 expression leading to worse outcomes [69] suggesting that EphA4 can be used as a useful clinical biomarker for patient prognosis. In addition, work by different investigators have suggested both opposite roles for both EphA5 and EphA7 proteins. Giaginis et al. demonstrated positive association of EphA5 and EphA7 with overall patient survival in NSCLC [69]. Conversely, work by Li and colleagues showed inhibition of lung cancer proliferation and metastasis following downregulation of EphA7 [70].

Enormous evidence in the literature is also available in support of tumor-suppressor roles for class B Eph receptors and ephrins. Overexpressing Eph receptor levels or stimulating receptor activity using soluble Fc fusion proteins have been attempted to elucidate their role in blocking tumor growth, proliferation, migration, and invasion in different cancers. This was evidenced in a study showing EphB3 overexpression enhanced cell-cell interaction and inhibited tumor growth in colon cancer cells [71]. We have observed similar results in a HNSCC model where forced EphB3 phosphorylation following treatment with an ephrin-B2-Fc fusion protein decreased cancer cell migration and growth, and enhanced response to PI3K inhibitor [72]. EphB4 is another Eph family protein that has been extensively studied in different tumor models and has been shown to have dichotomous functions. The EphB4 receptor has been shown to inhibit breast cancer cell tumorigenicity by regulating Abl-Crk pathway [73]. In contrast, overexpressing a dominant negative form of EphB4 lacking kinase domain required to transmit forward signals, has been shown to promote colorectal cancer proliferation and invasion [74]. Similar to Eph receptors, ephrin reverse signaling in cancer cells may in some cases also impart tumor suppressive function. This was reported in a study wherein ephrin-B reverse signaling may also inhibit the migratory and invasive potential of the cancer cells [5]. Ephrin-A5 was also demonstrated to exert negative regulation on growth promoting EGFR thus acting as a tumor suppressor in a glioma model [75]. Thus, considering the multi-faced role of Eph/ephrin signaling, careful design of agents targeting this pathway is warranted to ensure safe and efficient therapeutic intervention.

Therapeutic resistance mediated by Eph-ephrin signaling

Therapeutic resistance is a major roadblock that compromises the efficacy of commonly used drugs and treatment modalities. The resistance can be mediated by targets that are aberrantly regulated on either tumor cells, tumor microenvironment or both. Concerted research efforts are currently being directed at identifying such targets and understanding the underlying mechanisms associated with it. In this context, Eph-ephrin proteins have emerged as novel candidates that can be targeted to increase the potency of therapeutic agents. Studies from our laboratory and other research groups have shown that these proteins are aberrantly expressed on cancer cells, surrounding stroma including blood vessels. The functional role of Eph-ephrin proteins in mediating therapeutic resistance is discussed below.

Resistance against chemotherapeutic or targeted agents

It is increasingly evident from pre-clinical and clinical studies that despite advances in the field of cancer drug discovery and development, multidrug resistance sets in after an initial phase of response. Treatment resistance has become a major problem that blunts the therapeutic response to chemotherapeutic or targeted agents and identifying key pathways and molecular components that contributes to this phenomenon is of utmost importance. In addition to the various factors implicated in mediating treatment resistance including impaired drug uptake, resistance via multidrug efflux pumps, data has emerged from different groups showing that Eph-ephrins might play a critical role. In fact, Eph proteins have been shown to engage in a cross-talk with other tumor-promoting receptor tyrosine kinases and this presents another mechanism by which Eph receptors promote tumor development, progression, and resistance. A study published by Brantley-Sieders et al. [76], showed the involvement of EphA2 and how it cooperates with ErbB2 to promote mammary adenocarcinoma tumorigenesis and metastasis in a mouse model. In another study, it was demonstrated that elevated EphA2 levels confer both intrinsic and acquired resistance to a humanized monoclonal anti-HER2 antibody, trastuzumab, by interfering with AKT and MAPK pathways suggesting that EphA2 plays a role in therapeutic resistance [77]. Downregulated levels of EphA3 has been shown to contribute toward multidrug resistance by PI3K/BMX/STAT3 pathway [67]. EphA4 is another tyrosine kinase receptor associated with many cancers such as pancreatic ductal adenocarcinoma, gastric cancer, and multiple myeloma [78–80]. A recent study suggested the role of EphA4 in promoting cancer cell proliferation and cell-adhesion mediated drug resistance by stimulating the activation of AKT pathway in multiple myeloma [80]. EphA2, on the other hand, when overexpressed was found to promote a tumorigenic and metastatic phenotype with reduced sensitivity to tamoxifen in breast cancer [81, 82]. In an attempt to understand the mechanisms of acquired resistance to cisplatin in melanoma tumors, Yang et al. [83], conducted a genome-wide gene profiling analysis and found EphB4 to be elevated in the cisplatin resistant A375 melanoma xenografts. The tumors regained sensitivity to cisplatin following treatment with EphB4 inhibitor NVP-BHG712 by affecting PI3K and MAPK pathways [83]. EphB4-ephrin-B2 signaling was also implicated in mediating resistance to imatinib in chronic myeloid leukemia, which was prevented by silencing the expression of EphB4 receptor [84]. Thus, knowledge obtained from these studies revealed how Eph-ephrin axis can act as drivers of chemotherapeutic resistance and rationally designed agents targeting this gene family would be critical to improve patient outcomes.

Resistance against radiation therapy

Radiation therapy (RT) remains the mainstay of management of several different cancers including carcinoma of head and neck. Members of the Eph receptor and ephrin families, including EphA2, EphA5, EphB4, ephrin-A1, ephrin A3, and ephrin-B3 have previously been implicated in mediating radioresistance in different tumor models [85–88]. For instance, in a study examining NSCLC cell lines, ephrin-B3 emerged as one of the putative drivers of RT resistance in the gene expression analysis [88]. Moreover, ephrin-B3 downregulation was shown to affect MAPK and AKT pathways inducing RT-mediated G2/M cell cycle arrest, and apoptotic cell death suggesting enhanced radiosensitivity [88]. Staquicini and colleagues further interrogated clinical data and pre-clinical models to determine the role of EphA5 in SCLC radiation resistance suggesting that blockade of EphA5 activity might be a feasible approach to overcome radioresistance mechanism [89]. Our group has published studies demonstrating the role of EphB1, EphB2, and EphB4 knockdown in overcoming radiation resistance. We used medulloblastoma cell lines and genetically engineered mouse model to elucidate the role of EphB1 silencing in enhancing sensitivity toward radiation therapy [39]. The EphB1 silencing enhanced cellular radiosensitization by affecting medulloblastoma cell growth and migration [39]. Similar results were obtained in medulloblastoma cells following knockdown of EphB2 receptor [43]. Our data showed that combined EphB2 knockdown sensitizes medulloblastoma cells toward ionizing radiation by affecting cell viability and invasion [43]. Since EphB4-ephrin-B2 pathway is implicated in human malignancies, we investigated its role in imparting resistance toward RT in a human patient-derived xenograft model of HNSCC [90]. We observed that blockade of EphB4-ephrin-B2 signaling by sEphB4-HSA protein enhanced tumor regrowth and sensitivity toward RT partly via inducing apoptosis and inhibiting cell proliferation [90]. In a syngeneic model, EphB4-ephrin-B2 inhibition combined with RT sensitized the murine HNSCC tumors toward RT and resulted in significant tumor growth suppression by remodeling tumor immune microenvironment (Bhatia et al. 2019, Cancer Research, in press). We have also published that ephrin-B2 overexpression can serve as a prognostic indicator and predict response to therapy in solid tumors [23]. Furthermore, in another study we observed that locally advanced HNSCC patients treated with standard-of-care definitive chemo-RT have high levels of both EphB4 and ephrin-B2 after failure of treatment [91]. Treatment of HNSCC PDX tumors with EphB4-ephrin-B2 inhibitor resulted in a significant response toward an anti-EGFR antibody, cetuximab and RT combination improving survival in tumor-bearing mice [91]. Tumor growth suppression was accompanied by reduced levels of proliferation and pro-survival molecules and increased apoptosis [91]. Thus, in the current scenario, rational combination approaches targeting Eph-ephrin pathway in combination with RT seems to be a feasible strategy for the management of emerging tumor resistance.

Resistance against anti-angiogenic agents

Angiogenesis represent a critical mechanism that supports tumor growth and progression. This has led to the generation of anti-angiogenic therapies with an aim to counter direct VEGF signaling to starve tumors of the essential nutrient supply. But the enthusiasm in the field was dampened by the data that emerged from a series of pre-clinical and clinical studies. The beneficial effects from these angiogenic inhibitors lasted for a transient time and emerging drug resistance, off-target toxicities, increased invasion and metastasis led to further problems [92]. There is ample amount of evidence in the literature that underscores the reliance of tumors on alternate mechanisms to support vascularization that in turn compensate for and compromise the efficacy of anti-angiogenic agents. It is now well established that Eph-ephrins play a role in both physiological and pathological angiogenesis [16, 17, 93, 94] and research published by Wang et al. [17], and Sawamiphak et al. [16] elegantly illustrated the role of ephrin-B2 in this context. Thus, it is not surprising that dysregulated levels of EphB4-ephrin-B2 might be responsible for driving resistance against anti-angiogenic agents. In fact, elevated expression of EphB4 has been shown to contribute toward resistance to sunitinib in experimental glioma models by affecting vascular morphogenesis, pericyte coverage, and apoptosis [95]. The role of EphA2-ephrin-A1 signaling in overriding angiogenic pathways and contributing to drug resistance is well studied. One of the studies described how resistance to VEGFR1 and VEGFR2 blocking antibodies resulted in recurrent angiogenesis and tumor growth in pancreatic islet tumors [96]. Mechanistic analyses further confirmed the link between elevated levels of ephrin-A1 as well as pro-angiogenic factors and resistance to VEGF blockade [96]. In a study by Depner et al. [97], investigators observed that following treatment with bevacizumab, HIF-1α resulted in the stimulation of EMT repressor zinc finger E-box binding homeobox 2 (ZEB2), which in turn suppressed ephrin-B2. ZEB2-mediated suppression of ephrin-B2 was found to enhance tumor invasive potential and treatment resistance against bevacizumab in a murine glioma model [97]. However, in our laboratory, we have observed significant reduction in angiogenic markers following EphB4-ephrin-B2 inhibition in both head and neck cancer and pancreatic cancer patient-derived xenograft models (Bhatia et al. Cancer Research 2019, in press; Lennon et al. Clinical Cancer Research 2019, in press). The therapeutic potential of combining Dll4/Notch and Ephrin-B2/EphB4 inhibitors in disrupting tumor angiogenesis was evaluated by Djokovic et al. [98], and was found to be highly potent. Perhaps future studies should focus on the use of conditional Eph/ephrin knockout tumor models in addition to combination pharmacological inhibitory approaches, which would allow for the development of an improved generation of anti-angiogenic agents targeting the Eph/ephrin signaling axis that are more targeted and safer in their mode of action.

Resistance against immune checkpoint inhibitors

There are numerous clinical trials evaluating the benefit of immunotherapy in human cancer, including HNSCC. The objective response rate is at best 6%–20% [99] and data emerging from these studies clearly demonstrate that the vast majority of patients display either innate or adaptive resistance to immunotherapy [100–103]. In our preclinical HNSCC models, we have observed resistance to immune checkpoint inhibitor even in combination with RT [104] after a transient responsive phase [105], suggesting the involvement of alternate pathways that might be blunting the therapeutic immune response. While not much information is available on the involvement of EphB4-ephrin-B2 signaling in the cancer-related immune response, multiple reports have suggested that Eph/ephrin proteins modulate immune cell processes in inflammatory models, such as arteriosclerosis and wound healing [57–59, 106, 107]. Interactions between Eph-ephrin family of proteins have also been reported to regulate monocyte adhesion to the blood vessel wall [57], trans-endothelial migration [59, 60], T cell chemotaxis [108], activation, proliferation and apoptosis [109, 110], and mobilization of hematopoietic cells from bone marrow sinusoids [111, 112]. Recent data from our laboratory indicate that selective targeting of EphB4-ephrin-B2 pathway inhibits intratumoral immunosuppressive populations such as Tregs in both head and neck cancer and pancreatic cancer models (Bhatia et al. Cancer Research 2019, in press; Lennon et al. Clinical Cancer Research 2019, in press). In light of the emerging role of EphB4-ephrin-B2 in cancer immune response, one would speculate they may have some role to play in mediating resistance toward immune checkpoint inhibitors. We attempted to address this question in our syngeneic head and neck tumor models and found significant upregulation in both EphB4 and ephrin-B2 in anti-PDL1+RT treated tumors during the tumor regrowth phase suggesting that EphB4 and ephrin-B2 may be involved in imparting therapeutic resistance against immune checkpoint blockers (unpublished data). Detailed studies are currently underway to completely understand the molecular underpinnings of cellular immune response regulated by EphB4-ephrin-B2 interaction in head and neck cancers.

Enhancing therapeutic and radio-sensitivity by targeting Eph-ephrin signaling in cancers

Given the aberrant regulation of Eph/ephrins in cancers, several therapeutic strategies have been attempted or are currently under evaluation to target members of this gene family. Some of these approaches include the use of activating antibodies or ligand peptide mimetics to activate Eph kinase function through ligand stimulation. This is particularly beneficial in a scenario where Eph mediated forward signaling is known to stimulate tumor suppressive role. On the other hand, where Eph receptor activation promotes tumor growth and progression, use of kinase inhibitors or ephrin-blocking peptides or antibodies are exploited with an aim to inhibit active kinase signaling. Some of the Eph-ephrin inhibitors that are currently under pre-clinical or clinical development are described below and in Table 1. Key treatment strategies targeting Eph-ephrin pathway are illustrated in Fig. 3.

Proteins

sEphB4-HSA is a recombinant fusion protein that comprises of a full-length extracellular domain of human EphB4 and fused at its C-terminus, to full-length human serum albumin (HSA). sEphB4-HSA functions by inhibiting the interaction between EphB4 and ephrin-B2 by binding to ephrin-B2. Our laboratory and multiple other investigators have evaluated the efficacy of sEphB4-HSA in preclinical models of cancer in inhibiting different aspects of tumor growth and progression.

In a Kaposi sarcoma model, sEphB4-HSA efficiently inhibited cell adhesion, migration, and invasion. It was also shown to inhibit the activity of several growth factors, including VEGF, bFGF, EGF, and PDGF. A significant reduction in tumor growth partly associated with decreased vascular density and perfusion was also observed in vivo following sEphB4-HSA

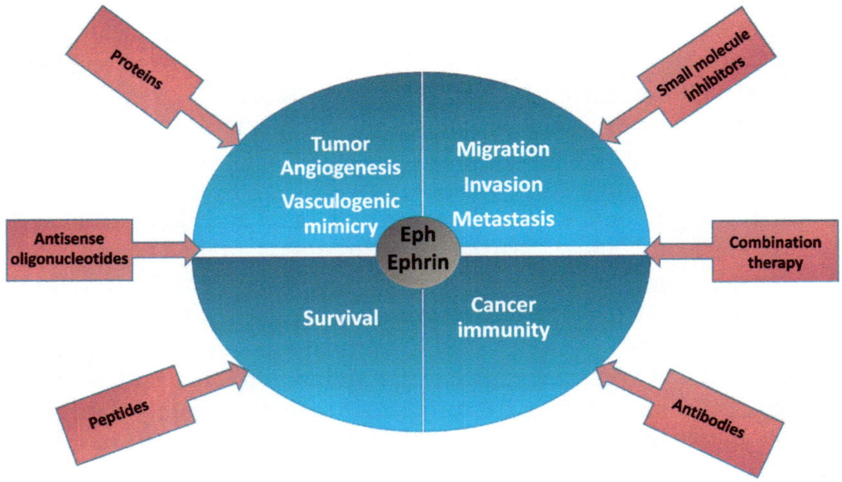

FIG. 3 Tumor promoting roles of Eph-ephrin signaling and therapeutic strategies for targeting Eph-ephrin axis.

treatment [113]. EphB4 has also been exploited as a therapeutic target in mesothelioma and sEphB4-HSA was shown to have significant tumor growth inhibitory activity alone or in combination with anti-VEGF antibody, bevacizumab in a xenograft model [114]. Similar results were evident when lung tumors implanted in a mouse model were treated with sEphB4-HSA protein [52].

Considering the fact that there has been limited progress in the field of head and neck cancer (HNSCC) over the past two decades and survival rates have remained dismal [115, 116], we have investigated the potency of sEphB4-HSA in combination with ionizing radiation. The definitive treatment of HNSCC patients comprises of cisplatin chemotherapy administered concurrently to radiotherapy that has been shown to improve progression-free survival (PFS), overall survival (OS), and organ preservation, but only approximately 50% of patients survive for more than 5 years [117]. For those patients with aggressive disease, particularly with heavy smoking history and HPV-negative disease, the outcomes are much worse, approximating a survival rate of 10% at 5 years [116]. Although high levels of epidermal growth factor receptor (EGFR) expression have been reported and correlated with poor prognosis in HNSCC [118], trials with EGFR inhibitors have shown only a modest increase in survival [119]. Attempts to match EGFR expression or gene copy number with treatments using EGFR inhibitors have failed [119] and EGFR mutations are rarely reported in HNSCCs [120–122]. The co-activation of multiple alternate receptor tyrosine kinases in HNSCCs is one mechanism of treatment resistance [123]. Based on the evidence available in the literature, Eph-ephrin families of proteins represent a viable target for therapeutic intervention. EphB4 and ephrin-B2 are highly amplified in HNSCCs [124], and act as prognostic indicators associated with poor survival outcomes [125]. Other groups have demonstrated that EphB4 also plays a key role in mediating cell survival, migration, and proliferation in HNSCC [50, 113]. Our data show that inhibition of the EphB4-EFNB2 interaction using sEphB4-HSA enhances radiosensitization of patient derived xenograft (PDX) models of HNSCC [90]. Our data also suggests that enhanced radiosensitization following EphB4-ephrin-B2 blockade is in part mediated by effects on tumor cell survival, DNA damage, and apoptosis [90]. In another study, we show that blocking EphB4-ephrin-B2 interaction using sEphB4-HSA enhanced response of HNSCC tumors to cetuximab-RT [91]. Tumor growth inhibition was further accompanied by reduction in the levels of proliferation and pro-survival molecules and increased apoptosis [91]. Work is currently underway in our laboratory to determine the effect of EphB4 and ephrin-B2 silencing on cancer cells and vasculature on tumor growth and progression in immunocompetent mouse models.

sEphB4-HSA has also been tested in clinical trials to determine dose-limiting toxicity, adverse reaction, pharmacokinetics, and efficacy in patients with malignant conditions. Published report on a phase I trial indicated that 1 out of 3 Kaposi Sarcoma patients had a complete response, 2 out of 17 head and neck cancer patients and 1 out of 8 hepatocellular carcinoma patient showed partial response, 50% had stable disease and disease progression was evident in remaining patients [126, 127]. In another pilot phase 1b trial currently underway, the efficacy of sEphB4-HSA will be evaluated in combination with standard chemotherapeutic regimens in advanced or metastatic solid tumors including NSCLC, pancreatic cancer, HNSCC, and gall bladder carcinoma (NCT02495896). Other phase II clinical trials are currently open to investigate sEphB4-HSA alone or in combination with Pembrolizumab in Kaposi Sarcoma (NCT02799485), in metastatic NSCLC, HNSCC (NCT03049618) or in urothelial carcinoma. The adjuvant effect of this recombinant protein is proposed to be evaluated in another pilot study before surgery in patients with bladder cancer, prostate cancer or kidney cancer (NCT02767921).

Peptides

Peptides targeting different members of Eph-ephrin families have also been generated and evaluated in multiple studies.

TNYL-RAW represents a 15 amino acid peptide that binds to the ephrin-binding pocket of EphB4 in a selective manner and blocks the interaction of the receptor with the ephrin-B2 ligand [128, 129]. This results in inhibition of tyrosine phosphorylation of both EphB4 and ephrin-B2 [128, 129]. Different forms of TNYL-RAW have been evaluated in pre-clinical studies [128]. Noberini et al. [128], determined the efficacy of PEGylated form of TNYL-RAW that showed increased binding affinity for EphB4 and minimal susceptibility to proteolytic degradation. It was further reported that PEG-TNYL-RAW was potent against blocking ephrin-B2 mediated EphB4 phosphorylation in melanoma cells and capillary-like tube formation and capillary sprouting in a co-culture system [128]. Our laboratory has also investigated plasmid form of TNYL-RAW conjugated to a Fc protein in in vivo models of HNSCC (Bhatia et al. 2019, Cancer Research). We found that TNYL-RAW-Fc effectively suppressed HNSCC tumor growth alone or in combination with radiation (Bhatia et al. 2019, Cancer Research). This was further accompanied by changes in immune landscape including significant reduction in immunosuppressive populations such as Tregs and tumor-associated macrophages (Bhatia et al. 2019, Cancer Research) suggesting that EphB4-ephrin-B2 inhibition not only affected tumor cells but also allowed remodeling of tumor immune microenvironment to affect tumor growth in a head and cancer mouse model. In another study, TNYL-RAW-Fc was reported to block tumor growth, hematopoietic cell mobilization and its infiltration into the mammary tumors [111].

Targeted drug delivery to the tissue of interest is another area that has garnered the attention of scientific community. This is particularly important to minimize undesirable systemic toxic effects induced by some of the chemotherapeutic drugs. To address the issue of non-specific targeting, conjugation of chemotherapeutic drugs to peptides has been attempted with an aim to target tumor-specific cell surface receptors [130]. In this context, a cyclic form of TNYL-RAW with enhanced in vivo stability has been tested in conjugation with doxorubicin-loaded nanosphere in a mouse tumor xenograft model [87]. The TNYL-RAW and doxorubicin complex resulted in complete regression of EphB4-positive tumors following local release of doxorubicin [87].

Peptides can also serve as a valuable function in molecular imaging of tumors in a non-invasive manner. This not only allows for early detection and diagnosis of tumors but also helps in evaluating the potential of anti-cancer agents. Due to the high retention ability and high EphB4 binding affinity, different derivatives of TNYL-RAW have been generated and tested in various imaging techniques. A study reported that tagging of the TNYL-RAW peptide with a radiometal chelator DOTA followed by loading with ^{64}Cu was successfully used as a radiotracer in PET-CT imaging of EphB4-positive prostate cancer cells and colon cancer cells implanted in a rodent model [131]. An alternate version of this peptide conjugate, Cy5.5-TNYL-RAWK-^{64}Cu-DOTA was successfully used to label both tumor cells and tumor vasculature in an orthotopic glioblastoma mouse model [132].

YSA is EphA2 agonist peptide that has been shown to inhibit pro-tumorigenic AKT-mTORc1 and Ras-ERK pathways [133]. It has also used in the form of PEGylated nanoparticles loaded with a fluorescent dye for imaging of lung cancer cells expressing high levels of EphA2 receptor [134]. Fluorescently-conjugated YSA peptide loaded in magnetic nanoparticles have been exploited to isolate EphA2-expressing ovarian cancer cells from peritoneal fluids and from human ascites fluid in mouse models and patients respectively [135, 136]. This approach can potentially be used for drug sensitivity analyses and analyze gene expression profiles and presence of mutations on extracted tumor cells.

SNEW represents antagonistic EphB2-binding peptide with ability to block signaling triggered by both EphB2 and its ephrin-B ligand [129]. SNEW was shown to inhibit COS and HUVEC retraction triggered as a result of EphB2 activation by ephrin-B2 [129]. Another study implicated the role of EphB2-ephrin-B2 blockade by SNEW in enhancing T-cell proliferation [110] thus suggesting that agents targeting this gene family can be used to modulate immune responses. Similar results were obtained by targeting EphB4 using TNYL-RAW peptide [110]. Derivatives of SNEW also underwent radiofluorination to find use for PET-imaging. However, data revealed that ^{18}F-labeled SNEW derivative had short retention time in the bloodstream and was rapidly eliminated from the system [137].

AzV36-NicL is another peptide derived from azurin that carries a radiosensitizing moiety nicotinamide. It has been shown to possess optimal binding specificity to EphA2, EphB2, and EphB4 receptors. The azurin derivative demonstrated effective radiosensitization effect in a metastasis model of Lewis Lung carcinoma increasing the efficacy of radiation treatment by about 13 fold compared to the control [138].

Studies have further exploited EphA2 for use in novel immunotherapeutic vaccines to stimulate immune response. For instance, Hatano et al. [62], have observed induction of an interferon response and cytotoxic T-lymphocyte (CTL) responses in a glioma model following administration of a vaccine that comprised of autologous dendritic cells with EphA2883-891 (TLADFDPRV), a synthetic EphA2 peptide.

Antibodies

Advances in the area of drug development has also led to the generation of antibodies targeting the Eph/ephrin system, some of which are already being tested in clinical trials.

Ifabotuzumab is a recombinant monoclonal antibody generated from the agonistic antibody IIIA4 that targets EphA3 receptor. EphA3 is known to be elevated in a variety of solid and hematological cancers. The EphA3 agonist was shown to induce activation of EphA3, followed by apoptosis of tumor-derived mesenchymal stem cells, disruption of tumor stroma and its vascular integrity ultimately resulting in tumor growth suppression in a prostate cancer model [139]. Significant tumor growth regression and enhanced survival was also observed following treatment with the radiolabeled form of IIIA4 antibody in a glioblastoma xenograft model [140].

Ifabotuzumab, formerly known as KB004, was also evaluated in a Phase I trial in patients with advanced hematological cancers. The study reported that KB004 was well tolerated with manageable toxicity profile. However, clinical efficacy was noted only in a few patients with remaining patients coming off the protocol due to disease progression or death [141].

MEDI-547 comprises of an agonistic antibody, 1C1 that targets EphA2, conjugated to a microtubule inhibitor monomethylauristatin. Treatment of cells with 1C1 conjugate triggered caspase 3/7 dependent apoptosis in EphA2-expressing prostate cancer PC3 cells and significant tumor growth inhibition in PC3 mouse xenograft and F98 rat syngeneic tumor models [142]. The conjugate did not show any adverse toxicities in these studies. MEDI-547 was also investigated in a phase I clinical trial in patients with relapsed or refractory solid tumors but the study was closed due to treatment-related bleeding and coagulation events [143].

In addition to Eph receptor tyrosine kinases, members of the ephrin ligand family have also been explored as candidates for targeting this gene family as described below.

PF-06647263 represents a humanized monoclonal antibody conjugated with a DNA-damaging agent calicheamicin. The potency of PF-06647263 was studied in ephrin-A4 expressing triple-negative breast cancer and ovarian cancer models [144]. The treatment achieved sustained tumor growth inhibition in both breast cancer and ovarian cancer PDX models [144]. PF-06647263 was also tested in patients with metastatic triple negative breast cancer in a phase I study (NCT02078752). The data presented at the American Society of Clinical Oncology 2017 meeting, confirmed that the conjugate was well-tolerated and some anti-tumor activity was evident in heavily pretreated patient cohort [145]. However, in the expansion cohort, no objective responses were reported in patients with triple negative breast cancer [145].

scFv B11 is a single-chain fragment variable (scFv) antibody targeted against ephrin-B2, which acts by preventing ephrin-B2 binding to EphB4 receptor thereby blocking the EphB4-ephrin-B2 downstream signaling [146]. Systemic administration of scFv B11 efficiently suppressed tumor growth in mice bearing pancreatic BxPC3 tumor xenografts [146]. This was partly attributed to increased apoptosis induced by blockade of EphB4-ephrin-B2 signaling in pancreatic tumors [146]. Treatment of BxPC3 tumors with B11 also affected angiogenesis and lymphangiogenesis indicated by significant reduction in the number of blood and lymphatic vessels [146]. Another study reported significant reduction in tumor growth following treatment with B11 antibody in a 4T1 mouse mammary tumor model [111]. In the same study, B11 treatment also resulted in a significant reduction in circulating white blood count and circulating colony-forming hematopoietic stem progenitor cells [111]. Perivascular invasion is a hallmark of Glioblastoma multiforme (GBM) tumors and is a critical factor contributing toward therapeutic resistance of these tumors. Krusche et al. [147], demonstrated that treatment of established glioblastoma tumors with B11 significantly inhibited tumor growth with complete tumor regression in 50% of the treated cohort alongwith marked reduction in tumor cell proliferation [147]. In our laboratory, we evaluated the efficacy of scFv B11 in combination with RT in a pancreatic tumor model. We observed that inhibition of EphB4-ephrin-B2 signaling with B11 radiosensitized pancreatic PDX tumors to a significant extent compared to single agent treatment (Lennon et al. 2019, Clinical Cancer Research, in press). Consistent with the previous findings, we observed significant decrease in CD31+ cells in the B11 treated group suggesting an effect on angiogenesis (Lennon et al. 2019, Clinical Cancer Research, in press). In addition, B11 treatment combined with RT resulted in significant changes in tumor-infiltrating immune cell populations including a marked reduction in Treg to Teffector cell ratio (Lennon et al. 2019, Clinical Cancer Research, in press). In our pancreatic xenograft and syngeneic tumor models, we observed that RT resulted in an increase in stromal fibrosis and combining B11 with RT significantly reduced it (Lennon et al. 2019, Clinical Cancer Research, in press) thus suggesting that the combined EphB4-ephrin-B2 inhibition with RT not only enhanced radiosensitivity of refractory pancreatic tumors but also modulated the tumor microenvironment to effectively inhibit tumor growth.

Small molecule inhibitors

Efforts in the field of drug discovery have led to the generation of kinase inhibitors that act by inhibiting the kinase domain of Eph receptors. Amongst the number of compounds tested, NVP-BHG712 surfaced as a highly selective and potent inhibitor of EphB4 kinase activity with excellent pharmacokinetic profile. The kinase inhibitor was reported to block EphB4 autophosphorylation in tissues and VEGF-driven angiogenesis in a growth factor implant model [148]. However, the issues related to lack of specificity, lack of access to their intracellular targets, and sub-optimal efficacy hindered the full exploitation potential of these kinase inhibitors [149] and led to the emergence of alternative strategies. This includes the use of protein-protein interaction (PPI) inhibitors that function by preventing the interaction between the extracellular ligand-binding domain of Eph receptors and their cognate ephrin ligands. Their ability to block both the forward and reverse signaling in Eph-expressing and ephrin-expressing cell, along with selective mode of action are some of the characteristic features due to which their use is favored over some of the available kinase inhibitors. Some of the PPI inhibitors are described below.

Lithocolic acid (LCA) is a competitive reversible inhibitor that was demonstrated to target protein-protein interaction with a greater affinity toward ephrin A subclass compared to ephrin B subclass. A study by Giorgio et al. [150], determined the efficacy of LCA in a cancer model and observed that LCA inhibited cell rounding and retraction in PC3 cells following EphA2 activation. This would be particularly relevant in tumor models where EphA2 activation drives tumor growth and progression.

UniPR129, the L-homo-Trp conjugate of lithocholic acid, is a competitive small molecule inhibitor that disrupts the interaction between EphA2-ephrin-A1 receptor-ligand pair. It was reported to block ephrin-A1-mediated phosphorylation of EphA2, and cell retraction in PC3 prostate carcinoma cells [151]. Furthermore, UniPR129 was shown to possess anti-angiogenic activity in functional assays and it effectively blocked cell migration in cultured human umbilical vein endothelial cells (HUVECs) without inducing any cytotoxic effects [151]. Although UniPR129 showed some promising results in vitro, its low oral bioavailability following administration in mice [152], posed a significant challenge for its future application.

UniPR1331 is an orally bioavailable small molecule inhibitor that has been reported to antagonize Eph-ephrin interaction both in vitro and in vivo [153]. UniPR1331 treatment resulted in a significant reduction of tube formation in HUVECs and vessel formation induced by $VEGF_{165}$ in an in vivo CAM assay [153]. The small molecule inhibitor was also shown to block EphA2 phosphorylation triggered by ephrin-A1 thus reducing EphA2 levels in GBM U87MG cells [153]. Treatment of GBM xenograft tumors resulted in a significant tumor growth inhibition and also delayed the time to disease progression in vivo [153]. In the same study, UniPR1331 treatment was found to enhance the efficacy of an anti-angiogenic drug,

Bevacizumab when administered as a combination therapy further suppressing the tumor growth in a xenograft model [153]. The combination treatment inhibited tumor growth by affecting pro-survival, growth and proliferation pathways as indicated by increased apoptosis, decreased vessel count, and decreased Ki67 levels [153]. Interestingly, UniPR1331 was shown to cross the blood-brain barrier and resulted in a potent anti-tumor activity when patient-derived glioblastoma propagating/undifferentiated/stem cells, TPC8, were orthotopically implanted in nude mice [153].

Antisense oligonucleotides or siRNAs

Antisense oligonucleotides or siRNAs targeting Eph/ephrin proteins have also been used to downregulate their expression and disrupt the downstream signaling activities. In this context, ephrin-B3 knockdown in combination with ionizing radiation has been documented to trigger cellular senescence, inhibiting MAPK ERK and p38 phosphorylation, and resulted in an elevated expression of p27(kip1) in a NSCLC model [88]. Studies published from our laboratory have also shown significant radiosensitization following EphB4 knockdown [90] in head and neck cancer and EphB2 knockdown [43] in medulloblastoma cells in vitro. In another study, investigators reported the efficacy of a multistage vector system loaded with liposomal EphA2 siRNA in enhancing therapeutic response to either paclitaxel or docetaxel in murine orthotopic models of ovarian cancer [154]. EphB4 silencing was also evaluated for its potent anti-tumor effects in combination with docetaxel in an ovarian cancer model where siRNA targeting of EphB4 inhibited tumor angiogenesis, and cell proliferation and resulted in enhanced apoptotic cell death by affecting PI3K signaling pathway [155] (Table 1).

TABLE 1 Therapeutic strategies targeting Eph-ephrins

Therapeutic agent	Target	Trial stage	ClinicalTrials.gov identifier, refs
Protein			
sEphB4-HSA	EphB4-Ephrin-B2	Phase Ib	NCT02495896; NCT02767921
		Phase II/IIa	NCT02799485; NCT03049618
		Preclinical	[52, 90, 91, 113, 114]
Peptides			
TNYL-RAW	EphB4-Ephrin-B2	Preclinical	[87, 111, 128, 129], Bhatia et al., Cancer Research, in press
YSA	EphA2	Preclinical	[133–135]
SNEW	EphB2	Preclinical	[110, 137]
AzV36-NicL	EphA2, EphB2, EphB4	Preclinical	[138]
EphA2883-891	EphA2	Phase I	NCT00390338
Antibodies			
Ifabotuzumab	EphA3		NCT03374943; NCT01211691 [139, 140]

TABLE 1 Therapeutic strategies targeting Eph-ephrins—Cont'd

Therapeutic agent	Target	Trial stage	ClinicalTrials.gov identifier, refs
		Phase I Preclinical	
MEDI-547	EphA2	Phase I Preclinial	NCT00796055 [142]
PF-06647263	Ephrin-A4	Phase I Preclinial	NCT02078752 [144]
scFv B11	EphB4-Ephrin-B2	Preclinical	[111, 146, 147], Lennon et al., Clinical Cancer Research, in press
Small molecule inhibitors			
NVP-BHG712	EphB4	Preclinical	[148]
Lithocolic acid (LCA)	EphA2-ephrin-A1	Preclinical	[150]
UniPR129	EphA2-ephrin-A1	Preclinical	[151, 152]
UniPR1331	EphA2-ephrin-A1	Preclinical	[153]
Antisense oligonucleotides or siRNAs			
siRNA	Ephrin-B3	Preclinical	[88]
	EphB4		[90, 155]
	EphB2		[43]
	EphA2		[154]

Challenges and future directions

Research over the past few years has considerably enhanced our current understanding on how aberrant signaling mediated by Eph-ephrins plays a role in tumorigenesis. There are several intriguing questions that still remain unexplored in the field and warrants further investigation. It has however become apparent from the wealth of information available in the literature that this gene family cannot be studied in isolation, mostly because Eph/ephrins engage in a complex cross-talk with multiple pathways to trigger a biological response. These direct or indirect regulatory interactions culminate in an agonist or antagonistic response and can be highly context dependent. It would be interesting to identify all the critical players in the signaling cascades that interact with Eph-ephrin proteins to unravel mechanistic insights and further formulate a rational therapeutic design.

Although several strategies including proteins, peptides, antibodies, and small molecule inhibitors have been evaluated to target Eph-ephrin proteins, their promiscuous activities and paradoxical roles in cancer presents a formidable challenge. It is well known that the Eph receptors initiate forward signaling cascade via their kinase domain and reverse signaling is triggered by cytoplasmic domain of transmembrane ephrin ligand. Thus, developing agents that selectively target a particular Eph/ephrin domain would uncover underlying mechanisms that drive diverse cellular activities by this family of proteins. Another challenge exists in developing strategies against Eph-ephrins because of their ability to orchestrate various cellular functions. The inhibitory effect of targeting a particular Eph receptor or ligand might be masked by a compensatory mechanism by other members of this gene family thus blunting the overall effect. Considerable amount of research has been done with Eph-ephrin targeted agents and most of these agents have been shown to possess manageable toxicities probably because of the reduced level of Eph-ephrin expression in the adult tissues. However, critical assessment of developmental toxicity is unquestionably justified given the imperative role of this gene family in the embryonic development.

Multiple studies have also shed light on the ability of some of the proteases to cleave Eph receptors and ephrins resulting in the shedding of their ectodomain in the extracellular milieu. This however, presents an exciting possibility on the ability of these paracrine cell-cell contact independent cues to define a biological outcome.

Another important aspect that needs to be prioritized in the field of Eph/ephrin cancer research is the need for improvement in animal models that are used in preclinical studies. Since complete ablation of some Eph and ephrins, such as EphB4 and ephrin B2, display perinatal lethality, the use of genetically engineered models with conditional knock-outs or knock-ins would help in understanding some of the pertinent questions related to Eph-ephrin biology. Eph-ephrins are undoubtedly attractive candidates for therapeutic intervention. However, in view of their broad cellular and tissue expression profile and established role in physiological functions, strategies aimed at blocking their activities should be carefully tailored to harness maximal therapeutic benefit in cancers.

Conclusion

The Eph receptor tyrosine kinases and their cognate ligands ephrins play a prominent role in tumorigenesis. Multiple reports exist correlating their dysregulated expression with a poor clinical prognosis in both solid tumors and hematological malignancies. Emerging data also suggest the involvement of Eph-ephrins in mediating therapeutic and radiation resistance. Therefore, ample opportunities exist to modulate Eph-ephrin axis to yield improved treatment outcomes in patients. In this context, a comprehensive understanding of Eph-ephrin mediated signaling mechanisms will be fundamental in selecting an optimal intervention to specifically target underlying malignant condition while minimizing undesirable effects on normal physiological functions.

Acknowledgments

The work published by Dr. Karam was supported by the Paul Calabresi Career Development Award for Clinical Oncology (CA086913), RSNA grant (RSD1713), Golfer's against Cancer, Cancer League of Colorado Grant, American Cancer Society Institutional Grant, Ruth L. Kirschstein National Research Service Award (T32CA17468), P30-CA046934 (University of Colorado Cancer Center Support Grant) and Marsico family endowment funds.

References

[1] Pasquale EB. Eph receptors and ephrins in cancer: bidirectional signalling and beyond. Nat Rev Cancer 2010; 10(3):165–80.

[2] Pasquale EB. Eph-ephrin bidirectional signaling in physiology and disease. Cell 2008;133(1):38–52.

[3] Himanen JP, Chumley MJ, Lackmann M, Li C, Barton WA, Jeffrey PD, Vearing C, Geleick D, Feldheim DA, Boyd AW, Henkemeyer M, Nikolov DB. Repelling class discrimination: ephrin-A5 binds to and activates EphB2 receptor signaling. Nat Neurosci 2004;7(5):501–9.

[4] Pasquale EB. Eph receptor signalling casts a wide net on cell behaviour. Nat Rev Mol Cell Biol 2005;6(6):462–75.

[5] Arvanitis D, Davy A. Eph/ephrin signaling: networks. Genes Dev 2008;22(4):416–29.

[6] Halford MM, Armes J, Buchert M, Meskenaite V, Grail D, Hibbs ML, Wilks AF, Farlie PG, Newgreen DF, Hovens CM, Stacker SA. Ryk-deficient mice exhibit craniofacial defects associated with perturbed Eph receptor crosstalk. Nat Genet 2000;25(4):414–8.

[7] Kasemeier-Kulesa JC, Bradley R, Pasquale EB, Lefcort F, Kulesa PM. Eph/ephrins and N-cadherin coordinate to control the pattern of sympathetic ganglia. Development 2006;133(24):4839–47.

[8] Klein R. Bidirectional modulation of synaptic functions by Eph/ephrin signaling. Nat Neurosci 2009;12 (1):15–20.

[9] Kandouz M. The Eph/Ephrin family in cancer metastasis: communication at the service of invasion. Cancer Metastasis Rev 2012;31(1–2):353–73.

[10] Barquilla A, Pasquale EB. Eph receptors and ephrins: therapeutic opportunities. Annu Rev Pharmacol Toxicol 2015;55:465–87.

[11] Shiuan E, Chen J. Eph receptor tyrosine kinases in tumor immunity. Cancer Res 2016;76(22):6452–7.

[12] Gerety SS, Wang HU, Chen ZF, Anderson DJ. Symmetrical mutant phenotypes of the receptor EphB4 and its specific transmembrane ligand ephrin-B2 in cardiovascular development. Mol Cell 1999;4(3):403–14.

[13] Heroult M, Schaffner F, Augustin HG. Eph receptor and ephrin ligand-mediated interactions during angiogenesis and tumor progression. Exp Cell Res 2006;312(5):642–50.

[14] Kuijper S, Turner CJ, Adams RH. Regulation of angiogenesis by Eph-ephrin interactions. Trends Cardiovasc Med 2007;17(5):145–51.

[15] Herbert SP, Huisken J, Kim TN, Feldman ME, Houseman BT, Wang RA, Shokat KM, Stainier DY. Arterial-venous segregation by selective cell sprouting: an alternative mode of blood vessel formation. Science 2009;326(5950):294–8.

[16] Sawamiphak S, Seidel S, Essmann CL, Wilkinson GA, Pitulescu ME, Acker T, Acker-Palmer A. Ephrin-B2 regulates VEGFR2 function in developmental and tumour angiogenesis. Nature 2010;465(7297):487–91.

[17] Wang Y, Nakayama M, Pitulescu ME, Schmidt TS, Bochenek ML, Sakakibara A, Adams S, Davy A, Deutsch U, Luthi U, Barberis A, Benjamin LE, Makinen T, Nobes CD, Adams RH. Ephrin-B2 controls VEGF-induced angiogenesis and lymphangiogenesis. Nature 2010;465(7297):483–6.

[18] Cheng N, Brantley DM, Chen J. The ephrins and Eph receptors in angiogenesis. Cytokine Growth Factor Rev 2002;13(1):75–85.

[19] Shekhar MP, Werdell J, Santner SJ, Pauley RJ, Tait L. Breast stroma plays a dominant regulatory role in breast epithelial growth and differentiation: implications for tumor development and progression. Cancer Res 2001;61 (4):1320–6.

[20] Brantley-Sieders DM, Fang WB, Hwang Y, Hicks D, Chen J. Ephrin-A1 facilitates mammary tumor metastasis through an angiogenesis-dependent mechanism mediated by EphA receptor and vascular endothelial growth factor in mice. Cancer Res 2006;66(21):10315–24.

[21] Ogawa K, Pasqualini R, Lindberg RA, Kain R, Freeman AL, Pasquale EB. The ephrin-A1 ligand and its receptor, EphA2, are expressed during tumor neovascularization. Oncogene 2000;19(52):6043–52.

[22] Shao Z, Zhang WF, Chen XM, Shang ZJ. Expression of EphA2 and VEGF in squamous cell carcinoma of the tongue: correlation with the angiogenesis and clinical outcome. Oral Oncol 2008;44(12):1110–7.

[23] Oweida A, Bhatia S, Hirsch K, Calame D, Griego A, Keysar S, Pitts T, Sharma J, Eckhardt G, Jimeno A, Wang XJ, Parkash G, Califano J, Karam SD. Ephrin-B2 overexpression predicts for poor prognosis and response to therapy in solid tumors. Mol Carcinog 2017;56(3):1189–96.

[24] Foo SS, Turner CJ, Adams S, Compagni A, Aubyn D, Kogata N, Lindblom P, Shani M, Zicha D, Adams RH. Ephrin-B2 controls cell motility and adhesion during blood-vessel-wall assembly. Cell 2006; 124(1):161–73.

[25] Salvucci O, Maric D, Economopoulou M, Sakakibara S, Merlin S, Follenzi A, Tosato G. EphrinB reverse signaling contributes to endothelial and mural cell assembly into vascular structures. Blood 2009;114(8):1707–16.

[26] Noren NK, Lu M, Freeman AL, Koolpe M, Pasquale EB. Interplay between EphB4 on tumor cells and vascular ephrin-B2 regulates tumor growth. Proc Natl Acad Sci U S A 2004;101(15):5583–8.

[27] Kirschmann DA, Seftor EA, Hardy KM, Seftor RE, Hendrix MJ. Molecular pathways: vasculogenic mimicry in tumor cells: diagnostic and therapeutic implications. Clin Cancer Res 2012;18(10):2726–32.

[28] Baeten CI, Hillen F, Pauwels P, de Bruine AP, Baeten CG. Prognostic role of vasculogenic mimicry in colorectal cancer. Dis Colon Rectum 2009;52(12):2028–35.

[29] Cao Z, Bao M, Miele L, Sarkar FH, Wang Z, Zhou Q. Tumour vasculogenic mimicry is associated with poor prognosis of human cancer patients: a systemic review and meta-analysis. Eur J Cancer 2013;49(18):3914–23.

[30] Wang W, Lin P, Han C, Cai W, Zhao X, Sun B. Vasculogenic mimicry contributes to lymph node metastasis of laryngeal squamous cell carcinoma. J Exp Clin Cancer Res 2010;29:60.

[31] Hess AR, Seftor EA, Gruman LM, Kinch MS, Seftor RE, Hendrix MJ. VE-cadherin regulates EphA2 in aggressive melanoma cells through a novel signaling pathway: implications for vasculogenic mimicry. Cancer Biol Ther 2006;5(2):228–33.

[32] Hess AR, Seftor EA, Gardner LM, Carles-Kinch K, Schneider GB, Seftor RE, Kinch MS, Hendrix MJ. Molecular regulation of tumor cell vasculogenic mimicry by tyrosine phosphorylation: role of epithelial cell kinase (Eck/EphA2). Cancer Res 2001;61(8):3250–5.

[33] Wang H, Lin H, Pan J, Mo C, Zhang F, Huang B, Wang Z, Chen X, Zhuang J, Wang D, Qiu S. Vasculogenic mimicry in prostate cancer: the roles of EphA2 and PI3K. J Cancer 2016;7(9):1114–24.

[34] Li G, Huang M, Cai Y, Ke Y, Yang Y, Sun X. miR141 inhibits glioma vasculogenic mimicry by controlling EphA2 expression. Mol Med Rep 2018;18(2):1395–404.

[35] Morrison LC, McClelland R, Aiken C, Bridges M, Liang L, Wang X, Di Curzio D, Del Bigio MR, Taylor MD, Werbowetski-Ogilvie TE. Deconstruction of medulloblastoma cellular heterogeneity reveals differences between the most highly invasive and self-renewing phenotypes. Neoplasia 2013;15(4):384–98.

[36] Kool M, Koster J, Bunt J, Hasselt NE, Lakeman A, van Sluis P, Troost D, Meeteren NS, Caron HN, Cloos J, Mrsic A, Ylstra B, Grajkowska W, Hartmann W, Pietsch T, Ellison D, Clifford SC, Versteeg R. Integrated genomics identifies five medulloblastoma subtypes with distinct genetic profiles, pathway signatures and clinicopathological features. PLoS One 2008;3(8):e3088.

[37] Fattet S, Haberler C, Legoix P, Varlet P, Lellouch-Tubiana A, Lair S, Manie E, Raquin MA, Bours D, Carpentier S, Barillot E, Grill J, Doz F, Puget S, Janoueix-Lerosey I, Delattre O. Beta-catenin status in paediatric medulloblastomas: correlation of immunohistochemical expression with mutational status, genetic profiles, and clinical characteristics. J Pathol 2009;218(1):86–94.

[38] Thompson MC, Fuller C, Hogg TL, Dalton J, Finkelstein D, Lau CC, Chintagumpala M, Adesina A, Ashley DM, Kellie SJ, Taylor MD, Curran T, Gajjar A, Gilbertson RJ. Genomics identifies medulloblastoma subgroups that are enriched for specific genetic alterations. J Clin Oncol 2006;24(12):1924–31.

[39] Bhatia S, Baig NA, Timofeeva O, Pasquale EB, Hirsch K, MacDonald TJ, Dritschilo A, Lee YC, Henkemeyer M, Rood B, Jung M, Wang XJ, Kool M, Rodriguez O, Albanese C, Karam SD. Knockdown of EphB1 receptor decreases medulloblastoma cell growth and migration and increases cellular radiosensitization. Oncotarget 2015;6(11):8929–46.

[40] Vindis C, Teli T, Cerretti DP, Turner CE, Huynh-Do U. EphB1-mediated cell migration requires the phosphorylation of paxillin at Tyr-31/Tyr-118. J Biol Chem 2004;279(27):27965–70.

[41] Vindis C, Cerretti DP, Daniel TO, Huynh-Do U. EphB1 recruits c-Src and p52Shc to activate MAPK/ERK and promote chemotaxis. J Cell Biol 2003;162(4):661–71.

[42] Han DC, Shen TL, Miao H, Wang B, Guan JL. EphB1 associates with Grb7 and regulates cell migration. J Biol Chem 2002;277(47):45655–61.

[43] Bhatia S, Hirsch K, Bukkapatnam S, Baig NA, Oweida A, Griego A, Calame D, Sharma J, Donson A, Foreman N, Albanese C, Venkataraman S, Vibhakar R, Karam SD. Combined EphB2 receptor knockdown with radiation decreases cell viability and invasion in medulloblastoma. Cancer Cell Int 2017;17:41.

[44] Sikkema AH, den Dunnen WF, Hulleman E, van Vuurden DG, Garcia-Manero G, Yang H, Scherpen FJ, Kampen KR, Hoving EW, Kamps WA, Diks SH, Peppelenbosch MP, de Bont ES. EphB2 activity plays a pivotal role in pediatric medulloblastoma cell adhesion and invasion. Neuro Oncol 2012;14(9):1125–35.

[45] Wang SD, Rath P, Lal B, Richard JP, Li Y, Goodwin CR, Laterra J, Xia S. EphB2 receptor controls proliferation/migration dichotomy of glioblastoma by interacting with focal adhesion kinase. Oncogene 2012;31(50):5132–43.

[46] Chukkapalli S, Amessou M, Dilly AK, Dekhil H, Zhao J, Liu Q, Bejna A, Thomas RD, Bandyopadhyay S, Bismar TA, Neill D, Azoulay L, Batist G, Kandouz M. Role of the EphB2 receptor in autophagy, apoptosis and invasion in human breast cancer cells. Exp Cell Res 2014;320(2):233–46.

[47] Nakada M, Anderson EM, Demuth T, Nakada S, Reavie LB, Drake KL, Hoelzinger DB, Berens ME. The phosphorylation of ephrin-B2 ligand promotes glioma cell migration and invasion. Int J Cancer 2010;126(5):1155–65.

[48] Ferguson BD, Tretiakova MS, Lingen MW, Gill PS, Salgia R. Expression of the EPHB4 receptor tyrosine kinase in head and neck and renal malignancies—implications for solid tumors and potential for therapeutic inhibition. Growth Factors 2014;32(6):202–6.

[49] Lv J, Xia Q, Wang J, Shen Q, Zhang J, Zhou X. EphB4 promotes the proliferation, invasion, and angiogenesis of human colorectal cancer. Exp Mol Pathol 2016;100(3):402–8.

[50] Masood R, Kumar SR, Sinha UK, Crowe DL, Krasnoperov V, Reddy RK, Zozulya S, Singh J, Xia G, Broek D, Schonthal AH, Gill PS. EphB4 provides survival advantage to squamous cell carcinoma of the head and neck. Int J Cancer 2006;119(6):1236–48.

[51] Yang NY, Pasquale EB, Owen LB, Ethell IM. The EphB4 receptor-tyrosine kinase promotes the migration of melanoma cells through Rho-mediated actin cytoskeleton reorganization. J Biol Chem 2006;281(43):32574–86.

[52] Ferguson BD, Liu R, Rolle CE, Tan YH, Krasnoperov V, Kanteti R, Tretiakova MS, Cervantes GM, Hasina R, Hseu RD, Iafrate AJ, Karrison T, Ferguson MK, Husain AN, Faoro L, Vokes EE Gill PS, Salgia R. The EphB4 receptor tyrosine kinase promotes lung cancer growth: a potential novel therapeutic target. PLoS One 2013;8(7):e67668.

[53] Yamashita T, Ohneda K, Nagano M, Miyoshi C, Kaneko N, Miwa Y, Yamamoto M, Ohneda O, Fujii-Kuriyama Y. Hypoxia-inducible transcription factor-2alpha in endothelial cells regulates tumor neovascularization through activation of ephrin A1. J Biol Chem 2008;283(27):18926–36.

[54] Pandey A, Shao H, Marks RM, Polverini PJ, Dixit VM. Role of B61, the ligand for the Eck receptor tyrosine kinase, in TNF-alpha-induced angiogenesis. Science 1995;268(5210):567–9.

[55] Hafner C, Schmitz G, Meyer S, Bataille F, Hau P, Langmann T, Dietmaier W, Landthaler M, Vogt T. Differential gene expression of Eph receptors and ephrins in benign human tissues and cancers. Clin Chem 2004;50(3):490–9.

[56] Zhang XQ, Takakura N, Oike Y, Inada T, Gale NW, Yancopoulos GD, Suda T. Stromal cells expressing ephrin-B2 promote the growth and sprouting of ephrin-B2(+) endothelial cells. Blood 2001;98(4):1028–37.

[57] Braun J, Hoffmann SC, Feldner A, Ludwig T, Henning R, Hecker M, Korff T. Endothelial cell ephrinB2-dependent activation of monocytes in arteriosclerosis. Arterioscler Thromb Vasc Biol 2011;31(2):297–305.

[58] Poitz DM, Ende G, Stutz B, Augstein A, Friedrichs J, Brunssen C, Werner C, Strasser RH, Jellinghaus S. EphrinB2/EphA4-mediated activation of endothelial cells increases monocyte adhesion. Mol Immunol 2015;68(2 Pt. C):648–56.

[59] Funk SD, Yurdagul Jr A, Albert P, Traylor Jr JG, Jin L, Chen J, Orr AW. EphA2 activation promotes the endothelial cell inflammatory response: a potential role in atherosclerosis. Arterioscler Thromb Vasc Biol 2012;32(3):686–95.

[60] Pfaff D, Heroult M, Riedel M, Reiss Y, Kirmse R, Ludwig T, Korff T, Hecker M, Augustin HG. Involvement of endothelial ephrin-B2 in adhesion and transmigration of EphB-receptor-expressing monocytes. J Cell Sci 2008;121(Pt. 22):3842–50.

[61] Tatsumi T, Herrem CJ, Olson WC, Finke JH, Bukowski RM, Kinch MS, Ranieri E, Storkus WJ. Disease stage variation in CD4+ and CD8+ T-cell reactivity to the receptor tyrosine kinase EphA2 in patients with renal cell carcinoma. Cancer Res 2003;63(15):4481–9.

[62] Hatano M, Eguchi J, Tatsumi T, Kuwashima N, Dusak JE, Kinch MS, Pollack IF, Hamilton RL, Storkus WJ, Okada H. EphA2 as a glioma-associated antigen: a novel target for glioma vaccines. Neoplasia 2005;7(8):717–22.

[63] Yamaguchi S, Tatsumi T, Takehara T, Sakamori R, Uemura A, Mizushima T, Ohkawa K, Storkus WJ, Hayashi N. Immunotherapy of murine colon cancer using receptor tyrosine kinase EphA2-derived peptide-pulsed dendritic cell vaccines. Cancer 2007;110(7):1469–77.

[64] Yamaguchi S, Tatsumi T, Takehara T, Sasakawa A, Hikita H, Kohga K, Uemura A, Sakamori R, Ohkawa K, Hayashi N. Dendritic cell-based vaccines suppress metastatic liver tumor via activation of local innate and acquired immunity. Cancer Immunol Immunother 2008;57(12):1861–9.

[65] Miao H, Gale NW, Guo H, Qian J, Petty A, Kaspar J, Murphy AJ, Valenzuela DM, Yancopoulos G, Hambardzumyan D, Lathia JD, Rich JN, Lee J, Wang B. EphA2 promotes infiltrative invasion of glioma stem cells in vivo through cross-talk with Akt and regulates stem cell properties. Oncogene 2015;34(5):558–67.
[66] Yang NY, Fernandez C, Richter M, Xiao Z, Valencia F, Tice DA, Pasquale EB. Crosstalk of the EphA2 receptor with a serine/threonine phosphatase suppresses the Akt-mTORC1 pathway in cancer cells. Cell Signal 2011;23 (1):201–12.
[67] Peng J, Wang Q, Liu H, Ye M, Wu X, Guo L. EPHA3 regulates the multidrug resistance of small cell lung cancer via the PI3K/BMX/STAT3 signaling pathway. Tumour Biol 2016;37(9):11959–71.
[68] Saintigny P, Peng S, Zhang L, Sen B, Wistuba II, Lippman SM, Girard L, Minna JD, Heymach JV, Johnson FM. Global evaluation of Eph receptors and ephrins in lung adenocarcinomas identifies EphA4 as an inhibitor of cell migration and invasion. Mol Cancer Ther 2012;11(9):2021–32.
[69] Giaginis C, Tsoukalas N, Bournakis E, Alexandrou P, Kavantzas N, Patsouris E, Theocharis S. Ephrin (Eph) receptor A1, A4, A5 and A7 expression in human non-small cell lung carcinoma: associations with clinicopathological parameters, tumor proliferative capacity and patients' survival. BMC Clin Pathol 2014;14(1):8.
[70] Li R, Sun Y, Jiang A, Wu Y, Li C, Jin M, Yan H, Jin H. Knockdown of ephrin receptor A7 suppresses the proliferation and metastasis of A549 human lung cancer cells. Mol Med Rep 2016;13(4):3190–6.
[71] Chiu ST, Chang KJ, Ting CH, Shen HC, Li H, Hsieh FJ. Over-expression of EphB3 enhances cell-cell contacts and suppresses tumor growth in HT-29 human colon cancer cells. Carcinogenesis 2009;30(9):1475–86.
[72] Bhatia S, Griego A, Lennon S, Oweida A, Sharma J, Rohmer C, Uyanga N, Bukkapatnam S, Van Court B, Raben D, Young C, Heasley L, Karam SD. Role of EphB3 receptor in mediating head and neck tumor growth, cell migration, and response to PI3K inhibitor. Mol Cancer Ther 2018;17:2049–59.
[73] Noren NK, Foos G, Hauser CA, Pasquale EB. The EphB4 receptor suppresses breast cancer cell tumorigenicity through an Abl-Crk pathway. Nat Cell Biol 2006;8(8):815–25.
[74] Dopeso H, Mateo-Lozano S, Mazzolini R, Rodrigues P, Lagares-Tena L, Ceron J, Romero J, Esteves M, Landolfi S, Hernandez-Losa J, Castano J, Wilson AJ, Ramon y Cajal S, Mariadason JM, Schwartz Jr S, Arango D. The receptor tyrosine kinase EPHB4 has tumor suppressor activities in intestinal tumorigenesis. Cancer Res 2009;69(18):7430–8.
[75] Li JJ, Liu DP, Liu GT, Xie D. EphrinA5 acts as a tumor suppressor in glioma by negative regulation of epidermal growth factor receptor. Oncogene 2009;28(15):1759–68.
[76] Brantley-Sieders DM, Zhuang G, Hicks D, Fang WB, Hwang Y, Cates JM, Coffman K, Jackson D, Bruckheimer E, Muraoka-Cook RS, Chen J. The receptor tyrosine kinase EphA2 promotes mammary adenocarcinoma tumorigenesis and metastatic progression in mice by amplifying ErbB2 signaling. J Clin Invest 2008;118 (1):64–78.
[77] Zhuang G, Brantley-Sieders DM, Vaught D, Yu J, Xie L, Wells S, Jackson D, Muraoka-Cook R, Arteaga C, Chen J. Elevation of receptor tyrosine kinase EphA2 mediates resistance to trastuzumab therapy. Cancer Res 2010;70 (1):299–308.
[78] Iiizumi M, Hosokawa M, Takehara A, Chung S, Nakamura T, Katagiri T, Eguchi H, Ohigashi H, Ishikawa O, Nakamura Y, Nakagawa H. EphA4 receptor, overexpressed in pancreatic ductal adenocarcinoma, promotes cancer cell growth. Cancer Sci 2006;97(11):1211–6.
[79] Miyazaki K, Inokuchi M, Takagi Y, Kato K, Kojima K, Sugihara K. EphA4 is a prognostic factor in gastric cancer. BMC Clin Pathol 2013;13(1):19.
[80] Ding L, Shen Y, Ni J, Ou Y, Ou Y, Liu H. EphA4 promotes cell proliferation and cell adhesion-mediated drug resistance via the AKT pathway in multiple myeloma. Tumour Biol 2017;39(3).
[81] Lu M, Miller KD, Gokmen-Polar Y, Jeng MH, Kinch MS. EphA2 overexpression decreases estrogen dependence and tamoxifen sensitivity. Cancer Res 2003;63(12):3425–9.
[82] Zelinski DP, Zantek ND, Stewart JC, Irizarry AR, Kinch MS. EphA2 overexpression causes tumorigenesis of mammary epithelial cells. Cancer Res 2001;61(5):2301–6.
[83] Yang X, Yang Y, Tang S, Tang H, Yang G, Xu Q, Wu J. EphB4 inhibitor overcome the acquired resistance to cisplatin in melanomas xenograft model. J Pharmacol Sci 2015;129(1):65–71.
[84] Li L, Xu N, Zhang JF, Xu LL, Zhou X, Huang BT, Li YL, Liu XL. EphB4/ephrinB2 contributes to imatinib resistance in chronic myeloid leukemia involved in cytoskeletal proteins. Int J Med Sci 2016;13(5):365–73.
[85] Li X, Wang L, Gu JW, Li B, Liu WP, Wang YG, Zhang X, Zhen HN, Fei Z. Up-regulation of EphA2 and down-regulation of EphrinA1 are associated with the aggressive phenotype and poor prognosis of malignant glioma. Tumour Biol 2010;31(5):477–88.

[86] Nojiri K, Iwakawa M, Ichikawa Y, Imadome K, Sakai M, Nakawatari M, Ishikawa K, Ishikawa A, Togo S, Tsujii H, Shimada H, Imai T. The proangiogenic factor ephrin-A1 is up-regulated in radioresistant murine tumor by irradiation. Exp Biol Med (Maywood) 2009;234(1):112–22.
[87] You J, Zhang R, Xiong C, Zhong M, Melancon M, Gupta S, Nick AM, Sood AK, Li C. Effective photothermal chemotherapy using doxorubicin-loaded gold nanospheres that target EphB4 receptors in tumors. Cancer Res 2012;72(18):4777–86.
[88] Stahl S, Kaminskyy VO, Efazat G, Hyrslova Vaculova A, Rodriguez-Nieto S, Moshfegh A, Lewensohn R, Viktorsson K, Zhivotovsky B. Inhibition of Ephrin B3-mediated survival signaling contributes to increased cell death response of non-small cell lung carcinoma cells after combined treatment with ionizing radiation and PKC 412. Cell Death Dis 2013;4:e454.
[89] Staquicini FI, Qian MD, Salameh A, Dobroff AS, Edwards JK, Cimino DF, Moeller BJ, Kelly P, Nunez MI, Tang X, Liu DD, Lee JJ, Hong WK, Ferrara F, Bradbury AR, Lobb RR, Edelman MJ, Sidman RL, Wistuba II, Arap W, Pasqualini R. Receptor tyrosine kinase EphA5 is a functional molecular target in human lung cancer. J Biol Chem 2015;290(12):7345–59.
[90] Bhatia S, Hirsch K, Sharma J, Oweida A, Griego A, Keysar S, Jimeno A, Raben D, Krasnoperov V, Gill PS, Pasquale EB, Wang XJ, Karam SD. Enhancing radiosensitization in EphB4 receptor-expressing head and neck squamous cell carcinomas. Sci Rep 2016;6:38792.
[91] Bhatia S, Sharma J, Bukkapatnam S, Oweida A, Lennon S, Phan A, Milner D, Uyanga N, Jimeno A, Raben D, Somerset H, Heasley L, Karam SD. Inhibition of EphB4-ephrin-B2 signaling enhances response to cetuximab-radiation therapy in head and neck cancers. Clin Cancer Res 2018;24(18):4539–50.
[92] Paez-Ribes M, Allen E, Hudock J, Takeda T, Okuyama H, Vinals F, Inoue M, Bergers G, Hanahan D, Casanovas O. Antiangiogenic therapy elicits malignant progression of tumors to increased local invasion and distant metastasis. Cancer Cell 2009;15(3):220–31.
[93] Salvucci O, Ohnuki H, Maric D, Hou X, Li X, Yoon SO, Segarra M, Eberhart CG, Acker-Palmer A, Tosato G. EphrinB2 controls vessel pruning through STAT1-JNK3 signalling. Nat Commun 2015;6:6576.
[94] Groppa E, Brkic S, Uccelli A, Wirth G, Korpisalo-Pirinen P, Filippova M, Dasen B, Sacchi V, Muraro MG, Trani M, Reginato S, Gianni-Barrera R, Yla-Herttuala S, Banfi A. EphrinB2/EphB4 signaling regulates non-sprouting angiogenesis by VEGF. EMBO Rep 2018;19(5):e45054.
[95] Uhl C, Markel M, Broggini T, Nieminen M, Kremenetskaia I, Vajkoczy P, Czabanka M. EphB4 mediates resistance to antiangiogenic therapy in experimental glioma. Angiogenesis 2018;21:873–81.
[96] Casanovas O, Hicklin DJ, Bergers G, Hanahan D. Drug resistance by evasion of antiangiogenic targeting of VEGF signaling in late-stage pancreatic islet tumors. Cancer Cell 2005;8(4):299–309.
[97] Depner C, Zum Buttel H, Bogurcu N, Cuesta AM, Aburto MR, Seidel S, Finkelmeier F, Foss F, Hofmann J, Kaulich K, Barbus S, Segarra M, Reifenberger G, Garvalov BK, Acker T, Acker-Palmer A. EphrinB2 repression through ZEB2 mediates tumour invasion and anti-angiogenic resistance. Nat Commun 2016;7:12329.
[98] Djokovic D, Trindade A, Gigante J, Badenes M, Silva L, Liu R, Li X, Gong M, Krasnoperov V, Gill PS, Duarte A. Combination of Dll4/Notch and Ephrin-B2/EphB4 targeted therapy is highly effective in disrupting tumor angiogenesis. BMC Cancer 2010;10:641.
[99] Szturz P, Vermorken JB. Immunotherapy in head and neck cancer: aiming at EXTREME precision. BMC Med 2017;15(1):110.
[100] AstraZeneca. AstraZeneca reports initial results from the ongoing MYSTIC trial in Stage IV lung cancer; 2017.
[101] LLC M. A phase 1b/2 safety and tolerability study of MEDI6469 in combination with therapeutic immune agents or monoclonal antibodies; 2017.
[102] Messenheimer DJ, Jensen SM, Afentoulis ME, Wegmann KW, Feng Z, Friedman DJ, Gough MJ, Urba WJ, Fox BA. Timing of PD-1 blockade is critical to effective combination immunotherapy with anti-OX40. Clin Cancer Res 2017;23:6165–77.
[103] Shrimali RK, Ahmad S, Verma V, Zeng P, Ananth S, Gaur P, Gittelman RM, Yusko E, Sanders C, Robins H, Hammond SA, Janik JE, Mkrtichyan M, Gupta S, Khleif SN. Concurrent PD-1 blockade negates the effects of OX40 agonist antibody in combination immunotherapy through inducing T-cell apoptosis. Cancer Immunol Res 2017;5(9):755–66.
[104] Oweida A, Hararah MK, Phan A, Binder D, Bhatia S, Lennon S, Bukkapatnam S, Van Court B, Uyanga N, Darragh L, Kim HM, Raben D, Tan AC, Heasley L, Clambey E, Nemenoff R, Karam SD. Resistance to radiotherapy and PD-L1 blockade is mediated by TIM-3 upregulation and regulatory T-cell infiltration. Clin Cancer Res 2018;24.

[105] Oweida A, Lennon S, Calame D, Korpela S, Bhatia S, Sharma J, Graham C, Binder D, Serkova N, Raben D, Heasley L, Clambey E, Nemenoff R, Karam SD. Ionizing radiation sensitizes tumors to PD-L1 immune checkpoint blockade in orthotopic murine head and neck squamous cell carcinoma. Oncoimmunology 2017;6(10):e1356153.
[106] Yu G, Luo H, Wu Y, Wu J. Ephrin B2 induces T cell costimulation. J Immunol 2003;171(1):106–14.
[107] Jellinghaus S, Poitz DM, Ende G, Augstein A, Weinert S, Stutz B, Braun-Dullaeus RC, Pasquale EB, Strasser RH. Ephrin-A1/EphA4-mediated adhesion of monocytes to endothelial cells. Biochim Biophys Acta 2013;1833 (10):2201–11.
[108] Sharfe N, Freywald A, Toro A, Dadi H, Roifman C. Ephrin stimulation modulates T cell chemotaxis. Eur J Immunol 2002;32(12):3745–55.
[109] Wohlfahrt JG, Karagiannidis C, Kunzmann S, Epstein MM, Kempf W, Blaser K, Schmidt-Weber CB. Ephrin-A1 suppresses Th2 cell activation and provides a regulatory link to lung epithelial cells. J Immunol 2004;172 (2):843–50.
[110] Nguyen TM, Arthur A, Hayball JD, Gronthos S. EphB and Ephrin-B interactions mediate human mesenchymal stem cell suppression of activated T-cells. Stem Cells Dev 2013;22(20):2751–64.
[111] Kwak H, Salvucci O, Weigert R, Martinez-Torrecuadrada JL, Henkemeyer M, Poulos MG, Butler JM, Tosato G. Sinusoidal ephrin receptor EPHB4 controls hematopoietic progenitor cell mobilization from bone marrow. J Clin Invest 2016;126(12):4554–68.
[112] Pfaff D, Fiedler U, Augustin HG. Emerging roles of the Angiopoietin-Tie and the ephrin-Eph systems as regulators of cell trafficking. J Leukoc Biol 2006;80(4):719–26.
[113] Scehnet JS, Ley EJ, Krasnoperov V, Liu R, Manchanda PK, Sjoberg E, Kostecke AP, Gupta S, Kumar SR, Gill PS. The role of Ephs, Ephrins, and growth factors in Kaposi sarcoma and implications of EphrinB2 blockade. Blood 2009;113(1):254–63.
[114] Liu R, Ferguson BD, Zhou Y, Naga K, Salgia R, Gill PS, Krasnoperov V. EphB4 as a therapeutic target in mesothelioma. BMC Cancer 2013;13:269.
[115] Ang KK, Zhang Q, Rosenthal DI, Nguyen-Tan PF, Sherman EJ, Weber RS, Galvin JM, Bonner JA, Harris J, El-Naggar AK, Gillison ML, Jordan RC, Konski AA, Thorstad WL, Trotti A, Beitler JJ, Garden AS, Spanos WJ, Yom SS, Axelrod RS. Randomized phase III trial of concurrent accelerated radiation plus cisplatin with or without cetuximab for stage III to IV head and neck carcinoma: RTOG 0522. J Clin Oncol 2014;32 (27):2940–50.
[116] Gillison ML, Zhang Q, Jordan R, Xiao W, Westra WH, Trotti A, Spencer S, Harris J, Chung CH, Ang KK. Tobacco smoking and increased risk of death and progression for patients with p16-positive and p16-negative oropharyngeal cancer. J Clin Oncol 2012;30(17):2102–11.
[117] Pignon JP, le Maitre A, Maillard E, Bourhis J, Group M-NC. Meta-analysis of chemotherapy in head and neck cancer (MACH-NC): an update on 93 randomised trials and 17,346 patients. Radiother Oncol 2009;92(1):4–14.
[118] Haddad RI, Shin DM. Recent advances in head and neck cancer. N Engl J Med 2008;359(11):1143–54.
[119] Specenier P, Vermorken JB. Cetuximab in the treatment of squamous cell carcinoma of the head and neck. Expert Rev Anticancer Ther 2011;11(4):511–24.
[120] Agrawal N, Frederick MJ, Pickering CR, Bettegowda C, Chang K, Li RJ, Fakhry C, Xie TX, Zhang J, Wang J, Zhang N, El-Naggar AK, Jasser SA, Weinstein JN, Trevino L, Drummond JA, Muzny DM, Wu Y, Wood LD, Hruban RH, Westra WH, Koch WM, Califano JA, Gibbs RA, Sidransky D, Vogelstein B, Velculescu VE, Papadopoulos N, Wheeler DA, Kinzler KW, Myers JN. Exome sequencing of head and neck squamous cell carcinoma reveals inactivating mutations in NOTCH1. Science 2011;333(6046):1154–7.
[121] Stransky N, Egloff AM, Tward AD, Kostic AD, Cibulskis K, Sivachenko A, Kryukov GV, Lawrence MS, Sougnez C, McKenna A, Shefler E, Ramos AH, Stojanov P, Carter SL, Voet D, Cortes ML, Auclair D, Berger MF, Saksena G, Guiducci C, Onofrio RC, Parkin M, Romkes M, Weissfeld JL, Seethala RR, Wang L, Rangel-Escareno C, Fernandez-Lopez JC, Hidalgo-Miranda A, Melendez-Zajgla J, Winckler W, Ardlie K, Gabriel SB, Meyerson M, Lander ES, Getz G, Golub TR, Garraway LA, Grandis JR. The mutational landscape of head and neck squamous cell carcinoma. Science 2011;333(6046):1157–60.
[122] Pickering CR, Zhang J, Yoo SY, Bengtsson L, Moorthy S, Neskey DM, Zhao M, Ortega Alves MV, Chang K, Drummond J, Cortez E, Xie TX, Zhang D, Chung W, Issa JP, Zweidler-McKay PA, Wu X, El-Naggar AK, Weinstein JN, Wang J, Muzny DM, Gibbs RA, Wheeler DA, Myers JN, Frederick MJ. Integrative genomic characterization of oral squamous cell carcinoma identifies frequent somatic drivers. Cancer Discov 2013;3 (7):770–81.
[123] Singleton KR, Kim J, Hinz TK, Marek LA, Casas-Selves M, Hatheway C, Tan AC, DeGregori J, Heasley LE. A receptor tyrosine kinase network composed of fibroblast growth factor receptors, epidermal growth factor

receptor, v-erb-b2 erythroblastic leukemia viral oncogene homolog 2, and hepatocyte growth factor receptor drives growth and survival of head and neck squamous carcinoma cell lines. Mol Pharmacol 2013;83(4):882–93.

[124] Gao J, Aksoy BA, Dogrusoz U, Dresdner G, Gross B, Sumer SO, Sun Y, Jacobsen A, Sinha R, Larsson E, Cerami E, Sander C, Schultz N. Integrative analysis of complex cancer genomics and clinical profiles using the cBioPortal. Sci Signal 2013;6(269):l1.

[125] Yavrouian EJ, Sinha UK, Rice DH, Salam MT, Gill PS, Masood R. The significance of EphB4 and EphrinB2 expression and survival in head and neck squamous cell carcinoma. Arch Otolaryngol Head Neck Surg 2008;134 (9):985–91.

[126] El-Khoueiry A, Gitlitz B, Cole S, Tsao-Wei D, Goldkorn A, Quinn D, Lenz HJ, Nieva J, Dorff T, Oswald M, Berg J, Menendez X, Karakozian K, Krasnoperov V, Liu R, Thomas J, Groshen S, Gill P. A first-in-human phase I study of sEphB4-HSA in patients with advanced solid tumors with expansion at the maximum tolerated dose (MTD) or recommended phase II dose (RP2D). Eur J Cancer 2016;69:S11.

[127] Hugen C, GM PG, El-Khoueiry A, Tsao-Wei D, Groshen S, Xu Y, Xu T, Kossari N, Kao B, Wei J, Rochefort H, Tai YC, Goldkorn A. Circulating tumor cell analysis in a first-in-human phase I study of sEphB4-HSA, an inhibitor of EphrinB2-EphB4 signaling. Eur J Cancer 2016;69:S126–7.

[128] Noberini R, Mitra S, Salvucci O, Valencia F, Duggineni S, Prigozhina N, Wei K, Tosato G, Huang Z, Pasquale EB. PEGylation potentiates the effectiveness of an antagonistic peptide that targets the EphB4 receptor with nanomolar affinity. PLoS One 2011;6(12):e28611.

[129] Koolpe M, Burgess R, Dail M, Pasquale EB. EphB receptor-binding peptides identified by phage display enable design of an antagonist with ephrin-like affinity. J Biol Chem 2005;280(17):17301–11.

[130] Reubi JC, Waser B. Concomitant expression of several peptide receptors in neuroendocrine tumours: molecular basis for in vivo multireceptor tumour targeting. Eur J Nucl Med Mol Imaging 2003;30(5):781–93.

[131] Xiong C, Huang M, Zhang R, Song S, Lu W, Flores 2nd L, Gelovani J, Li C. In vivo small-animal PET/CT of EphB4 receptors using 64Cu-labeled peptide. J Nucl Med 2011;52(2):241–8.

[132] Huang M, Xiong C, Lu W, Zhang R, Zhou M, Huang Q, Weinberg J, Li C. Dual-modality micro-positron emission tomography/computed tomography and near-infrared fluorescence imaging of EphB4 in orthotopic glioblastoma xenograft models. Mol Imaging Biol 2014;16(1):74–84.

[133] Koolpe M, Dail M, Pasquale EB. An ephrin mimetic peptide that selectively targets the EphA2 receptor. J Biol Chem 2002;277(49):46974–9.

[134] Patel AR, Chougule M, Singh M. EphA2 targeting pegylated nanocarrier drug delivery system for treatment of lung cancer. Pharm Res 2014;31(10):2796–809.

[135] Scarberry KE, Dickerson EB, McDonald JF, Zhang ZJ. Magnetic nanoparticle-peptide conjugates for in vitro and in vivo targeting and extraction of cancer cells. J Am Chem Soc 2008;130(31):10258–62.

[136] Scarberry KE, Dickerson EB, Zhang ZJ, Benigno BB, McDonald JF. Selective removal of ovarian cancer cells from human ascites fluid using magnetic nanoparticles. Nanomedicine 2010;6(3):399–408.

[137] Pretze M, Kuchar M, Bergmann R, Steinbach J, Pietzsch J, Mamat C. An efficient bioorthogonal strategy using CuAAC click chemistry for radiofluorinations of SNEW peptides and the role of copper depletion. ChemMedChem 2013;8(6):935–45.

[138] Micewicz ED, Jung C, Schaue D, Luong H, McBride WH, Ruchala P. Small azurin derived peptide targets ephrin receptors for radiotherapy. Int J Pept Res Ther 2011;17:247.

[139] Vail ME, Murone C, Tan A, Hii L, Abebe D, Janes PW, Lee FT, Baer M, Palath V, Bebbington C, Yarranton G, Llerena C, Garic S, Abramson D, Cartwright G, Scott AM, Lackmann M. Targeting EphA3 inhibits cancer growth by disrupting the tumor stromal microenvironment. Cancer Res 2014;74(16):4470–81.

[140] Day BW, Stringer BW, Al-Ejeh F, Ting MJ, Wilson J, Ensbey KS, Jamieson PR, Bruce ZC, Lim YC, Offenhauser C, Charmsaz S, Cooper LT, Ellacott JK, Harding A, Leveque L, Inglis P, Allan S, Walker DG, Lackmann M, Osborne G, Khanna KK, Reynolds BA, Lickliter JD, Boyd AW. EphA3 maintains tumorigenicity and is a therapeutic target in glioblastoma multiforme. Cancer Cell 2013;23(2):238–48.

[141] Swords RT, Greenberg PL, Wei AH, Durrant S, Advani AS, Hertzberg MS, Jonas BA, Lewis ID, Rivera G, Gratzinger D, Fan AC, Felsher DW, Cortes JE, Watts JM, Yarranton GT, Walling JM, Lancet JE. KB004, a first in class monoclonal antibody targeting the receptor tyrosine kinase EphA3, in patients with advanced hematologic malignancies: Results from a phase 1 study. Leuk Res 2016;50:123–31.

[142] Jackson D, Gooya J, Mao S, Kinneer K, Xu L, Camara M, Fazenbaker C, Fleming R, Swamynathan S, Meyer D, Senter PD, Gao C, Wu H, Kinch M, Coats S, Kiener PA, Tice DA. A human antibody-drug conjugate targeting EphA2 inhibits tumor growth in vivo. Cancer Res 2008;68(22):9367–74.

[143] Annunziata CM, Kohn EC, LoRusso P, Houston ND, Coleman RL, Buzoianu M, Robbie G, Lechleider R. Phase 1, open-label study of MEDI-547 in patients with relapsed or refractory solid tumors. Invest New Drugs 2013;31(1):77–84.
[144] Damelin M, Bankovich A, Park A, Aguilar J, Anderson W, Santaguida M, Aujay M, Fong S, Khandke K, Pulito V, Ernstoff E, Escarpe P, Bernstein J, Pysz M, Zhong W, Upeslacis E, Lucas J, Lucas J, Nichols T, Loving K, Foord O, Hampl J, Stull R, Barletta F, Falahatpisheh H, Sapra P, Gerber HP, Dylla SJ. Anti-EFNA4 calicheamicin conjugates effectively target triple-negative breast and ovarian tumor-initiating cells to result in sustained tumor regressions. Clin Cancer Res 2015;21(18):4165–73.
[145] Ignacio Garrido-Laguna IEK, Burris H, Hamilton EP, Braiteh FS, Weise A, Abu-Khalaf MM, Zopf C, Lakshminarayanan M, Holland JS, Baffa R, Hong DS, Hassan R. A phase I study of PF-06647263, a novel EFNA4-ADC, in patients with metastatic triple negative breast cancer. J Clin Oncol 2017;35(15 suppl):2511.
[146] Abengozar MA, de Frutos S, Ferreiro S, Soriano J, Perez-Martinez M, Olmeda D, Marenchino M, Canamero M, Ortega S, Megias D, Rodriguez A, Martinez-Torrecuadrada JL. Blocking ephrinB2 with highly specific antibodies inhibits angiogenesis, lymphangiogenesis, and tumor growth. Blood 2012;119(19):4565–76.
[147] Krusche B, Ottone C, Clements MP, Johnstone ER, Goetsch K, Lieven H, Mota SG, Singh P, Khadayate S, Ashraf A, Davies T, Pollard SM, De Paola V, Roncaroli F, Martinez-Torrecuadrada J, Bertone P, Parrinello S. EphrinB2 drives perivascular invasion and proliferation of glioblastoma stem-like cells. Elife 2016;5:e14845.
[148] Martiny-Baron G, Holzer P, Billy E, Schnell C, Brueggen J, Ferretti M, Schmiedeberg N, Wood JM, Furet P, Imbach P. The small molecule specific EphB4 kinase inhibitor NVP-BHG712 inhibits VEGF driven angiogenesis. Angiogenesis 2010;13(3):259–67.
[149] Noberini R, Lamberto I, Pasquale EB. Targeting Eph receptors with peptides and small molecules: progress and challenges. Semin Cell Dev Biol 2012;23(1):51–7.
[150] Giorgio C, Hassan Mohamed I, Flammini L, Barocelli E, Incerti M, Lodola A, Tognolini M. Lithocholic acid is an Eph-ephrin ligand interfering with Eph-kinase activation. PLoS One 2011;6(3):e18128.
[151] Hassan-Mohamed I, Giorgio C, Incerti M, Russo S, Pala D, Pasquale EB, Zanotti I, Vicini P, Barocelli E, Rivara S, Mor M, Lodola A, Tognolini M. UniPR129 is a competitive small molecule Eph-ephrin antagonist blocking in vitro angiogenesis at low micromolar concentrations. Br J Pharmacol 2014;171(23):5195–208.
[152] Incerti M, Russo S, Callegari D, Pala D, Giorgio C, Zanotti I, Barocelli E, Vicini P, Vacondio F, Rivara S, Castelli R, Tognolini M, Lodola A. Metadynamics for perspective drug design: computationally driven synthesis of new protein-protein interaction inhibitors targeting the EphA2 receptor. J Med Chem 2017;60(2):787–96.
[153] Festuccia C, Gravina GL, Giorgio C, Mancini A, Pellegrini C, Colapietro A, Delle Monache S, Maturo MG, Sferra R, Chiodelli P, Rusnati M, Cantoni A, Castelli R, Vacondio F, Lodola A, Tognolini M. UniPR1331, a small molecule targeting Eph/ephrin interaction, prolongs survival in glioblastoma and potentiates the effect of antiangiogenic therapy in mice. Oncotarget 2018;9(36):24347–63.
[154] Shen H, Rodriguez-Aguayo C, Xu R, Gonzalez-Villasana V, Mai J, Huang Y, Zhang G, Guo X, Bai L, Qin G, Deng X, Li Q, Erm DR, Aslan B, Liu X, Sakamoto J, Chavez-Reyes A, Han HD, Sood AK, Ferrari M, Lopez-Berestein G. Enhancing chemotherapy response with sustained EphA2 silencing using multistage vector delivery. Clin Cancer Res 2013;19(7):1806–15.
[155] Spannuth WA, Mangala LS, Stone RL, Carroll AR, Nishimura M, Shahzad MM, Lee SJ, Moreno-Smith M, Nick AM, Liu R, Jennings NB, Lin YG, Merritt WM, Coleman RL, Vivas-Mejia PE, Zhou Y, Krasnoperov V, Lopez-Berestein G, Gill PS, Sood AK. Converging evidence for efficacy from parallel EphB4-targeted approaches in ovarian carcinoma. Mol Cancer Ther 2010;9(8):2377–88.

CHAPTER

10

Manipulating redox active metal metabolism to enhance cancer therapy

M.S. Petronek, B.G. Allen

Department of Radiation Oncology, The University of Iowa, Iowa City, IA, United States

Abstract

Redox active metals, including iron, play essential roles in regulating cellular growth and metabolism. Approximately 2% of cellular iron is unbound and can participate in redox reactions. This labile iron pool (LIP) is tightly regulated by cells as excess LIP can participate in Fenton chemistry reactions generating reactive oxygen species. Many malignancies have increased iron uptake resulting in increased LIPs that contribute to cancer proliferation. Iron chelation therapy reduces LIPs and slows cancer progression. Alternatively, treatment with pharmacological ascorbate further increases cancer cell LIPs thereby increasing oxidative stress. Currently multiple clinical trials are combining pharmacological ascorbate with radiation and chemotherapy in a variety of cancers. Drugs containing redox active catalytic metals are also reducing radiation and chemotherapy-induced toxicity. The superoxide dismutase mimetic, GC4419, contains a catalytically active Mn^{2+} ion that dismutes superoxide radicals and protects against radiation and chemotherapy induced mucositis in head and neck cancer patients.

Abbreviations

$Asc^{\bullet -}$	ascorbate radical
$AscH^-$	ascorbate monoanion; ascorbate
$AscH_2$	ascorbic acid; vitamin C
$AscH^{2-}$	ascorbate dianion
DFO	desferrioxamine
ETC	electron transport chain
Fpn-1	ferroportin-1
Ft	ferritin
H&N	head and neck
IRE	iron response element
IRP-1	iron responsive protein-1
IRP-2	iron responsive protein-2
MGMT	O^6-Methylguanine-DNA methyltransferase

https://doi.org/10.1016/B978-0-12-817868-3.00010-X

NB	neuroblastoma
OM	oral mucositis
RT	radiation therapy
SOD	superoxide dismutase
Tf	transferrin
TfR1	transferrin receptor 1
TfR2	transferrin receptor 2

Conflict of interest

No potential conflicts of interest were disclosed.

Basic science of redox active metal metabolism and cancer

Head and neck (H&N) cancer is frequently treated with combinations of radiation therapy (RT), chemotherapy, and/or surgery. Patients with local or regionally advanced head and neck cancer are often recommended to receive multimodality therapy including radiation and chemotherapy with encouraged participation in clinical trials. Clinical trials which modulate reduction-oxidation (redox) active metal metabolism provide a novel mechanism to improve the therapeutic ratio between cancer and normal tissues and can potentially enhance standard of care anticancer therapy. Redox active pathways currently being investigated in H&N cancer to improve patient outcomes include iron and manganese metabolism.

Iron is an essential element that is central to many different biological processes critical to sustain life. Iron is redox active in that it can serve as either an electron donor (Fe^{2+} to Fe^{3+}) or electron acceptor (Fe^{3+} to Fe^{2+}) and is involved in many different cellular reactions due to its ability to gain and lose electrons. The three major classes of iron containing proteins are: iron-sulfur containing proteins, heme-containing proteins, and iron enzymes that do not contain heme or iron-sulfur clusters [1]. Due to its reactive nature, iron levels and metabolism are tightly regulated by the cell.

In the human body, there is approximately 3–4 g of total iron. Most of the iron in the body is stored as hemoglobin in erythrocytes, which contain approximately 2–3 g of the body's total iron. In contrast to erythrocytes, plasma contains 2–3 mg of iron bound to the plasma iron carrier, transferrin (Tf) [2]. Tf delivers ferric iron (Fe^{3+}) to cells where it is taken in via receptor mediated endocytosis to be utilized. During endocytosis, Tf binds to the membrane bound receptor, transferrin receptor 1 or 2 (TfR1/2), and the complex is internalized into the cell as an endosome where ferrous iron (Fe^{2+}) is released under acidic conditions and utilized in either heme-containing proteins, iron sulfur (Fe-S) clusters, stored as ferritin, or contributed to the labile iron pool (LIP) [3].

Iron is centrally involved in various metabolic processes including iron-sulfur clusters in the electron transport chain (ETC) complexes and in the tricarboxylic acid cycle. The iron-sulfur clusters in the ETC function as electron carriers as they participate in redox reactions moving electrons through the *ETC*. ETC complex I contains 9 Fe-S clusters, complex II contains 3 Fe-S clusters, and complex III contains 1 Fe-S cluster [4–6]. The enzyme aconitase also contains an Fe-S cluster. In the tricarboxylic acid cycle, aconitase converts citrate into isocitrate; however, in the cytoplasm, the same protein can have either aconitase activity if the cell has sufficient iron stores or act as iron regulatory protein 1 (IRP1) regulating cellular iron concentrations in iron deplete states based on the unstable nature of its Fe-S cluster. When iron stores are low, IRP1 inhibits the formation of ferritin and increases transferrin receptor generation to increase cellular iron stores. Fe-S clusters are also incorporated into several proteins that are essential for DNA replication, synthesis, and repair [7]. When intracellular iron levels exceeds storage capacity or utilization needs, the labile iron pool (LIP) is formed [8]. The LIP pool, which is ferrous in nature, makes up a small percentage of cellular iron (<5%) and is able to rapidly diffuse throughout the cell [9,10]. It is believed that the LIP contributes significantly to the oxidative stress of the cell through the generation of reactive oxygen species (ROS) via Fenton chemistry (Eqs. 1–3) and has been implicated in numerous diseases including cancer [10–12]. Because of its strong contribution as a catalyst of ROS generation, the LIP concentration acts as a strong trigger for intracellular iron metabolism.

$$Fe^{3+} + O_2^{\bullet -} \longrightarrow Fe^{2+} + O_2 \quad (1)$$

$$Fe^{2+} + H_2O_2 \longrightarrow Fe^{3+} + OH^- + HO^\bullet \quad (2)$$

$$\text{Net reaction}: O_2^{\bullet -} + H_2O_2 \xrightarrow{Fe} O_2 + OH^- + HO^\bullet \quad (3)$$

Due to its importance in a variety of cellular processes and its implications in disease, it is imperative that intracellular iron levels are tightly regulated. There are two primary methods that cells use to regulate iron levels: (1) systemic hormonal regulation and (2) posttranscriptional regulation of iron uptake and storage proteins [13]. Iron is regulated systemically by the hormone hepcidin. Hepcidin is a 25 amino acid long peptide that originates from an 84 amino acid pro-peptide following its furin cleavage [13,14]. Hepcidin is secreted by hepatocytes and binds to α2-macroglobulin in the plasma where it is circulated throughout the body [15]. Hepcidin acts as a negative regulator for cellular iron levels by regulating the degradation of ferroportin-1 (Fpn-1) [2,13–16]. Fpn-1 is a transmembrane protein that controls the export of intracellular iron into the plasma for circulation throughout the body. Once hepcidin is secreted by the liver, it binds to Fpn-1 on the basolateral surface of the cell. Binding of hepcidin to Fpn-1 results in the internalization and degradation of the Fpn-1/hepcidin complex [15,17]. The result of this internalization/degradation of the complex is the trapping of Fe^{2+} within the cell [18] limiting the available iron for circulation throughout the body. The secretion of hepcidin by the hepatocytes of the liver is mediated by total serum iron levels. In circumstances with high serum iron levels, hepcidin is secreted to promote the degradation of Fpn-1 and the sequestering of Fe^{2+} within the cell. Conversely, low serum iron inhibits the secretion of hepcidin and allows for an increase in Fe^{3+} excretion into circulation [16]. In this regard, hepcidin acts as negative feedback control mechanism for controlling Fe levels in circulation. A summary of cellular iron regulation may be found in Fig. 1.

Iron is taken into cells by receptor mediated endocytosis of diferric transferrin bound to transferrin receptors on the cell membrane where the iron is then released into the cell for utilization or storage as ferritin [11,13,19]. TfR1 and ferritin expressions are primarily regulated by posttranscriptional modifications of their mRNA, by iron responsive proteins 1 and 2 (IRP1/2). IRP1/2 are responsive to the intracellular LIPs and are effective posttranscriptional regulators of intracellular iron. IRP1/2 share many similarities including conserved sequences and binding to iron response elements (IREs) on the mRNA of TfR1, Fpn-1, and

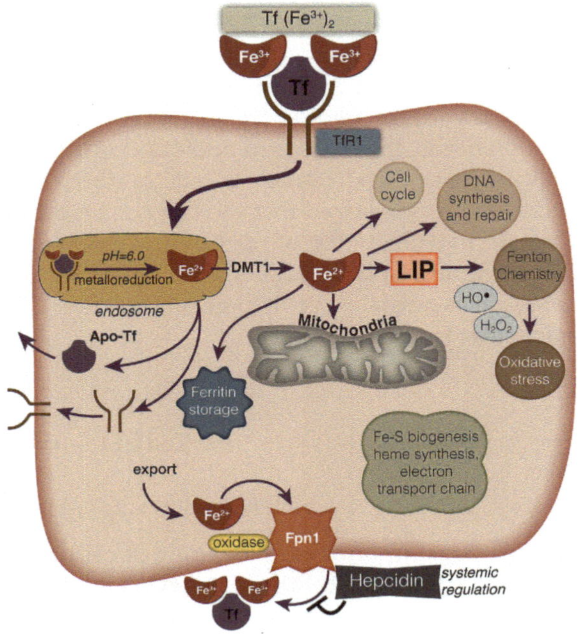

FIG. 1 Basic schematic of intracellular iron metabolism. Iron is taken into the cell *via* endocytosis of di-ferric-transferrin bound to transferrin receptors. When released from the endosome, ferrous iron makes up the labile iron pool which may be utilized for various cellular processes, stored in ferritin, or exported following oxidation by ferroportin. Systemically, ferroportin is regulated by the hormone hepcidin.

ferritin. Differences between IRP1/2 are that IRP1 also contains cytosolic aconitase activity while IRP2 does not. Additionally, IRP1 is regulated by a [4Fe-4S] cluster at its core that regulates its ferritin and transferrin receptor mRNA binding activity [20]. Conversely, IRP2 does not contain any Fe-S clusters and does not directly sense changes in intracellular iron levels but instead undergoes proteasomal degradation mediated by the iron sensitive SKP1-CUL1-FBXL5 E3 ubiqutin ligase complex [21]. When under conditions of low intracellular iron, IRP1/2 stabilizes TfR1 mRNA by binding to IREs at the 3′ end of the transcript to allow for an increase in iron uptake via Tf-TfR1 endocytosis. In conditions of low iron, IRP1/2 will bind to IREs at the 5′ end of the Fpn-1/ferritin transcript and repress mRNA ferritin translation thereby decreasing iron export/storage. When the cell is in a state of high iron (i.e. significant increase in the LIP), IRP1's Fe-S cluster will inhibit its binding to the respective IREs and IRP2 will release and be degraded. This allows for RNAase to degrade TfR1 mRNA, limiting cellular iron uptake. Conversely, the release of IRP1/2 from the IREs of Fpn-1/ferritin will allow for the translation of the proteins for increased iron storage and cellular export of iron [3,13,16,19,20]. In this manner, the cell can tightly and efficiently regulate the LIP and reduce cellular oxidative stress (Fig. 2).

Iron regulation and metabolism are significantly altered in cancer cells. Relative to normal cells, cancer cells have increased expression of TfR1 and reduced ferritin expression [16,22–25]. Additionally, as TfR1 has been shown to be regulated posttranscriptionally by the IRP-IRE mechanism, TfR1 is also a downstream target of the proto-oncogene transcription factor, c-Myc [16,26] which is linked to a variety of cancer initiation and progression events. Previous studies have shown that colorectal cancer progression is associated with the

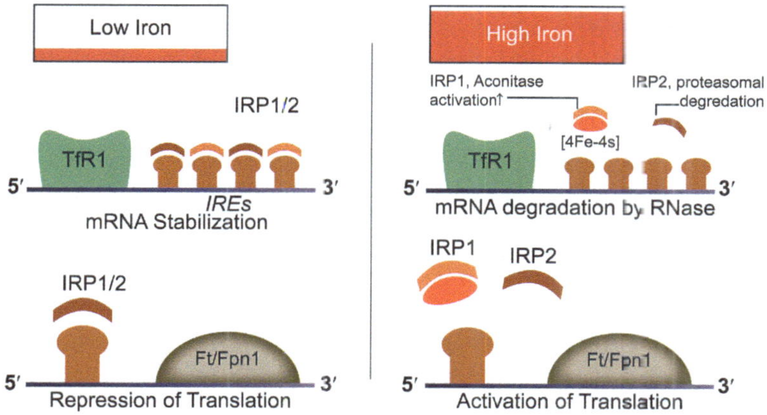

FIG. 2 Posttranscriptional modifications of key iron metabolic proteins. Under low iron conditions iron responsive proteins 1/2 bind at the 5′ end of transferring receptor and the 3′ end of ferritin and ferroportin transcripts. This stabilizes TfR mRNA and allows for translation to occur, while it represses translation of ferritin/Fpn1. In doing so, the cell is able to bring more iron into the cell without storing or exporting it in order to maintain homeostatic balance. Under conditions in which the LIP is high, IRP 1/2 are released from their Iron responsive elements on the mRNA allowing for TfR mRNA to be degraded and the translation of ferritin/Fpn1. The result of this is decreased intracellular import of iron and increased iron storage and export in order to stabilize the LIP.

upregulation of iron import proteins, such as TfR1 and TfR2,as well as the down regulation of iron export proteins, such as Fpn1 [27]. Along with TfR1, the expression of TfR2, has been shown to be upregulated in neoplastic cells [28–30] and increase in vivo tumor growth in nude mice when compared to those tumors overexpressing TfR1 alone [28]. Nonsmall cell lung cancer cells have reduced expression of ferritin and increased expression of TfR1 relative to normal bronchial epithelial cells [25]. Serum ferritin levels are also used in neuroblastoma patients as a prognostic indicator with high levels of ferritin indicating a poor prognosis and low levels of ferritin indicating a good prognosis [16,31].

With the growing evidence that iron metabolism is altered in cancer, it appears to be a prime target for anticancer therapy. Research groups have aimed to induce tumor cell death by targeting iron metabolism in two ways: chelation therapy to hamper the increased cellular intake of iron that is necessary for proliferation and therapies targeted to take advantage of the increased cellular iron to increase the LIP and generate increased iron catalyzed ROS causing DNA, protein and lipid damage.

Iron chelation cancer therapy

Because neoplastic cells have an increased reliance on iron uptake relative to normal cells [16], iron chelators have been assessed as a possible anticancer therapy. Desferrioxamine (DFO), a hexadentate iron chelator that is capable of binding iron in a 1:1 ratio, was initially used for the treatment of β-thalassemia associated iron overload [32–34]. DFO has a relatively short half-life and is hydrophilic rather than lipophilic leading to limited membrane permeability. Due to its rapid metabolism and poor membrane permeability, it is unable to be ingested orally and must be delivered in multiple subcutaneous injections [16,34]. In 1988, Becton and Bryles showed the antitumor potential for DFO in neuroblastoma (NB) cells [35]. NB cells collected from two untreated children with Evans Stage IV disease demonstrated growth inhibition and cytotoxic effects after treatment with DFO in a time- and dose-dependent manner [35]. Following this study, DFO showed similar antitumor effects in ovarian and cervical carcinomas as well as in leukemia cells [36–38].

Following the preclinical success of DFO in the treatment of cancer cells, multiple DFO clinical trials were conducted. In 1990, a phase II clinical trial containing 9 NB patients received a 5-day course of DFO at 150 mg/kg/day. Treatment responses were observed in 7 of 9 patients with no associated drug toxicity and one patient exhibited a 48% decrease in tumor mass [39]. Another trial in 57 NB patients combined DFO with cyclophosphamide, etoposide, carboplatin and thiotepa. This study found a complete response in 24 patients, partial response in 26 patients, 3 minor responses, and 4 patients showed progressive disease [40]. DFO anticancer success was further exhibited in a case of neonatal acute leukemia where DFO treatment increased growth of hematopoetic progenitor cells with no associated growth of peripheral blood blast cells. In 2011, a clinical study in 10 patients with chemorefractory hepatocellular carcinoma showed a 20% partial response rate with DFO treatment with no associated grade 4 adverse events [41]. One patient, who had presented with a large hepatocellular tumor with lung metastases demonstrated a complete response following DFO treatment [41].

While several studies have shown great promise for iron chelation therapy, there have been some conflicting results in the literature. In 1994, a clinical trial of 10 children with NB tumors treated with DFO failed to produce a partial or complete response. Along with its ineffectiveness in this trial, DFO also demonstrated significant toxicities including blurred vision, dizziness, and leg cramping [42]. With conflicting efficacy, difficult delivery, and rapid metabolism of DFO more recent efforts have gone into generating more efficient iron chelators to use in therapy. Two more modern chelation therapies have been introduced to the clinic, deferiprone and deferasirox, are currently under investigation and may potentially be more universally effective in iron targeted cancer therapy than DFO [43,44].

Pharmacological ascorbate cancer therapy

Ascorbic acid ($AscH_2$), commonly referred to as vitamin C, is a ketolactone that has two ionizable hydroxyl groups at the 2 and 3 positions, respectively. The hydroxyl groups have pK_a's at 4.2 and 11.6; thus at a pH of 7.4, the most prevalent form is the ascorbate monoanion ($AscH^-$) [45]. Ascorbate can act as a reducing agent capable of undergoing a reversible Michaelis oxidation to dehydroascorbic acid (DHA) (Eqs. 4, 5) with a radical intermediate (ascorbate radical; $Asc^{\bullet -}$) [45,46]. The ascorbate radical is rapidly dismutated to DHA and is generally considered an unreactive radical species because of the resonance stabilization of its unpaired electron at pH = 7.0 ($k_{obs} = 2 \times 10^5\,M^{-1}\,s^{-1}$) [46].

$$AscH_2 \leftrightarrow Asc^{\bullet -} + 2H^+ \quad (4)$$

$$2Asc^{\bullet -} + H^+ \leftrightarrow AscH^- + DHA \quad (5)$$

Ascorbate can undergo autooxidation via the formation of the ascorbate dianion ($AscH^{2-}$) (Eq. 6) and is pH dependent [45,46]. A one unit increase in pH induces a ten-fold increase in $AscH^{2-}$, resulting in a ten-fold increase in the rate of autooxidation [47].

$$AscH^{2-} + O_2 \rightarrow Asc^{\bullet-} + O_2^{\bullet-} \quad (6)$$

$$k \approx 3 \times 10^{-1} M^{-1} s^{-1}$$

Therefore, while ascorbate is able to be oxidized to produce superoxide without the aid of catalytic metals, it occurs very slowly and is not likely to occur significantly at physiological pH due to the limited amount of the $AscH^{2-}$ present.

Ascorbate oxidation occurs more readily in the presence of catalytic metals [48,49]. In the presence of iron, ascorbate is able to act as a one-electron reducing agent that can reduce ferric (Fe^{3+}) to ferrous (Fe^{2+}) while producing an ascorbate radical (Eq. 7). In the presence of oxygen, the ferrous iron resulting from the reaction with ascorbate is able to perform the converse reaction to Eq. (1) to generate superoxide (Eq. 8). The superoxide radical is able to be dismuted by superoxide dismutase (SOD) to produce H_2O_2 and O_2 (Eq. 9).

$$AscH^- + Fe^{3+} \rightarrow Fe^{2+} + Asc^{\bullet-} \quad (7)$$

$$Fe^{2+} + O_2 \rightarrow Fe^{3+} + O_2^{\bullet-} \quad (8)$$

$$2O_2^{\bullet-} + 2H^+ \xrightarrow{SOD} H_2O_2 + O_2 \quad (9)$$

H_2O_2 is a byproduct of the high doses of ascorbate delivered over the course of treatment and is central determinant to the cytotoxic effects of ascorbate. When ascorbate exists in millimolar levels, the autooxidation of ascorbate (Eq. 6) is able to produce considerable amounts of H_2O_2 without the aid of catalytic metals. Therefore, an ascorbate concentration of 20 mM will consist of $\approx 1\,\mu M$ Asc^{2-} (pH = 7.4) and may result in a flux of H_2O_2 at a rate of approximately $10\,nM\,s^{-1}$ [45].

The altered iron metabolism of neoplastic cells may be a driving force behind the effectiveness of high doses of ascorbate as an anticancer agent when combined with traditional cancer therapies. In cancer cells, the upregulation of transferrin receptors and reduced ferritin expression may lead to more available catalytic iron available to oxidize ascorbate and participate in other oxidant reactions [25].

An additional property of ascorbate that enhances its pro-oxidant property is its ability to effectively labilize iron in cancerous cells. With an increased labile iron pool, more iron is available within the cell to participate in oxidation reactions with ascorbate and oxygen to produce toxic ROS and participate in classic Fenton chemistry (Eqs. 7–9). Moser et al. demonstrated that ascorbate increased the labile iron pool in MIA PaCa-2 tumor homogenates removed from nude athymic mice using EPR spectroscopy and that combining ascorbate with ionizing radiation had an additive effect in increasing labile iron pools [50]. Schoenfeld et al. showed that ascorbate selectively increases the labile iron pool in nonsmall cell lung cancer and glioblastoma cells in vitro relative to normal cells and that the increase in labile iron is necessary for the cytotoxic effects of ascorbate [25]. One hypothesis is that ascorbate generated H_2O_2 mediates the disruption of Fe-S containing proteins, such as complex I and II of the electron transport chain as well as aconitase. Additionally, in vitro knockdown of TfR and overexpression of ferritin was rescued cells from the cytotoxic effects of ascorbate providing further evidence for its dependence on the LIP [25]. Along with the release of iron from Fe-S clusters, ascorbate may release iron stored in ferritin. Ferritin is able to store up to 4500 atoms of iron and reducing agents, such as ascorbate, thiols, and reduced flavins, may labilize iron from ferritin [51,52] to potentially increase oxidative stress and form $HO^{\bullet}$ via Fenton chemistry [53,54].

Gram doses ascorbate given intravenously (pharmacological ascorbate; achieving mM plasma ascorbate concentrations) were initially determined to have value as a cancer treatment in the 1970s as multiple trials showed the safety and efficacy of the ascorbate in the treatment of a variety of terminal patients [55–58]. But, interest in ascorbate as an anticancer agent waned following the completion of two randomized, double-blind clinical trials when high doses of oral ascorbate failed to provide benefit relative to placebo [59,60].

Upon further investigation into the pharmacokinetics of pharmacological ascorbate, a significant difference in the delivery of ascorbate between the successful and unsuccessful trial emerged. Initial trials were delivered using both intravenous and oral routes while it was only delivered orally in those trials showing a lack of efficacy as an anticancer therapy. It was later discovered that oral ascorbate administration does not achieve mM plasma ascorbate concentrations. The steady-state concentration of ascorbate in the plasma is $\approx 80\,\mu M$ with a given dose of 200 mg. As doses exceeded 200 mg, absorption decreases and excretion increases, with plasma saturation occurring at approximately 1000 mg [61,62]. It was then determined that when given orally, ascorbate plasma concentrations saturate at $\approx 220\,\mu M$ with a dose of 3 g every 6 h. Using an intravenous delivery of 50 g, plasma concentrations reached 13.4 mM [63]. A pilot clinical trial using high dose ascorbate delivered intravenously showed that when given 10 g, plasma concentrations reach an average of 1.1 mM [64]. Since the beginning of the use of ascorbate in cancer therapy, it has been determined that best way to effectively deliver pharmacological doses of ascorbate is intravenously.

After determining efficacious pharmacological ascorbate delivery approaches, there has been an increasing interest in re-evaluating the potential of pharmacological ascorbate as an anticancer agent. Multiple phase I clinical trials have been completed and demonstrate that it is pharmacologically safe and may improve patient outcomes when combined with standard of care radiation and chemotherapy. In 2004, Rioridan et al. found pharmacological ascorbate treatments to be safely tolerated in cases of renal cell carcinoma, colorectal cancer, pancreatic cancer, non-Hodgkin's lymphoma, and breast cancer [65]. In 2005, Rioridan et al. performed a pilot study in which intravenous ascorbate was delivered in continuous infusions of 150 to 710 $mg/kg\,day^{-1}$ for up to 8 weeks in 24 late-stage, terminally ill cancer patients. All of the patients on the study were vitamin C deficient prior to beginning the trial and following the infusion of pharmacological ascorbate the mean plasma concentrations reached 1.1 mM. The most common adverse events that this study had reported included minor nausea, edema, and dry mouth/skin. One patient experienced the development of a kidney stone 13 days into treatment but had a history of renal caliculi [64]. In 2008, a phase I clinical trial including both solid tumor and hematologic malignancies dosed with 0.4, 0.6, 0.9, and 1.5 $g\,kg^{-1}$ ascorbate three times per week demonstrated no significant toxicity but also failed to show any anticancer responses and concluded that although pharmacological ascorbate may be well tolerated in patients, the promise of ascorbate as an anticancer therapy may lie in its use with other cytotoxic therapeutic agents [66]. A subsequent phase I clinical trial was completed in 2013 in patients with biopsy-proven advanced or metastatic solid tumors treated with pharmacological ascorbate infusions of 1 $g\,min^{-1}$, for 4 days $week^{-1}$, for 4 weeks with ascorbate doses of 30 $g\,m^{-2}$ initially followed by 20 g increases in order to find a maximally tolerable dose. They discovered that doses at 70, 90, and 110 $g\,m^{-2}$ maintained plasma levels of ascorbate of 10–20 mM for 5–6 h, with a maximal blood plasma concentration of 49 mM. This study recommended an ascorbate dose of 70–80 $g\,m^{-2}$ with a 1 $g\,min^{-1}$ infusion for 4 weeks to maintain tolerance [67]. In 2012, Monti et al. completed a phase I clinical trial combining pharmacological ascorbate with gemcitabine and erlotinib for patients with stage IV pancreatic ductal carcinoma. Over the course of 8 weeks, 14 patients received ascorbate 3 times per week in combination with gemcitabine and erlotinib with 9 of the 14 patients completing the trial. In the patients who completed the trial, 7 had stable disease while 2 showed progression according the RESIST criteria. The study reported 15 nonserious adverse events and 8 serious adverse events, which the investigators believed were attributed to erlotinib and gemcitabine [68]. In 2014, a phase I clinical trial of 14 stage IV pancreas cancer patients combining pharmacological ascorbate with gemcitabine showed that pharmacological ascorbate is safe when combined with gemcitabine and suggested that pharmacological ascorbate may improve patient outcomes. The study was a dose-escalation trial designed to recruited 14 patients to receive 15–125 g of ascorbate per infusion with two infusions per week during 4-week cycles with the target plasma ascorbate concentrations ≥20 mM. Subjects continued to receive ascorbate infusions until experiencing a grade 3 dose limiting toxicity or disease progression (as defined by RESIST criteria). Of the 14 subjects, 9 completed at least two treatment cycles with the most common side effects being dry mouth, nausea, and diarrhea. No dose limiting toxicities were observed. In addition, the trial showed that as of August 2014, the average overall survival was 15 ± 2 months and average time to progression was 26 ± 7 weeks which was a dramatic improvement relative to treatment with gemcitabine alone where historical overall survival was 6 months and historical average time to progression of 9 weeks. From this study, the authors were able to suggest that not only is ascorbate well tolerated in combination with cytotoxic chemotherapy agents, but pharmacological ascorbate has the potential to significantly enhance pancreas cancer treatment efficacy [69]. A phase 1 clinical trial in 11 glioblastoma multiforme (GBM) subjects received combined pharmacological ascorbate concurrently with radiation and temozolomide chemotherapy. Intravenous ascorbate was administered three times a week while receiving radiation therapy and temozolomide (~28 weeks during the adjuvant phase) and twice a week following the completion of radiation therapy in combination with temozolomide (~6 months). Similar to the pancreas trial, this was also a dose escalation trial (15–125 g of ascorbate per infusion) during radiation phase with ascorbate doses escalated to achieve a target ascorbate plasma level of ≥20 mM. Subjects who received 87.5 g ascorbate per infusion consistently achieved 20 mM ascorbate plasma concentrations. Pharmacological ascorbate was considered safe and tolerable in all 13 subjects who participated in the trial, showing minimal grade 3 and 4 toxicities. During the radiation phase, nonhematological toxicities observed included the following: grade 2/3 fatigue, grade 2/3 nausea, grade 2 infection, and grade 3 vomiting. During the 28-week adjuvant phase following radiotherapy, toxicities observed included: grade 2 fatigue, grade 2 nausea, grade 1 vomiting, grade 3 leukopenia, and grade 3 neutropenia. Importantly, during the entirety of the course of treatment no serious adverse events were able to be ascribed to treatment with ascorbate [25]. Average progression-free survival was shown to be 13.3 months (median of 9.4 months) with an average overall survival was 21.5 months (median of 18.2 months) [25,70]. Stupp et al. had previously shown that patients receiving temozolomide and radiation alone who lacked methylation of the O^6-methylguanine-DNA methyltransferase (MGMT) promoter region had significantly reduced survival (median OS = 12.7 vs. 21.7 months) [70]. In the phase I trial combining pharmacological ascorbate with radiation and temozolomide, 73% of patients lacked MGMT methylation and this group of patients, median overall survival was 23.0 months. Interestingly, at the time of publication, 3 of 8 patients lacking MGMT promoter methylation remained alive and had no evidence of disease based on imaging [25]. These data demonstrating that ascorbate is safe when combined with radiation and chemotherapy are very encouraging and suggest that ascorbate may be extended to additional disease sites including head and neck cancer.

Superoxide dismutase mimetic, GC4419, prevents and reduces severity of oral mucositis in head and neck cancer patients

GC4419 is a small molecule redox active manganese (Mn) containing superoxide dismutase mimetic developed by Galera Therapeutics Inc. In the presence of superoxide, GC4419's Mn^{2+} is oxidized to Mn^{3+} thereby reducing superoxide to hydrogen peroxide. Hydrogen peroxide can then be reduced to molecular oxygen and water via catalase, glutathione peroxidases, peroxiredoxins or thioredoxins. Because radiation induces bursts of superoxide which are associated with the development of oral mucositis (OM), GC4419 has great promise in preventing and reducing the severity of oral mucositis (OM) in head and neck cancer patients treated with radiation and platinum chemotherapy [71]. A phase 1b/2a clinical trial combining GC4419 with concurrent radiation and cisplatin in patients with oropharyngeal carcinoma demonstrated that GC441 was well tolerated and decreased both the incidence and severity of severe oral mucositis. Observed dose-limiting toxicities included grade 3 gastroenteritis and vomiting with hyponatremia. Severe oral mucositis (World Health Organization OM grades 3 and 4) incidence was 29% with a median duration of only 2.5 days [71]. Following the phase 1b/2a study, a phase 2b trial of 223 patients with oropharyngeal carcinoma receiving 70 Gy/35 fx with concurrent cisplatin were randomized 1:1:1 to receive 30 mg, 90 mg GC4419, or placebo via intravenous infusion prior to each radiation fraction. It was recently reported that 90 mg doses of GC4419 reduced OM duration to a median of 1.5 days as compared to the placebo arm with a median of 19 days. Along with the reduced OM duration, 90 mg GC4419 also significantly delayed the onset of OM from 39 days to 61 days and incidence of OM through 60 Gy from 58% to 36%. Additionally, 90 mg GC4419 reduced the incidence of grade 4 OM from 30% to 16%. The phase 2b study found no significant toxicities related to the delivery of GC4419 indicating that the SOD mimic provides a safe and efficacious approach to reduce and prevent OM in regards to incidence, duration, and severity [72]. Due to the promising results in early stage clinical trials, GC4419 was recently granted fast track designation by the FDA for use in the reduction of incidence and severity of OM associated with radiation and chemotherapy in the treatment of head and neck cancer [73].

TABLE 1 Active clinical trials manipulating redox active metals to improve cancer therapy

Clinical trial #	Phase	Condition(s)	Interventions	Status
NCT03689712	Phase III	Oral Mucositis	GC4419, Placebo	Recruiting
NCT03659084	Observational	Myelodysplastic syndrome, Myeloid leukemia	EXJADE (Deferasirox)	Recruiting
NCT02341495	Phase II	Acute myeloid leukemia	Deferasirox, Cholecalciferol, Azacitidine	Recruiting
NCT01852890	Phase I	Pancreatic cancer	Ascorbate, Gemcitabine, Radiation Therapy	Active, not recruiting
NCT03541486	Phase II	Pancreatic cancer	Ascorbate, Gemcitabine, Radiation Therapy	Not yet recruiting
NCT02905578	Phase II	Pancreatic cancer	Ascorbate, Gemcitabine, Nab-paclitaxel	Not yet recruiting
NCT02420314	Phase II	Nonsmall cell lung carcinoma	Ascorbate, Paclitaxel, Carboplatin	Recruiting
NCT02905591	Phase II	Nonsmall cell lung carcinoma	Ascorbate, Paclitaxel, Carboplatin, Radiation Therapy	Not yet recruiting
NCT03468075	Phase II	Sarcoma	Ascorbate, Gemcitabine	Recruiting
NCT03508726	Phase Ib/II	Sarcoma	Ascorbate, Radiation Therapy	Not yet recruiting
NCT02344355	Phase II	Glioblastoma- Multiforme	Ascorbate, Temozolomide, Radiation	Recruiting
NCT03602235	Phase I	Multiple myeloma	Ascorbate, Melphalan	Not yet recruiting
NCT02341495	Phase II	Acute Myeloid Leukemia	Deferasirox, Cholecalciferol, Azacitidine	Recruiting
NCT03387475	Phase II	Myelodysplastic Syndrome	Deferasirox	Recruiting
NCT02943668	Phase II	Anemia, Myelodysplastic Syndrome	Deferasirox	Recruiting
NCT03340974	Phase I/II	Pancreatic Cancer	GC4419, Placebo, Radiation Therapy	Recruiting
NCT03689712	Phase III	Oral Mucositis	GC4419, Placebo	Recruiting

Conclusion/summary

Manipulating redox active metals has the potential to enhance cancer therapy by making cancer cells more sensitive to radiation and chemotherapy as well as protecting against radiation and chemotherapy toxicity. Promising ongoing research continues in chelation studies, pharmacological ascorbate trials, and superoxide dismutase mimetic trials (Table 1). Clinical trials to date have shown a great deal of promise for ascorbate to further enhance current standard of care therapies at a number of different tumor sites including the pancreas, brain, lung, and head and neck.

Acknowledgments

The authors would like to acknowledge Gareth Smith for the production of the high-quality figures used in this chapter.

References

[1] Kaplan J, Ward DM. The essential nature of iron usage and regulation. Curr Biol 2013;23(15):R642–6.
[2] Nemeth E, Ganz T. The role of hepcidin in iron metabolism. Acta Haematol 2009;122(2–3):78–86.
[3] Arredondo M, Núñez MT. Iron and copper metabolism. Mol Aspects Med 2005;26(4–5):313–27.
[4] Hinchliffe P, Sazanov LA. Organization of iron-sulfur clusters in respiratory complex I. Science 2005 Jul 29;309 (5735):771–4.
[5] Cecchini G. Function and structure of complex II of the respiratory chain. Annu Rev Biochem 2003;72:77–109.
[6] Rich PR, Maréchal A. The mitochondrial respiratory chain. Essays Biochem 2010;47:1–23.
[7] Zhang C. Essential functions of iron-requiring proteins in DNA replication, repair and cell cycle control. Protein Cell 2014;5(10):750–60.
[8] Mills E, Dong X, Wang F, Xu H. Mechanisms of brain iron transport: insight into neurodegeneration and CNS disorders. Future Med Chem 2010;2(1):51.
[9] Breuer W, Shvartsman M, Cabantchik ZI. Intracellular labile iron. Int J Biochem Cell Biol 2008;40(3):350–4.
[10] Kruszewski M. Labile iron pool: the main determinant of cellular response to oxidative stress. Mutat Res 2003;531(1–2):81–92.
[11] Piñero DJ, Connor JR. Iron in the brain: an important contributor in normal and diseased states, iron in the brain: an important contributor in normal and diseased states. Neuroscientist 2000;6(6):435–53.
[12] Silva B, Faustino P. An overview of molecular basis of iron metabolism regulation and the associated pathologies. Biochim Biophys Acta 2015;1852(7):1347–59.
[13] Hentze MW, Muckenthaler MU, Galy B, Camaschella C. Two to tango: regulation of mammalian iron metabolism. Cell 2010;142(1):24–38.
[14] Papanikolaou G, Pantopoulos K. Iron metabolism and toxicity. Toxicol Appl Pharmacol 2005;202(2):199–211.
[15] Peslova G, Petrak J, Kuzelova K, Hrdy I, Halada P, Kuchel PW, et al. Hepcidin, the hormone of iron metabolism, is bound specifically to α-2-macroglobulin in blood. Blood 2009;113(24):6225–36.
[16] Richardson DR, Kalinowski DS, Lau S, Jansson PJ, Lovejoy DB. Cancer cell iron metabolism and the development of potent iron chelators as anti-tumour agents. Biochim Biophys Acta 2009;1790(7):702–17.
[17] Ramey G, Deschemin J-C, Durel B, Canonne-Hergaux F, Nicolas G, Vaulont S. Hepcidin targets ferroportin for degradation in hepatocytes. Haematologica 2009;95(3):501–4.
[18] Oates PS. The role of hepcidin and ferroportin in iron absorption. Histol Histopathol 2007;22(7):791–804.
[19] Pantopoulos K. Iron metabolism and the IRE/IRP regulatory system: an update. Ann N Y Acad Sci 2004;1012:1–13.
[20] Cairo G, Pietrangelo A. Iron regulatory proteins in pathobiology. Biochem J 2000;352(Pt 2):241–50.
[21] Takahashi-Makise N, Ward DM, Kaplan J. On the mechanism of iron sensing by IRP2: new players, new paradigms. Nat Chem Biol 2009;5(12):874–5.
[22] Sutherland R, Delia D, Schneider C, Newman R, Kemshead J, Greaves M. Ubiquitous cell-surface glycoprotein on tumor cells is proliferation-associated receptor for transferrin. Proc Natl Acad Sci U S A 1981;78(7):4515–9.
[23] Trowbridge IS, Lopez F. Monoclonal antibody to transferrin receptor blocks transferrin binding and inhibits human tumor cell growth in vitro. Proc Natl Acad Sci U S A 1982;79(4):1175–9.
[24] Taetle R, Honeysett JM. Effects of monoclonal anti-transferrin receptor antibodies on in vitro growth of human solid tumor cells. Cancer Res 1987;47(8):2040–4.
[25] Schoenfeld JD, Sibenaller ZA, Mapuskar KA, Wagner BA, Cramer-Morales KL, Furqan M, et al. O(2)(•−) and H(2)O(2)-Mediated disruption of Fe metabolism causes the differential susceptibility of NSCLC and GBM cancer cells to pharmacological ascorbate. Cancer Cell 2017;31(4):487–500.e8.

[26] O'Donnell KA, Yu D, Zeller KI, Kim J, Racke F, Thomas-Tikhonenko A, et al. Activation of transferrin receptor 1 by c-Myc enhances cellular proliferation and tumorigenesis. Mol Cell Biol 2006 Mar;26(6):2373–86.

[27] Brookes MJ, Hughes S, Turner FE, Reynolds G, Sharma N, Ismail T, et al. Modulation of iron transport proteins in human colorectal carcinogenesis. Gut 2006;55(10):1449–60.

[28] Kawabata H, Germain RS, Vuong PT, Nakamaki T, Said JW, Koeffler HP. Transferrin receptor 2-alpha supports cell growth both in iron-chelated cultured cells and in vivo. J Biol Chem 2000;275(22):16618–25.

[29] Nakamaki T, Kawabata H, Saito B, Matsunawa M, Suzuki J, Adachi D, et al. Elevated levels of transferrin receptor 2 mRNA, not transferrin receptor 1 mRNA, are associated with increased survival in acute myeloid leukaemia. Br J Haematol 2004;125(1):42–9.

[30] Smilevska T, Stamatopoulos K, Samara M, Belessi C, Tsompanakou A, Paterakis G, et al. Transferrin receptor-1 and 2 expression in chronic lymphocytic leukemia. Leuk Res 2006;30(2):183–9.

[31] Hann HW, Levy HM, Evans AE. Serum ferritin as a guide to therapy in neuroblastoma. Cancer Res 1980; 40(5):1411–3.

[32] Wong C, Richardson DR. β-Thalassaemia: emergence of new and improved iron chelators for treatment. Int J Biochem Cell Biol 2003;35(7):1144–9.

[33] Kalinowski DS, Richardson DR. The evolution of iron chelators for the treatment of iron overload disease and cancer. Pharmacol Rev 2005;57(4):547–83.

[34] Bedford MR, Ford SJ, Horniblow RD, Iqbal TH, Tselepis C. Iron chelation in the treatment of cancer: a new role for deferasirox? J Clin Pharmacol 2013;53(9):885–91.

[35] Becton DL, Bryles P. Deferoxamine inhibition of human neuroblastoma viability and proliferation. Cancer Res 1988;48(24 Pt 1):7189–92.

[36] Simonart T, Boelaert JR, Mosselmans R, Andrei G, Noel J-C, De Clercq E, et al. Antiproliferative and apoptotic effects of iron chelators on human cervical carcinoma cells. Gynecol Oncol 2002;85(1):95–102.

[37] Becton DL, Roberts B. Antileukemic effects of deferoxamine on human myeloid leukemia cell lines. Cancer Res 1989;49(17):4809–12.

[38] Brard L, Granai CO, Swamy N. Iron chelators deferoxamine and diethylenetriamine pentaacetic acid induce apoptosis in ovarian carcinoma. Gynecol Oncol 2006;100(1):116–27.

[39] Donfrancesco A, Deb G, Dominici C, Pileggi D, Castello MA, Helson L. Effects of a single course of deferoxamine in neuroblastoma patients. Cancer Res 1990;50(16):4929–30.

[40] Donfrancesco A, Bernardi BD, Carli M, Mancini A, Nigro M, Sio LD, et al. Deferoxamine followed by cyclophosphamide, etoposide, carboplatin, thiotepa, induction regimen in advanced neuroblastoma: preliminary results. Eur J Cancer 1995;31(4):612–5.

[41] Yamasaki T, Terai S, Sakaida I. Deferoxamine for advanced hepatocellular carcinoma. N Engl J Med 2011; 365(6):576–8.

[42] Blatt J. Deferoxamine in children with recurrent neuroblastoma. Anticancer Res 1994;14(5B):2109–12.

[43] Tury S, Assayag F, Bonin F, Chateau-Joubert S, Servely J-L, Vacher S, et al. The iron chelator deferasirox synergises with chemotherapy to treat triple-negative breast cancers. J Pathol 2018;246(1):103–14.

[44] Simões RV, Veeraperumal S, Serganova IS, Kruchevsky N, Varshavsky J, Blasberg RG, et al. Inhibition of prostate cancer proliferation by Deferiprone. NMR Biomed 2017;30(6). https://doi.org/10.1002/nbm.3712.

[45] Du J, Cullen JJ, Buettner GR. Ascorbic acid: chemistry, biology and the treatment of cancer. Biochim Biophys Acta 2012;1826(2):443–57.

[46] BHJ B. Chemistry of ascorbic acid radicals. Adv Chem Ser (United States) 1982;200: Available from: https://www.osti.gov/biblio/6200707.

[47] Buettner GR. In the absence of catalytic metals ascorbate does not autoxidize at pH 7: ascorbate as a test for catalytic metals. J Biochem Biophys Methods 1988;16(1):27–40.

[48] Frei B, Lawson S. Vitamin C and cancer revisited. Proc Natl Acad Sci U S A 2008;105(32):11037–8.

[49] Buettner GR, Jurkiewicz BA. Catalytic metals, ascorbate and free radicals: combinations to avoid. Radiat Res 1996;145(5):532–41.

[50] Moser JC, Rawal M, Wagner BA, Du J, Cullen JJ, Buettner GR. Pharmacological ascorbate and ionizing radiation (IR) increase labile iron in pancreatic cancer. Redox Biol 2013;2:22–7.

[51] Boyer RF, McCleary CJ. Superoxide ion as a primary reductant in ascorbate-mediated ferritin iron release. Free Radic Biol Med 1987;3(6):389–95.

[52] Melman G, Bou-Abdallah F, Vane E, Maura P, Arosio P, Melman A. Iron release from ferritin by flavin nucleotides. Biochim Biophys Acta 2013;1830(10):4669–74.

[53] Keyer K, Imlay JA. Superoxide accelerates DNA damage by elevating free-iron levels. Proc Natl Acad Sci U S A 1996;93(24):13635–40.
[54] Biemond P, van Eijk HG, Swaak AJ, Koster JF. Iron mobilization from ferritin by superoxide derived from stimulated polymorphonuclear leukocytes. Possible mechanism in inflammation diseases. J Clin Invest 1984;73 (6):1576–9.
[55] Cameron E, Campbell A. The orthomolecular treatment of cancer. II. Clinical trial of high-dose ascorbic acid supplements in advanced human cancer. Chem Biol Interact 1974;9(4):285–315.
[56] Cameron E, Campbell A, Jack T. The orthomolecular treatment of cancer. III. Reticulum cell sarcoma: double complete regression induced by high-dose ascorbic acid therapy. Chem Biol Interact 1975;11(5):387–93.
[57] Cameron E, Pauling L. Supplemental ascorbate in the supportive treatment of cancer: Prolongation of survival times in terminal human cancer. Proc Natl Acad Sci U S A 1976;73(10):3685–9.
[58] Cameron E, Pauling L. Supplemental ascorbate in the supportive treatment of cancer: reevaluation of prolongation of survival times in terminal human cancer. Proc Natl Acad Sci U S A 1978;75(9):4538–42.
[59] Creagan ET, Moertel CG, O'Fallon JR, Schutt AJ, O'Connell MJ, Rubin J, et al. Failure of high-dose vitamin C (ascorbic acid) therapy to benefit patients with advanced cancer. N Engl J Med 1979;301(13):687–90.
[60] Creagan ET, Moertel CG, O'Fallon JR, Schutt AJ, O'Connell MJ, Rubin J, et al. Failure of high-dose vitamin C (ascorbic acid) therapy to benefit patients with advanced cancer. A controlled trial. N Engl J Med 1979; 301(13):687–90.
[61] Graumlich JF, Ludden TM, Conry-Cantilena C, Cantilena LR, Wang Y, Levine M. Pharmacokinetic model of ascorbic acid in healthy male volunteers during depletion and repletion. Pharm Res 1997;14(9):1133–9.
[62] Levine M, Conry-Cantilena C, Wang Y, Welch RW, Washko PW, Dhariwal KR, et al. Vitamin C pharmacokinetics in healthy volunteers: evidence for a recommended dietary allowance. Proc Natl Acad Sci U S A 1996; 93(8):3704–9.
[63] Padayatty SJ, Sun H, Wang Y, Riordan HD, Hewitt SM, Katz A, et al. Vitamin C pharmacokinetics: implications for oral and intravenous use. Ann Intern Med 2004;140(7):533–7.
[64] Riordan HD, Casciari JJ, González MJ, Riordan NH, Miranda-Massari JR, Taylor P, et al. A pilot clinical study of continuous intravenous ascorbate in terminal cancer patients. P R Health Sci J 2005;24(4):269–76.
[65] Riordan HD, Riordan NH, Jackson JA, Casciari JJ, Hunninghake R, González MJ, et al. Intravenous vitamin C as a chemotherapy agent: a report on clinical cases. P R Health Sci J 2004;23(2):115–8.
[66] Hoffer LJ, Levine M, Assouline S, Melnychuk D, Padayatty SJ, Rosadiuk K, et al. Phase I clinical trial of i.v. ascorbic acid in advanced malignancy. Ann Oncol 2008;19(11):1969–74.
[67] Stephenson CM, Levin RD, Spector T, Lis CG. Phase I clinical trial to evaluate the safety, tolerability, and pharmacokinetics of high-dose intravenous ascorbic acid in patients with advanced cancer. Cancer Chemother Pharmacol 2013;72(1):139–46.
[68] Monti DA, Mitchell E, Bazzan AJ, Littman S, Zabrecky G, Yeo CJ, et al. Phase I evaluation of intravenous ascorbic acid in combination with gemcitabine and erlotinib in patients with metastatic pancreatic cancer. PLoS One 2012;7(1):e29794. Available from: https://www.ncbi.nlm.nih.gov/pmc/articles/PMC3260161/.
[69] Welsh JL, Wagner BA, van't Erve TJ, Zehr PS, Berg DJ, Halfdanarson TR, et al. Pharmacological ascorbate with gemcitabine for the control of metastatic and node-positive pancreatic cancer (PACMAN): results from a phase I clinical trial. Cancer Chemother Pharmacol 2013;71(3):765–75.
[70] Stupp R, Mason WP, van den Bent MJ, Weller M, Fisher B, Taphoorn MJB, et al. Radiotherapy plus concomitant and adjuvant temozolomide for glioblastoma. N Engl J Med 2005;352(10):987–96.
[71] Anderson CM, Sonis ST, Lee CM, Adkins D, Allen BG, Sun W, et al. Phase 1b/2a trial of the superoxide dismutase mimetic GC4419 to reduce chemoradiotherapy-induced oral mucositis in patients with oral cavity or oropharyngeal carcinoma. Int J Radiat Oncol Biol Phys 2018;100(2):427–35.
[72] Anderson CM, Lee CM, Saunders D, Curtis A, Dunlap N, Nangia CS, et al. Results of a randomized, placebo (PBO) controlled, double-blind P2b trial of GC4419 (avisopasem manganese) to reduce duration, incidence and severity and delay onset of severe radiation-related oral mucositis (SOM) in patients (pts) with locally advanced squamous cell cancer of the oral cavity (OC) or oropharynx (OP). JCO 2018;36(15 suppl):6006.
[73] FDA grants fast track designation to GC4419 for oral mucositis. [Internet]. January 8. Available from: https://www.healio.com/hematology-oncology/head-neck-cancer/news/online/%7bba1cfa0f-4473-454b-9fc3-ef567c22dec2%7d/fda-grants-fast-track-designation-to-gc4419-for-oral-mucositis; 2016. Accessed 6 October 2018.

CHAPTER

11

Fibroblast growth factor receptors as therapeutic targets in head and neck squamous cell carcinomas

Gopika Senthilkumar, David M. Francis, Andrew M. Baschnagel

Department of Human Oncology, University of Wisconsin Carbone Cancer Center, University of Wisconsin School of Medicine and Public Health, University of Wisconsin, Madison, WI, United States

Abstract

The fibroblast growth factor receptor family and their ligands regulate cell proliferation, invasion, and numerous other pathways essential for tumorigenesis. Aberrations in FGFRs have been reported to activate signaling pathways involved in cell survival and therapy resistance in tumors, thereby making FGFRs notable therapeutic targets. In head and neck squamous cell carcinomas (HNSCC) FGFR expression and FGFR1 amplification are associated with invasive disease, increased aggressiveness of disease, and poor overall survival. FGFR inhibition in HNSCC has demonstrated strong in vitro efficacy. The few clinical trials that have been reported suggest that FGFR inhibition can lead to tumor shrinkage. Multiple studies have found that FGFR-inhibition helps overcome therapeutic resistance to current standard of care therapies. The role of the FGFRs as therapeutic targets and biomarkers in HNSCC will be discussed in this chapter. We will highlight pre-clinical and clinical studies that investigate FGFR inhibition as a means for enhancing antitumor effects in HNSCC.

Abbreviations

FGFR	fibroblast growth factor receptor
HNSCC	head and neck squamous cell carcinoma
EGFR	epithelial growth factor receptor
HPV	human papillomavirus
MAPK	Mitogen activated protein kinase
PI3K	phospoinositide-3-kinase
STAT	signal transducers and activators of transcription
PLCγ	phospholipase C-γ

Improving the Therapeutic Ratio in Head and Neck Cancer
https://doi.org/10.1016/B978-0-12-817868-3.00011-1

TCGA the cancer genome atlas
MTOR mammalian target of rapamycin
FISH fluorescence in situ hybridization
PMOL potentially malignant oral lesions
OSCC oral squamous cell carcinoma
SNP small nucleotide polymorphism

Conflict of interest

No potential conflicts of interest were disclosed.

Introduction

Molecularly targeted therapy in HNSCC has mainly focused on targeting the epidermal growth factor receptor (EGFR) [1, 2]. Therapies targeting other cell signaling pathways have had limited success in HNSCC, despite their proven usefulness in other cancers [3–5]. Large-scale genomic profiling studies have identified the fibroblast growth factor receptor (FGFR) family as potential biomarkers and therapeutic targets in HNSCC [6, 7]. FGFRs are overexpressed and amplified in both human papillomavirus (HPV) positive and HPV negative tumors. There are currently multiple FGFR small molecular inhibitors being evaluated in both the preclinical and clinical setting. Several groups have demonstrated that FGFR inhibition, in conjunction with EGFR inhibition or radiation therapy, can synergistically inhibit proliferation of HNSCC cell lines, thereby suggesting that FGFR could play a role in resistance of HNSCC to current standard of care therapies [8–10].

The family of fibroblast growth factor receptors

Fibroblast growth factor receptors (FGFR) are receptor tyrosine kinases that regulate cell proliferation and differentiation, embryogenesis, angiogenesis, survival, organogenesis, migration, and wound repair [11–16]. The FGFR family consists of four members (FGFR1, FGFR2, FGR3 and FGFR4) and one receptor (FGFR5, also known as FGFRL1) that binds fibroblast growth factor (FGF) ligands but lacks an intracellular kinase domain. The receptors differ based on their ligand-specificities and tissue distribution. All four tyrosine kinase receptors contain a ligand-binding extracellular domain, a single transmembrane domain, and a tyrosine kinase cytoplasmic domain (Fig. 1) that are encoded by 20 exons of the FGFR genes [17, 18]. The extracellular domain consists of two or three immunoglobulin (Ig) domains with eight acidic residues (i.e., the acidic box) located between the IgI and IgII domains [19]. The IgI domain (amino-terminal) and the acidic box are involved with auto-inhibition of the receptor [19]. The most proximal Ig domain to the cell membrane (IgIII or IgIII in some) determines the ligand-binding specificity of the receptors [18]. Alternative splicing and inclusion of exons 7, 8, and 9 generates IgIIIa/b/c, all three of which are seen in FGFR1-3 [20–23]. The FGFR4 gene only codes for IgIIIc [22]. The various isoforms of IgIII lead to differential specificities of the receptors to various FGF ligands such as the FGF1 ligand being able to bind to IgIIIb and IgIIIc while the FGF2 ligand specifically binds to IgIIIc [18]. Moreover, the IgIIb isoform is more highly expressed in epithelial cells while the IgIIc is found more commonly in mesenchymal tissue—highlighting the tissue specificity of the receptor isoforms [24, 25].

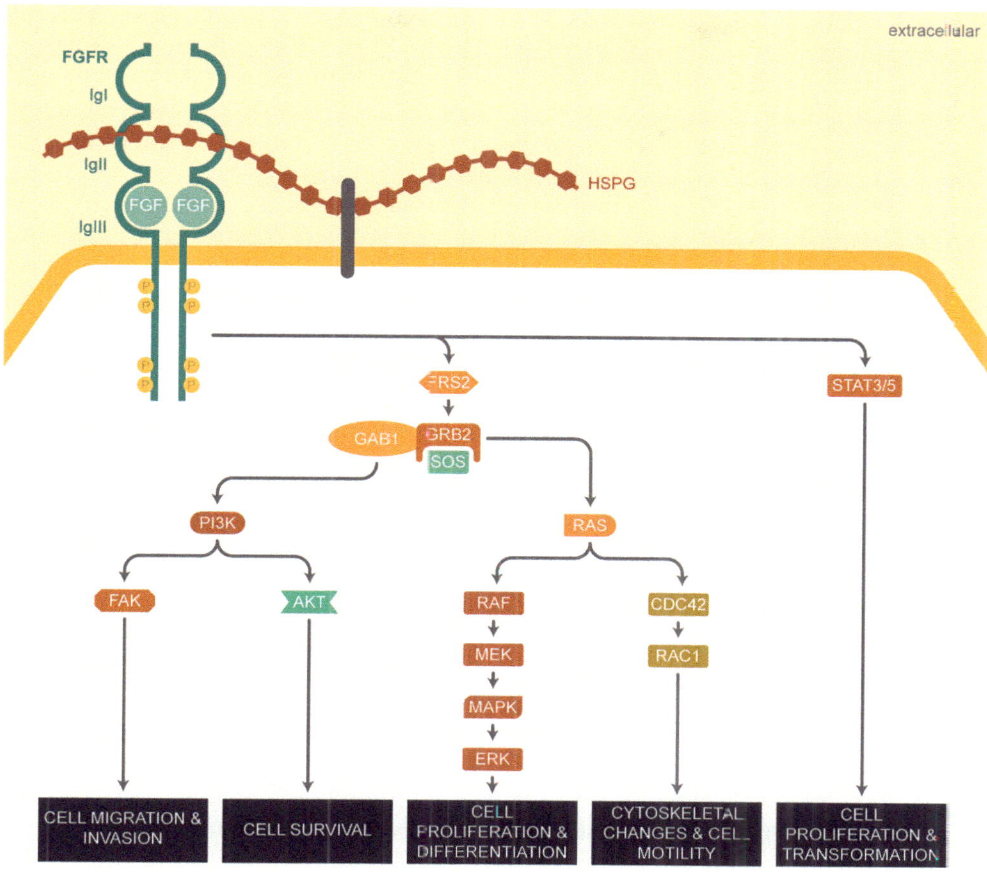

FIG. 1 The FGFR signaling pathway. Activation of the four receptors (FGFR1, FGFR2, FGFR3, and FGFR4) by the 20 different FGF ligands activates downstream pathways that promote and regulate tumor growth.

A FGFR-like protein, termed FGFR5, has recently been discovered. It lacks the tyrosine kinase domain and is thought to act as a decoy receptor by binding to FGFs without causing activation of downstream signaling [26].

The FGF ligands involve over 20 proteins whose interactions with FGFRs mediate Ras-dependent Mitogen Activated Protein Kinase (MAPK), Phospoinositide-3-Kinase (PI3K-AKT), Phospholipase C-γ (PLCγ), and Signal Transducers and Activators of Transcription (STAT) signaling pathways [27]. The ligands can be divided into 7 subfamilies. The canonical (i.e., secreted or paracrine) FGFs become sequestered by heparan sulfate proteoglycans (HSPs) to the cell surface and extracellular matrix. The HSPs protect the ligands from protease-mediated degradation, thereby stabilizing FGFR interaction with the ligand [27, 28]. The endocrine FGFs interact with Klotho family proteins instead of HSPs, and the intracellular FGFs act as co-factors for voltage-gated sodium channels [16, 28]. Once FGFs bind to the receptors, the tyrosine kinase domains dimerize and become autophosphorylated. This leads to a multistep process of receptor activation and signaling [29–31]. First, adaptor proteins dock with the phosphorylated tyrosine group. This activates the MAPK and PI3K-AKT pathways which then utilize PLCγ to amplify MAPK signaling. FGFR activation can also promote STAT signaling. These activated pathways coordinate control of cell proliferation and differentiation, migration and invasion, survival, shape and motility, and transformation.

Dysregulation of the FGFR signaling cascade has been associated with numerous cancers [32, 33]. Dysregulation can occur via multiple alterations. Receptor and ligand can be overexpressed. The genomic locus encoding the gene can be amplified. Changes in posttranscriptional processing can affect protein stability. Mutations can result in constitutive receptor activation, lowered sensitivity to ligand-binding, formation of fusion proteins, and isoform switching. Specific mutations and amplifications in FGFRs have been reported to cause aberrant activation of signaling pathways leading to therapeutic resistance, suppression of apoptotosis, and mesenchymal transformation in tumors, thereby making FGFRs notable biomarkers and therapeutic targets [28, 32–35]. The currently proposed mechanisms for FGFR influenced oncogenesis include neoangiogenesis, acquired resistance to existing cancer treatments, and mutations that lead to cell growth and survival (i.e., driver mutations) [32, 36]. A growing number of preclinical studies report promising antiapoptotic and/or antiproliferative tumor responses following FGFR inhibition. Importantly, these effects appear to result in reversal of resistance to existing therapeutics as well [28, 34, 37–39].

Genomic profiling and patient tissue analysis of HNSCC samples have identified FGFR family members as potential therapeutic targets [6, 7, 9, 40–43]. A study analyzing 951 patient samples reported that 73% of oral squamous cell carcinomas (OSCC) and 85% of oropharynx squamous cell carcinomas highly co-express two or more FGFR family members [44]. The same group also reported decreased proliferation following FGFR-inhibition in a HNSCC cell line expressing high levels of FGFR [8]. They also reported that a nonresponsive high-FGFR expressing cell line demonstrated resistance mediated via the EGFR pathway [8]. This cell line was sensitive to combined EGFR and FGFR inhibition [8]. Another recent study reported that the overexpression of FGFR1 mRNA and protein was related to more invasive disease, local and/or nodal recurrence, and poor prognosis [45]. Overexpression of FGFR3 mRNA and protein has been linked to worse overall survival [45]. Numerous other studies highlight the association between FGFR-aberrations and outcome in HNSCC suggesting that FGFR alterations could be a prognostic biomarker [8, 44, 46–49]. In addition, multiple studies using preclinical models of HNSCC demonstrate antitumor effects following FGFR-inhibition alone or in combination with other therapies [8, 49–53].

FGFR as a prognostic marker for HNSCC

Numerous studies have investigated the prognostic value of FGFR family members in HNSCC (Table 1). Each family member is discussed separately, below.

FGFR1

FGFR1 is the most frequently altered family member in HNSCC. The most common alterations include overexpression and amplification. Mutations in FGFR1 are seen infrequently in

TABLE 1 Studies assessing the prognostic impact of FGFR on HNSCC

	First author, year	Primary tumor site	# of patient samples	Prognostic marker	Biomarker technique	Outcome and prognostic impact
FGFR 1-4						
[44]	Koole et al. 2016	oral and oropharynx	951	FGFR1, 2, 3 protein expression	IHC	73% of oral squamous cell carcinomas and 85% of oropharynx squamous cell carcinomas show overexpression of two or more FGFR family member proteins. FGFR family protein expression was not correlated with overall survival. FGFR1 is more frequently overexpressed in HPV negative oral and oropharynx cancers, while FGFR2 and 3 are overexpressed in HPV positive tumors.
[45]	Starska et al. 2018	larynx	137	FGFR1, 3 mRNA and protein expression	qPCR, western blot, IHC	Overexpression of FGFR1 mRNA and protein was related to higher invasion rates, local/nodal recurrence rate, and poor prognosis. Overexpression of FGFR3 mRNA and protein was related to shorter survival time. High FGFR1 protein levels and FGFR 1 or 3 mRNA levels in combination were associated with lower overall survival.
[61]	Brands et al. 2017	HNSCC	530	FGFR1-4 mRNA expression, mutations, amplification, gains	RNA microarray from TCGA, Aanalysis, sqRT-PCR	46% of the patient samples had changes in FGFR. 31% of the patients had FGFR1 amplification of gain, 21% had FGFR3 amplification, mutation, or gain, 9% had FGFR2 mutations, and 8% had FGFR4 mutations. No correlation was found between overall survival and FGFR alterations.
FGFR 1						
[8]	Koole et al. 2016	Oral and oropharynx	436	FGFR1 protein expression and gene copy number	IHC, FISH	FGFR1 was overexpressed in 82% of HPV+ samples and 75% of HPV− samples. FGFR1 overexpression is related to poor overall survival and disease-free survival in HPV-negative HNSCC.

Continued

TABLE 1 Studies assessing the prognostic impact of FGFR on HNSCC—Cont'd

	First author, year	Primary tumor site	# of patient samples	Prognostic marker	Biomarker technique	Outcome and prognostic impact
[48]	Goke et al. 2013	hypopharynx, oropharynx, oral cavity, larynx, and other unknown	555	FGFR1 copy number (i.e., amplification)	FISH	FGFR1 amplification was associated with consumption of nicotine and alcohol, and poor patient outcomes. Amplification was also more common in primary hypopharynx and larynx squamous cell carcinomas, and less common in HPV-infection related oropharynx and oral cavity tumors. FGFR1 amplification was also correlated with higher T-stages of tumors and presence of lymphovascular invasion (i.e., more aggressive tumors). There was no significant correlation between FCFR1 amplification and overall survival.
[40]	Freier et al. 2007	Oral	92 FISH, 178 IHC	FGFR1 amplification and protein expression	FISH, IHC	High FGFR1 expression leads to carcinogenesis at early stage of oral carcinoma development. There was no correlation between overall survival and amplification of FGFR1 or protein overexpression.
[41]	Hase et al. 2006	oral	61	FGFR1 expression	IHC	FGFR1 expression in fibroblasts was higher in high-invasive and lymph node metastatic tumors, and it was correlated with patient prognosis.
[49]	Nguyen et al. 2013	HNSCC, oral	100	FGFR1 expression	IHC	FGFR1 expression was correlated with nuclear polymorphism, pattern of invasion, histological differentiation, and increased mitoses. Nuclear FGFR1 expression was associated with the same characteristics in Oral SCC.
[56]	Young et al. 2013	oral tongue	123	FGFR1 amplification	FISH	FGFR1 amplification was associated with smoking but not with T-stage, alcohol history, and nodal status. No significant correlation between FGFR1 amplification and survival were found.

FGFR 2,3						
[60]	Nayak et al. 2015	oral- OSCC and potentially malignant oral lesions (PMOL)	71 PMOL, 108 OSCC, 52 Healthy	FRFR 2,3 expression	IHC, RT-PCR	FGFR-2 overexpression was associated with malignant transformation of PMOL to OSCC.
FGFR 4						
[80]	Choi et al. 2013	oral	24	FGFR4 small nucleotide polymorphism	PCR, direct sequence	FGFR4 Arg/Arg or Arg/Gly at amino acid 388 was related to advanced N stage disease progression compared to Gly/Gly.
[46]	Azad et al. 2012	tongue, gum, mouth, oropharynx, hypopharynx, pharynx, larynx	531	FGFR4 small nucleotide polymorphism	PCR	No significant correlation between FGFR4 Gly/Arg388 genotype and overall survival or disease-free survival was found.
[80]	Chou et al. 2017	oral	955	FGFR4 genetic polymorphisms	PCR	FGFR4 rs351855 could be involved with susceptibility to oral squamous cell carcinoma development. Within that SNP, the patients with genotypes Gly/Arg or Gly/Arg in combination with Arg/Arg showed high risk of OSCC, while the homozygous Arg/Arg genotype showed low risk of advanced stage OSCC.
[76]	Tanuma et al. 2010	oral	150	FGFR4 genetic polymorphisms	PCR, PCR-single-stranded conformation polymorphism	FGFR4 Arg388 allele and TP53 mutations are related to poor patient prognosis. During a 60 month observation time post radical surgery, majority of the patients with the Arg388 allele and TP53 died while over 90% of those with Gly388 allele and wild-type TP53 survived. The combination of both is related to worse prognosis.

Continued

TABLE 1 Studies assessing the prognostic impact of FGFR on HNSCC—Cont'd

	First author, year	Primary tumor site	# of patient samples	Prognostic marker	Biomarker technique	Outcome and prognostic impact
[71]	Ansell et al. 2008	HNSCC	110	FGFR4 gene and RNA	PCR-RFLP	Males with FGFR4 Gly(388) allele have an increased risk of developing HNSCC. In cell lines, the Arg388 allele was not found to play a role in radiosensitivity, but it did slightly increase cell line sensitivity to cisplatin.
[77]	De Costa Andrade et al. 2007	HNSCC	75	FGFR4 Arg(388) allele	PCR-RFLP and DNA sequencing	The presence of at least one FGFR4 Arg388 allele was correlated with reduced overall survival and increased mortality risk.
[78]	Farnebo et al. 2013	HNSCC	40	FGFR4 small nucleotide polymorphism	IHC, PCR-RFLP	FGFR4 Gly388Arg single nucleotide polymorphism was associated with longer overall survival.
[79]	Dutra et al. 2012	oral and oropharynx	122	FGFR4 expression, Gly388Arg genotype	IHC, PCR-RFLP	Presence of FGFR4 Arg388 genotype was associated with lymphatic embolization and premature death related to disease. Low expression of FGFR4 was associated with lymph node positivity, premature disease relapse, and death related to disease.
[81]	Shi et al. 2015	nasopharynx	55	FGFR4 expression	IHC, western blot	Overexpression of FGFR4 was correlated with high proliferation and clinical stage, and poor prognosis. It might also be associated with increased tumor migration.
[75]	Streit et al. 2004	oral, oropharynx	104	FGFR4 expression, Arg388 allele	IHC, PCR	Overexpression of FGFR4 with Arg388 allele was associated with reduced overall survival and advanced tumor stage. However, the overexpression of FGFR4 Gly388 did not affect disease progression. The overexpression of FGFR4 and the Arg388 allele was related to poor clinical outcomes.

HSNCC [54]. FGFR1 is overexpressed in 75%–85% of HPV positive and HPV negative HNSCCs and amplification of FGFR1, at the 8p12 locus is seen in 15–20% of HPV negative tumors [40, 48].

FGFR1 amplification or overexpression is associated with higher T-stage of disease, poor overall survival, and more invasive disease [8, 40, 41, 45, 48, 49]. A study with 336 patient samples reported that FGFR1 was overexpressed in 82% of HPV positive samples and 75% of HPV negative samples, and FGFR1 overexpression was related to poor overall survival and disease-free survival in HPV-negative HNSCC [8]. The same group also reported that FGFR1 was more frequently overexpressed in HPV negative oral and oropharynx cancers, while FGFR2 and 3 are more frequently overexpressed in HPV positive tumors [8, 44]. A study involving 353 patient samples and 279 samples from The Cancer Genome Atlas (TCGA) found that FGFR1 mRNA expression was similar in both HPV positive and negative samples [54]. In the TCGA dataset, FGFR1 amplification (determined by SNP array) was reported in 10% of HPV negative tumors and not found in any of the HPV positive tumors. In addition, FGFR2 and FGFR3 amplification was uncommon (<2%) and only seen in HPV negative tumors [54]. It should be noted that the TCGA dataset included very few HPV positive tumors [50, 55]. Another study with 555 patient samples analyzed FGFR1 copy number determined by fluorescence in situ hybridization (FISH) and found that FGFR1 amplification was present in 15% of all patients, more common in primary hypopharynx and larynx squamous cell carcinomas, and less common in HPV positive oropharynx and oral cavity tumors. FGFR1 amplification was also correlated with higher T-stage and the presence of lymphovascular invasion (i.e., more aggressive tumors) [48]. However, this group did not find a significant correlation between FGFR1 amplification and overall survival, confirming the results published by other groups [48, 56]. Overall, FGFR1 expression seems to serve as a prognostic marker in HPV negative oral and oropharynx cancers, and FGFR1 amplification seems to predict the aggressiveness of hypopharynx and larynx squamous cell carcinomas. These studies also highlight the relationship and importance of HPV status in determining the prognostic impact of FGFR1, FGFR2, and FGFR3. It also must be noted that the clinically relevant cutoff levels for amplification remains to be determined. The method of determining FGFR1 amplification either with FISH or with various sequencing techniques can also influence the results given the heterogeneity of the 8p12 amplicon. FISH is not as sensitive or specific; thus, focal amplification of FGFR1 is likely to be more frequent with FISH techniques than when estimated by SNP arrays or sequencing. [57].

FGFR2

The Human Protein Atlas analyzed 499 patient samples and reported that high expression of FGFR2, seen in 60% (299 of 499) of patients, is associated with favorable outcome in HNSCC; with a 52% five-year survival rate in patients with high-FGFR2 expression compared to a 40% five-year survival r(57)ate in low FGFR2 expressing patients [58]. This association was not replicated in another study that analyzed 951 patient samples: no significant correlation was seen between FGFR2 protein expression and overall survival in oral cavity and oropharyngeal squamous cell carcinomas [44]. Using a preclinical animal model of chemically induced OSCC in hamsters, Vairaktaris and colleagues demonstrated that FGFR2 and

FGFR3 promoted dysplasia and early invasion [59]. Interestingly, a study that analyzed 71 potentially malignant oral lesions (PMOL) reported that FGFR2 and FGF2 overexpression was associated with malignant transformation of PMOL to OSCC [60]. The contradicting results and the limited number of studies highlights the need for further exploring FGFR2's prognostic role in HNSCC, with a focus on specific FGFR2 aberrations rather than overall mutational or expressional profiles.

While FGFR2 protein overexpression is common, mutations and amplification of the FGFR2 gene found infrequently. A 2017 study that analyzed 530 patient samples found that FGFR2 mutations were found only in 9% of HNSCC patients, and there was no correlation between FGFR2 mutations and overall survival [61]. The TCGA analysis found FGFR2 to be altered in only 2% of HPV negative samples with two patients showing amplification and two patients showing mutations. There were no FGFR2 alterations reported in the HPV positive cohort [55]. In addition, Seiwert et al. reported two FGFR2 mutations (4%), N569D and N569K, in their cohort of HPV positive tumors and no mutations in HPV negative tumors. The authors of this paper suggest that these mutations may have oncogenic potential [7]. A sequencing study from Singapore examining 128 nasopharyngeal carcinomas found two FGFR2 mutations: N549K and L20P [62].

FGFR3

Amplification of FGFR3 is not common within the HNSCC TCGA dataset, but overexpression of mRNA and protein are seen [55]. The Human Protein Atlas found that FGFR3 was highly expressed in three quarters of patient samples (362 of 499), but FGFR3 was not prognostic of outcome [58]. Brands and colleagues analyzed 530 HNSCC patient samples and found that 21% had FGFR3 amplification, mutation, or gain, but no correlation between FGFR3 aberrations and overall survival was observed [61]. Additionally, the study conducted by Nayak, et al. on PMOLs found that 37% of untransformed PMOLs and 60% of transformed PMOLs had upregulated FGFR3 expression [60]. Interestingly, upregulation did not correlate with malignant transformation of PMOLs to OSCC [60]. A 2018 study with 137 squamous laryngeal cancer patient samples found that overexpression of FGFR3 mRNA and protein was related to shorter survival time, and FGFR3 mRNA levels were inversely correlated with overall survival [45]. Koole et al. analyzed 522 HNSCC TCGA patient samples and found that only 3% of the samples showed upregulation of FGFR3 mRNA. However, 48% of oral and 59% of oropharyngeal squamous cell carcinomas overexpressed FGFR3 protein; a finding that was not correlated with overall or disease-free survival [63]. They also found no difference in protein overexpression between HPV positive (53%) and HPV negative (59%) tumors. Note that only 19% of the samples analyzed in this study were HPV positive [63]. In another study, the same group reported that FGFR2/3 were highly co-expressed in HPV positive oropharyngeal squamous cell carcinomas, however the co-expression was not statistically significant [44].

Somatic mutations in FGFR3 are found in 1% in HPV negative tumors and 12% of HPV positive tumors. Stewart et al., identified six S249C FGFR3 mutations, all in HPV positive tumors, as well as one K413N somatic mutation in a HPV negative tumor [7]. The FGFR3 S249C mutation has also been reported in one HPV positive tumor in the TCGA HNSCC cohort [6] and in the TCGA lung squamous cell and bladder carcinoma cohorts [55, 64]. The S249C mutation may cause the receptor to form ligand-independent dimers through intermolecular disulfide bonds, which results in constitutive receptor activation [65–67] or structural changes of the dimers [68]. Preclinical reports suggest that combination FGFR and EGFR therapies may be important in mediating resistance to FGFR-targeting in S249C FGFR3-activating mutated cancers [69] and preliminary phase I clinical trial data indicate that FGFR3-activating mutations in bladder cancers may be responsive to the pan-FGFR inhibitor, BGJ398 [70].

FGFR4

Compared to the other FGF receptors, FGFR4 is not highly overexpressed or amplified in HNSCC [6]. Base alterations in FGFR4 have been described including the presence of a Glycine (Gly) to Arginine (Arg) small nucleotide polymorphism (SNP) at position 388 within the transmembrane domain of FGFR4 [71]. The presence of this Arg388 SNP has been previously associated with worse prognosis in many cancers such as breast, colon, prostate, soft tissue carcinoma, and HNSCC [71–75]. A study with nearly one thousand OSCC patient samples identified FGFR4 rs351855 (a specific SNP) as a susceptibility locus for HNSCC development. Within that SNP, they found that the patients with genotypes Gly/Arg or Gly/Arg in combination with Arg/Arg showed high risk of OSCC, while the homozygous Arg/Arg genotype showed low risk of advanced stage OSCC [47]. Numerous other studies broadly concluded that the presence/overexpression of FGFR4 with Arg388 was associated with reduced overall survival and advanced tumor stage, while the presence of FGFR4 Gly388 either did not affect disease progression or improved overall survival [47, 75–80]. Data on this SNP is not conclusive. Another group failed to identify a correlation between FGFR4 Gly/Arg388 and either disease-free or overall survival in a study including over 500 patients [46]. Yet another group studying just over one hundred patient samples reached exactly the opposite conclusion: males with FGFR4 Gly388 allele had an increased risk of developing HNSCC [71]. Of the two studies that analyzed the overall FGFR4 expression, one found that low expression of FGFR4 was associated with lymph node positivity, premature disease relapse, and death related to disease in oral and oropharynx squamous cell carcinomas [79]. Another study found that the overexpression of FGFR4 was correlated with high proliferation and clinical stage, and poor prognosis in nasopharyngeal carcinoma [81]. This suggests that the implications of overall expression of FGFR4 might be disease-site specific even within HNSCC, and that presence of Arg388 vs. Gly388 in FGFR4 might be a valuable prognostic marker in HNSCC.

FGF ligands

While numerous studies have analyzed the prognostic implications of the FGFRs in HNSCC, only a few studies to date have analyzed the FGF ligand's role in HNSCC. FGF-2 ligand (i.e., basic FGF) expression is nearly 12 fold higher in HNSCC tissue compared to noncancerous tissues [82]. The FGF-2 ligand has been broadly associated with tumor invasion and proliferation of per-tumoral fibroblasts [34]. In HNSCC, the prognostic implications of FGF-2 have remain unclear due to conflicting results among various studies [42, 59, 82, 83].

In summary, overexpression and amplification of FGFR1, and the presence of the Arg388 allele in FGFR4 seem to be reliable predictors of poor prognosis and more aggressive disease in HNSCC. More research on FGFR2, FGFR3, and the numerous FGFR ligands is required. Moreover, while correlations have been reported between specific FGFR alterations and overall survival, groups have found no correlation between overall FGFR-family member expression and overall survival [44, 56, 61]. Thus, as more work is done to further analyze the prognostic implications of FGFR on HNSCC, a focus on the role of specific aberrations in each member of the FGFR family might be more effective than aggregate correlations.

Preclinical inhibition of FGFR in HNSCC

Given the frequency of FGFR alterations and their potential prognostic implications, numerous groups have explored the preclinical benefits of FGFR-inhibition in HNSCC (Table 2). BGJ398, a specific pan-FGFR kinase inhibitor that acts on FGFR1-3, has been shown to increase apoptosis and reduced basal level activation of downstream targets (pERK1/2) in cell lines expressing FGFR1 [50]. They also found that cell-line sensitivity to FGFR-inhibition was correlated with FGFR1 mRNA and protein expression, but not copy number gain or FGF2 or FGF3 mRNA levels [50]. This group used two in vivo xenografts with high FGFR1 amplification and expression and reported that one of them showed tumor growth control following BGJ398 treatment [50]. Another group used BGJ398 to demonstrate that FGFR signaling was implicated in HNSCC resistance to cisplatin [84]. In the UM-SCC-22B ALDH and CD44 high cell line (i.e., cancer stem cells resistant to cisplatin), combination therapy of cisplatin and BGJ398 decreased proliferation of the resistant cells [84]. Likewise, in UM-SCC-1 ALDH and CD44 high cells, BGJ298 alone was enough to decrease cisplatin cell proliferation [84]. Other groups have also reported that FGFR inhibition sensitized resistant HNSCC tumors to bevacizumab [85–87]. These studies show promising preclinical antitumor and antitherapeutic resistance effects of FGFR- inhibition in HNSCC.

A study that utilized six HNSCC cell lines reported that AZD4547, a potent and selective tyrosine kinase inhibitor of FGFR1-3, resulted in reduced cell proliferation and FGFR signaling in a high FGFR1 expressing cell line [8]. The cell lines that did not show high FGFR1 expression did not respond to treatment. The cell lines that showed high expression of FGFR1 but were resistant to FGFR-inhibition benefitted from combined EGFR inhibition suggesting that EGFR signaling may be a potential mechanism for FGFR-treatment resistance [8]. Another group further characterized cell lines into three categories: FGFR-dependent, EGFR and FGFR co-dependent, and EGFR-dependent. They demonstrated that the cell lines that were both EGFR and FGFR dependent showed significantly lowered cell growth following combination therapy of FGFR and EGFR inhibitors, than either inhibitor alone [9]. This result reflects the role of receptor tyrosine kinase co-activations in therapeutic resistance, whereby multiple receptors that are involved in oncogenic signaling can compensate for the few that are inhibited [9, 88].

Another study used parallel RNA interference genetic screens to show that upregulation of EGFR signaling led to resistance to FGFR inhibition in FGFR3-mutant and translocated cell lines [33]. They also reported increased antitumor effects when both FGFR and EGFR were inhibited in vitro and in vivo [33]. These studies highlight that FGFR-inhibition is prone to

TABLE 2 Preclinical studies of FGFR inhibitors in HNSCC

	First author, year	FGFR inhibitor	Drug target	Cell line	in vivo xenograft	Effect
[8]	Koole et al. 2016	AZD4547	pan-FGFR (FGFR1/2/3 and weakly FGFR4)	UTSCC9, UTSCC12A, UTSCC40, SCC147, BICR16, CCL30	none	Only CCL30 (high FGFR1-expressing cell line) responded to AZD4547 treatment. UTSCC9, UTSCC12A, and UTSCC40 did not express FGFR, which could explain the unresponsiveness to treatment. SCC147 and BICR16 showed high expression of FGFR but showed resistance to inhibition. These FGFR-inhibition resistant tumors benefitted from combination therapy with EGFR inhibition (gefitinib).
[50]	Goke et al. 2015	NVP-BGJ398	pan-FGFR kinase inhibitor; predominantly active on FGFR1-3	CCL30, 584-A2, SCC9, Cal27, UMSCC25, HN6, UMSCC8, JHU011, HN12, UMSCC1, HN, FaDu	CUHN015, CUHN072	BGJ398 increased apoptosis and reduced basal level activation of downstream targets (pERK1/2) in cell lines expressing FGFR1. FGFR inhibition sensitivity is also correlated with FGFR1 mRNA and protein expression but not copy number gain, and FGFR2 and FGFR3 mRNA levels were not correlated. Although CUHNO72 xenograft had high FGFR1 amplification and expression, there was no tumor control from BGJ298 administration. However, the CUHNO15 xenograft had significant tumor growth control.
[49]	Nguyen et al. 2013	PD173074	FGFR1 ATP pocket and kinase activity	Ho-1-N-1, Ho-1-U-1, HOC313, HOC621, KOSCC33A, KOSCC25B	none	FGFR1 was overexpressed in Epithelial-mesenchymal transition (EPT) HNSCC cell lines compared to non-EMT cell lines. In HOC313 cells, decreased proliferation, invasion, and phosphorylation of ERK1/2 and p38 following treatment with PD173074 was observed. Their morphologies were also changed from spindle-

Continued

TABLE 2 Preclinical studies of FGFR inhibitors in HNSCC—Cont'd

	First author, year	FGFR inhibitor	Drug target	Cell line	in vivo xenograft	Effect
						shape to cobble-stone shape. This drug inhibits the MAPK pathway which regulates AP-1 and induced MET, which then suppresses cancer growth and invasion.
[52]	Singleton et al. 2013	AZ8010	FGFR-specific tyrosine kinase	UMSCC25, 584-A2, CCL30	none	ERBB2 and/or MET inhibition in conjunction with FGFR inhibition decreased growth relative to FGFR inhibition alone with the inhibition of all three being the most effective in inducing apoptosis and inhibiting growth.
[51]	Singleton et al. 2016	AZD4547	pan-FGFR (FGFR1/2/3 and weakly FGFR4)	584-A2, CCL30	Lung cancer cells	Inhibition of MTOR or AKT along with FGFR resulted in synergistic in vitro tumor growth suppression than FGFR-inhibitor monotherapy.
[53]	Sweeny et al. 2012	PD173074	small molecule tyrosine kinase inhibitor of FGFR1/2/3	FADU, OSC19, Cal27, SCC1, SCC5, SCC22A, HS27	SCC-5	FGFR inhibition in vitro resulted in decreased proliferation in the fibroblast cell line (HS27) and endothelial cells but not HNSCC cells. In HNSCC and fibroblast coculture, FGFR inhibition suppressed fibroblast-enhanced tumor growth. In vivo, treatment with PD173074 led to significant growth inhibition. The treated tumors contained smaller stromal components, showed decreased proliferation, and had increased apoptosis.

[85]	Gyanchandani et al. 2018	PD173074	small molecule tyrosine kinase inhibitor of FGFR1/2/3	Tu-138, bevacizumab-resistant isogenic clones	Tu-138, bevacizumab-resistant isogenic clones	FGFR inhibition sensitized resistant tumors to bevacizumab.
[84]	McDermott et al. 2018	BGJ398	pan-FGFR kinase inhibitor	UM-SCC-1, UM-SCC-22B	None	FGFR signaling is implicated in HNSCC resistance to cisplatin therapy. In the UM-SCC-22B ALDH and CD44 high cell line (i.e., cancer stem cells resistant to cisplatin), combination therapy of cisplatin and BGJ398 decreased proliferation of the resistant cells, while in UM-SCC-1 ALDH and CD44 high cell line, BGJ298 alone was enough to decrease cisplatin resistant-cell proliferation.
[9]	Marshall et al. 2011	RO4383596, AZ12908010, FP-1039	KDR, FGFR, and PDGFR receptor tyrosine kinase (RO4…), FGFR 1-3 (AZ1…), FGFR1 (FP…)	UMSCC8, UMSCC19, UMSCC25, HN4, HN31, CCL30, Detroit562, Ca9-22, 584-A2	none	FP-1039 inhibited growth in FGF2 expressing cell lines, suggesting that FGF-2 is an autocrine growth factor. UMSCC8, Ca9-22 and UMSCC2 respond to both EGFR and FGFR inhibition, 584-A2, CCL30 and Detroit 562 respond solely to FGFR inhibition, and UMSCC19, HN4, HN31 respond to EGFR inhibition (measured by decreased in phospho-ERK and cell growth). Both AZ12908010 and RO4383596 decreased signals to the same level, suggesting that FGFRs are driving ERK-signaling in the three cell lines solely dependent on FGFR signaling. In the cell-lines that responded to both treatments, combination therapy resulted in significantly lower cell growth than either treatment alone.

resistance and that inhibition of EGFR is one potential mechanism for overcoming that resistance. A different group that used AZD4547 showed that inhibition of mammalian target of rapamycin (MTOR) along with FGFR resulted in synergistic in vitro tumor growth suppression compared to FGFR-inhibitor monotherapy [51]. They concluded that parallel activation of the MTOR/AKT pathway affected the FGFR-dependent cell lines' sensitivity to FGFR inhibition [51]. The same group also found that ErbB2 and/or MET inhibition in conjunction with FGFR inhibition decreased tumor growth relative to FGFR inhibition alone, with the inhibition of all three being the most effective in inducing apoptosis and inhibiting growth [52]. These results are not surprising, as amplification of MET following FGFR-inhibition has been thought to lead to resistance via the ErbB3 pathway [89]. Kim et al. used clonal selection to isolate squamous cell lung carcinoma cell lines that were resistant to AZD4547 or the FGFR inhibitor BAY1163877 and found overexpression and activation of MET in the resistant lines [89]. When the cell lines were treated with AZD4547, they found gene amplification of MET which was thought to cause resistance via activation of ErbB3 [89]. Moreover, when they forced overexpression of MET in the regular cell lines, the cells showed resistance to FGFR inhibitors; when treated with both MET and FGFR inhibitors, the investigators found a synergistic reduction in cell growth [89]. These studies reflect the therapeutic potential of inhibiting FGFR in combination with pathways that could be involved in FGFR-therapy resistance, to achieve desired tumor control in HNSCC.

Numerous other studies have tested PD173074, a FGFR1 ATP pocket and kinase inhibitor, and reported decreased proliferation, and invasion both in vitro [49] and in vivo [53]. FGFR inhibition in vitro resulted in decreased proliferation in the fibroblast cell line (HS27) and endothelial cells but not HNSCC cells. In HNSCC cells with fibroblast coculture, FGFR inhibition suppressed fibroblast-enhanced tumor growth [53]. In vivo, treatment with PD173074 led to significant growth inhibition and the treated tumors contained smaller stromal components, showed decreased proliferation, and demonstrated increased apoptosis [53]. This study highlights the role of the tumor-microenvironment on drug-response. The first study analyzed the mechanism via which PD173074 acts and concluded that the drug induced MET by blocking the MAPK pathway, which then suppresses cancer growth and invasion [49]. These studies further highlight the differences in the mechanisms through which the varying specific FGFR-inhibitors act, thereby showcasing the need to further preclinical trials that assess: (i) the tumor-response following drug treatment in cells with varying FGFR aberrations as discussed previously; (ii) the mechanism through which the drugs leads to tumor-suppression; and, (iii) the resistance mechanisms that need to be inhibited in conjunction with the drug.

FGFR inhibition in combination with radiation

A few studies have also explored the efficacy of inhibiting FGFR as a mechanism for overcoming resistance to radiation in numerous cancers [85–87]. FGFR inhibition in glioblastoma radiosensitized resistant cells by increasing mitotic cell death, decreasing FGFR1 availability on the cell membrane, increasing receptor ubiquitylation, inhibiting activated RhoB, and modulating the level of hypoxia inducible factor 1α [86]. FGFR-inhibition did not affect survival of radiosensitive cells, suggesting that the FGFR pathway was truly implicated in the radioresistance mechanism in brain glioblastomas [86]. Moreover, the mice treated with drug before radiation showed greater neurological deficit free survival [86]. These results are consistent with the radiosensitizing effects of FGFR-inhibition others have found in human squamous cell carcinomas [87]. However, in colorectal cancer, researchers did not see a radiosensitizing effect of FGFR-inhibition [90]. The only other published study on HNSCC utilized a single cell line and a nonspecific, multikinase inhibitor, and concluded that FGFR inhibition does not radiosensitize HNSCC [91]. We have seen promising radiosensitizing effects of FGFR-inhibition using AZD4547 in both HNSCC and nonsmall cell lung cancers both in vitro and in vivo using multiple cell lines and FGFR-expressing patient-derived xenografts [10, 92]. Interestingly, both groups that studied radiosensitization in SCC conducted transcriptome studies of HNSCC cell lines and reported that FGFR-signaling pathway could be a potential target for radiosensitization [87, 91]. Moreover, others have shown that administration of FGF2 along with radiation can protect epithelial cells from radiation-induced apoptosis. In vivo, they found that FGF2 protected the mice from fatal radiation pneumonitis [93, 94]. Others have also reported that the administration of FGF2 to gastrointestinal crypt cells and hematopoietic tissues reduces radiation-induced apoptosis, progenitor cell differentiation, epithelial proliferation, and stem cell renewal, further suggesting that the FGF/FGFR signaling cascade could be involved in the regulation of cellular response following radiation therapy and is a potential mechanism through which tumors show radioresistance [95–97]. Thus, the inhibition of the FGFR signaling axis in conjunction with radiation is a logical combination therapy to explore in HNSCC. The limited number of studies in HNSCC and widely reported radiosensitization potential of FGFR-inhibition highlights the need for further preclinical research on radiation combination studies.

In summary, the existing preclinical data suggest that FGFR-inhibition could lead to antitumor effects in FGFR-expressing HNSCC and can help overcome resistance to existing treatments. Furthermore, combination therapy with additional molecularly targeted agents (e.g., MTOR, EGFR, MET, see appropriate chapters) and radiation should be further explored to understand the synergistic benefit versus the added toxicities.

Clinical inhibition of FGFR in HNSCC

Based on the promising preclinical data described above, a few clinical studies have begun exploring the potential of FGFR inhibition in HNSCC. The only study with published data that is specific for HNSCC patients involved three patients and did not report antitumor response. Two out of three patients had noticeable toxicities [98]. The other studies are nonspecific to HNSCC patients or ongoing or still recruiting as summarized in the Table 3.

The FGFR inhibitors that are currently being tested in HNSCC both in preclinical and clinical settings have previously been tested in patients with numerous other cancers [32]. Some of the more promising clinical studies on the specific-FGFR inhibitors are discussed below. Drugs such as nintedanib, abemaciclib and ponatinib are nonspecific FGFR inhibitors that have been tested in patients with FGFR aberrations [32], but this chapter will not focus on those.

AZD4547 is a potent small molecule inhibitor of FGFR 1-3 [32]. In a Phase I trial of AZD4547 (NCT00979134) in patients with advanced solid malignancies, investigators reported that at

TABLE 3 Clinical studying assessing the effects of FGFR inhibition (HNSCC)

	First author, year	NCT #	Phase and status of trial	Conditions	FGFR inhibitor	Drug target	Study design	# of patients	Effect
[99]	Peter S. Hammerman, 2014	NCT01761747	II, terminated	HNSCC presence	Ponatinib (AP24534)	FGFR kinase	Single group assignment; Ponatinib by mouth daily for 4 weeks	3	All three patients showed FGFR amplification/mutation but none of the patients showed any response to ponatinib treatment and 2/3 reported toxicities.
[100]	Tanguy Seiwert, 2018	NCT02706691	II, recruiting	FGFR1,2 gene amplification and mutation in HNSCC, HPV status, recurrent HNSCC or nasopharynx carcinoma or oro-pharyngeal SCC	Infigratinib (BGJ398)	FGFR1-3	Single group assignment; 125 mg drug by mouth one daily for 3 weeks on and 1 week off—repeats every 28 days until disease progression or unacceptable toxicity	20 (estimated)	N/A
[101]	Myung-Ju Ahn, 2015	NCT02558387	II, unknown	recurrent or metastatic salivary gland HNSCC	Nintedanib (BIBF1120)	PDGFR/FGFR 1-2, VEGFR 1-3 tyrosine kinase	Single Group Assignment, 200 mg drug delivered orally twice daily until progression or unacceptable toxicity.	20	Enrollment was stopped after 20 patients as no responders were observed at the end of Stage I. No partial response was seen, but 75% disease control rate was reported. At 6 months, progression-free survival was 60%, and grade 3 adverse events included liver enzyme elevation, nausea/vomiting. Five of the 20 patients required dose reduction.

[102]	Hwan Jung Yun, 2017	NCT03292250	II, recruiting	HNSCC	BYL719, Poziotinib, Nintedanib, Abemaciclib, Durvalumab, Tremelimumab	Ninte-danib PDGFR/FGFR 1-2, VEGFR 1-3 tyrosine kinase	Nonrandomized, parallel assignment based on molecular characterization of tumors during or after palliative platinum chemotherapy	259	N/A
[103]	Bayer, 2013	NCT01976741	I, active not recruiting	any type of solid tumor with FGFR expression/mutation level. Expansion only to sqNSCLC, HNSCC, bladder cancer, and lung adeno-carcinoma	Rogaratinib (BAY1163877)	pan-FGFR inhibitor	Single group assignment, starting with 100 mg per patient, the drug will be orally administered twice daily	168	N/A
[104]	Bayer, 2017	NCT02592785	I, completed	Japanese subjects with refractory, locally advanced, or metastatic solid tumors that had high FGFR mRNA expression.	Rogaratinib (BAY11638)	Pan-FGFR inhibitor	Single group assignment, cohort 1 (600 mg), cohort 2 (800 mg), orally administered.	9	All patients had adverse events (commonly hyperphosphatemia with grade 1 or 2); 1 epipharynx cancer patient had partial response, five patients had stable disease. 800 mg BID is safe and tolerated in Japanese patients.

80 mg PO BID, pharmacologic target inhibition could be achieved. However, dose limiting toxicities included mucositis, renal failure, hyperphosphatemia, and elevated liver enzymes [32, 105]. In an expansion cohort with 13 patients that had advanced gastric and gastroesophageal cancer, 6 patients reported fatigue and nausea, 8 reported vomiting, 7 had decreased appetite and diarrhea, and 4 patients had hyperphosphatemia, constipation, and dry eyes [106]. A Phase II proof of concept study (NCT01795768) that tested AZD4547 in patients with HER2 negative breast cancer and nonsmall cell lung cancer that have high FGFR1 expression and gastroesophageal cancer with FGFR2 amplification showed that AZD4547 had greater activity in patients with FGFR2 amplified gastroesophageal cancer than the other FGFR1 overexpressed cancers [107]. The researchers also noted that the patients with gastroesophageal cancer that showed partial response to AZD4547 had elevated FGFR2 copy number in their cell free plasma DNA, providing a potential biomarker for screening in future patients [32, 107]. While AZD4547 has been tested in HNSCC cell lines and in vivo xenografts, it is not currently being tested in clinical trials specifically for HNSCC.

BGJ398 (Novartis) is a potent, pan-FGFR inhibitor (targets FGFR1-3) that is currently being testing in HNSCC (NCT02706691) [32]. In an ongoing dose escalation clinical trial (NCT01004224) with 94 enrolled patients, preliminary results show partial responses in four FGFR3 mutated bladder cancer patients and 2 FGFR1 amplified squamous cell lung cancer patients [108]. In an expansion cohort, the adverse effects reported include reversible hyperphosphatemia, stomatitis, alopecia, fatigue, and decreased appetite [109]. In a Phase I trial for patients with advanced solid tumors that showed FGFR alterations determined by FISH, treatment with BGJ398 was reported to show antitumor activity in patients with FGFR1-amplified squamous nonsmall cell lung cancer and FGFR3-mutant bladder or urothelial cancers. The study also reported the aforementioned adverse effects and concluded that their dosing had tolerable and manageable safety profiles [110].

Rogaratinib (BAY1163877) is a selective, potent small-molecule pan-FGFR (FGFR1-4) inhibitor [111]. The Bayer group conducted two phase I clinical studies in patients with solid tumors (including HNSCC patients), and the study completed with 9 patients (all solid tumors, not specific to HNSCC) reported a disease control rate of 56%—1 patient with partial response and 5 with stable disease (NCT01976741, NCT02592785) [103, 104]. The same group analyzed the results specific to the HNSCC patients enrolled in both of those studies (10 patients administered 800 mg bid), and found that two of them had partial response, six had stable disease, and two patients had progressive disease. Seven out of the 10 HNSCC patients assessed, including the two patients that had partial response, overexpressed FGFR3 mRNA. The common adverse effects in these patients were hyperphosphataemia and diarrhea. Although the sample size is limited, the group concluded that Rogaratinib treatment should be further investigated in HNSCC patients with high FGFR3 mRNA expression [112].

These clinical studies, although limited, support the need for more in-depth clinical analysis of FGFR inhibition in patients with FGFR-expressing/amplified tumors. In both of the clinical studies that reported results in HNSCC, the FGFR biomarker selected either did not translate to clinical response from FGFR-inhibition (NCT01761747) or led to partial response in a small subset of the patients expressing the biomarker (NCT01976741, NCT02592785) [98]. Thus, one of the biggest barriers towards further clinical implementation remains the need for accurate, predictive biomarkers that can be used for selecting patients for different FGFR-targeted therapies [113]. Given the preclinical studies demonstrating potential synergy between FGFR inhibition and other anticancer treatments, clinical studies might see more success from combination therapies that target and overcome therapeutic resistance [8–10, 84, 88].

Conclusion

Ultimately, given the limited number of effective molecularly targeted therapies in HNSCC and the high rate of disease recurrence, there is a need for biomarker driven therapies that can overcome tumor resistance and improve patient prognosis [114–117]. Based on the current evidence, the FGFR pathway could be a promising target to meet this need. FGFR signaling is involved with cell proliferation, survival, migration, invasion, and numerous other pathways essential for tumorigenesis. Specific aberrations in FGFRs have been reported to activate pathways involved with therapy resistance, antiapoptosis, mitogenesis, and mesenchymal response in tumors, thereby making FGFRs notable biomarkers and therapeutic targets [28, 32–35]. Biomarker analysis in HNSCC by various groups has shown that specific FGFR expression and amplifications are common and are associated with poor overall survival, higher invasion rates, and overall increased aggressiveness of disease [8, 40, 41, 45, 48, 49]. Numerous preclinical studies report antiapoptotic and antigrowth tumor response following FGFR inhibition, and many have found that FGFR-inhibition helps overcome therapeutic resistance to current standard of care therapies [28, 34, 37–39] While clinical studies in HNSCC that test the efficacy of FGFR inhibition are limited, the promising preclinical data supports the need for further clinical investigation. Moreover, the growing number of biomarker driven umbrella trials and improvements in biomarker screening/sequencing technologies could accelerate further investigation of FGFR-inhibition in HNSCC patients [32, 113]. This trend towards precision medicine provides a realistic method for implementing FGFR-inhibitors as part of the standard of care in HNSCC patients with FGFR aberrations.

Acknowledgments

This project was supported in part by the Specialized Program of Research Excellence (SPORE) program, through the NIH National Institute for Dental and Craniofacial Research (NIDCR) and National Cancer Institute (NCI), grant P50DE026787. The content is solely the responsibility of the authors and does not necessarily represent the official views of the NIH.

References

[1] Bonner JA, Harari PM, Giralt J, Cohen RB, Jones CU, Sur RK, et al. Radiotherapy plus cetuximab for locoregionally advanced head and neck cancer: 5-year survival data from a phase 3 randomised trial, and relation between cetuximab-induced rash and survival. Lancet Oncol 2010;11(1):21–8.

[2] Vermorken JB, Mesia R, Rivera F, Remenar E, Kawecki A, Rottey S, et al. Platinum-based chemotherapy plus cetuximab in head and neck cancer. N Engl J Med 2008;359(11):1116–27.

[3] Luke JJ, Flaherty KT, Ribas A, Long GV. Targeted agents and immunotherapies: optimizing outcomes in melanoma. Nat Rev Clin Oncol 2017;14(8):463–82.

[4] Pilié PG, Tang C, Mills GB, Yap TA. State-of-the-art strategies for targeting the DNA damage response in cancer. Nat Rev Clin Oncol 2019;16(2):81–104.

[5] Recondo G, Facchinetti F, Olaussen KA, Besse B, Friboulet L. Making the first move in EGFR-driven or ALK-driven NSCLC: first-generation or next-generation TKI? Nat Rev Clin Oncol 2018;15(11):694–708.
[6] Cancer Genome Atlas Network. Comprehensive genomic characterization of head and neck squamous cell carcinomas. Nature 2015;517(7536):576–82.
[7] Seiwert TY, Zuo Z, Keck MK, Khattri A, Pedamallu CS, Stricker T, et al. Integrative and comparative genomic analysis of HPV-positive and HPV-negative head and neck squamous cell carcinomas. Clin Cancer Res 2015;21(3):632–41.
[8] Koole K, Brunen D, van Kempen PM, Noorlag R, de Bree R, Lieftink C, et al. FGFR1 is a potential prognostic biomarker and therapeutic target in head and neck squamous cell carcinoma. Clin Cancer Res 2016;22(15):3884–93.
[9] Marshall ME, Hinz TK, Kono SA, Singleton KR, Bichon B, Ware KE, et al. Fibroblast growth factor receptors are components of autocrine signaling networks in head and neck squamous cell carcinoma cells. Clin Cancer Res 2011;17(15):5016–25.
[10] Gopika SenthilKumar MM, Fisher M, Brennan S, Kaushik S, Able L, Harari PM, … Andrew MB, editors. Radiosensitization of head and neck cancer by FGFR inhibition. 110th annual meeting of the American Association for cancer research; Atlanta, GA; 2019.
[11] Haugsten EM, Wiedlocha A, Olsnes S, Wesche J. Roles of fibroblast growth factor receptors in carcinogenesis. Mol Cancer Res 2010;8(11):1439–52.
[12] Thisse B, Thisse C. Functions and regulations of fibroblast growth factor signaling during embryonic development. Dev Biol 2005;287(2):390–402.
[13] Wesche J, Haglund K, Haugsten EM. Fibroblast growth factors and their receptors in cancer. Biochem J 2011;437(2):199–213.
[14] Neilson KM, Friesel R. Ligand-independent activation of fibroblast growth factor receptors by point mutations in the extracellular, transmembrane, and kinase domains. J Biol Chem 1996;271(40):25049–57.
[15] Beenken A, Mohammadi M. The FGF family: biology, pathophysiology and therapy. Nat Rev Drug Discov 2009;8(3):235–53.
[16] Ornitz DM, Itoh N. The fibroblast growth factor signaling pathway. Wiley Interdiscip Rev Dev Biol 2015;4(3):215–66.
[17] Tiong KH, Mah LY, Leong CO. Functional roles of fibroblast growth factor receptors (FGFRs) signaling in human cancers. Apoptosis 2013;18(12):1447–68.
[18] Wang JF, Shen M, Fong GH, Hill DJ. A soluble fibroblast growth factor receptor is released from HL-60 promyelocytic leukemia cells: implications for paracrine growth control. Growth Factors 2000;17(3):203–14.
[19] Knights V, Cook SJ. De-regulated FGF receptors as therapeutic targets in cancer. Pharmacol Ther 2010;125(1):105–17.
[20] Johnson DE, Lu J, Chen H, Werner S, Williams LT. The human fibroblast growth factor receptor genes: a common structural arrangement underlies the mechanisms for generating receptor forms that differ in their third immunoglobulin domain. Mol Cell Biol 1991;11(9):4627–34.
[21] Dell KR, Williams LT. A novel form of fibroblast growth factor receptor 2. Alternative splicing of the third immunoglobulin-like domain confers ligand binding specificity. J Biol Chem 1992;267(29):21225–9.
[22] Chellaiah AT, McEwen DG, Werner S, Xu J, Ornitz DM. Fibroblast growth factor receptor (FGFR) 3. Alternative splicing in immunoglobulin-like domain III creates a receptor highly specific for acidic FGF/FGF-1. J Biol Chem 1994;269(15):11620–7.
[23] Werner S, Duan DS, de Vries C, Peters KG, Johnson DE, Williams LT. Differential splicing in the extracellular region of fibroblast growth factor receptor 1 generates receptor variants with different ligand-binding specificities. Mol Cell Biol 1992;12(1):82–8.
[24] Eswarakumar VP, Lax I, Schlessinger J. Cellular signaling by fibroblast growth factor receptors. Cytokine Growth Factor Rev 2005;16(2):139–49.
[25] Johnson DE, Williams LT. Structural and functional diversity in the FGF receptor multigene family. Adv Cancer Res 1993;60:1–41.
[26] Gong SG. Isoforms of receptors of fibroblast growth factors. J Cell Physiol 2014;229(12):1887–95.
[27] Harmer NJ, Ilag LL, Mulloy B, Pellegrini L, Robinson CV, Blundell TL. Towards a resolution of the stoichiometry of the fibroblast growth factor (FGF)-FGF receptor-heparin complex. J Mol Biol 2004;339(4):821–34.
[28] Porta R, Borea R, Coelho A, Khan S, Araujo A, Reclusa P, et al. FGFR a promising druggable target in cancer: molecular biology and new drugs. Crit Rev Oncol Hematol 2017;113:256–67.

[29] Gotoh N. Regulation of growth factor signaling by FRS2 family docking/scaffold adaptor proteins. Cancer Sci 2008;99(7):1319–25.

[30] Lew ED, Furdui CM, Anderson KS, Schlessinger J. The precise sequence of FGF receptor autophosphorylation is kinetically driven and is disrupted by oncogenic mutations. Sci Signal 2009;2(58):ra6.

[31] Peng WC, Lin X, Torres J. The strong dimerization of the transmembrane domain of the fibroblast growth factor receptor (FGFR) is modulated by C-terminal juxtamembrane residues. Protein Sci 2009;18(2):450–9.

[32] Chae YK, Ranganath K, Hammerman PS, Vaklavas C, Mohindra N, Kalyan A, et al. Inhibition of the fibroblast growth factor receptor (FGFR) pathway: the current landscape and barriers to clinical application. Oncotarget 2017;8(9):16052–74.

[33] Dienstmann R, Rodon J, Prat A, Perez-Garcia J, Adamo B, Felip E, et al. Genomic aberrations in the FGFR pathway: opportunities for targeted therapies in solid tumors. Ann Oncol 2014;25(3):552–63.

[34] Brooks AN, Kilgour E, Smith PD. Molecular pathways: fibroblast growth factor signaling: a new therapeutic opportunity in cancer. Clin Cancer Res 2012;18(7):1855–62.

[35] Turner N, Grose R. Fibroblast growth factor signalling: from development to cancer. Nat Rev Cancer 2010;10 (2):116–29.

[36] Touat M, Ileana E, Postel-Vinay S, Andre F, Soria JC. Targeting FGFR signaling in cancer. Clin Cancer Res 2015;21(12):2684–94.

[37] Manchado E, Weissmueller S, JPt M, Chen CC, Wullenkord R, Lujambio A, et al. A combinatorial strategy for treating KRAS-mutant lung cancer. Nature 2016;534(7609):647–51.

[38] Lieu C, Heymach J, Overman M, Tran H, Kopetz S. Beyond VEGF: inhibition of the fibroblast growth factor pathway and antiangiogenesis. Clin Cancer Res 2011;17(19):6130–9.

[39] Turkington RC, Longley DB, Allen WL, Stevenson L, McLaughlin K, Dunne PD, et al. Fibroblast growth factor receptor 4 (FGFR4): a targetable regulator of drug resistance in colorectal cancer. Cell Death Dis 2014;5:e1046.

[40] Freier K, Schwaenen C, Sticht C, Flechtenmacher C, Mühling J, Hofele C, et al. Recurrent FGFR1 amplification and high FGFR1 protein expression in oral squamous cell carcinoma (OSCC). Oral Oncol 2007;43(1):60–6.

[41] Hase T, Kawashiri S, Tanaka A, Nozaki S, Noguchi N, Kato K, et al. Correlation of basic fibroblast growth factor expression with the invasion and the prognosis of oral squamous cell carcinoma. J Oral Pathol Med 2006;35 (3):136–9.

[42] Wakulich C, Jackson-Boeters L, Daley TD, Wysocki GP. Immunohistochemical localization of growth factors fibroblast growth factor-1 and fibroblast growth factor-2 and receptors fibroblast growth factor receptor-2 and fibroblast growth factor receptor-3 in normal oral epithelium, epithelial dysplasias, and squamous cell carcinoma. Oral Surg Oral Med Oral Pathol Oral Radiol Endod 2002;93(5):573–9.

[43] Drugan CS, Paterson IC, Prime SS. Fibroblast growth factor receptor expression reflects cellular differentiation in human oral squamous carcinoma cell lines. Carcinogenesis 1998;19(6):1153–6.

[44] Koole K, Clausen MJ, van Es RJ, van Kempen PM, Melchers LJ, Koole R, et al. FGFR family members protein expression as prognostic markers in oral cavity and oropharyngeal squamous cell carcinoma. Mol Diagn Ther 2016;20(4):363–74.

[45] Starska K, Forma E, Lewy-Trenda I, Stasikowska-Kanicka O, Skóra M, Bryś M. Fibroblast growth factor receptor 1 and 3 expression is associated with regulatory PI3K/AKT kinase activity, as well as invasion and prognosis, in human laryngeal cancer. Cell Oncol (Dordr) 2018;41(3):253–68.

[46] Azad AK, Bairati I, Samson E, Cheng D, Mirshams M, Qiu X, et al. Validation of genetic sequence variants as prognostic factors in early-stage head and neck squamous cell cancer survival. Clin Cancer Res 2012;18 (1):196–206.

[47] Chou CH, Hsieh MJ, Chuang CY, Lin JT, Yeh CM, Tseng PY, et al. Functional FGFR4 Gly388Arg polymorphism contributes to oral squamous cell carcinoma susceptibility. Oncotarget 2017;8(56):96225–38.

[48] Göke F, Bode M, Franzen A, Kirsten R, Goltz D, Göke A, et al. Fibroblast growth factor receptor 1 amplification is a common event in squamous cell carcinoma of the head and neck. Mod Pathol 2013;26(10):1298–306.

[49] Nguyen PT, Tsunematsu T, Yanagisawa S, Kudo Y, Miyauchi M, Kamata N, et al. The FGFR1 inhibitor PD173074 induces mesenchymal-epithelial transition through the transcription factor AP-1. Br J Cancer 2013;109(8):2248–58.

[50] Göke F, Franzen A, Hinz TK, Marek LA, Yoon P, Sharma R, et al. FGFR1 expression levels predict BGJ398 sensitivity of FGFR1-dependent head and neck squamous cell cancers. Clin Cancer Res 2015; 21(19):4356–64.

[51] Singleton KR, Hinz TK, Kleczko EK, Marek LA, Kwak J, Harp T, et al. Kinome RNAi screens reveal synergistic targeting of MTOR and FGFR1 pathways for treatment of lung cancer and HNSCC. Cancer Res 2015;75 (20):4398–406.

[52] Singleton KR, Kim J, Hinz TK, Marek LA, Casás-Selves M, Hatheway C, et al. A receptor tyrosine kinase network composed of fibroblast growth factor receptors, epidermal growth factor receptor, v-erb-b2 erythroblastic leukemia viral oncogene homolog 2, and hepatocyte growth factor receptor drives growth and survival of head and neck squamous carcinoma cell lines. Mol Pharmacol 2013;83(4):882–93.

[53] Sweeny L, Liu Z, Lancaster W, Hart J, Hartman YE, Rosenthal EL. Inhibition of fibroblasts reduced head and neck cancer growth by targeting fibroblast growth factor receptor. Laryngoscope 2012;122(7):1539–44.

[54] Cancer Genome Atlas Network. Comprehensive genomic characterization of head and neck squamous cell carcinomas. Nature 2015;517(7536):576–82.

[55] Cancer Genome Atlas Research Network, Weinstein JN, Collisson EA, Mills GB, Shaw KR, Ozenberger BA, et al. The cancer genome atlas pan-cancer analysis project. Nat Genet 2013;45(10):1113–20.

[56] Young RJ, Lim AM, Angel C, Collins M, Deb S, Corry J, et al. Frequency of fibroblast growth factor receptor 1 gene amplification in oral tongue squamous cell carcinomas and associations with clinical features and patient outcome. Oral Oncol 2013;49(6):576–81.

[57] Weiss J, Sos ML, Seidel D, Peifer M, Zander T, Heuckmann JM, et al. Frequent and focal FGFR1 amplification associates with therapeutically tractable FGFR1 dependency in squamous cell lung cancer. Sci Transl Med 2010;2(62).

[58] Uhlen M, Zhang C, Lee S, Sjostedt E, Fagerberg L, Bidkhori G, et al. A pathology atlas of the human cancer transcriptome. Science 2017;357(6352). https://doi.org/10.1126/science.aan2507.

[59] Vairaktaris E, Ragos V, Yapijakis C, Derka S, Vassiliou S, Nkenke E, et al. FGFR-2 and -3 play an important role in initial stages of oral oncogenesis. Anticancer Res 2006;26(6B):4217–21.

[60] Nayak S, Goel MM, Makker A, Bhatia V, Chandra S, Kumar S, et al. Fibroblast growth factor (FGF-2) and its receptors FGFR-2 and FGFR-3 may be putative biomarkers of malignant transformation of potentially malignant oral lesions into oral squamous cell carcinoma. PLoS One 2015;10(10):e0138801.

[61] Brands RC, Knierim LM, De Donno F, Steinacker V, Hartmann S, Seher A, et al. Targeting VEGFR and FGFR in head and neck squamous cell carcinoma in vitro. Oncol Rep 2017;38(3):1877–85.

[62] Lin DC, Meng X, Hazawa M, Nagata Y, Varela AM, Xu L, et al. The genomic landscape of nasopharyngeal carcinoma. Nat Genet 2014;46(8):866–71.

[63] Koole K, van Kempen PM, Swartz JE, Peeters T, van Diest PJ, Koole R, et al. Fibroblast growth factor receptor 3 protein is overexpressed in oral and oropharyngeal squamous cell carcinoma. Cancer Med 2016;5(2):275–84.

[64] Cancer Genome Atlas Research Network. Comprehensive genomic characterization of squamous cell lung cancers. Nature 2012;489(7417):519–25.

[65] d'Avis PY, Robertson SC, Meyer AN, Bardwell WM, Webster MK, Donoghue DJ. Constitutive activation of fibroblast growth factor receptor 3 by mutations responsible for the lethal skeletal dysplasia thanatophoric dysplasia type I. Cell Growth Differ 1998;9(1):71–8.

[66] Tomlinson DC, Hurst CD, Knowles MA. Knockdown by shRNA identifies S249C mutant FGFR3 as a potential therapeutic target in bladder cancer. Oncogene 2007;26(40):5889–99.

[67] Del Piccolo N, Placone J, Hristova K. Effect of thanatophoric dysplasia type I mutations on FGFR3 dimerization. Biophys J 2015;108(2):272–8.

[68] Meric-Bernstam F, Frampton GM, Ferrer-Lozano J, Yelensky R, Perez-Fidalgo JA, Wang Y, et al. Concordance of genomic alterations between primary and recurrent breast cancer. Mol Cancer Ther 2014;13(5):1382–9.

[69] Herrera-Abreu MT, Pearson A, Campbell J, Shnyder SD, Knowles MA, Ashworth A, et al. Parallel RNA interference screens identify EGFR activation as an escape mechanism in FGFR3-mutant cancer. Cancer Discov 2013;3(9):1058–71.

[70] Sequist LV, Cassier P, Varga A, Tabernero J, Schellens JHM, Delord J-P, … Wolf J, editors. Phase I study of BGJ398, a selective pan-FGFR inhibitor in genetically preselected advanced solid tumors. Proceedings of the 105th annual meeting of the American Association for cancer research,San Diego, CA; 2014.

[71] Ansell A, Farnebo L, Grénman R, Roberg K, Thunell LK. Polymorphism of FGFR4 in cancer development and sensitivity to cisplatin and radiation in head and neck cancer. Oral Oncol 2009;45(1):23–9.

[72] Bange J, Prechtl D, Cheburkin Y, Specht K, Harbeck N, Schmitt M, et al. Cancer progression and tumor cell motility are associated with the FGFR4 Arg(388) allele. Cancer Res 2002;62(3):840–7.

[73] Wang J, Stockton DW, Ittmann M. The fibroblast growth factor receptor-4 Arg388 allele is associated with prostate cancer initiation and progression. Clin Cancer Res 2004;10(18 Pt. 1):6169–78.

[74] Morimoto Y, Ozaki T, Ouchida M, Umehara N, Ohata N, Yoshida A, et al. Single nucleotide polymorphism in fibroblast growth factor receptor 4 at codon 388 is associated with prognosis in high-grade soft tissue sarcoma. Cancer 2003;98(10):2245–50.

[75] Streit S, Bange J, Fichtner A, Ihrler S, Issing W, Ullrich A. Involvement of the FGFR4 Arg388 allele in head and neck squamous cell carcinoma. Int J Cancer 2004;111(2):213–7.

[76] Tanuma J, Izumo T, Hirano M, Oyazato Y, Hori F, Umemura E, et al. FGFR4 polymorphism, TP53 mutation, and their combinations are prognostic factors for oral squamous cell carcinoma. Oncol Rep 2010;23(3):739–44

[77] da Costa Andrade VC, Parise Jr O, Hors CP, de Melo Martins PC, Silva AP, Garicochea B. The fibroblast growth factor receptor 4 (FGFR4) Arg388 allele correlates with survival in head and neck squamous cell carcinoma. Exp Mol Pathol 2007;82(1):53–7.

[78] Farnebo L, Tiefenböck K, Ansell A, Thunell LK, Garvin S, Roberg K. Strong expression of survivin is associated with positive response to radiotherapy and improved overall survival in head and neck squamous cell carcinoma patients. Int J Cancer 2013;133(8):1994–2003.

[79] Dutra RL, de Carvalho MB, Dos Santos M, Mercante AM, Gazito D, de Cicco E, et al. FGFR4 profile as a prognostic marker in squamous cell carcinoma of the mouth and oropharynx. PLoS One 2012;7(11):e50747.

[80] Choi KY, Rho YS, Kwon KH, Chung EJ, Kim JH, Park IS, et al. ECRG1 and FGFR4 single nucleotide polymorphism as predictive factors for nodal metastasis in oral squamous cell carcinoma. Cancer Biomark 2012;12 (3):115–24.

[81] Shi S, Li X, You B, Shan Y, Cao X, You Y. High expression of FGFR4 enhances tumor growth and metastasis in nasopharyngeal carcinoma. J Cancer 2015;6(12):1245–54.

[82] Dellacono FR, Spiro J, Eisma R, Kreutzer D. Expression of basic fibroblast growth factor and its receptors by head and neck squamous carcinoma tumor and vascular endothelial cells. Am J Surg 1997;174(5):540–4.

[83] Janot F, el-Naggar AK, Morrison RS, Liu TJ, Taylor DL, Clayman GL. Expression of basic fibroblast growth factor in squamous cell carcinoma of the head and neck is associated with degree of histologic differentiation. Int J Cancer 1995;64(2):117–23.

[84] McDermott SC, Rodriguez-Ramirez C, McDermott SP, Wicha MS, Nör JE. FGFR signaling regulates resistance of head and neck cancer stem cells to cisplatin. Oncotarget 2018;9(38):25148–65.

[85] Gyanchandani R, Ortega Alves MV, Myers JN, Kim S. A proangiogenic signature is revealed in FGF-mediated bevacizumab-resistant head and neck squamous cell carcinoma. Mol Cancer Res 2013;11(12):1585–96.

[86] Ader I, Delmas C, Skuli N, Bonnet J, Schaeffer P, Bono F, et al. Preclinical evidence that SSR128129E—a novel small-molecule multi-fibroblast growth factor receptor blocker—radiosensitises human glioblastoma. Eur J Cancer 2014;50(13):2351–9.

[87] Uzawa K, Ishigami T, Fushimi K, Kawata T, Shinozuka K, Kasamatsu A, et al. Targeting fibroblast growth factor receptor 3 enhances radiosensitivity in human squamous cancer cells. Oncogene 2011;30(43):4447–52.

[88] Xu AM, Huang PH. Receptor tyrosine kinase coactivation networks in cancer. Cancer Res 2010;70(10):3857–60.

[89] Kim SM, Kim H, Yun MR, Kang HN, Pyo KH, Park HJ, et al. Activation of the Met kinase confers acquired drug resistance in FGFR-targeted lung cancer therapy. Oncogenesis 2016;5(7):e241

[90] Verstraete M, Debucquoy A, Gonnissen A, Dok R, Isebaert S, Devos E, et al. In vitro and in vivo evaluation of the radiosensitizing effect of a selective FGFR inhibitor (JNJ-42756493) for rectal cancer. BMC Cancer 2015;15:946.

[91] Skinner HD, Giri U, Yang L, Woo SH, Story MD, Pickering CR, et al. Proteomic profiling identifies PTK2/FAK as a driver of radioresistance in HPV-negative head and neck cancer. Clin Cancer Res 2016;22(18):4643–50.

[92] Margot Miller MF, Senthilkumar G, Kaushik S, Able L, Brennan S, Iyer G, Kimple R, Baschnagel AM, editors. Evaluation of the pan-FGFR inhibitor AZD4547 with radiation in non-small cell lung cancer. 110th annual meeting of the American Association for cancer research; Atlanta, GA; 2019.

[93] Haimovitz-Friedman A, Balaban N, McLoughlin M, Ehleiter D, Michaeli J, Vlodavsky I, et al. Protein kinase C mediates basic fibroblast growth factor protection of endothelial cells against radiation-induced apoptosis. Cancer Res 1994;54(10):2591–7.

[94] Fuks Z, Persaud RS, Alfieri A, McLoughlin M, Ehleiter D, Schwartz JL, et al. Basic fibroblast growth factor protects endothelial cells against radiation-induced programmed cell death in vitro and in vivo. Cancer Res 1994;54(10):2582–90.

[95] Houchen CW, George RJ, Sturmoski MA, Cohn SM. FGF-2 enhances intestinal stem cell survival and its expression is induced after radiation injury. Am J Phys 1999;276(1 Pt. 1):G249–58.
[96] Zhang L, Sun W, Wang J, Zhang M, Yang S, Tian Y, et al. Mitigation effect of an FGF-2 peptide on acute gastrointestinal syndrome after high-dose ionizing radiation. Int J Radiat Oncol Biol Phys 2010;77(1):261–8.
[97] Kim BH, Jung HW, Seo SH, Shin H, Kwon J, Suh JM. Synergistic actions of FGF2 and bone marrow transplantation mitigate radiation-induced intestinal injury. Cell Death Dis 2018;9(3):383.
[98] NCT01761747. Available from: https://clinicaltrials.gov.
[99] Hammerman P. Ponatinib for squamous cell lung and head and neck cancers. ClinicalTrials.gov.; 2019.
[100] Tanguy S. BGJ398 in treating patients with FGFR positive recurrent head and neck cancer. ClinicalTrials.gov.; 2019.
[101] Kim Y, Lee SJ, Park K, Lee S-H, Sun JM, Keam B, et al. Phase II trial of nintedanib in patients with recurrent or metastatic salivary gland cancer: a multicenter phase II study. J Clin Oncol 2017;34(15 suppl). https://doi.org/10.1200/JCO.2016.34.15_suppl.6090.
[102] ICH GCP—Clinical Trials Registry. A biomarker-driven, open label, single arm, multicentre phase II study of abemaciclib in patients with recurrent or metastatic head and neck squamous cell carcinoma who failed to platinum-based therapy. Seoul National University Hospital; 2019. Available from: https://ichgcp.net/clinical-trials-registry/NCT03356587.
[103] Joerger M, Cassier P, Penel N, Cathomas R, Richly H, Schostak M, et al. Rogaratinib treatment of patients with advanced urothelial carcinomas prescreened for tumor FGFR mRNA expression. J Clin Oncol 2018;36(6 suppl). https://doi.org/10.1200/JCO.2018.36.6_suppl.494.
[104] Makoto Tahara ST, Ocker M, Ince S, Nogai H, Zhang C, Sei F, Ellinghaus P, editors. Safety, tolerability, pharmacokinetics, and efficacy of pan-fibroblast growth factor receptor inhibitor rogaratinib in Japanese patients with refractory, locally advanced metastatic solid tumors selected with a novel, mRNA-based patient selection strategy. Proceedings of the AACR-NCI-EORTC International conference: molecular targets and cancer therapeutics, Philadelphia, PA; 2017 [Abstract].
[105] Paik PK, Shen R, Ferry D, Soria J-C, Mathewson A, Kilgour E, et al., editors. A phase 1b open-label multicenter study of AZD4547 in patients with advanced squamous cell lung cancers: preliminary antitumor activity and pharmacodynamic data. ASCO annual meeting; 2014.
[106] Saggese M, Arkenau H-T, Hollebecque A, Alastair M, Rose LC, Donal L, et al., editors. A phase 1 expansion cohort of the fibroblast growth factor receptor (FGFR) inhibitor AZD4547 in patients (pts) with advanced gastric (GC) and gastroesophageal (GOJ) cancer. ASCO annual meeting; 2014.
[107] Smyth EC, Turner NC, Peckitt C, Pearson A, Brown G, Chua S, Cunningham D, editors. Phase II multicenter proof of concept study of AZD4547 in FGFR amplified tumours. ASCO annual meeting; 2015.
[108] Nogova L, Sequist LV, Garcia JMP, Andre F, Delord J-P, Hidalgo M, et al. Evaluation of BGJ398, a fibroblast growth factor receptor 1-3 kinase inhibitor, in patients with advanced solid tumors harboring genetic alterations in fibroblast growth factor receptors: results of a global phase I, dose-escalation and dose-expansion study. J Clin Oncol 2017;35(2):157–65.
[109] Nogova L, Sequist LV, Cassier PA, Hidalgo M, Delord J-P, Schuler MH, et al. Targeting FGFR1-amplified lung squamous cell carcinoma with the selective pan-FGFR inhibitor BGJ398. J Clin Oncol 2014;32(15 suppl):8034.
[110] Nogova L, Sequist LV, Perez Garcia JM, Andre F, Delord JP, Hidalgo M, et al. Evaluation of BGJ398, a fibroblast growth factor receptor 1-3 Kinase inhibitor, in patients with advanced solid tumors harboring genetic alterations in fibroblast growth factor receptors: results of a global phase I, dose-escalation and dose-expansion study. J Clin Oncol 2017;35(2):157–65.
[111] Collin MP, Lobell M, Hubsch W, Brohm D, Schirok H, Jautelat R, et al. Discovery of rogaratinib (BAY 1163877): a pan-FGFR inhibitor. ChemMedChem 2018;13(5):437–45.
[112] Joerger M, Takahashi S, Sayehli CM, Slosarczyk S, Mendivil AN, Cassier PA, … Tahara M, editors. Phase I experience with rogaratinib in patients with head and neck cancer selected based on FGFR mRNA overexpression. European Society for Medical Oncology 2018 Congress; 2018.
[113] Galot R, Le Tourneau C, Guigay J, Licitra L, Tinhofer I, Kong A, et al. Personalized biomarker-based treatment strategy for patients with squamous cell carcinoma of the head and neck: EORTC position and approach. Ann Oncol 2018;29(12):2313–27.
[114] Boyle PLB. World cancer report 2008. Lyon, France: International Agency for Research on Cancer; 2008.
[115] Bourhis J, Le Maitre A, Baujat B, Audry H, Pignon JP, Meta-Analysis of Chemotherapy in Head NCCG, et al. Individual patients' data meta-analyses in head and neck cancer. Curr Opin Oncol 2007;19(3):188–94.

[116] Brockstein B, Haraf DJ, Rademaker AW, Kies MS, Stenson KM, Rosen F, et al. Patterns of failure, prognostic factors and survival in locoregionally advanced head and neck cancer treated with concomitant chemoradiotherapy: a 9-year, 337-patient, multi-institutional experience. Ann Oncol 2004;15(8):1179–86.

[117] Ferlay J, Steliarova-Foucher E, Lortet-Tieulent J, Rosso S, Coebergh JW, Comber H, et al. Cancer incidence and mortality patterns in Europe: estimates for 40 countries in 2012. Eur J Cancer 2013;49(6):1374–403.

CHAPTER

12

The MET receptor as a therapeutic target in head and neck squamous cell carcinomas

David M. Francis, Gopika Senthilkumar, Huaising C. Ko, Andrew M. Baschnagel

Department of Human Oncology, University of Wisconsin Carbone Cancer Center, University of Wisconsin School of Medicine and Public Health, University of Wisconsin, Madison, WI, United States

Abstract

Head and neck squamous cell carcinomas (HNSCC) are a heterogeneous group of tumors with mortality rates approaching 50%. Resistance to standard therapies has led to the investigation of novel molecular targets that drive cellular growth and proliferation. The mesenchymal-epithelial transition factor (MET) is a tyrosine kinase receptor that is overexpressed and dysregulated in HNSCC. MET activation leads to upregulation of downstream signaling through the PI3K/AKT and RAS/MAPK pathways, which in turn promote cellular metabolism, growth, and proliferation. In this chapter, we discuss the role of MET in signaling pathways that drive HNSCC, highlighting preclinical and emerging clinical investigations of MET inhibition as a means of enhancing antitumor treatment strategies.

Abbreviations

ALDH aldehyde dehydrogenase 1
ATP adenosine triphosphate
CSC cancer stem cells
EGF epidermal growth factor
EGFR epidermal growth factor receptor
FISH fluorescence in situ hybridization
HGF hepatocyte growth factor
HNSCC head and neck squamous cell carcinoma
HPV human papillomavirus
JAK/STAT3 janus kinase/signal transducer and activator of transcription
MAPK mitogen activated protein kinase

Improving the Therapeutic Ratio in Head and Neck Cancer
https://doi.org/10.1016/B978-0-12-817868-3.00012-3

MET	mesenchymal-epithelial transition factor
NF-κB	nuclear factor-kappa B
PI3K	phosphatidylinositol-3 kinase
TCGA	the cancer genome atlas
TGF-α	transforming growth factor α

Conflict of interest

No potential conflicts of interest were disclosed.

Introduction

Despite advances in surgery, chemotherapy, and radiation, locally advanced HNSCC represent a challenging disease to treat, with mortality rates approaching 50% [1, 2]. A strong rationale exists to identify specific biologic pathways that drive tumorigenesis and modulate resistance to standard therapy. Such aberrant nodes represent attractive targets that can be inhibited, thereby increasing the therapeutic efficacy while potentially mitigating treatment-related toxicity. Increasing our understanding of how unique genomic alterations augment signaling pathways can lead to the development of new therapies specifically tailored to individual tumors.

The most well-known dysregulated receptor tyrosine kinase in HNSCC is the epidermal growth factor receptor (EGFR). EGFR is a transmembrane protein that is activated by binding of extracellular ligand growth factors, including epidermal growth factor (EGF) and transforming growth factor α (TGF-α). EGFR has been found to be abnormally activated and overexpressed in up to 90% of HNSCC. Two large randomized trials have shown the benefit of targeting EGFR with cetuximab, an IgG1 monoclonal antibody against the ligand-binding domain of EGFR [3, 4]. One randomized study demonstrated an overall survival advantage of adding weekly cetuximab with definitive radiotherapy in patients with locally advanced HNSCC [3] and another study found a survival benefit of adding cetuximab to chemotherapy in first-line treatment of recurrent or metastatic HNSCC [4].

While EGFR directed therapy has shown clinical benefit in two randomized studies it has also failed to improve outcomes in multiple other studies [5, 6]. Outcomes of patients with HNSCC still remain poor and new therapies that target novel targets are desperately needed. The mesenchymal-epithelial transition factor (MET also known as c-MET) is a promising potential therapeutic target. It is highly overexpressed and dysregulated in HNSCC and has been found to be associated with poor prognosis and radiation resistance [7, 8].

Overview of the MET pathway

The MET tyrosine kinase receptor plays a key role in cellular proliferation, embryogenesis, tissue development, and wound healing [9, 10]. The MET heterodimer is comprised of a 45 kDa extracellular subunit connected to a 145 kDa transmembrane subunit via a disulfide link [11]. The transmembrane subunit contains an intracellular catalytic region that modulates kinase activity. The hepatocyte growth factor (HGF) also known as scatter factor, serves as the ligand for MET, and is widely secreted from mesenchymal cells throughout the extracellular matrix of most tissues as an inactive precursor (pro-HGF) [12]. HGF consists of six domains with a high affinity MET binding site in the α chain and a low affinity MET binding site in the β chain [13]. Several cytokines including tumor necrosis factor-α (TNF-α) and transforming growth factor-β (TGF-β) increase the transcription of both MET and HGF in response to tissue damage and inflammation [14].

Cleavage of pro-HGF by the urokinase plasminogen activator results in an activated heterodimer that binds to MET resulting in receptor dimerization and autophosphorylation of tyrosine residues Tyr1234 and Tyr1235 within the intracellular catalytic subunit. Subsequent phosphorylation of tyrosine residues Tyr1349 and Tyr1356 serve as docking sites for the recruitment of signaling molecules and transducers that mediate downstream signaling pathways such as: the mitogen activated protein kinase (MAPK) cascade, phosphatidylinositol-3 kinase (PI3K)/AKT axis, Janus kinase/signal transducer and activator of transcription (JAK/STAT3), and Nuclear factor-kappa B NF-κB complex [15–18]. Functional crosstalk between MET and other signaling pathways regulated by the EGFR and WNT–β-catenin further contribute to intracellular networking that initiates cellular transformation and differentiation [19]. Aberrant activation of the MET receptor in cancer triggers signaling through downstream pathways, including the PI3K/AKT and the RAS/MAPK pathways, which promote metastasis, invasion, angiogenesis, and proliferation.

Dysregulation of MET in HNSCC

Sustained proliferative signaling via increased receptor expression, upregulation of ligand production, or avoidance of feedback inhibition represents a unique biologic capability that is acquired during the process of human tumorigenesis [20]. The constitutive activation of the MET pathway through somatic mutation, copy number alterations, receptor overexpression, feedback augmentation, chromosomal rearrangements, or aberrant intracellular crosstalk has been shown to result in malignant transformation of multiple different cancers, including HNSCC [21, 22]. Malignant extracellular stimuli that are exacerbated in cancer cells populations such as increased hypoxia, release of inflammatory cytokines, or pro-angiogenic factors have been shown to transcriptionally activate MET resulting in enhanced cellular proliferation and tumor growth [23, 24].

In HNSCC, increased activation of the HGF/MET signaling pathway results from a variety of aberrations including overexpression of both MET and HGF proteins, amplification of the MET gene, and MET mutations. In addition, paracrine signaling, overexpression of ligand and autocrine loop formation can also lead to sustained MET activation. Overexpression of the MET protein from transcriptional upregulation in the absence of gene amplification is the most frequently observed MET alteration, being present in >80% of HNSCC tumors [7, 8, 25–29]. MET amplification, on the other hand, has been reported to be seen in approximately 13% of HNSCC [30]. Both autocrine and paracrine mechanisms can also trigger MET signaling. HNSCC tumor cells typically do not produce HGF. HGF is secreted by tumor-associated fibroblasts and is involved in paracrine activation [24]. HGF is overexpressed in about 50% of HNSCC associated extracellular stroma [24].

MET mutations are uncommon and occur in a limited number of HNSCCs [31, 32]. The Cancer Genome Atlas Network (TCGA) observed MET exon 14 skipping mutations in approximately 2% of HPV-negative tumors [32]. Exon 14 deletion mutations were also observed in a cohort of 10 HNSCC patients randomized to receive docetaxel with or without the EGFR-inhibitor gefitinib [31]. On this trial, MET mutations tended to predict poorer overall survival regardless of treatment received. In addition to cellular proliferation, MET mutations have been implicated in malignant invasion and metastasis. Di Renzo and colleagues identified MET oncogene somatic mutations in four patient lymph node metastases out of 15 HNSCC cases examined [33]. They identified two somatic mutations: the Y1230C in one patient and the Y1235D in three patient samples which lead to constitutive activation of MET kinase activity and were strongly associated with lymph node metastases as opposed to primary tumors, suggesting unique clonal selection during HNSCC progression and supporting a role of MET in driving HNSCC metastasis. Further preclinical analysis from Tao and colleagues demonstrated that knocking down MET in orthotopic HNSCC mouse models resulted in impaired activation of downstream MAPK signaling and reduced tumor dissemination to regional lymph nodes, supporting the notion of MET signaling in HNSCC locoregional spread [34]. Tan and colleagues discovered MET alterations in 11% of 66 squamous cell carcinomas arising from the oral tongue, second only to mutations in p53. Using somatic mutation profiling, they identified a specific MET N375S variant that had a poorer outcome than wild-type MET counterparts with median locoregional recurrence free survival rates of only 11 months versus 90 months in patients without MET N357S variant ($P = .008$) [35]. These data suggest a mechanism by which novel activating MET alterations drive rare molecular HNSCC subtypes that perform worse compared with non-MET associated HNSCC.

MET activation has also been thought to drive the development of cancer stem cells (CSC) in the tumorigenesis of some tumor types, including HNSCC. Lim and colleagues observed that treating HNSCC stem-like cells in vitro with HGF led to increased sphere-forming capacity and promoted the expression of stem cell markers Oct4, Sox2, and CD44 [36]. In cells that exhibited high levels of aldehyde dehydrogenase 1 (ALDH), a marker for HNSCC stem cells, transcriptional levels of MET were increased. Furthermore, knockdown of MET attenuated the cells' ability to form spheres, express stem cell markers, and even decreased their sensitivity to cisplatin chemotherapy via suppression of an ABCG2 transporter gene. These data were recapitulated by Sun and colleagues using mouse HNSCC xenograft models of varying MET expression levels. They observed that tumor cells with MET positivity exhibited high colony-forming ability in vitro, upregulated self-renewal pathways, and resistance to cisplatin [37]. These cells increased their expression of self-renewal pathways and were responsible for mediating metastasis compared with non-MET expressing HNSCC control cells. In a clinical cohort of HNSCC patient samples, Baschnagel et al., have shown high CD44 expression is associated with high MET expression and combined expression of these two markers predicts for poor prognosis in HNSCC patients treated with chemoradiation [38]. Taken together, these data indicate how MET may contribute to the development of a CSC-like phenotype that mediates resistance to traditional cytotoxic chemotherapy and drives both locoregional progression and distant failure in HNSCC.

The interaction and crosstalk of MET with other intracellular signaling pathways may also contribute to HNSCC tumorigenesis and potentiate therapeutic resistance to molecular targeted inhibition. Wheeler and colleagues initially identified upregulation of MET in addition with EGFR, HER2, and HER3 in HNSCC cell models that were highly resistant to cetuximab [39]. Using immunoprecipitation techniques, they observed preferential EGFR binding with HER2, HER3, and MET compared with cetuximab-sensitive parental control cells [39]. In lung cancer cell lines, focal amplification of the MET proto-oncogene has been associated with resistance to the EGFR inhibitor gefitinib by driving HER3 dependent

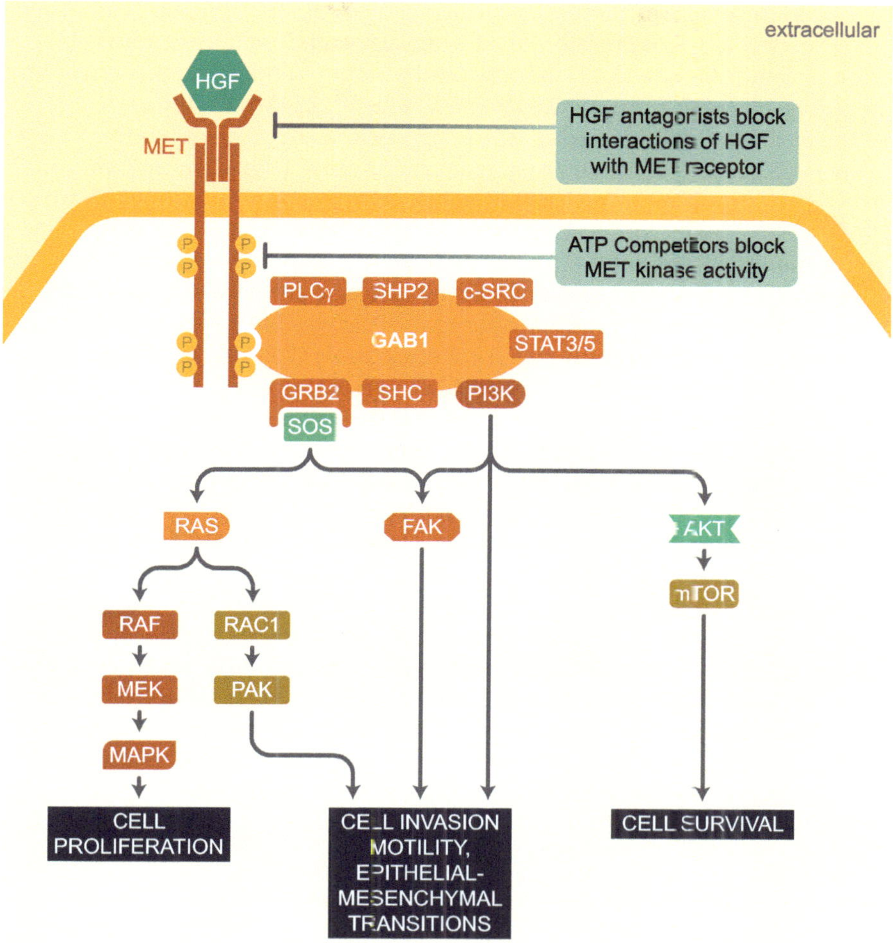

FIG. 1 MET signaling pathway. Activation the MET receptor by HGF ligand activates downstream pathways that promote and regulate tumor growth.

activation of the PI3K pathway [40]. Recent data from the Chicago Head and Neck Cancer Genomic Cohort elucidated the ability of HGF to restore viability in HNSCC tumor tissue treated initially with gefitinib and how such mediators impair the response of tumor cells to EGFR inhibition [41] (Fig. 1).

MET as prognostic marker for HNSCC

While genomic analyses of HNSCC suggest that the mutation rate of MET is relatively rare, pathologic MET receptor upregulation and HGF ligand overexpression in the extracellular stroma have been observed in the vast majority of HNSCC [22, 25, 30]. Numerous observations have correlated MET expression level with advanced primary tumor stage, involved lymph nodes, resistance to standard therapy, and poorer overall survival across several solid tumors, including HNSCC from a variety of subsites [8, 25–27, 42–45].

Some of the earliest data illustrating this prognostic association arose from Sawatsubashi and colleagues in 82 cases of primary laryngeal carcinoma. They demonstrated that MET expression in cancerous epithelium was significantly correlated with involved lymph node status and Ki-67 labeling index, suggesting that overexpression of MET corresponds to a growth advantage for malignant cells [44]. These data were recapitulated in a larger series from Baschnagel and colleagues who examined archival tissue from 107 HNSCC patients arising from the oral cavity, oropharynx, hypopharynx, or larynx who received definite radiation combined with concurrent platinum-based chemotherapy (78%) or cetuximab (22%). In their tissue microarray, 93% of patients expressed MET [25]. On univariate analysis, high MET expression predicted for worse locoregional control, distant metastasis, disease-free survival, and overall survival. On multivariate analysis, after adjustment for site, T-stage, tobacco history, and EGFR status, high MET expression predicted for worse disease-free survival ($P=.011$) [25]. This correlation existed in both EGFR-positive ($P=.032$) and -negative ($P=.008$) patients [25]. In p16-negative patients, those with high MET expression had worse disease-free survival than did those with low MET expression ($P=.036$) [25]. High MET expression did not correlate with outcomes in p16-positive patients. A multitude of other studies (Table 1) including at least four separate metaanalyses (Table 2) have confirmed overexpression of MET to be associated with higher clinical stage and worse prognosis [9, 46–48]. These four metaanalyses included up 2000 patients each treated with surgery, radiation or chemotherapy with MET expression determined mainly by immunohistochemistry (IHC). Consistently, MET has been shown to be associated with both worse overall survival and disease-free survival independent of treatment and disease site.

TABLE 1 Studies assessing the prognostic impact of MET on HNSCC

First author, year	Primary tumor	No. of patients	Copy number RNA expression protein expression	Biomarker technique	Outcome or prognostic impact
Fountzilas, 2009 [51]	oral, oropharynx, hypopharynx, larynx	37	protein and gene expression	IHC, FISH	There is no association between MET gene/protein expression and complete response following concomitant radiation treatment with cisplatin and cetuximab.
Choe, 2012 [42]	oral, oropharynx, hypopharynx, larynx	82	protein expression and gene amplification	IHC, FISH	Protein expression of MET was higher in oral and oropharynx tumors than nasopharynx and nasal cavity samples. Expression was also associated with lymph node metastasis but not well-differentiated histologic type.

TABLE 1 Studies assessing the prognostic impact of MET on HNSCC—Cont'd

First author, year	Primary tumor	No. of patients	Copy number RNA expression protein expression	Biomarker technique	Outcome or prognostic impact
Lacroix, 2014 [52]	oropharynx	143	mutation, amplification and protein overexpression	FISH, Sanger sequencing, IHC	There was no correlation between MET mutation, amplification, protein expression and survival.
Brauswetter, 2016 [43]	oral, oropharynx, hypopharynx, larynx	152	copy number	FISH	Increase in MET copy number was associated with decreased disease-specific survival and decreased levels of EGFR.
Sawatsubashi, 1998 [44]	larynx	82	protein expression	IHC	MET expression was associated with lymph node status and increased proliferation. There was no correlation between expression and histological type, T-category, distant metastasis, and clinical stage.
Aebersold, 2001 [26]	oropharynx	97	protein expression	IHC	MET expression was associated with poor rates of complete primary tumor remission and decreased local failure-free survival, disease-free survival, and overall survival.
Lo Muzio, 2004 [53]	oral	73	protein expression	IHC	MET expression was associated with higher lymph node metastases and patients with low/no MET expression had higher survival rates.
Endo, 2006 [54]	oral	99	protein expression	IHC	MET expression was associated with local and distant metastatic recurrence, decreased disease-free survival.
Kim, 2006 [55]	hypopharynx	40	protein expression	IHC	MET expression was associated with lymph node metastasis.
Lo Muzio, 2006 [27]	oral	84	protein expression	IHC	MET expression was associated with lower overall survival.

Continued

TABLE 1 Studies assessing the prognostic impact of MET on HNSCC—Cont'd

First author, year	Primary tumor	No. of patients	Copy number RNA expression protein expression	Biomarker technique	Outcome or prognostic impact
Kim, 2010 [7]	oral	61	protein expression	IHC	MET expression was associated with lymph node metastasis, tumor classification, recurrence, and decreased survival.
Zhao, 2011 [56]	oral	76	protein expression	IHC	High MET expression was correlated with higher clinical stage, expression of vascular endothelial growth factor-C, incidence of peritumoral lymphatic invasion, and positive lymph node status, and lower overall and disease-free survival.
Lim, 2012 [8]	oral	71	protein expression	IHC	MET overexpression was associated with neck metastasis and lower survival. Constitutive activation of MET was associated with migration and invasion in vitro and promoted tongue tumor growth in vitro and in vivo.
Baschnagel, 2014 [25]	oral, oropharynx, hypopharynx, larynx	107	protein expression	IHC	High MET expression predicted for worse prognosis overall In p16-negative patients MET was associated with reduced disease-free survival. No correlation between MET expression and survival in p16 positive patients. MET expression predicted worse outcomes regardless of EGFR status.
Madoz-Gurpide, 2015 [50]	oral, oropharynx, hypopharynx, larynx	57	protein expression, phosphorylation, gene status, mutations	IHC	Overexpression of MET in recurrent or metastatic tumors and MET phosphorylation was associated with lower progression-free and overall survival.

TABLE 1 Studies assessing the prognostic impact of MET on HNSCC—Cont'd

First author, year	Primary tumor	No. of patients	Copy number RNA expression protein expression	Biomarker technique	Outcome or prognostic impact
Chen, 2004 [57]	oral	90	protein expression	IHC	MET expression was associated with increased tumor staging, lymph node metastasis, and clinical staging
Freudlsperger, 2010 [58]	oral	211	protein expression	IHC	No correlation between MET expression and tumor size, lymph node status, distant metastasis, or survival rates. A negative correlation between MET expression and tumor stage was found.
Qian, 2016 [45]	oropharynx	78	protein expression	IHC	Overexpression of MET was associated with HPV+ status and tumor stage. In patients with T3/T4 disease, MET overexpression was correlated with HPV+ status and low p53 levels. Overexpression was also slightly associated with poor disease-free survival in HPV + patients.

TABLE 2 Metaanalyses assessing the prognostic impact of MET expression in HNSCC patients

First author, year	Number of studies	No. of patients	Biomarker technique	Outcome or prognostic impact
Szturz, 2017 [9]	28	2019	IHC	MET IHC score 3+ or above was associated with inferior survival outcomes and advanced disease.
Kim, 2017 [46]	16	1948	IHC	high MET expression is significantly associated with worse pathological features and prognosis
Vsiansky, 2018 [47]	17	1724	IHC	High levels of MET expression predict unfavorable prognosis associated with common clinicopathological features
Li, 2019 [48]	19	2419	IHC or IF	MET expression was associated with N classification and poorer prognosis

One reason for observed inferior outcomes in HNSCC overexpressing MET may be resistance to radiotherapy. Akervall and colleagues evaluated RNA expression using cDNA microarray analysis from 38 HNSCC patients who were treated with radiation [49]. They identified five biomarkers (VEGF, BCL-2, CLAUDIN-4, YAP-1 and MET) and subsequently tested these markers on a second cohort of 86 HNSCC patients who received radiation. Among these patients, MET staining was found to predict for incomplete (partial or no) response to therapy ($P < .01$) [49]. Furthermore, when YAP-1 and MET expression levels were combined the prediction of radio-resistance was increased, suggesting a synergistic relationship that may impact treatment decisions.

MET overexpression also appears to correlate with resistance to EGFR inhibition. A recent retrospective analysis was performed in 57 patients with recurrent or metastatic HNSCC who were treated with cetuximab [50]. Of these, MET overexpression was detected in 58% of samples. Univariate testing showed a significantly worse outcome in MET-overexpressing patients for progression free survival and overall survival. Moreover, p-MET expression was also significantly associated with a poor clinical outcome for overall survival. Patients with p-MET overexpression had worse prognosis (median progression free survival of 15 months; median overall survival of 18 months) compared with p-MET negative/low expression cases (median progression survival of 37 months; median overall survival of 48 months) [50]. Multivariate analysis confirmed the independent prognostic significance of p-MET for progression free survival overall survival. As such, these data implicate HGF/MET pathway activation with worse outcome in recurrent/metastatic HNSCC patients.

It should be noted that one challenge in cumulatively interpreting these data is the intrinsic heterogeneity across HNSCC subsites and the broad variety of methods that were used to detect and score MET expression. However, the consistent correlation of MET overexpression with unfavorable pathologic tumor characteristics and poor outcomes across multiple institutions and head and neck subsites highlights the therapeutic opportunity of targeting MET in developing novel anticancer strategies.

MET as therapeutic target

The abundance of data on aberrant MET signaling in HNSCC and its role in promoting both metastasis and resistance to standard therapies provides a strong rationale for identifying and developing novel agents that downregulate the MET signaling pathway as an anticancer strategy. Indeed, multiple preclinical studies have shown antitumor success through suppressing MET signaling with monoclonal antibodies or inhibiting the tyrosine kinase domain of the MET protein itself in HNSCC models (Table 3). MET inhibition has also been shown to result in synergistic or addictive effects with cytotoxic chemotherapy and other molecular targeted inhibitors.

One of the earliest series from Seiwert and colleagues evaluated the use of small-molecule MET inhibitors SU11274 and PF-2341066 (crizotinib) as a means of abrogating MET signaling in HNSCC cell models [30]. SU11274 is a first generation MET inhibitor that acts as an adenosine triphosphate (ATP)-competitive small molecule that inhibits MET catalytic activity, thereby inducing cell cycle arrest and eventual apoptosis. PF-2341066 (crizotinib) is a water-soluble, orally bioavailable, and ATP-competitive dual inhibitor of MET and anaplastic lymphoma kinase (ALK). Cell lines that were treated with SU11274 led to suppression of cell motility and migration despite HGF stimulation, compared with untreated control cells. Additionally, PF-2341066 (crizotinib) inhibited HGF-dependent MET phosphorylation in a dose-dependent manner in colony formation assays. Xenograft tumor models that were treated

TABLE 3 Preclinical studies of MET inhibitors in head and neck squamous cell carcinomas

First author	MET inhibitor	Cell line	Xenograft	Effect
Seiwert, 2009 [30]	SU11274/ PF-2341066 (crizotinib)	SCC9, SCC15, SCC178, SQ20B, SCC28, SCC61, HN31, MSK921, OSS3, SCC35, HN5	OSCC3, SCC35	Inhibition of MET using SU22274 decreased cell viability, MET signaling, and migration in vitro. Crizotinib treatment in vitro decreased colony formation and inhibited HGF-depended MET phosphorylation in a few of the cell lines. In vivo, it decreased angiogenesis and proliferation. In conjunction with cisplatin and erlotinib treatment, SU11274 reduced in vitro cell viability synergistically.
Knowles, 2009 [24]	SU22274/ PF-2341066 (crizotinib)	PCI-37A, UM-22B, 1483	UM-22B	MET inhibition using SU22274 blocked interleukin-8 release, HGF-induced signaling, and wound healing in vitro. Treatment with PF-2341066 in vitro also inhibited cell growth and in UM-22B, inhibited would healing. PF-2341066 was used in vivo, and it caused a 50% tumor growth inhibition, lowered proliferation, and increased apoptosis.
Liu, 2011 [64]	foretinib	SCC15, HN5	N/A	Inhibition of MET using foretinib and HER using erlotinib/lapatinib synergistically decreased cell growth and increased apoptosis.
Sen, 2011 [59]	PHA665752, PF02341066 (crizotinib)	Tu167, TR146, UMSCC14a, Osc19, Tu167R2, Tu167R1, Tu138	Osc-19	PHA665752 inhibited MET kinase activity and when used in conjunction with dasatinib (SFK inhibitor), it increased apoptosis. In vivo treatment with dasatinib and crizotinib resulted in decreased tumor size and nodal metastasis.
Xu, 2011 [61]	SU11274/ PF-2341066 (crizotinib)	PCI-15B, UM-22A, UM-22B, Cal-33, OSC-19	UM-22B	Both drugs inhibited cell growth in vivo but crizotinib had better effects. Crizotinib treatment in conjunction with gefitinib decreased cell proliferation, invasion, and wound healing. In vivo, the combination of drugs decreased tumor volume and increased apoptosis.
Stabile, 2013 [65]	PF04217903	686LN	686LN	The combination therapy with erlotinib decreased cell viability in vitro and decreased tumor volume and increased apoptosis in vivo.

Continued

TABLE 3 Preclinical studies of MET inhibitors in head and neck squamous cell carcinomas—Cont'd

First author	MET inhibitor	Cell line	Xenograft	Effect
Lim, 2014 [36]	SU11274	K3,K4,K5	none	Drug treatment reduced sphere formation and expression of Oct4 and Sox2 stem cell markers.
Sun, 2014 [66]	SU22274/ PF-2341066 (crizotinib)	HN-0-22, HN-0-29, HN-0-46	HN-0-22, HN-0-29, HN-0-46	Treatment with both drugs reduced sphere formation in vitro. In vivo treatment of crizotinib in conjunction with docetaxel/cisplatin decreased tumor growth in all the xenografts and decreased sphere formation and number of stem-like cells in HN022. Crizotinib with docetaxel treatment reduced secondary tumor formation. In HN-0-29 and HN-0-46, crizotinib treatment with docetaxel lowered metastasis and increased mice survival along with the crizotinib-only treatment.
Xi, 2015 [67]	Tivantinib	Cal27, Tca8113	none	Treatment with drug decreased cell growth and colony formation regardless of the inhibition of HGF/c-MET axis inhibition.
Baschnagel, 2015 [63]	PF-2341006 (crizotinib)	UT-SCC-5, UT-SCC-14, UT-SCC-15, U-SCC-16A	UT-SCC-14, UT-SCC-15	Crizotinib inhibited p-MET but not downstream proteins, decreased cell growth, and radiosensitized UT-SCC-16A but not UT-SCC-14 and UT-SCC-15. In vivo, there was a small reduction in tumor growth and no radiosensitization effects.

with PF-2341066 (crizotinib) exhibited marked angiogenesis and proliferation suppression compared with control vehicle-treated mice in vivo. Moreover, they observed synergistic inhibition of cell growth when treating cell lines with SU11274 and EGFR inhibitors cisplatin or erlotinib.

These results were recapitulated by Knowles et al. who also examined the use of SU11274 and PF-2341066 (crizotinib) in HNSCC models [24]. They confirmed that pretreatment inhibition via SU11274 completely prevented MET phosphorylation which coincided with subsequent loss of AKT and p44/p42 MAPK phosphorylation, all of which normally occurs in response to HGF ligand activation. This inhibition prevented both cellular migration as well as proliferation, both of which were confirmed using in vivo models treated with PF- 2341066 (crizotinib). Treatment of gastric carcinoma and glioblastoma murine xenograft models with PF-2341066 (crizotinib) have also resulted in tumor volume inhibition and increased numbers of apoptotic cells compared with mice that received vehicle control injection.

Sen and colleagues examined the use of PHA665752 in oral cavity cell lines [59]. PHA665752 is a second-generation small-molecule ATP-competitive MET inhibitor that shares an indolinone core structure with SU11274 and exhibits higher selectivity for MET than other tyrosine kinases [60]. They observed that blocking c-Src with dasatinib resulted in MET inhibition in sensitive HNSCC cell lines. This was confirmed using small interfering RNA to knockdown c-Src, which led to substantial inhibition of MET. Interestingly, MET was not inhibited in cell lines that were resistant to Src inhibition, suggesting that persistent MET activation might mediate such resistance. Both PHA665752 and dasatinib inhibited tumor cell invasion and migration, and the combination was more effective than the single agent treatment. Furthermore, they observed a trend toward decreased nodal metastases in mice treated with both crizotinib and dasatinib, compared with mice that received control or single agents alone.

MET inhibition has been shown to result in synergistic or additive effects with cytotoxic chemotherapy and other molecularly targeted inhibitors. Xu and colleagues examined the potential of combining MET inhibition with EGFR inhibition in HNSCC models [61]. They confirmed that treating HNSCC cells with crizotinib and gefitinib led to diminished cell proliferation, invasion, and wound healing compared with crizotinib or gefitinib alone. They achieved maximal inhibition of both downstream MAPK and AKT phosphorylation when treating HNSCC cells with crizotinib plus gefitinib followed by stimulation with HGF. Partial inhibition was observed in cells treated with either inhibitor alone. The synergistic in vitro effects were confirmed in HNSCC xenograft models. Mice that received gefitinib plus crizotinib for 5 days per week for 3 weeks experienced decreased tumor volume compared with mice that received vehicle control or crizotinib alone. The number of apoptotic cells in the combination group was significantly higher than the mean number of apoptotic cells in the crizotinib and gefitinib treatment groups. Recently, PF-04217903 was introduced as a novel ATP-competitive small-molecule inhibitor of MET kinase. It demonstrated >1000-fold selectivity for MET compared with >150 kinases, making it one of the most selective MET inhibitors described to date [62].

Inhibiting MET in combination with radiation

Baschnagel and colleagues examined the potential of enhancing the antitumor effect of radiation with crizotinib, a potent ATP-competitive inhibitor of the MET receptor, in preclinical HNSCC models [63]. Using clonogenic survival assays, they observed that certain HNSCC cells exposed to crizotinib prior to irradiation experienced in vitro radiosensitization, although this was not recapitulated across all cell lines. Additionally, they did not observe an additive inhibitory effect of crizotinib to radiation in HNSCC xenograft models, despite MET being overexpressed in these cell lines. Mice that received both crizotinib and radiation actually experienced increased tumor growth compared with mice that received radiation alone ($P = .07$). One reason postulated for the lack of radiosensitization observed in this model was activation of alternative pathways by other RTKs. As such, they recommended that future radiosensitizing studies focus on inhibiting combinations of RTKs, such as of EGFR along with MET, and inhibiting common downstream proteins such as PI3K/AKT/mTOR.

Clinical MET inhibition

To date, MET inhibition has yet to demonstrate any clinically significant survival advantage in patients with HNSCC. Seiwert and colleagues conducted a 2-stage Phase II trial evaluating the use of single-agent foretinib in patients with recurrent or metastatic HNSCC [68]. Foretinib (also known as GSK1363089 or XL880) is an oral multikinase ATP-competitive inhibitor primarily targeting MET and VEGF receptor-2 signaling pathways by binding in the adenosine triphosphate pocket. Foretinib also targeted additional RTKs including MET-related recepteur d'origine nantais (RON) receptor and AXL and TIE-2, both of which modulate malignant angiogenesis [70]. Of the 14 patients they enrolled in the first stage of their study, 50% (7/14) showed stable disease, 43% (6/14) experienced tumor shrinkage and two patients had prolonged disease stabilization for ≥13 months. Foretinib was administered as 240 mg orally for 5 consecutive days of a 14-day cycle. The most common adverse events were fatigue, constipation and hypertension, which were managed with additional medication or dosing adjustments. Unfortunately, the study did not meet criteria for continuing to the second stage.

TABLE 4 Ongoing clinical trials of MET inhibitors

First author, year	NCT #	Phase and status of trial	MET inhibitor	Study design	# of patients	Effect
Seiwert, 2013 [68]	NCT00725764	II, completed	Foretinib	Nonrandomized	14	50% (7/14) showed stable disease, 43% (6/14) experienced tumor shrinkage and two patients had prolonged disease stabilization for ≥13 months
Vokes, 2015 [69]	NCT01696955	II, completed	Tivantinib	Randomized to cetuximab plus tivantinib versus cetuximab alone	78	The response rate in the combination arm was 8.5% compared with 5.3% in the cetuximab alone arm (NS). The median progression free survival in both arms was 4 months and the median overall survival was 8 months (NS).
Novartis pharmaceuticals	NCT02205398	IB, terminated	Capmatinib	Nonrandomized	13	N/A
Eli Lilly	NCT01285037	I, completed	Merestinib	Single Group Assignment	190	N/A

Preliminary results from Vokes et al. were presented on patients treated with tivantinib (ARQ197), an oral small-molecule nonspecific MET antagonist [69]. They evaluated tivantinib in combination with cetuximab in 78 patients with recurrent or metastatic HNSCC and compared them with patients who received cetuximab alone. Tivantinib was administered orally at a dose of 360 mg twice daily and cetuximab was given every 2 weeks at a dose of 500 mg/m^2. The response rate in the combination arm was 8.5% compared with 5.3% in the cetuximab alone arm, which was not statistically significant. The response rate in HPV-positive oropharyngeal HNSCC patients was 0% in both arms. The response rate in HPV-negative pts was 12.5% in the combination group and 8.5% in the cetuximab alone group. The median progression free survival in both arms was 4 months and the median overall survival was 8 months. Both treatments were well tolerated with 12.5% of patients experiencing grade 3 leucopenia. Collectively, tivantinib given with cetuximab did not significantly improve the response rate and survival in this cohort of HNSCC patients.

Additional phase I trials (Table 4) are currently evaluating the use of novel small molecule inhibitors with a varying selectivity against MET. One Phase Ib trial (NCT02205398) is investigating the use of INC280 (capmatinib), an ATP-competitive antagonist of MET, in combination with cetuximab in MET positive metastatic HNSCC patients who already received cetuximab. Another Phase I study (NCT01285037) exploring the use of merestinib (LY2801653) with or without cetuximab in patients who failed platinum-based chemotherapy has completed recruitment.

Conclusion

MET is highly overexpressed in a majority of HNSCC and is associated with advanced disease and therapeutic resistance. High MET expression is also strongly linked with poor prognosis including worse overall survival and disease-free survival. Preclinical data utilizing MET inhibition in HNSCC models have shown promising additive and synergistic effects with other treatment strategies. Translation into meaningful clinical benefit has yet to be demonstrated. Identifying and selecting active agents that integrate effectively and safely into current treatment paradigms through well-designed randomized trials are crucial. MET inhibition represents a potential avenue for combination treatment protocols involving other molecular targeted agents in HNSCC.

Acknowledgments

This project was supported in part by the Specialized Program of Research Excellence (SPORE) program, through the NIH National Institute for Dental and Craniofacial Research (NIDCR) and National Cancer Institute (NCI), grant P50DE026787. The content is solely the responsibility of the authors and does not necessarily represent the official views of the NIH.

References

[1] Global Burden of Disease Cancer Collaboration, Fitzmaurice C, Allen C, Barber RM, Barregard L, Bhutta ZA, Brenner H, et al. Global, regional, and national cancer incidence, mortality, years of life lost, years lived with disability, and disability-adjusted life-years for 32 Cancer groups, 1990 to 2015: a systematic analysis for the global burden of disease study. JAMA Oncol 2017;3(4):524–48.

[2] Torre LA, et al. Global cancer statistics, 2012. CA Cancer J Clin 2015;65(2):87–108.
[3] Bonner JA, et al. Radiotherapy plus cetuximab for locoregionally advanced head and neck cancer: 5-year survival data from a phase 3 randomised trial, and relation between cetuximab-induced rash and survival. Lancet Oncol 2010;11(1):21–8.
[4] Vermorken JB, et al. Platinum-based chemotherapy plus cetuximab in head and neck cancer. N Engl J Med 2008;359(11):1116–27.
[5] Vermorken JB, Stöhlmacher-Williams J, Davidenko I, Licitra L, Winquist E, Villanueva C, Foa P, Rottey S, Skladowski K, Tahara M, Pai VR, Faivre S, Blajman CR, Forastiere AA, Stein BN, Oliner KS, Pan Z, Bach BA. Cisplatin and fluorouracil with or without panitumumab in patients with recurrent or metastatic squamous-cell carcinoma of the head and neck (SPECTRUM): an open-label phase 3 randomised trial. Lancet Oncol 2013;14(8):697–710.
[6] Gillison ML, et al. Radiotherapy plus cetuximab or cisplatin in human papillomavirus-positive oropharyngeal cancer (NRG oncology RTOG 1016): a randomised, multicentre, non-inferiority trial. Lancet 2019;393(10166):40–50.
[7] Kim CH, et al. C-Met expression as an indicator of survival outcome in patients with oral tongue carcinoma. Head Neck 2010;32(12):1655–64.
[8] Lim YC, et al. Overexpression of c-Met promotes invasion and metastasis of small oral tongue carcinoma. Oral Oncol 2012;48(11):1114–9.
[9] Szturz P, et al. Prognostic value of c-MET in head and neck cancer: a systematic review and meta-analysis of aggregate data. Oral Oncol 2017;74:68–76.
[10] Trusolino L, Bertotti A, Comoglio PM. MET signalling: principles and functions in development, organ regeneration and cancer. Nat Rev Mol Cell Biol 2010;11(12):834–48.
[11] Hartmann S, Bhola NE, Grandis JR. HGF/Met signaling in head and neck cancer: impact on the tumor microenvironment. Clin Cancer Res 2016;22(16):4005–13.
[12] Sonnenberg E, et al. Scatter factor hepatocyte growth-factor and its receptor, the C-Met tyrosine kinase, can mediate a signal exchange between mesenchyme and epithelia during mouse development. J Cell Biol 1993;123(1):223–35.
[13] Basilico C, et al. A high affinity hepatocyte growth factor-binding site in the immunoglobulin-like region of Met. J Biol Chem 2008;283(30):21267–77.
[14] Hoot KE, et al. HGF upregulation contributes to angiogenesis in mice with keratinocyte-specific Smad2 deletion. J Clin Investig 2010;120(10):3606–16.
[15] Ponzetto C, et al. A multifunctional docking site mediates signaling and transformation by the hepatocyte growth factor/scatter factor receptor family. Cell 1994;77(2):261–71.
[16] Johnson GL, Lapadat R. Mitogen-activated protein kinase pathways mediated by ERK, JNK, and p38 protein kinases. Science 2002;298(5600):1911–2.
[17] Zhang YW, et al. Requirement of Stat3 signaling for HGF/SF-Met mediated tumorigenesis. Oncogene 2002;21(2):217–26.
[18] Fan S, et al. Role of NF-kappaB signaling in hepatocyte growth factor/scatter factor-mediated cell protection. Oncogene 2005;24(10):1749–66.
[19] Gherardi E, et al. Targeting MET in cancer: rationale and progress. Nat Rev Cancer 2012;12(2):89–103.
[20] Hanahan D, Weinberg RA. Hallmarks of cancer: the next generation. Cell 2011;144(5):646–74.
[21] Szturz P, et al. Understanding c-MET signalling in squamous cell carcinoma of the head & neck. Crit Rev Oncol Hematol 2017;111:39–51.
[22] Peruzzi B, Bottaro DP. Targeting the c-Met signaling pathway in cancer. Clin Cancer Res 2006;12(12):3657–60.
[23] Pennacchietti S, et al. Hypoxia promotes invasive growth by transcriptional activation of the met protooncogene. Cancer Cell 2003;3(4):347–61.
[24] Knowles LM, et al. HGF and c-Met participate in paracrine tumorigenic pathways in head and neck squamous cell cancer. Clin Cancer Res 2009;15(11):3740–50.
[25] Baschnagel AM, et al. C-Met expression is a marker of poor prognosis in patients with locally advanced head and neck squamous cell carcinoma treated with chemoradiation. Int J Radiat Oncol Biol Phys 2014;88(3):701–7.
[26] Aebersold DM, et al. Involvement of the hepatocyte growth factor/scatter factor receptor c-met and of Bcl-xL in the resistance of oropharyngeal cancer to ionizing radiation. Int J Cancer 2001;96(1):41–54.
[27] Lo Muzio L, et al. Effect of c-Met expression on survival in head and neck squamous cell carcinoma. Tumour Biol 2006;27(3):115–21.

[28] Marshall DD, Kornberg LJ. Overexpression of scatter factor and its receptor (c-met) in oral squamous cell carcinoma. Laryngoscope 1998;108(9):1413–7.

[29] Kim YJ, et al. Immunohistochemical study identifying prognostic biomolecular markers in nasopharyngeal carcinoma treated by radiotherapy. Head Neck 2010;33(10):1458–66.

[30] Seiwert TY, et al. The MET receptor tyrosine kinase is a potential novel therapeutic target for head and neck squamous cell carcinoma. Cancer Res 2009;69(7):3021–31.

[31] Argiris A, et al. Phase III randomized, placebo-controlled trial of docetaxel with or without gefitinib in recurrent or metastatic head and neck cancer: an eastern cooperative oncology group trial. J Clin Oncol 2013;31(11):1405–14.

[32] Cancer Genome Atlas Network. Comprehensive genomic characterization of head and neck squamous cell carcinomas. Nature 2015;517(7536):576–82.

[33] Di Renzo MF, et al. Somatic mutations of the MET oncogene are selected during metastatic spread of human HNSC carcinomas. Oncogene 2000;19(12):1547–55.

[34] Tao X, Hill KS, Gaziova I, Sastry SK, Qui S, Szaniszlo P, et al. Silencing Met receptor tyrosine kinase signaling decreased oral tumor growth and increased survival of nude mice. Oral Oncol 2014;50:104–12.

[35] Tan DS, et al. Tongue carcinoma infrequently harbor common actionable genetic alterations. BMC Cancer 2014;14:679.

[36] Lim YC, Kang HJ, Moon JH. C-Met pathway promotes self-renewal and tumorigenecity of head and neck squamous cell carcinoma stem-like cell. Oral Oncol 2014;50(7):633–9.

[37] Sun S, Wang Z. Head neck squamous cell carcinoma c-Met(+) cells display cancer stem cell properties and are responsible for cisplatin-resistance and metastasis. Int J Cancer 2011;129(10) 2337–48.

[38] Baschnagel AM, et al. Combined CD44, c-MET, and EGFR expression in p16-positive and p16-negative head and neck squamous cell carcinomas. J Oral Pathol Med 2017;46(3):208–13.

[39] Wheeler DL, et al. Mechanisms of acquired resistance to cetuximab: role of HER (ErbB) family members. Oncogene 2008;27(28):3944–56.

[40] Engelman JA, et al. MET amplification leads to gefitinib resistance in lung cancer by activating ERBB3 signaling. Science 2007;316(5827):1039–43.

[41] Tepper SR, et al. Growth factor expression mediates resistance to EGFR inhibitors in head and neck squamous cell carcinomas. Oral Oncol 2016;56:62–70.

[42] Choe JY, et al. Expression of c-Met is different along the location and associated with lymph node metastasis of head and neck carcinoma. Korean J Pathol 2012;46(6):515–22.

[43] Brauswetter D, Dános K, Gurbi B, Félegyházi ÉF, Birtalan E, Meggyesházi N, Krenács T, Tamás L, Peták I. Copy number gain of PIK3CA and MET is associated with poor prognosis in head and neck squamous cell carcinoma. Virchows Arch 2016;468(5):579–87.

[44] Sawatsubashi M, et al. Expression of c-Met in laryngeal carcinoma. Virchows Arch 1998;432(4):331–5.

[45] Qian GQ, et al. Human papillomavirus oncoprotein E6 upregulates c-Met through p53 downregulation. Eur J Cancer 2016;65:21–32.

[46] Kim JH, Kim BJ, Kim HS. Clinicopathological impacts of high c-Met expression in head and neck squamous cell carcinoma: a meta-analysis and review. Oncotarget 2017;8(68):113120–8.

[47] Vsiansky V, et al. Prognostic role of c-Met in head and neck squamous cell cancer tissues: a meta-analysis. Sci Rep 2018;8(1).

[48] Li L, et al. Role of c-Met expression on prognosis of head and neck cancer: a literature review and meta-analysis. Head Neck 2019;41:1999–2006.

[49] Akervall J, et al. A novel panel of biomarkers predicts radioresistance in patients with squamous cell carcinoma of the head and neck. Eur J Cancer 2014;50(3):570–81.

[50] Juan Madoz-Gúrpide SZ, Chamizo C, Casado V, Caramés C, Gavín E, Cristóbal I, García-Foncillas J, Rojo F. Activation of MET pathway predicts poor outcome to cetuximab in patients with recurrent or metastatic head and neck cancer. J Transl Med 2015;13(282):1–13.

[51] Fountzilas G, et al. MMP9 but not EGFR, MET, ERCC1, P16, and P-53 is associated with response to concomitant radiotherapy, cetuximab, and weekly cisplatin in patients with locally advanced head and neck cancer. J Oncol 2009;2009:1–17.

[52] Lacroix L, et al. MET genetic abnormalities unreliable for patient selection for therapeutic intervention in oropharyngeal squamous cell carcinoma. Plos One 2014;9(1):1–8.

[53] Lo Muzio L, Leonardi R, Mignogna MD, Pannone G, Rubini C, Pieramici T, Trevisiol L, Ferrari F, Serpico R, Testa N, De Rosa G, Staibano S. Scatter factor receptor (c-Met) as possible prognostic factor in patients with oral squamous cell carcinoma. Anticancer Res 2004;24:1063–9.
[54] Endo K, et al. Prognostic value of cell motility activation factors in patients with tongue squamous cell carcinoma. Hum Pathol 2006;37(8):1111–6.
[55] Kim CH, et al. Expression of hepatocyte growth factor and c-Met in hypopharyngeal squamous cell carcinoma. Acta Otolaryngol 2006;126(1):88–94.
[56] Zhao D, Wang SH, Feng Y, Hua CG, Zhao J, Tang XF. Intratumoral c-Met expression is associated with vascular endothelial growth factor C expression, lymphangiogenesis, and lymph node metastasis in oral squamous cell carcinoma: implications for use as a prognostic marker. Hum Pathol 2011;10:1514–23.
[57] Chen YS, Wang JT, Chang YF, Liu BY, Wang YP, Sun A, Chiang CP. Expression of hepatocyte growth factor and c-met protein is significantly associated with the progression of oral squamous cell carcinoma in Taiwan. J Oral Pathol Med 2004;33(4):209–17.
[58] Freudlsperger C, et al. Prognostic value of c-Met expression in oral squamous cell carcinoma. Exp Ther Med 2010;1(1):69–72.
[59] Sen B, et al. Distinct interactions between c-Src and c-Met in mediating resistance to c-Src inhibition in head and neck cancer. Clin Cancer Res 2011;17(3):514–24.
[60] Christensen JG, et al. A selective small molecule inhibitor of c-Met kinase inhibits c-Met-dependent phenotypes in vitro and exhibits cytoreductive antitumor activity in vivo. Cancer Res 2003;63(21):7345–55.
[61] Xu H, et al. Dual blockade of EGFR and c-Met abrogates redundant signaling and proliferation in head and neck carcinoma cells. Clin Cancer Res 2011;17(13):4425–38.
[62] Zou HY, et al. Sensitivity of selected human tumor models to PF-04217903, a novel selective c-Met kinase inhibitor. Mol Cancer Ther 2012;11(4):1036–47.
[63] Baschnagel AM, et al. Crizotinib fails to enhance the effect of radiation in head and neck squamous cell carcinoma xenografts. Anticancer Res 2015;35(11):5973–82.
[64] Liu L, et al. Synergistic effects of foretinib with HER-targeted agents in MET and HER1- or HER2-coactivated tumor cells. Mol Cancer Ther 2011;10(3):518–30.
[65] Stabile LP, et al. C-Src activation mediates erlotinib resistance in head and neck cancer by stimulating c-Met. Clin Cancer Res 2013;19(2):380–92.
[66] Sun S, et al. Targeting the c-Met/FZD8 signaling axis eliminates patient-derived cancer stem-like cells in head and neck squamous carcinomas. Cancer Res 2014;74(24):7546–59.
[67] Xi WH, et al. Tivantinib (ARQ-197) exhibits anti-tumor activity with down-regulation of FAK in oral squamous cell carcinoma. Biochem Biophys Res Commun 2015;457(4):723–9.
[68] Seiwert T, et al. Phase II trial of single-agent foretinib (GSK1363089) in patients with recurrent or metastatic squamous cell carcinoma of the head and neck. Invest New Drugs 2013;31(2):417–24.
[69] Vokes EE, et al. A randomized phase II trial of the MET inhibitor tivantinib plus cetuximab versus cetuximab alone in patients with recurrent/metastatic head and neck cancer. J Clin Oncol 2015;33(15).
[70] Eder JP, et al. A phase I study of foretinib, a multi-targeted inhibitor of c-Met and vascular endothelial growth factor receptor 2. Clin Cancer Res 2010;16(13):3507–16.

C H A P T E R

13

Autophagy in head and neck cancer therapy

Yong-Syu Lee[a], *Jaimee C. Eckers*[b], *Randall J. Kimple*[c]

[a]Department of Human Oncology, University of Wisconsin School of Medicine and Public Health, University of Wisconsin, Madison, WI, United States [b]Department of Chemical & Biological Sciences, Rockford University, Rockford, IL, United States [c]Department of Human Oncology, University of Wisconsin Carbone Cancer Center, University of Wisconsin School of Medicine and Public Health, University of Wisconsin, Madison, WI, United States

Abstract

Autophagy is an evolutionarily conserved cellular process that functions in the degradation and recycling of damaged cellular components in response to cell stress. Autophagy plays context-dependent roles in tumorigenesis. Multiple anti-cancer therapies can lead to activation of autophagy within the tumor including chemotherapy and radiation therapy. This chapter will review the mechanisms by which these treatments activate autophagy, available drugs which can be used to modulate autophagy, and how modulation of autophagy can aid in overcoming resistance to treatments commonly used in patients with head and neck cancer.

Abbreviations

3-MA	3-methyladenine
Akt.	protein kinase B
AMBRA1	autophagy and Beclin 1 regulator 1
AMPK	AMP-activated protein kinase
ANE	areca nut extract
ATG	autophagy-related gene
Bcl-2	B-cell lymphoma 2
Bcl-xL	B-cell lymphoma-extra large
Bif-1	Bax interacting factor 1
CMA	chaperone-mediated autophagy
CQ	chloroquine
CRT	chemoradiation
DAPK	death-associated protein kinase
EGFR	epidermal growth factor receptor
ESCRT	the endosomal sorting complexes required for transport machinery

https://doi.org/10.1016/B978-0-12-817868-3.00013-5

FIP200	FAK family-interacting protein of 200 kDa
HCQ	hydroxychloroquine
HFK	human foreskin keratinocyte
HNC	head and neck cancer
HNSCC	head and neck squamous cell carcinoma
HO-1	heme oxygenase-1
HPV	human papillomavirus
Hsp70	heat shock protein 70
IP_3R	inositol 1,4,5-triphosphate receptor
JNK	c-JUN N-terminal kinase
LAMP2A	lysosome-associated membrane protein 2A
LC3	microtubule-associated protein 1A/1B-light chain 3
MOMP	mitochondrial outer membrane permeabilization
mTOR	mechanistic target of rapamycin
NOX4	NADPH oxidase 4
NSCLC	non-small cell lung cancer
PCa	prostate cancer cell
PDX	patient-derived xenograft
PE	phosphatidylethanolamine
PI3K	phosphoinositide 3-kinase
PIK3C3	phosphatidylinositol 3-kinase catalytic subunit type 3
PINK1	PTEN-induced kinase 1
pRB	retinoblastoma protein
Rab7	Ras-related protein 7
RB1CC1	RB1 coiled-coil 1
ROS	reactive oxygen species
RT	radiotherapy
SNARE	soluble NSF attachment protein receptor
SQSTM1	aka Sequestosome 1
Ulk	Unc-51 like autophagy activating kinase
UVRAG	UV radiation resistance-associated gene protein
Vps34	vacuolar protein sorting-associated protein 34

Conflict of interest

No potential conflicts of interest were disclosed.

Autophagy is an evolutionary mechanism that delivers proteins, organelles or other components to autolysosomes/lysosomes for degradation and recycling in response to cellular stresses or diverse signals (Fig. 1A) [1]. The process of autophagy is composed of several sequential steps, from sequestration to degradation, and each step is strictly regulated by multiple genes [2]. The feature of multi-step regulation involves autophagy in various physiological and pathological phenotypes in cells [3], such as anti-aging [4], developmental processes [5], cell death, starvation adaption [6], oncogenesis, and other diverse diseases [7].

Types of autophagy and their functions

Three different types of autophagy have been identified—macroautophagy, microautophagy and chaperone-mediated autophagy (CMA) [8]. Macroautophagy is the most well-understood type of autophagy and commonly abbreviated as the term "autophagy" or "bulk autophagy" (Hereafter, we will refer to it as "autophagy" and focus on the discussion

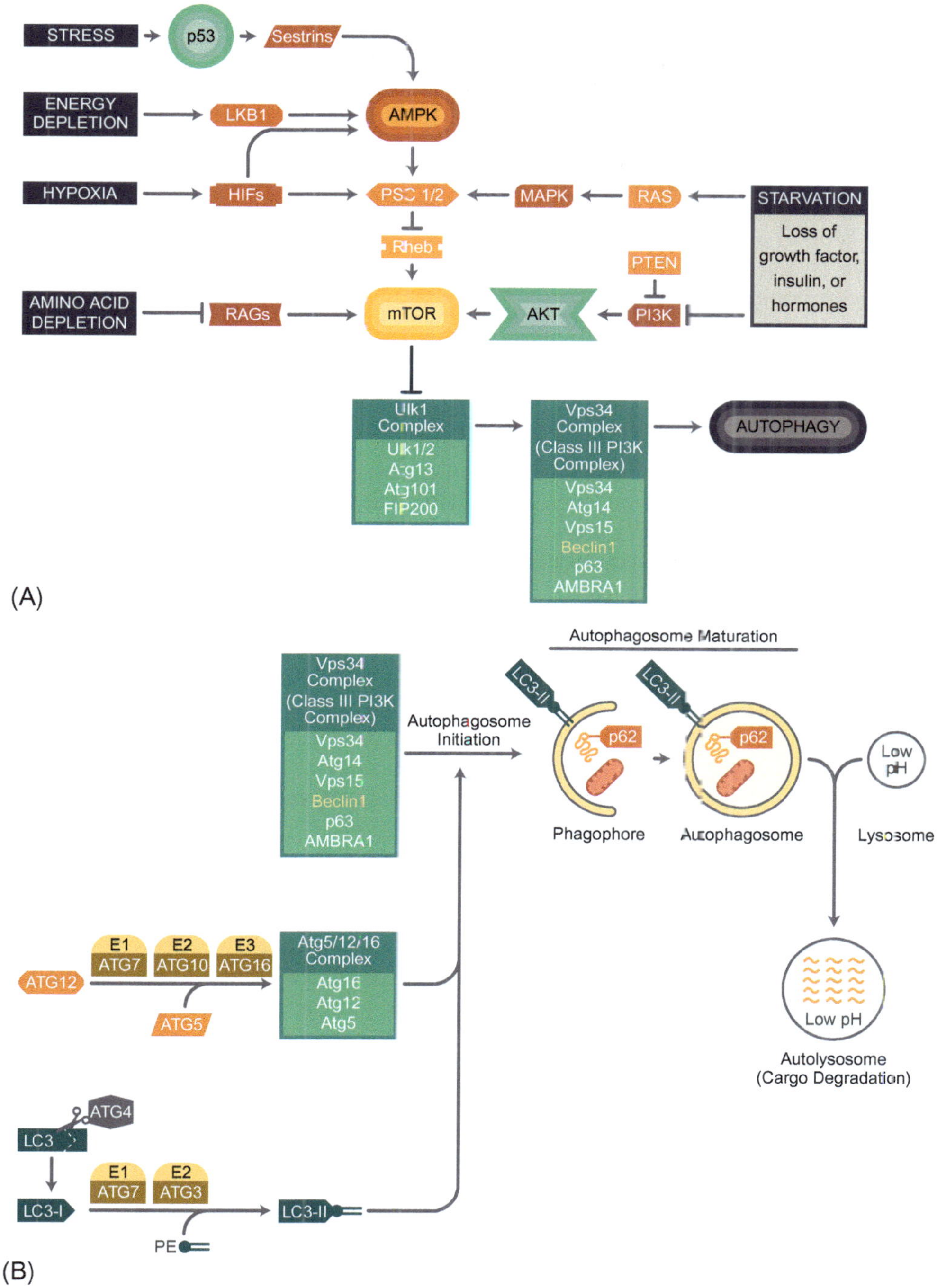

FIG. 1 Causes of autophagy and the autophagy signaling pathway. (A) Various cellular stresses (starvation, hypoxia, amino acid depletion, energy depletion or stress) mediate the AMPK-mTOR pathway to induce autophagy. mTOR is the "switch" for autophagy activation. Upon mTOR inhibition, the Ulk1 complex becomes activated and promotes activation of the Vps34 complex. (B) The Vps34 complex mediates autophagosome maturation, together with the Atg5/12/16 and LC3II complex. The autophagosome can engulf cargo that has been labeled by p62-labeled (e.g., proteins or organelles) during its maturation. Fusion of the autophagosome with the lysosome leads to degradation of the autophagosome contents.

of this type of autophagy in this chapter). It involves the engulfment of organelles and proteins through double membrane vesicles which are known as autophagosomes. These autophagosomes eventually fuse with lysosomes to form autolysosomes in order to degrade the organelles or proteins and recycle free amino acids for metabolic use [9]. However, this sequestration step in autophagy is non-selective relative to proteasome-mediated degradation, and excessive destruction of cellular components may lead to cell death [6].

Another type of autophagy, microautophagy, has been categorized into two subtypes—non-selective and selective. Non-selective microautophagy refers to the direct engulfment of cellular components or cytoplasm by tubular membrane invagination into lysosomes. By contrast, selective microautophagy directly targets specific organelles via lysosomes, such as macropexophagy (peroxisomes) and micromitophagy (mitochondria). Research has associated microautophagy with several neurodegenerative diseases but a direct link to cancer has yet to be discovered [10].

The third type of autophagy is chaperone-mediated autophagy (CMA). CMA involves a process in which specific proteins, containing an exposed amino acid motif (KFERQ), are directly recognized by the heat shock protein 70 (Hsp70) and then delivered to lysosomes by the lysosome-associated membrane protein 2A (LAMP2A) [11]. Notably, there is evidence indicating that CMA is associated with tumor growth and progression [12], making it a potential target for clinical therapy in the future.

Signaling pathways of autophagy

Autophagy-related genes (Atgs) are evolutionarily conserved for their highly precise regulation of autophagy [13]. Atgs are involved in each stage of autophagy—initiation, nucleation of autophagosome, elongation of the autophagosome membrane, fusion with the lysosome, and cargo degradation [14].

Both nutrient starvation and the AMP-activated protein kinase (AMPK) signaling pathway are able to initiate autophagy by means of the activation of the Unc-51 like autophagy activating kinase 1 (Ulk1) complex including Ulk1/Atg1, Ulk2, Atg13, Atg101, and FAK family-interacting protein of 200 kDa (FIP200, also known as RB1 coiled-coil 1, RB1CC1) (Fig. 1B) [13]. This Ulk1 complex can then further activate the class III phosphoinositide 3-kinase (PI3K) complex consisting of vacuolar protein sorting-associated protein 34 (Vps34, also known as phosphatidylinositol 3-kinase catalytic subunit type 3, PIK3C3), Beclin-1 (Atg6), Atg14 (BARKOR), Vps15 (p150), p63 (also known as UV radiation resistance-associated gene protein, UVRAG), and autophagy and Beclin 1 regulator 1 (AMBRA1). Beclin-1 serves as a scaffold protein to gather all complex members together in order to interact with each other [15]. The interaction between Beclin-1 and the lipid kinase Vps34 plays an essential role in autophagosome formation. The repression of autophagosome formation has been observed by means of the inhibition of Vps34 activity through administration of wortmannin [16] or 3-methyladenine (3-MA) [17], and the inhibition of Beclin-1 or Vps34 phosphorylation through Ulk1 inhibitor, SBI-0206965 [18]. Another important autophagy signaling regulator, mechanistic target of rapamycin (mTOR), which is highly activated in HPV-positive HNC and related to cell survival, can directly bind to ULK1 for phosphorylation, thereby reducing the kinase activity of phosphorylated ULK1 and inhibiting autophagy

[19]. Targeting the mTOR signaling pathway has been considered as one of the most important anti-cancer strategies in HNC therapy. In addition, evidence indicates that the binding of anti-apoptotic proteins B-cell lymphoma 2 (Bcl-2) or B-cell lymphoma-extra large (Bcl-xL) to Beclin-1 leads to the inhibition of autophagy [20]. Inositol 1,4,5-triphosphate receptor (IP_3R) also has the ability to interact with Beclin-1 to contribute to the repression of autophagy, which could be compromised by the IP_3R antagonist, xestospongin B [21].

The expansion of the autophagosome membrane requires two ubiquitin-like conjugation systems (Fig. 1B) [13]. The first conjugation, Atg5-Atg12 complex, begins with the activation of Atg12 by Atg7 followed by Atg10 replacing Atg7 for further activation. Atg12 is eventually conjugated with the internal lysine residue of Atg5 through its C-terminal glycine, and linked to Atg16 as the Atg5-Atg12-Atg16 complex for autophagosome membrane elongation, which dissociates from the completed autophagosome membrane [22]. Another conjugation involves microtubule-associated protein 1A/1B-light chain 3 (LC3, also known as Atg8) and phosphatidylethanolamine (PE). In order to assist in autophagosome membrane elongation, the precursor form LC3 is cleaved by the protease Atg4 [23] and forms LC3-I. LC3-I is subsequently conjugated with PE to form the LC3-I-PE complex, also known as LC3-II, a process which is mediated by Atg7 and Atg3. In comparison to the Atg5-Atg12-Atg16 complex, LC3-II not only helps the expansion of the autophagosome membrane but also stays on the completed membrane until it is degraded in the autolysosome, which allows it to be used as a marker of autophagy [24]. Interestingly, a cross-talk between the two conjugation systems has been demonstrated [25, 26] and the blockage of each conjugation system results in the inhibition of autophagosome membrane formation [27].

During the maturation of the autophagosome, the elongating autophagosome membrane fuses with endosomes in order to engulf specific cellular components with p62 (aka Sequestosome 1, SQSTM1). p62 non-covalently interacts with ubiquitin and mediates the transfer of the cargo to autophagosomes [28]. Ultimately, the autophagosomes fuse with lysosomes and create autolysosomes. There are several required proteins involved in mediating the fusion process as well as providing GTPase activity, such as the endosomal sorting complexes required for the transport (ESCRT) machinery, soluble NSF attachment protein receptors (SNAREs), Ras-related protein 7 (Rab7), and the class C Vps proteins. Loss of these required proteins negatively affects the autophagosome maturation [29–31]. LC3-II and p62 are eventually degraded in the autolysosome; a process that has been commonly used to measure the autophagic flux [24].

Autophagy and apoptosis

Autophagy and apoptosis regulate the turnover of cellular components by self-eating and self-killing, respectively [32]. The interplay between these two mechanisms is intricate. They share some regulators including p53, BH3-only proteins (BAD, BID, etc.), death-associated protein kinase (DAPK), and c-JUN N-terminal kinase (JNK) which activate or inhibit the processes [33].

In response to nutrient deficiency or DNA damage, autophagy is commonly induced in cells to enhance survival. Upon detection of cellular stress, some p53 proteins enter into the nucleus and turn on pro-autophagic genes, including *AMPK*, in order to induce autophagy [34]. Beclin-1 is also released from Bcl-2 by the interference of BH3-only proteins (BAD, BID, PUMA, NOXA) to initiate autophagosome formation [35]. These BH-3 only proteins are also associated with apoptosis [36]. DAPK and JNK target Beclin-1 and Bcl-2, respectively, in response to cellular stress leading to the release of Beclin-1 from Bcl-2 and induction of autophagy [37, 38].

In cancer cells, autophagy provides the ability to evade apoptosis through degradation of proapoptotic proteins and inactivation of the mitochondrial outer membrane permeabilization (MOMP). Evidence has shown that inhibition of autophagy enhances caspase 8 activity in colon cancer cells [39]. Mitophagy is induced by Parkin and PTEN-induced kinase 1 (PINK1) to degrade damaged mitochondria so that cells are able to avoid apoptosis induction [40]. Conversely, apoptosis is also able to inhibit autophagy. Bax-mediated cleavage of Beclin-1 leads to the impaired interaction between Beclin-1 and Vps34, which inhibits autophagy [41].

Role of autophagy in human cancer

The association of autophagy with cancers

Autophagy is considered not only a mechanism of cell protection due to its ability to mitigate cell stresses and recycle metabolic precursors, but also as a driver of cell death in autophagy deficient cells or through excessive destruction of cellular components [42]. This paradoxical feature is highly associated with tumor formation in different stages [7].

At the early stage of tumorigenesis, autophagy serves as a tumor suppressor. The heterozygous deletion of Beclin-1 is significantly associated with the tumorigenesis of breast cancer (50–70%), ovarian cancer (75%), and prostate cancer (40%). This association is in line with the in vivo studies showing that deficient Beclin-1 increases spontaneous malignancies [43, 44]. By contrast, overexpression of Beclin-1 compromises the tumorigenesis in a MCF7 xenograft model by the inhibition of cell proliferation [45]. Moreover, evidence indicates that deficiency of Atg4 favors fibrosarcoma formation with the increased sensitivity to chemical carcinogens [46]. Deletion of Atg5 or Atg7 promotes the development of benign liver tumors in Atg5 or Atg7 knock-out mice [47]. Mutations of Atg2B, Atg5, Atg9B, Atg12 and UVRAG contribute to the development of gastric and colorectal cancers [48, 49]. Loss of Bax interacting factor 1 *(Bif-1)*, a gene interacting with Beclin-1 to enhance autophagy, is associated with colon adenocarcinomas due to the inhibition of programmed cell death [50]. Interestingly, the tumor suppressor gene *p53*, the most commonly mutated gene in human cancers, plays a dual role in the regulation of autophagy. Wild type p53 induces autophagy through its transcriptional function in the nucleus, but inhibits autophagy in the cytoplasm by means of its interaction with FIP200 [51, 52]. Mutated p53 is preferentially located in the cytoplasm and works to inhibit autophagy [53].

Conversely, autophagy is considered a mechanism of cell protection in terms of tumor progression. When a tumor grows in size, there are several cellular stresses that activate autophagy in order to cope with the stress and maintain homeostasis. These include insufficient supplies of nutrients and oxygens, the accumulation of cellular waste, loss of growth factors, and production of reactive oxygen species (ROS) [45]. There are growing data indicating that autophagy may be critical for tumor progression. Although the monoallelic deletion of Beclin-1 is pivotal for tumorigenesis [45], the remaining Beclin-1 allele implies that tumor cells require autophagy for malignant transformation [54]. This evidence also accounts for the finding that deletion of Atg5 or Atg7 leads to benign liver tumors in mouse models [47]. Inhibition and deletion of Atg5, respectively, interferes with tumor growth of human pancreatic ductal adenocarcinomas and RAS-transformed mouse kidney epithelial cells in mouse models [55, 56], which further demonstrates the importance of autophagy at the late stage of tumorigenesis and tumor progression [7].

Role of HNSCC risk factors in autophagy induction

As the sixth leading cancer in the world, head and neck cancer ranges from nose and paranasal sinuses to the oral cavity, larynx, hypopharynx and oropharynx [57]. Most head and neck cancer patients (90%) have squamous cell carcinoma (HNSCC), which account for 3% of the cancer incidence and 1.6% of the cancer death in the United States, and has a 40–50% 5-year survival rate worldwide [58]. There are several risk factors associated with HNSCC, such as tobacco, alcohol, areca nut extract (ANE), human papillomaviruses (HPV16 and HPV18), and several other factors [59, 60]. In terms of gene alternations, overexpression of the epidermal growth factor receptor (EGFR), mutation of p53, and inactivation of p16 have been reported as the three major changes related to HNSCC tumorigenesis [61–63].

Because autophagy plays a role in both tumor suppression (in the early stage of tumorigenesis) and tumor protection (in cancer progression), the modulation of autophagy by different risk factors for HNSCC can be considered as a potential therapeutic target. Cigarette smoke contains several toxins (carbon monoxide, etc.) and carcinogens (superoxide, nitrogen oxide, nicotine, etc.), which contribute to some negative biological responses in the human body such as inhibition of immune cell proliferation and formation of DNA adducts [64, 65]. Moreover, there is accumulating data suggesting that smoking induces autophagy through ROS and SIRT-1-PARP-dependent pathways [66, 67]. Although tobacco smoking contributes to oxidative stresses activating apoptotic signaling, cells can ameliorate apoptosis by heme oxygenase-1 (HO-1) induced expression of LC3B, thereby activating autophagy to inhibit apoptosis [68]. ANE has been associated with chronic inflammation by ROS, and it confers cisplatin resistance on oral squamous cell carcinoma (OSCC) through autophagy induction [60].

HPV is considered as a major risk factor in HNSCC due to its two oncogenic proteins E6 and E7. The complex of E6 and its associated proteins can target p53 for degradation. E7 is able to target the retinoblastoma protein (pRB, tumor suppressor gene) leading to persistent activation of E2F transcription factors, which contributes to the aberration of the cell cycle and tumorigenesis [69]. Knockdown of autophagy related genes or inhibition of autophagy significantly increases HPV16 infectivity [70]. Moreover, the protein kinase B (Akt) and mTOR signaling pathways have been activated by HPV16 pseudo virions in the early stage of host cell infection; these changes can result in the inhibition of autophagy [71]. HPV16 E6 protein maintains mTOR pathway activation, which blocks autophagy, during nutrient loss in primary human foreskin keratinocytes (HFKs) [72]. In contrast, HPV16 E7 protein can activate autophagy by enhancing the lipidation of LC3 in nutrient-deficient human keratinocytes in order to induce autophagy-mediated cell death [73]. These results indicate that the balance between E6 and E7 can regulate autophagy in HPV-infected cells.

Autophagy modulation in head and neck cancer therapy

Currently, some anti-cancer therapies, such as radiotherapy and chemotherapy, have been shown to induce autophagy in cancer cells [74–76] indicating that the modulation of autophagy has potential as an anticancer therapy. As previously discussed, autophagy plays a dual role in the determination of cancer cell fate. In the early stage of tumorigenesis, autophagy acts as a tumor suppressor but it functions oppositely in the later stages of cancer development [7]. Understanding the crosstalk between autophagy and current anticancer drugs is important for the improvement of current treatments.

Effects of current chemotherapy on autophagy

EGFR is overexpressed in over 80% of HNSCCs [77] and one of the main downstream signaling pathways is PI3K/Akt/mTOR. Activation of this pathway is seen in up to 47% of HNSCCs [78]. These EGFR-initiated proteins are highly expressed in the early stages leading to tumorigenesis and activation of the PI3K/Akt/mTOR pathway successfully blocks autophagy induction, indicating that both the EGFR and PI3K/Akt/mTOR pathways may be potential targets for HNSCC therapy. Recent studies revealed that current chemotherapies induce autophagy [57], indicating a potential underlying resistance mechanism in HNSCCs.

Gefitinib (aka ZD1839 or Iressa) inactivates EGFR by blocking the binding of ATP to the active site of the kinase domain thus repressing the major downstream signaling pathways: Akt/ERK/Stat3 [79]. EGFR mutations in exons 18, 19 and 21 allow cancer cells to be significantly sensitive to this tyrosine kinase inhibitor. Clinically, 10–15% of HNSCC patients respond to gefitinib treatment [80]. Gefitinib can simultaneously induce apoptosis and autophagy by itself or in combination with other treatments such as MK-2206 (Akt inhibitor) [81], docetaxel (anti-microtubule agent) [82], or radiation. Inhibition of autophagy significantly enhances the efficacy of gefitinib, underscoring the benefit of modulating autophagy for anticancer therapy [83].

Erlotinib (aka OSI-774 or Tarceva) is a specific EGFR tyrosine kinase inhibitor that reversibly binds to the ATP binding site of EGFR to inactivate receptor activity. A single dose of erlotinib leads to a 29% response rate from HNSCC patients [84]. Accumulating data has shown that NADPH oxidase 4 (NOX4) mediates an increase in oxidative stress in erlotinib-treated cells, indicating that a combination of erlotinib with traditional cancer therapies that induce oxidative stress may be more effective [85]. Low doses of erlotinib also greatly induces autophagy to protect cancer cells from cell death, but autophagic cell death has been found in those cells treated with high dose of erlotinib. Furthermore, inhibition of autophagy enhances cytotoxicity of erlotinib [86].

Saracatinib (aka AZD0530) selectively inhibits Src family proteins (c-Src, c-Yes, Lck and Bcr-Abl) which play a role in cancer cell proliferation, angiogenesis, survival, invasion and metastasis [87]. Overexpression of Src is associated with metastases through the Vimentin/Snail pathway in variety of cancers including HNSCC. Blockade of Src kinase activity or knockdown of Src successfully represses invasion and metastases in HNSCC cells in vitro and in vivo [88]. However, results from a phase II clinical trial indicate a lack of efficacy of saracatinib alone in metastatic HNSCC patients [89], suggesting the need for combination treatment. Autophagy protects non-small cell lung cancer (NSCLC) cells from saracatinib-induced apoptosis and inhibition of the autophagy pathway enhances the cytotoxicity of saracatinib [90]. In addition, saracatinib induces autophagy in prostate cancer cells (PCa), which is also observed when Src is knocked down in PCa cells. The combination of saracatinib and the autophagy inhibitor chloroquine (CQ) leads to increased apoptosis and a 64% inhibition of cell growth [91]. Taken together, the inhibition of both autophagy and Src may lead to better tumor control.

Cetuximab (aka IMC-225 or Erbitux) is a recombinant chimeric IgG1 antibody (65% human and 35% murine) which targets the EGFR ligand binding domain to inactivate EGFR, promote EGFR internalization, and causes cell cycle arrest and eventual apoptosis. Cetuximab is widely used in HNSCC treatment due to the high frequency of EGFR overexpression; however, only 50% of patients respond to this antibody [92]. Cetuximab can induce the translocation of EGFR to the nucleus, which confers cetuximab resistance on HNSCC cells. Interestingly, dasatinib (Src family proteins inhibitor) shows the capacity for a blockade of cetuximab-induced EGFR translocation [93], suggesting that the combination of cetuximab with dasatinib may benefit HNSCC cancer treatment. The combination of cetuximab and cisplatin (a DNA crosslinker) leads to a promising outcome in metastatic HNSCC treatment, which increases response from 20% to 36% [94, 95]. Cetuximab also induces autophagy, as indicated by the observation of an increase in LC3 conversion, autophagosome formation, and inhibition of autolysosome-induced LC3II accumulation. However, co-treatment with rapamycin (mTOR inhibitor), which is known as an autophagy inducer, increases apoptosis in cetuximab-resistant HNSCC cells [96], suggesting that autophagy serves a dual role in HNSCC response to EGFR-targeted therapy.

Panitumumab (aka ABX-EGF or Vectibix) is a human IgG2 antibody that blocks the binding of EGF to EGFR by targeting the ectodomain. Clinically, panitumumab possesses higher affinity in the HNSCC cell (UM-SCC-22B) than other antibodies [97] and improved progression-free survival in the phase III clinical trial on HNSCC patients while showing low toxicity [98]. As an IgG2 antibody, panitumumab rarely causes antibody-dependent cellular cytotoxicity (ADCC) compared to IgG1 antibodies such as cetuximab. Panitumumab also shows an ability to sensitize HNSCC cells to radiotherapy by increasing DNA damage and blocking radiation-induced EGFR phosphorylation [99]. One autophagy marker, Beclin-1, is increased in panitumumab-treated cells. In vitro data indicates panitumumab inhibits cell proliferation through autophagy independently of K-Ras and EGFR in colon cancer cells [100]. More research needs to be done to fully determine the connection between panitumumab and autophagy.

Effect of radiation therapy on autophagy

Radiotherapy is frequently used to treat cancer patients, and it significantly increases autophagy in cancer cells [57]. Some studies suggest that autophagy enhances programmed cell death in irradiated cancer cells [74]. This hypothesis is supported by the evidenced that rapamycin sensitizes MCF-7 breast cancer cells to radiotherapy [101]. Moreover, rapamycin also enhances cell death in apoptosis-deficient cells treated with radiation [102]. In human oral squamous cell carcinoma (OSCC) cell line, OC3, co-treatment of rapamycin with radiation leads to a further reduction in cell viability, suggesting the synergistic effects of radiation and autophagy induction in OC3 cells [103]. On the other hand, several reports indicate a protective role of autophagy in irradiated cancer cells [104]. Pretreatment with autophagy inhibitor 3-MA or CQ reduces cell viability more than radiotherapy alone [105]. However, knockdown of different autophagy related genes results in differential effects on survival depending on the gene and the dose of radiation in different cancer cell lines [106], indicating that the role of autophagy in radiotherapy remains unclear and needs to be further determined.

Induction of autophagy in head and neck cancer

Since modulation of autophagy has been implicated as a potential anticancer therapy, there are several new drugs developed to modulate autophagy in different fashions: (1) induction, and (2) inhibition. Most autophagy inducers target the mTOR/Akt signaling pathway due to the ability of this pathway to negatively regulate autophagy induction [33]. It is important to note that at this time, none of the drugs used to modulate autophagy in patients were developed for that purpose.

Rapamycin (aka sirolimus or Rapamune) is an mTOR inhibitor which forms a complex with FKBP12 to bind with mTOR leading to restricted access of mTOR substrates, which ultimately inhibits mTOR activity [107]. Rapamycin induces autophagy based on the inhibition of mTOR and activation of the Akt/mTOR pathway is associated with 80% of HNSCC lesions [108]. Several studies prove rapamycin's anti-cancer effect through its promotion of apoptosis in HNSCC-tumor xenograft models. In vitro studies also indicate that rapamycin enhances the effects of other anti-cancer drugs including cisplatin, carboplatin, doxorubicin, paclitaxel, topotecan and mitoxantrone in HNSCC cell lines [109]. In terms of rapamycin-induced autophagy, inhibition of autophagy compromises rapamycin cytotoxicity, suggesting that rapamycin-induced autophagy functions as an alternative programmed cell death pathway [57].

Everolimus (aka RAD001 or Afinitor) is a derivative of rapamycin and similarly functions to inhibit mTOR. Everolimus has been tested for its anti-cancer effect in a variety of cancer cells [110] and autophagy-mediated programmed cell death induction has been found in everolimus-treated cells. Everolimus-induced autophagy also improves cell sensitivity to doxorubicin and radiation. In addition, everolimus-induced autophagy contributes to Met inactivation, which is involved in cell proliferation, migration, angiogenesis and survival [111].

Temsirolimus (aka CCI-779 or Torisel) is another derivative of rapamycin that inhibits mTOR activity and effectively, but not completely, decreases the activation of downstream proteins p70S6K and 4E-BP1 in HNSCC cells. Temsirolimus has growth-inhibitory effect on the cells that only depend on PI3K/Akt/mTOR pathway. Clinically, temsirolimus enhances the anti-proliferative effect of traditional treatments when used in conjunction with radiotherapy or chemotherapy that targets EGFR [112, 113]. Temsirolimus has been shown to induce autophagy and arrest the cell cycle at G_1 by downregulation of p21 in lymphoma cells. Prolonged exposure to this drug may cause massive autophagy leading to autophagy-induced programmed cell death in cancer cells [114].

Inhibition of autophagy in head and neck cancer

Several drugs which can block autophagy have been developed. As was describe for autophagy inducers, the compounds used clinically were not developed as specific autophagy inhibitors, but rather this was found to be an off-target effect. The only two autophagy inhibitors that have been used clinically remain chloroquine (CQ) and hydroxychloroquine (HCQ) [15]. Other autophagy inhibitors are in preclinical development stages and are primarily used to study autophagy in the lab.

Chloroquine (CQ) can block autophagic degradation by not only deacidifying lysosomes but interfering with the fusion of autophagosomes to lysosomes [115]. Co-treatment of CQ and other anti-cancer drugs including cisplatin, erlotinib, and saracatinib increases cell death in HNSCC cells [86, 91, 116]. Clinical trials demonstrate CQ improves the median survival of glioblastoma patients both alone [117] and in combination with radiation and the alkylating agent, temozolomide [118]. CQ also has other anti-cancer effects which are independent of autophagy, such as vessel normalization for restricting invasion and metastases in cancer cells [119].

Hydroxychloroquine (HCQ, also known as Plaquenil) functions similarly to CQ and has been shown to augment the anticancer effect of CYT997 (tubulin polymerization inhibitor), by causing ROS-induced apoptosis in HNSCC cells [120]. It has had success alone as well as in combination with other chemotherapies in treating canine lymphoma. Barnard and colleagues demonstrated the combination was safe resulted in a 93% objective response rate [121]. The dose of HCQ necessary for combination therapy in human is well understood [122]. Preclinical data shows that patient-derived xenograft (PDX) models of pancreatic cancer responded well to HCQ monotherapy while 61% of pancreatic adenocarcinoma patients treated with combined treatment of HCQ and gemcitabine (cytosine analogue) have been found to decrease the detection of the cancer marker CA 19-9. It was also observed that the patients with higher autophagy have increased progression-free survival and overall survival [123].

The challenges of modulating autophagy in cancer therapy

One of the primary challenges of modulating autophagy in cancer therapy is that, to date, none of the clinically used drugs was developed specifically as an autophagy modulator. In head and neck cancer, there are currently multiple open trials testing drugs which modulate autophagy in combination with other systemic therapies (Table 1). It been hypothesized that chemotherapy and radiation therapy can be combined with autophagy modulation (induction or inhibition) to overcome therapeutic resistance and improve tumor control. Additional preclinical research is needed to determine whether a specific subtype of autophagy should

TABLE 1 Clinical trails testing drugs which modulate autophagy in combination with other systemic therapies

Drug class/ mechanism	Agent names	Title	Suggested patient population	Phase	Enrollment	Status	Identifier
mTOR inhibitor	RAD001/ cisplatin	RAD001 With Weekly Cisplatin and Radiation Therapy in Patients With Locally Advanced Head and Neck Cancer	Locally advanced head and neck cancer	I	3	Terminated	NCT01058408
mTOR inhibitor	RAD001/ cisplatin	Radiation Therapy (IMRT) + RAD001 (Everolimus) + Cisplatin for Patients With Head and Neck Cancer	Head and neck cancer	I	13	Completed	NCT00858663
mTOR inhibitor	RAD001 (everolimus)/ placebo	Everolimus Versus Placebo in Head and Neck Cancer	Head and neck cancer	II	160	Active, not recruiting	NCT01111058
mTOR inhibitor	RAD001/ cetuximab/ carboplatin	A Study of Carboplatin, Cetuximab and RAD001 in Advanced Head and Neck Cancer	Advanced head and neck cancer	I/II	20	Completed	NCT01283334
mTOR inhibitor	RAD001 (everolimus)/ docetaxel	Everolimus and Docetaxel in Treating Patients With Recurrent, Locally Advanced, or Metastatic Head and Neck Cancer	Recurrent, locally advanced or metastatic head and neck cancer	I/II	4	Terminated	NCT01313390
mTOR inhibitor	RAD001 (everolimus)	Study in Locally Advanced Squamous Cell Carcinoma of Head and Neck	Locally advanced head and neck cancer	II	80	Active, not recruiting	NCT01133678

mTOR inhibitor	Temsirolimus	Efficacy Study of Temsirolimus to Treat Head and Neck Cancer	Head and neck cancer	II	42	Completed	NCT01172769
mTOR inhibitor	Temsirolimus/ paclitaxel/ carboplatin	Temsirolimus + Weekly Paclitaxel + Carboplatin for Recurrent or Metastatic Head and Neck Squamous Cell Cancer (HNSCC)	Recurrent, locally advanced or metastatic head and neck squamous cell carcinoma	I/II	48	Completed	NCT01016769

Drug class/ mechanism	Intervention	Title	Suggested patient population	Phase	Enrollment	Status	Identifier
mTOR inhibitor	Temsirolimus/ cisplatin	Temsirolimus With or Without Cetuximab in Patients With Recurrent and/or Metastatic Head and Neck Cancer Who Did Not Respond to Previous Therapy Recurrent	Recurrent and/or metastatic head and neck cancer who did not respond to previous therapy recurrent	II	86	Completed	NCT01256385
mTOR inhibitor	Rapamycin (sirolimus)	Rapamycin Therapy in Head and Neck Squamous Cell Carcinoma	Head and neck squamous cell carcinoma	I/II	37	Completed	NCT01195922

be targeted and/or whether targeting at a specific point in the autophagy pathway is preferred. Further research is clearly needed to identify how and in which patients autophagy modulators are most likely to improve outcomes when combined with standard anti-cancer treatments.

Acknowledgments

This project was supported in part by funding from the American Cancer Society (RSG-16-091-01-TBG and MBG-18-205-01-COUN) and the Specialized Program of Research Excellence (SPORE) program, through the NIH National Institute for Dental and Craniofacial Research (NIDCR) and National Cancer Institute (NCI), grant P50DE026787. The content is solely the responsibility of the authors and does not necessarily represent the official views of the NIH.

References

[1] Codogno P, Meijer AJ. Autophagy and signaling: their role in cell survival and cell death. Cell Death Differ 2005;12(Suppl. 2):1509–18.
[2] Meijer AJ, Codogno P. Regulation and role of autophagy in mammalian cells. Int J Biochem Cell Biol 2004;36(12):2445–62.
[3] Mizushima N, Klionsky DJ. Protein turnover via autophagy: implications for metabolism. Annu Rev Nutr 2007;27:19–40.
[4] Rubinsztein DC, Marino G, Kroemer G. Autophagy and aging. Cell 2011;146(5):682–95.
[5] Levine B, Klionsky DJ. Development by self-digestion: molecular mechanisms and biological functions of autophagy. Dev Cell 2004;6(4):463–77.
[6] Mizushima N. The pleiotropic role of autophagy: from protein metabolism to bactericide. Cell Death Differ 2005;12(Suppl. 2):1535–41.
[7] Rubinsztein DC, Codogno P, Levine B. Autophagy modulation as a potential therapeutic target for diverse diseases. Nat Rev Drug Discov 2012;11(9):709–30.
[8] Galluzzi L, Baehrecke EH, Ballabio A, Boya P, Bravo-San Pedro JM, Cecconi F, Choi AM, Chu CT, Codogno P, Colombo MI, Cuervo AM, Debnath J, Deretic V, Dikic I, Eskelinen EL, Fimia GM, Fulda S, Gewirtz DA, Green DR, Hansen M, Harper JW, Jaattela M, Johansen T, Juhasz G, Kimmelman AC, Kraft C, Ktistakis NT, Kumar S, Levine B, Lopez-Otin C, Madeo F, Martens S, Martinez J, Melendez A, Mizushima N, Munz C, Murphy LO, Penninger JM, Piacentini M, Reggiori F, Rubinsztein DC, Ryan KM, Santambrogio L, Scorrano L, Simon AK, Simon HU, Simonsen A, Tavernarakis N, Tooze SA, Yoshimori T, Yuan J, Yue Z, Zhong Q, Kroemer G. Molecular definitions of autophagy and related processes. EMBO J 2017;36(13):1811–36.
[9] Mizushima N. Methods for monitoring autophagy. Int J Biochem Cell Biol 2004;36(12):2491–502.
[10] Li WW, Li J, Bao JK. Microautophagy: lesser-known self-eating. Cell Mol Life Sci 2012;69(7):1125–36.
[11] Thorburn A, Thamm DH, Gustafson DL. Autophagy and cancer therapy. Mol Pharmacol 2014;85(6):830–8.
[12] Kon M, Kiffin R, Koga H, Chapochnick J, Macian F, Varticovski L, Cuervo AM. Chaperone-mediated autophagy is required for tumor growth. Sci Transl Med 2011;3(109).
[13] Ravikumar B, Sarkar S, Davies JE, Futter M, Garcia-Arencibia M, Green-Thompson ZW, Jimenez-Sanchez M, Korolchuk VI, Lichtenberg M, Luo S, Massey DC, Menzies FM, Moreau K, Narayanan U, Renna M, Siddiqi FH, Underwood BR, Winslow AR, Rubinsztein DC. Regulation of mammalian autophagy in physiology and pathophysiology. Physiol Rev 2010;90(4):1383–435.
[14] Mizushima N, Yoshimori T, Ohsumi Y. The role of Atg proteins in autophagosome formation. Annu Rev Cell Dev Biol 2011;27:107–32.
[15] Levy JMM, Towers CG, Thorburn A. Targeting autophagy in cancer. Nat Rev Cancer 2017;17(9):528–42.
[16] Blommaart EF, Krause U, Schellens JP, Vreeling-Sindelarova H, Meijer AJ. The phosphatidylinositol 3-kinase inhibitors wortmannin and LY294002 inhibit autophagy in isolated rat hepatocytes. Eur J Biochem 1997;243(1–2):240–6.
[17] Seglen PO, Gordon PB. 3-methyladenine: specific inhibitor of autophagic/lysosomal protein degradation in isolated rat hepatocytes. Proc Natl Acad Sci U S A 1982;79(6):1889–92.

[18] Egan DF, Chun MG, Vamos M, Zou H, Rong J, Miller CJ, Lou HJ, Raveendra-Panickar D, Yang CC, Sheffler DJ, Teriete P, Asara JM, Turk BE, Cosford ND, Shaw RJ. Small molecule inhibition of the autophagy kinase ULK1 and identification of ULK1 substrates. Mol Cell 2015;59(2):285–97.

[19] Wong PM, Puente C, Ganley IG, Jiang X. The ULK1 complex: sensing nutrient signals for autophagy activation. Autophagy 2013;9(2):124–37.

[20] Pattingre S, Tassa A, Qu X, Garuti R, Liang XH, Mizushima N, Packer M, Schneider MD, Levine B. Bcl-2 antiapoptotic proteins inhibit Beclin 1-dependent autophagy. Cell 2005;122(6):927–39.

[21] Vicencio JM, Ortiz C, Criollo A, Jones AW, Kepp O, Galluzzi L, Joza N, Vitale I, Morselli E, Tailler M, Castedo M, Maiuri MC, Molgo J, Szabadkai G, Lavandero S, Kroemer G. The inositol 1,4,5-trisphosphate receptor regulates autophagy through its interaction with Beclin 1. Cell Death Differ 2009;16(7):1006–17.

[22] Mizushima N, Sugita H, Yoshimori T, Ohsumi Y. A new protein conjugation system in human. The counterpart of the yeast Apg12p conjugation system essential for autophagy. J Biol Chem 1998;273(51):33889–92.

[23] Webb JL, Ravikumar B, Rubinsztein DC. Microtubule disruption inhibits autophagosome-lysosome fusion: implications for studying the roles of aggresomes in polyglutamine diseases. Int J Biochem Cell Biol 2004;36 (12):2541–50.

[24] Klionsky DJ, Abeliovich H, Agostinis P, Agrawal DK, Aliev G, Askew DS, Baba M, Baehrecke EH, Bahr BA, Ballabio A, Bamber BA, Bassham DC, Bergamini E, Bi X, Biard-Piechaczyk M, Blum JS, Bredesen DE, Brodsky JL, Brumell JH, Brunk UT, Bursch W, Camougrand N, Cebollero E, Cecconi F, Chen Y, Chin L-S, Choi A, Chu CT, Chung J, Clark RSB, Clarke PGH, Clarke SG, Clave C, Cleveland JL, Codogno P, Colombo MI, Coto-Montes A, Cregg JM, Cuervo AM, Debnath J, Dennis PB, Dennis PA, Demarchi F, Deretic V, Devenish RJ, Di Sano F, Dice JF, Distelhorst CW, Dinesh-Kumar SP, Eissa NT, DiFiglia M, Djavaheri-Mergny M, Dorsey FC, Dröge W, Dron M, Dunn JWA, Duszenko M, Elazar Z, Esclatine A, Eskelinen E-L, Fésüs L, Finley KD, Fuentes JM, Fueyo-Margareto J, Fujisaki K, Galliot B, Gao F-B, Gewirtz DA, Gibson SB, Gohla A, Goldberg AL, Gonzalez R, González-Estévez C, Gorski SM, Gottlieb RA, Häussinger D, He Y-W, Heidenreich K, Hill JA, Høyer-Hansen M, Hu X, Huang W-P, Iwasaki A, Jäättelä M, Jackson WT, Jiang X, Jin SV, Johansen T, Jung JU, Kadowaki M, Kang C, Kelekar A, Kessel DH, Kiel JAKW, Kim HP, Kimchi A, Kinsella TJ, Kiselyov K, Kitamoto K, Knecht E, Komatsu M, Kominami E, Kondo S, Kovács AL, Kroemer G, Kuan C-Y, Kumar R, Kundu M, Landry J, Laporte M, Le W, Lei H-Y, Levine B, Lieberman AP, Lim K-L, Lin F-C, Liou W, Liu LF, Lopez-Berestein G, López-Otín C, Lu B, Macleod KF, Malorni W, Martinet W, Matsuoka K, Mautner J, Meijer AJ, Meléndez A, Michels P, Miotto G, Mistiaen WP, Mizushima N, Mograbi B, Moore MN, Moreira PI, Moriyasu Y, Motyl T, Münz C, Murphy LO, Naqvi NI, Neufeld TP, Nishino I, Nixon RA, Noda T, Nürnberg B, Ogawa M, Oleinick NL, Olsen LJ, Ozpolat B, Paglin S, Palmer GE, Papassideri IS, Parkes M, Perlmutter DH, Perry G, Piacentini M, Pinkas-Kramarski R, Prescott M, Proikas-Cezanne T, Raben N, Rami A, Reggiori F, Rohrer B, Rubinsztein DC, Ryan KM, Sadoshima J, Sakagami H, Sakai Y, Sandri M, Sasakawa C, Sass M, Schneider C, Seglen PO, Seleverstov O, Settleman J, Shacka JJ, Shapiro IM, Sibirny AA, Silva-Zacarin ECM, Simon H-U, Simone C, Simonsen A, Smith MA, Spanel-Borowski K, Srinivas V, Steeves M, Stenmark H, Stromhaug PE, Subauste CS, Sugimoto S, Sulzer D, Suzuki T, Swanson MS, Tabas I, Takeshita F, Talbot NJ, Tallóczy Z, Tanaka K, Tanaka K, Tanida I, Taylor GS, Taylor JP, Terman A, Tettamanti G, Thompson CB, Thumm M, Tolkovsky AM, Tooze SA, Truant R, Tumanovska LV, Uchiyama Y, Ueno T, Uzcátegui NL, van der Klei IJ, Vaquero EC, Vellai T, Vogel MW, Wang H-G, Webster P, Xi Z, Xiao G, Yahalom J, Yang J-M, Yap GS, Yin X-M, Yoshimori T, Yue Z, Yuzaki M, Zabirnyk O, Zheng X, Zhu X, Deter RL. Guidelines for the use and interpretation of assays for monitoring autophagy in higher eukaryotes. Autophagy 2014;4(2):151–75.

[25] Fujita N, Itoh T, Omori H, Fukuda M, Noda T, Yoshimori T. The Atg16L complex specifies the site of LC3 lipidation for membrane biogenesis in autophagy. Mol Biol Cell 2008;19(5):2092–100.

[26] Hanada T, Noda NN, Satomi Y, Ichimura Y, Fujioka Y, Takao T, Inagaki F, Ohsumi Y. The Atg12-Atg5 conjugate has a novel E3-like activity for protein lipidation in autophagy. J Biol Chem 2007;282(52):37298–302.

[27] Fujita N, Hayashi-Nishino M, Fukumoto H, Omori H, Yamamoto A, Noda T, Yoshimori T. An Atg4B mutant hampers the lipidation of LC3 paralogues and causes defects in autophagosome closure. Mol Biol Cell 2008;19 (11):4651–9.

[28] Birgisdottir AB, Lamark T, Johansen T. The LIR motif—crucial for selective autophagy. J Cell Sci 2013;126(Pt 15):3237–47.

[29] Rusten TE, Vaccari T, Lindmo K, Rodahl LM, Nezis IP, Sem-Jacobsen C, Wendler F, Vincent JP, Brech A, Bilder D, Stenmark H. ESCRTs and Fab1 regulate distinct steps of autophagy. Curr Biol 2007;17(20):1817–25.

[30] Lee JA, Beigneux A, Ahmad ST, Young SG, Gao FB. ESCRT-III dysfunction causes autophagosome accumulation and neurodegeneration. Curr Biol 2007;17(18):1561–7.

[31] Filimonenko M, Stuffers S, Raiborg C, Yamamoto A, Malerod L, Fisher EM, Isaacs A, Brech A, Stenmark H, Simonsen A. Functional multivesicular bodies are required for autophagic clearance of protein aggregates associated with neurodegenerative disease. J Cell Biol 2007;179(3):485–500.

[32] Marino G, Niso-Santano M, Baehrecke EH, Kroemer G. Self-consumption: the interplay of autophagy and apoptosis. Nat Rev Mol Cell Biol 2014;15(2):81–94.

[33] Bhat P, Kriel J, Shubha Priya B, Basappa, Shivananju NS, Loos B. Modulating autophagy in cancer therapy: advancements and challenges for cancer cell death sensitization. Biochem Pharmacol 2018;147:170–82.

[34] Gao W, Shen Z, Shang L, Wang X. Upregulation of human autophagy-initiation kinase ULK1 by tumor suppressor p53 contributes to DNA-damage-induced cell death. Cell Death Differ 2011;18(10):1598–607.

[35] Elgendy M, Sheridan C, Brumatti G, Martin SJ. Oncogenic Ras-induced expression of Noxa and Beclin-1 promotes autophagic cell death and limits clonogenic survival. Mol Cell 2011;42(1):23–35.

[36] Lomonosova E, Chinnadurai G. BH3-only proteins in apoptosis and beyond: an overview. Oncogene 2008;27(Suppl. 1):S2–19.

[37] Zalckvar E, Berissi H, Mizrachy L, Idelchuk Y, Koren I, Eisenstein M, Sabanay H, Pinkas-Kramarski R, Kimchi A. DAP-kinase-mediated phosphorylation on the BH3 domain of beclin 1 promotes dissociation of beclin 1 from Bcl-XL and induction of autophagy. EMBO Rep 2009;10(3):285–92.

[38] Wei Y, Pattingre S, Sinha S, Bassik M, Levine B. JNK1-mediated phosphorylation of Bcl-2 regulates starvation-induced autophagy. Mol Cell 2008;30(6):678–88.

[39] Hou W, Han J, Lu C, Goldstein LA, Rabinowich H. Autophagic degradation of active caspase-8: a crosstalk mechanism between autophagy and apoptosis. Autophagy 2010;6(7):891–900.

[40] Youle RJ, Narendra DP. Mechanisms of mitophagy. Nat Rev Mol Cell Biol 2011;12(1):9–14.

[41] Luo S, Rubinsztein DC. Apoptosis blocks Beclin 1-dependent autophagosome synthesis: an effect rescued by Bcl-xL. Cell Death Differ 2010;17(2):268–77.

[42] Patil S, Rao RS, Raj AT. Dual role of autophagy in oral cancer. J Int Oral Health 2015;7(6):i–ii.

[43] Liang XH, Jackson S, Seaman M, Brown K, Kempkes B, Hibshoosh H, Levine B. Induction of autophagy and inhibition of tumorigenesis by beclin 1. Nature 1999;402(6762):672–6.

[44] Aita VM, Liang XH, Murty VV, Pincus DL, Yu W, Cayanis E, Kalachikov S, Gilliam TC, Levine B. Cloning and genomic organization of beclin 1, a candidate tumor suppressor gene on chromosome 17q21. Genomics 1999;59(1):59–65.

[45] Nagelkerke A, Bussink J, Geurts-Moespot A, Sweep FC, Span PN. Therapeutic targeting of autophagy in cancer. Part II: pharmacological modulation of treatment-induced autophagy. Semin Cancer Biol 2015;31:99–105.

[46] Marino G, Salvador-Montoliu N, Fueyo A, Knecht E, Mizushima N, Lopez-Otin C. Tissue-specific autophagy alterations and increased tumorigenesis in mice deficient in Atg4C/autophagin-3. J Biol Chem 2007;282(25):18573–83.

[47] Takamura A, Komatsu M, Hara T, Sakamoto A, Kishi C, Waguri S, Eishi Y, Hino O, Tanaka K, Mizushima N. Autophagy-deficient mice develop multiple liver tumors. Genes Dev 2011;25(8):795–800.

[48] Panda PK, Mukhopadhyay S, Das DN, Sinha N, Naik PP, Bhutia SK. Mechanism of autophagic regulation in carcinogenesis and cancer therapeutics. Semin Cell Dev Biol 2015;39:43–55.

[49] Kang MR, Kim MS, Oh JE, Kim YR, Song SY, Kim SS, Ahn CH, Yoo NJ, Lee SH. Frameshift mutations of autophagy-related genes ATG2B, ATG5, ATG9B and ATG12 in gastric and colorectal cancers with microsatellite instability. J Pathol 2009;217(5):702–6.

[50] Coppola D, Khalil F, Eschrich SA, Boulware D, Yeatman T, Wang HG. Down-regulation of Bax-interacting factor-1 in colorectal adenocarcinoma. Cancer 2008;113(10):2665–70.

[51] Tasdemir E, Maiuri MC, Morselli E, Criollo A, D'Amelio M, Djavaheri-Mergny M, Cecconi F, Tavernarakis N, Kroemer G. A dual role of p53 in the control of autophagy. Autophagy 2014;4(6):810–4.

[52] Morselli E, Shen S, Ruckenstuhl C, Bauer MA, Marino G, Galluzzi L, Criollo A, Michaud M, Maiuri MC, Chano T, Madeo F, Kroemer G. p53 inhibits autophagy by interacting with the human ortholog of yeast Atg17, RB1CC1/FIP200. Cell Cycle 2011;10(16):2763–9.

[53] Morselli E, Tasdemir E, Maiuri MC, Galluzzi L, Kepp O, Criollo A, Vicencio JM, Soussi T, Kroemer G. Mutant p53 protein localized in the cytoplasm inhibits autophagy. Cell Cycle 2008;7(19):3056–61.

[54] Edinger AL, Thompson CB. Defective autophagy leads to cancer. Cancer Cell 2003;4(6):422–4.

[55] Yang S, Wang X, Contino G, Liesa M, Sahin E, Ying H, Bause A, Li Y, Stommel JM, Dell'antonio G, Mautner J, Tonon G, Haigis M, Shirihai OS, Doglioni C, Bardeesy N, Kimmelman AC. Pancreatic cancers require autophagy for tumor growth. Genes Dev 2011;25(7):717–29.
[56] Guo JY, Chen HY, Mathew R, Fan J, Strohecker AM, Karsli-Uzunbas G, Kamphorst JJ, Chen G, Lemons JM, Karantza V, Coller HA, Dipaola RS, Gelinas C, Rabinowitz JD, White E. Activated Ras requires autophagy to maintain oxidative metabolism and tumorigenesis. Genes Dev 2011;25(5):460–70.
[57] Rikiishi H. Autophagic action of new targeting agents in head and neck oncology. Cancer Biol Ther 2012;13(11):978–91.
[58] Siegel RL, Miller KD, Jemal A. Cancer statistics, 2017. CA Cancer J Clin 2017;67(1):7–30.
[59] Sannigrahi MK, Singh V, Sharma R, Panda NK, Khullar M. Role of autophagy in head and neck cancer and therapeutic resistance. Oral Dis 2015;21(3):283–91.
[60] Xu Z, Huang CM, Shao Z, Zhao XP, Wang M, Yan TL, Zhou XC, Jiang EH, Liu K, Shang ZJ. Autophagy induced by areca nut extract contributes to decreasing cisplatin toxicity in oral squamous cell carcinoma cells: roles of reactive oxygen species/AMPK signaling. Int J Mol Sci 2017;18(3):524.
[61] Sano T, Oyama T, Kashiwabara K, Fukuda T, Nakajima T. Expression status of p16 protein is associated with human papillomavirus oncogenic potential in cervical and genital lesions. Am J Pathol 1998;153(6):1741–8.
[62] Kalyankrishna S, Grandis JR. Epidermal growth factor receptor biology in head and neck cancer. J Clin Oncol 2006;24(17):2666–72.
[63] Langendijk JA, Psyrri A. The prognostic significance of p16 overexpression in oropharyngeal squamous cell carcinoma: implications for treatment strategies and future clinical studies. Ann Oncol 2010;21(10):1931–4.
[64] Stampfli MR, Anderson GP. How cigarette smoke skews immune responses to promote infection, lung disease and cancer. Nat Rev Immunol 2009;9(5):377–84.
[65] Harrigan JA, Vezina CM, McGarrigle BP, Ersing N, Box HC, Maccubbin AE, Olson JR. DNA adduct formation in precision-cut rat liver and lung slices exposed to benzo[a]pyrene. Toxicol Sci 2004;77(2):307–14.
[66] Hwang JW, Chung S, Sundar IK, Yao H, Arunachalam G, McBurney MW, Rahman I. Cigarette smoke-induced autophagy is regulated by SIRT1-PARP-1-dependent mechanism: implication in pathogenesis of COPD. Arch Biochem Biophys 2010;500(2):203–9.
[67] Lin MH, Hsieh WF, Chiang WF, Hong WZ, Hsu YR, Cheng YC, Chen TC, Hsu KC, Lin PY, Liu SY, Liu YC. Autophagy induction by the 30-100kDa fraction of areca nut in both normal and malignant cells through reactive oxygen species. Oral Oncol 2010;46(11):822–8.
[68] Was H, Dulak J, Jozkowicz A. Heme oxygenase-1 in tumor biology and therapy. Curr Drug Targets 2010;11(12):1551–70.
[69] Chen SF, Yu FS, Chang YC, Fu E, Nieh S, Lin YS. Role of human papillomavirus infection in carcinogenesis of oral squamous cell carcinoma with evidences of prognostic association. J Oral Pathol Med 2012;41(1):9–15.
[70] Griffin LM, Cicchini L, Pyeon D. Human papillomavirus infection is inhibited by host autophagy in primary human keratinocytes. Virology 2013;437(1):12–9.
[71] Surviladze Z, Sterk RT, DeHaro SA, Ozbun MA. Cellular entry of human papillomavirus type 16 involves activation of the phosphatidylinositol 3-kinase/Akt/mTOR pathway and inhibition of autophagy. J Virol 2013;87(5):2508–17.
[72] Spangle JM, Munger K. The human papillomavirus type 16 E6 oncoprotein activates mTORC1 signaling and increases protein synthesis. J Virol 2010;84(18):9398–407.
[73] Zhou X, Munger K. Expression of the human papillomavirus type 16 E7 oncoprotein induces an autophagy-related process and sensitizes normal human keratinocytes to cell death in response to growth factor deprivation. Virology 2009;385(1):192–7.
[74] Tam SY, Wu VW, Law HK. Influence of autophagy on the efficacy of radiotherapy. Radiat Oncol 2017;12(1) 57.
[75] Kondo Y, Kanzawa T, Sawaya R, Kondo S. The role of autophagy in cancer development and response to therapy. Nat Rev Cancer 2005;5(9):726–34.
[76] Sui X, Chen R, Wang Z, Huang Z, Kong N, Zhang M, Han W, Lou F, Yang J, Zhang Q, Wang X, He C, Pan H. Autophagy and chemotherapy resistance: a promising therapeutic target for cancer treatment. Cell Death Dis 2013;4.
[77] Leemans CR, Braakhuis BJ, Brakenhoff RH. The molecular biology of head and neck cancer. Nat Rev Cancer 2011;11(1):9–22.

[78] Pedrero JM, Carracedo DG, Pinto CM, Zapatero AH, Rodrigo JP, Nieto CS, Gonzalez MV. Frequent genetic and biochemical alterations of the PI 3-K/AKT/PTEN pathway in head and neck squamous cell carcinoma. Int J Cancer 2005;114(2):242–8.
[79] Wakeling AE, Guy SP, Woodburn JR, Ashton SE, Curry BJ, Barker AJ, Gibson KH. ZD1839 (Iressa): an orally active inhibitor of epidermal growth factor signaling with potential for cancer therapy. Cancer Res 2002;62(20):5749–54.
[80] Pernas FG, Allen CT, Winters ME, Yan B, Friedman J, Dabir B, Saigal K, Mundinger GS, Xu X, Morris JC, Calvo KR, Van Waes C, Chen Z. Proteomic signatures of epidermal growth factor receptor and survival signal pathways correspond to gefitinib sensitivity in head and neck cancer. Clin Cancer Res 2009;15(7):2361–72.
[81] Cheng Y, Zhang Y, Zhang L, Ren X, Huber-Keener KJ, Liu X, Zhou L, Liao J, Keihack H, Yan L, Rubin E, Yang JM. MK-2206, a novel allosteric inhibitor of Akt, synergizes with gefitinib against malignant glioma via modulating both autophagy and apoptosis. Mol Cancer Ther 2012;11(1):154–64.
[82] Chu Q, Amano O, Kanda Y, Kunii S, Wang Q, Sakagami H. Tumor-specific cytotoxicity and type of cell death induced by gefitinib in oral squamous cell carcinoma cell lines. Anticancer Res 2009;29(12):5023–31.
[83] Han W, Pan H, Chen Y, Sun J, Wang Y, Li J, Ge W, Feng L, Lin X, Wang X, Wang X, Jin H. EGFR tyrosine kinase inhibitors activate autophagy as a cytoprotective response in human lung cancer cells. PLoS ONE 2011;6(6).
[84] Vergez S, Delord JP, Thomas F, Rochaix P, Caselles O, Filleron T, Brillouet S, Canal P, Courbon F, Allal BC. Preclinical and clinical evidence that Deoxy-2-[18F]fluoro-D-glucose positron emission tomography with computed tomography is a reliable tool for the detection of early molecular responses to erlotinib in head and neck cancer. Clin Cancer Res 2010;16(17):4434–45.
[85] Orcutt KP, Parsons AD, Sibenaller ZA, Scarbrough PM, Zhu Y, Sobhakumari A, Wilke WW, Kalen AL, Goswami P, Miller Jr FJ, Spitz DR, Simons AL. Erlotinib-mediated inhibition of EGFR signaling induces metabolic oxidative stress through NOX4. Cancer Res 2011;71(11):3932–40.
[86] Eimer S, Belaud-Rotureau MA, Airiau K, Jeanneteau M, Laharanne E, Veron N, Vital A, Loiseau H, Merlio JP, Belloc F. Autophagy inhibition cooperates with erlotinib to induce glioblastoma cell death. Cancer Biol Ther 2011;11(12):1017–27.
[87] Summy JM, Gallick GE. Treatment for advanced tumors: SRC reclaims center stage. Clin Cancer Res 2006;12(5):1398–401.
[88] Lang L, Shay C, Xiong Y, Thakkar P, Chemmalakuzhy R, Wang X, Teng Y. Combating head and neck cancer metastases by targeting Src using multifunctional nanoparticle-based saracatinib. J Hematol Oncol 2018;11 (1):85.
[89] Fury MG, Baxi S, Shen R, Kelly KW, Lipson BL, Carlson D, Stambuk H, Haque S, Pfister DG. Phase II study of saracatinib (AZD0530) for patients with recurrent or metastatic head and neck squamous cell carcinoma (HNSCC). Anticancer Res 2011;31(1):249–53.
[90] Rothschild SI, Gautschi O, Batliner J, Gugger M, Fey MF, Tschan MP. MicroRNA-106a targets autophagy and enhances sensitivity of lung cancer cells to Src inhibitors. Lung Cancer 2017;107:73–83.
[91] Wu Z, Chang PC, Yang JC, Chu CY, Wang LY, Chen NT, Ma AH, Desai SJ, Lo SH, Evans CP, Lam KS, Kung HJ. Autophagy blockade sensitizes prostate cancer cells towards Src family kinase inhibitors. Genes Cancer 2010;1(1):40–9.
[92] Tejani MA, Cohen RB, Mehra R. The contribution of cetuximab in the treatment of recurrent and/or metastatic head and neck cancer. Biologics 2010;4:173–85.
[93] Li C, Iida M, Dunn EF, Wheeler DL. Dasatinib blocks cetuximab- and radiation-induced nuclear translocation of the epidermal growth factor receptor in head and neck squamous cell carcinoma. Radiother Oncol 2010;97(2):330–7.
[94] Vermorken JB, Mesia R, Rivera F, Remenar E, Kawecki A, Rottey S, Erfan J, Zabolotnyy D, Kienzer HR, Cupissol D, Peyrade F, Benasso M, Vynnychenko I, De Raucourt D, Bokemeyer C, Schueler A, Amellal N, Hitt R. Platinum-based chemotherapy plus cetuximab in head and neck cancer. N Engl J Med 2008;359(11):1116–27.
[95] de Mello RA, Geros S, Alves MP, Moreira F, Avezedo I, Dinis J. Cetuximab plus platinum-based chemotherapy in head and neck squamous cell carcinoma: a retrospective study in a single comprehensive European cancer institution. PLoS ONE 2014;9(2).
[96] Li X, Lu Y, Pan T, Fan Z. Roles of autophagy in cetuximab-mediated cancer therapy against EGFR. Autophagy 2010;6(8):1066–77.

[97] Foon KA, Yang XD, Weiner LM, Belldegrun AS, Figlin RA, Crawford J, Rowinsky EK, Dutcher JP, Vogelzang NJ, Gollub J, Thompson JA, Schwartz G, Bukowski RM, Roskos LK, Schwab GM. Preclinical and clinical evaluations of ABX-EGF, a fully human anti-epidermal growth factor receptor antibody. Int J Radiat Oncol Biol Phys 2004;58(3):984–90.
[98] Vermorken JB, Stohlmacher-Williams J, Davidenko I, Licitra L, Winquist E, Villanueva C, Foa P, Rottey S, Skladowski K, Tahara M, Pai VR, Faivre S, Blajman CR, Forastiere AA, Stein BN, Oliner KS, Pan Z, Bach BA, SPECTRUM investigators. Cisplatin and fluorouracil with or without panitumumab in patients with recurrent or metastatic squamous-cell carcinoma of the head and neck (SPECTRUM): an open-label phase 3 randomised trial. Lancet Oncol 2013;14(8):697–710.
[99] Kruser TJ, Armstrong EA, Ghia AJ, Huang S, Wheeler DL, Radinsky R, Freeman DJ, Harari PM. Augmentation of radiation response by panitumumab in models of upper aerodigestive tract cancer. Int J Radiat Oncol Biol Phys 2008;72(2):534–42.
[100] Giannopoulou E, Antonacopoulou A, Matsouka P, Kalofonos HP. Autophagy: novel action of panitumumab in colon cancer. Anticancer Res 2009;29(12):5077–82.
[101] Paglin S, Lee NY, Nakar C, Fitzgerald M, Plotkin J, Deuel B, Hackett N, McMahill M, Sphicas E, Lampen N, Yahalom J. Rapamycin-sensitive pathway regulates mitochondrial membrane potential, autophagy, and survival in irradiated MCF-7 cells. Cancer Res 2005;65(23):11061–70.
[102] Kim KW, Mutter RW, Cao C, Albert JM, Freeman M, Hallahan DE, Lu B. Autophagy for cancer therapy through inhibition of pro-apoptotic proteins and mammalian target of rapamycin signaling. J Biol Chem 2006;281(48):36883–90.
[103] Wu SY, Liu YW, Wang YK, Lin TH, Li YZ, Chen SH, Lee YR. Ionizing radiation induces autophagy in human oral squamous cell carcinoma. J BUON 2014;19(1):137–44.
[104] Maycotte P, Thorburn A. Autophagy and cancer therapy. Cancer Biol Ther 2011;11(2):127–37.
[105] Chaachouay H, Ohneseit P, Toulany M, Kehlbach R, Multhoff G, Rodemann HP. Autophagy contributes to resistance of tumor cells to ionizing radiation. Radiother Oncol 2011;99(3):287–92.
[106] Apel A, Herr I, Schwarz H, Rodemann HP, Mayer A. Blocked autophagy sensitizes resistant carcinoma cells to radiation therapy. Cancer Res 2008;68(5):1485–94.
[107] Hausch F, Kozany C, Theodoropoulou M, Fabian AK. FKBPs and the Akt/mTOR pathway. Cell Cycle 2013;12(15):2366–70.
[108] Molinolo AA, Hewitt SM, Amornphimoltham P, Keelawat S, Rangdaeng S, Meneses Garcia A, Raimondi AR, Jufe R, Itoiz M, Gao Y, Saranath D, Kaleebi GS, Yoo GH, Leak L, Myers EM, Shintani S, Wong D, Massey HD, Yeudall WA, Lonardo F, Ensley J, Gutkind JS. Dissecting the Akt/mammalian target of rapamycin signaling network: emerging results from the head and neck cancer tissue array initiative. Clin Cancer Res 2007;13(17):4964–73.
[109] Aissat N, Le Tourneau C, Ghoul A, Serova M, Bieche I, Lokiec F, Raymond E, Faivre S. Antiproliferative effects of rapamycin as a single agent and in combination with carboplatin and paclitaxel in head and neck cancer cell lines. Cancer Chemother Pharmacol 2008;62(2):305–13.
[110] Vignot S, Faivre S, Aguirre D, Raymond E. mTOR-targeted therapy of cancer with rapamycin derivatives. Ann Oncol 2005;16(4):525–37.
[111] Lin CI, Whang EE, Donner DB, Du J, Lorch J, He F, Jiang X, Price BD, Moore Jr FD, Ruan DT. Autophagy induction with RAD001 enhances chemosensitivity and radiosensitivity through Met inhibition in papillary thyroid cancer. Mol Cancer Res 2010;8(9):1217–26.
[112] Bozec A, Etienne-Grimaldi MC, Fischel JL, Sudaka A, Toussan N, Formento P, Milano G. The mTOR-targeting drug temsirolimus enhances the growth-inhibiting effects of the cetuximab-bevacizumab-irradiation combination on head and neck cancer xenografts. Oral Oncol 2011;47(5):340–4.
[113] Wang Z, Valera JC, Zhao X, Chen Q, Silvio GJ. mTOR co-targeting strategies for head and neck cancer therapy. Cancer Metastasis Rev 2017;36(3):491–502
[114] Yazbeck VY, Buglio D, Georgakis GV, Li Y, Iwado E, Romaguera JE, Kondo S, Younes A. Temsirolimus downregulates p21 without altering cyclin D1 expression and induces autophagy and synergizes with vorinostat in mantle cell lymphoma. Exp Hematol 2008;36(4):443–50.
[115] Yang YP, Hu LF, Zheng HF, Mao CJ, Hu WD, Xiong KP, Wang F, Liu CF. Application and interpretation of current autophagy inhibitors and activators. Acta Pharmacol Sin 2013;34(5):625–35.

[116] New J, Arnold L, Ananth M, Alvi S, Thornton M, Werner L, Tawfik O, Dai H, Shnayder Y, Kakarala K, Tsue TT, Girod DA, Ding WX, Anant S, Thomas SM. Secretory autophagy in cancer-associated fibroblasts promotes head and neck cancer progression and offers a novel therapeutic target. Cancer Res 2017;77(23):6679–91.

[117] Briceno E, Reyes S, Sotelo J. Therapy of glioblastoma multiforme improved by the antimutagenic chloroquine. Neurosurg Focus 2003;14(2).

[118] Sotelo J, Briceno E, Lopez-Gonzalez MA. Adding chloroquine to conventional treatment for glioblastoma multiforme: a randomized, double-blind, placebo-controlled trial. Ann Intern Med 2006;144(5):337–43.

[119] Maes H, Kuchnio A, Peric A, Moens S, Nys K, De Bock K, Quaegebeur A, Schoors S, Georgiadou M, Wouters J, Vinckier S, Vankelecom H, Garmyn M, Vion AC, Radtke F, Boulanger C, Gerhardt H, Dejana E, Dewerchin M, Ghesquiere B, Annaert W, Agostinis P, Carmeliet P. Tumor vessel normalization by chloroquine independent of autophagy. Cancer Cell 2014;26(2):190–206.

[120] Gao L, Zhao X, Lang L, Shay C, Andrew Yeudall W, Teng Y. Autophagy blockade sensitizes human head and neck squamous cell carcinoma towards CYT997 through enhancing excessively high reactive oxygen species-induced apoptosis. J Mol Med (Berl) 2018;96(9):929–38.

[121] Barnard RA, Wittenburg LA, Amaravadi RK, Gustafson DL, Thorburn A, Thamm DH. Phase I clinical trial and pharmacodynamic evaluation of combination hydroxychloroquine and doxorubicin treatment in pet dogs treated for spontaneously occurring lymphoma. Autophagy 2014;10(8):1415–25.

[122] Rangwala R, Chang YC, Hu J, Algazy KM, Evans TL, Fecher LA, Schuchter LM, Torigian DA, Panosian JT, Troxel AB, Tan KS, Heitjan DF, DeMichele AM, Vaughn DJ, Redlinger M, Alavi A, Kaiser J, Pontiggia L, Davis LE, O'Dwyer PJ, Amaravadi RK. Combined MTOR and autophagy inhibition: phase I trial of hydroxychloroquine and temsirolimus in patients with advanced solid tumors and melanoma. Autophagy 2014;10(8):1391–402.

[123] Boone BA, Bahary N, Zureikat AH, Moser AJ, Normolle DP, Wu WC, Singhi AD, Bao P, Bartlett DL, Liotta LA, Espina V, Loughran P, Lotze MT, Zeh 3rd HJ. Safety and biologic response of pre-operative autophagy inhibition in combination with gemcitabine in patients with pancreatic adenocarcinoma. Ann Surg Oncol 2015;22(13):4402–10.

CHAPTER

14

Combining Chk1/2 inhibition with radiation in head and neck cancer

Anatoly Nikolaev, Eddy S. Yang

Department of Radiation Oncology, Comprehensive Cancer Center, University of Alabama-Birmingham School of Medicine, Birmingham, AL, United States

Abstract

Despite the effectiveness of nonsurgical standard treatments, the survival rates of high risk head and neck cancer patients still remains <50%. Therefore, new targeted therapeutic approaches are needed to improve the outcomes of patients with advanced high risk head and neck cancer. The majority of head and neck cancers harbor amplification or overexpression of EGFR, and over 50% have mutations in TP53. These alterations can impact the DNA damage response. ATR/Chk1 and ATM/Chk2 are the two main pathways that control cellular responses to DNA damage by engaging G2/M and G1/S cell cycle checkpoints. Chk1/2 inhibitors sensitize head and neck cancer tumor cells to DNA damaging agents, such as cisplatin, cetuximab, and ionizing radiation, and potentiate DNA damage in vitro and reduce tumor growth in vivo in head and neck models. The safety and efficacy of Chk1/2 inhibition in combination with first line chemotherapy and radiation is currently being evaluated in clinical trials for advanced and metastatic head and neck cancers. In this review we will highlight the emerging molecular therapeutics targeting DNA damage response and repair checkpoints and review the available data from pre-clinical studies and on-going clinical trials.

Abbreviations

DNA-PK	DNA-dependent protein kinase
DSB	double strand break
EBV	Epstein-Barr virus
EGFR	epidermal growth factor receptor
HNSCC	head and neck squamous cell carcinoma
HPV	human papillomavirus
HRR	homologous recombination repair
NHEJ	non-homologous end joining
PDX	patient-derived xenografts
RPA	recurrent partition analysis
RT	radiation therapy
SCC	squamous cell carcinoma

Improving the Therapeutic Ratio in Head and Neck Cancer
https://doi.org/10.1016/B978-0-12-817868-3.00014-7

Conflict of interest

Dr. Yang has received research funding from Eli Lilly, is a consultant for AstraZeneca and is on the advisory board for Strata Oncology.

Introduction to head and neck cancers

Molecular pathogenesis of head and neck cancer

Head and neck is a molecularly heterogeneous group of malignancies with multiple different underlying etiologies [1–3]. While the majority of oropharyngeal cancer has been linked to human papillomavirus HPV infection, oral cavity and laryngeal cancers are associated with smoking and alcohol consumption, and nasopharyngeal cancers with Epstein-Barr virus (EBV) infection [4–10]. Oral leukoplakia and erythroplakia are the known precursor lesions of oral cavity SCC [7]. Mucosal damage caused by tobacco smoking contributes to activation of epidermal growth factor receptor (EGFR) and cyclin D1 overexpression, leading to dysregulated proliferation. Mutagenic effects of tobacco smoke are the likely contributors to genetic changes that include 3p and 9p LOH, CDKN2A (p16) inactivation by promoter methylation, EGFR amplification, and mutations in TP53 [7].

The HPV-positive oropharyngeal head and neck cancers represent the majority of oropharyngeal cancers. The most common high risk HPV type associated with oropharyngeal head and neck cancer is HPV16 [9, 10]. The two HPV genes directly involved in pathogenesis of head and neck cancer are *E6* and *E7*. HPV viral protein forms a complex with p53 and a cellular ubiquitin ligase E6AP, resulting in p53 ubiquitination and proteasome degradation. This functional inactivation of p53 has been linked to uncontrolled cell proliferation and genomic instability. In addition, E6 was shown to target DNA repair pathways via p53-dependent and p53-independent mechanisms. Indeed, we and others have shown DNA repair deficiencies in HPV+ head and neck cancer cell models [11–14]. For instance, HPV-positive head and neck cancer cells display increased sensitivity to radiation in clonogenic survival assays, and a reduced double strand break (DBS) repair capacity as compared to HPV-negative cells [11–14]. The observed significant delay in repair of DSBs in HPV-positive cells as compared to HPV-negative cells was presumably due to a deficiency in both non-homologous end joining (NHEJ) and in homologous recombination repair (HRR) [11–15]. Alterations in expression of DNA damage repair genes were also detected in HPV-positive cancers, including *XRCC1*, *ERCC2*, and *ERCC1* [16]. Recent evidence indicates that E6 can inhibit XRCC1 function and targets MGMT for proteasomal degradation [16].

Viral protein E7 inactivates Rb, leading to release of the transcription factor E2F, p16 overexpression, and cell cycle progression [9, 10]. Since both E6 and E7 inhibit post-mitotic checkpoints, they are believed to promote genomic instability by inducing polyploidy in cancer cells [16]. In addition, HPV infection had been shown to contribute to genomic instability via ectopic expression of viral E1 and E2 proteins. These viral proteins drive HPV replication from integrated viral genomes ultimately causing accumulation of chromosomal defects [16]. Given the exquisite sensitivity of HPV-positive head and neck cancer cells to DNA damaging agents due to their DNA repair deficiencies, which correlates well with improved treatment outcomes, the current approach to treating these cancers with concurrent chemoradiation with cisplatin seems appropriate (Table 1) [11–15, 18, 19].

TABLE 1 Risk stratification of head and neck cancer based on recurrent partition analysis (RPA) for OS [17]

Risk category	HPV status, pack-years of tobacco smoking, nodal and/or tumor stage	3-Year OS
Low risk	HPV-positive and ≤10 pack-years, or HPV-positive, >10 pack-years and N0-2a	93.0%
Intermediate risk	HPV-positive, >10 pack-years and N2b-3, or HPV-negative, ≤10 pack-years,T2-3	70.8%
High risk	HPV-negative, ≤10 and T4 or >10 pack-years	46.2%

TP53 is the most frequently mutated gene in HPV-negative head and neck cancer, with reported p53 mutation rates of 75–85% in HPV-negative head and neck squamous cell carcinoma (HNSCC) cases [20–22]. Mutated p53 status is associated with decreased overall survival and resistance to conventional therapy [23]. The anatomical sites with the most prevalent p53 mutations in head and neck cancer are larynx and hypopharynx (83.5%), followed by tongue and oral cavity (75.6%). Oropharyngeal HNSCCs, including tonsil and base of the tongue (BOT), have the lowest p53 mutation rates of approximately 28.6% [20–23]. Loss of TP53 function results in G1/S checkpoint deficiency, uncontrolled tumor cell proliferation, invasion and metastasis, and is thought to contribute to genetic instability [24, 25]. In addition, TP53 mutations render cells deficient in G1/S checkpoint and dependent on the remaining G2/M checkpoint to repair their DNA. Therefore, abrogation of the remaining G2/M checkpoint was hypothesized to sensitize these tumor cells to DNA damaging agents such as chemotherapy and radiation. The ATR/Chk1 pathway plays a key role on the G2/M checkpoint and consequently, inhibition of Chk1 activity results in the loss of this checkpoint and DNA repair defect. These may lead to a premature entry into mitosis with unrepaired DNA and cell death by mitotic catastrophe. Therefore, abrogation of the G2/S checkpoint with a Chk1 inhibitor is expected to sensitize head and neck cancers to ionizing radiation and other DNA damaging agents such as cisplatin [26].

In addition to TP53 mutations, EGFR amplification/overexpression was found in the majority of HPV-negative head and neck cancers and had been linked to replicative stress [11–15, 27–31]. Furthermore, an inverse correlation was observed between the HPV status (as assessed by p16 expression) and EGFR overexpression, indicating that HPV-positive cancers may be driven by a different biology and may not respond to EGFR-targeting therapies [28–31]. Importantly, EGFR signaling is induced by chemotherapy and radiation in cancer cells, and results in an enhanced capacity for repair of double strain breaks (DSBs) [11–15, 32]. Conversely, inhibition of EGFR signaling with cetuximab elicits a DNA repair deficiency in two major pathways of DNA repair, homologous recombination repair (HRR) and non-homologous end joining (NHEJ) [11–14]. Consequently, inhibition of EGFR with cetuximab combined with a G2/S checkpoint inhibitor and DNA damaging agent would force cancer cells with unrepaired DNA into premature mitosis, thus leading to cell death by mitotic catastrophe. Therefore, a combinatorial therapy with Chk1 inhibitor, cetuximab, and ionizing radiation may represent a promising therapeutic approach in HPV-negative head and neck cancer with evidence of the EGFR pathway activation.

Standard treatments and risk stratification

The first line treatment for the high risk locoregionally advanced head and neck cancer is a combination of radiation with concurrent chemotherapy [33]. The importance of adding concurrent chemotherapy (high dose cisplatin) to fractionated radiation therapy was established in a number of phase III randomized clinical trials [34]. Adelstein et al. showed that the addition of high dose bolus cisplatin to conventionally fractionated radiation treatments improves 3-year overall survival of patients locoregionally advanced unresectable HNSCC (overall survival of 23% with radiation alone, vs 37% for chemo-radiation) [34]. Likewise, the addition of concurrent carboplatin and 5-FU to definitive conventionally fractionated radiation treatment improved 5-year overall survival (22% vs 16% with radiation alone), disease-free survival (DFS, 27% vs 15% with radiation alone), and loco-regional control (LRC, 48% vs 25% with radiation alone) [35]. According to MACH-NC meta-analysis of chemo-radiation trials, the addition of chemotherapy to radiation treatment resulted in a 4.5% overall survival (OS) benefit, with concurrent chemo-radiation having an absolute OS benefit of 6.5% at 5 years [36, 37]. The highest 5-year absolute OS benefit with concomitant chemo-radiation was for oral cavity (8.9%) and oropharyngeal cancers (8.1%) [38].

A retrospective analysis of RTOG 0129 indicated that HPV-positive head and neck cancers behave differently compared to HPV negative. Indeed, the 3-year overall survival was dramatically improved for HPV-positive patients (OS, 82.4% vs 57.1%), with a 58% reduction in the risk of death [17]. The 3-year progression-free survival (PFS) was 73.7% in HPV-positive group vs 43.4% in HPV-negative group [17]. Consequently, a recurrent partition analysis for OS (RPA) divided head and neck cancer patients into three classes based on their HPV status, smoking, and T and N-stages. The low-risk patients were HPV-positive and had $\leq$10 pack year history of smoking, or HPV-positive with >10 pack-year history of smoking and had nodal stages N0-2a [17]. Intermediate-risk patients were either HPV-positive with >10 pack year smoking history and N2b-3, or HPV-negative with $\leq$10 pack year smoking history and T2-3 primary. High risk patients were either HPV-negative with $\leq$10 pack year smoking history and T4 primary, or HPV-negative with >10 pack-year smoking history [17] (Table 1). The 3-year OS was 93.0% in the low-risk group, 70.8% in the intermediate-risk group, and 46.2% in the high-risk group [17] (Table 1).

Targeted therapies for head and neck cancer

The role of biologics in the treatment of head and neck cancer is evolving. EGFR was found to be overexpressed in 90% of head and neck cancer cases. EGFR overexpression was shown to be of prognostic significance [39]. Indeed, high levels of EGFR expression are associated with reduced OS and DFS as compared to low EGFR expression cases [39]. Furthermore, EGFR overexpression was found to be inversely correlated with HPV status, and the HPV-negative/EGFR-positive combination was associated with poor response to therapy and reduced overall survival [28–31].

We and others have previously shown that activation of the EGFR pathway confers resistance to radiation by enhancing cellular DSB repair capacity [11–15]. The two major pathways of DNA repair responsible to resistance of tumor cells to chemotherapy and ionizing radiation are NHEJ and HRR [32]. While the HRR is a high fidelity DNA repair pathway that operates in the late S and G2 phases of the cell cycle, NHEJ is error-prone and is thought to be active in the G1 and early S-phases [32]. The NHEJ pathway is orchestrated by DNA-dependent protein kinase (DNA-PK), Ku70/Ku80, and DNA ligase XRCC4. In contrast, HRR involves multiple proteins including Rad51, ATM and BRCA1/2 [32]. As we and others have shown, the EGFR pathway is involved in increasing tumor cell DNA repair capacity by engaging both the NHEJ and HRR pathways [11–15, 32]. Indeed, chemotherapy (cisplatin) and radiation induce nuclear import of EGFR, leading to increased nuclear kinase activity of DNA-PK and an enhanced NHEJ [14].

Cetuximab, a chimeric monoclonal antibody to EGFR, reversed nuclear import of EGFR, blocked DNA repair, and sensitized tumor cells to radiation [13, 14]. Specifically, inhibition of EGFR suppressed HRR by attenuating DNA damage-induced Rad51 foci and cytoplasmic retention of BRCA1, a key protein responsible for HRR [15]. Cetuximab also reduced DNA-PK activation, a key player in NHEJ. These results indicated that cetuximab causes a deficiency in the two major DNA repair pathways, HRR and NHEJ [32]. Additionally, cetuximab may engage the immune system responses by triggering antibody-dependent cell-mediated toxicity (ADCC) and other effector functions [27]. These data provide mechanistic insight into the synergies between cetuximab and radiation, which was verified in a prospective randomized phase 3 trial evaluating the combination of cetuximab with concurrent radiation therapy (RT) for locoregionally advanced head and neck cancer [40, 41].

This trial showed a significant 5-year overall survival benefit compared to RT alone (OS of 45.6% vs 36.4%), and improved locoregional control (LRC, 24.4 vs 14.9 months) [40, 41]. However, RTOG 0522 phase 3 trial testing the hypothesis that the addition of cetuximab to radiation-cisplatin would improve progression free survival failed to show any improvement in outcomes with this combination, concluding that it should not be prescribed routinely [42]. Furthermore, the final results of RTOG1016 directly comparing cisplatin-radiation to cetuximab-radiation in HPV-positive locoregionally advanced head and neck cancer was reported recently [43]. Radiotherapy plus cetuximab did not meet the non-inferiority criteria for overall survival (HR 1.45, $P = .5056$ for non-inferiority). Importantly, a 5-year progression-free survival was significantly lower in the cetuximab group compared with the cisplatin group: 67.3% vs 78.4% (HR 1.72, $P = .0002$). The 5-year locoregional failure was significantly higher in the cetuximab group compared with the cisplatin group: 17.3% vs 9.9% [43]. The failure to demonstrate a non-inferiority of cetuximab may be explained by an inverse correlation between the HPV status and EGFR overexpression in tumor cells as discussed above. Indeed, HPV-positive head and neck cancers may be driven by viral oncogenic proteins E1, E2, E6 and E7, and not by EGFR signaling.

Despite the improved outcomes of combining radiation with either concurrent chemotherapy (cisplatin) or cetuximab compared to radiation alone in locoregionally advanced head and neck cancers, the relapses still occur, especially in HPV-negative high risk groups, with a 3-year OS <50% [17] (Table 1). Therefore, a new therapy is needed, especially for locoregionally advanced HPV-negative patients in the intermediate and high risk groups.

DNA damage response pathways

The ATM/Chk2 and ATR/Chk1 signaling pathways play a crucial role in cellular responses to genotoxic stress [26, 44, 45] (Figs. 1 and 2). Chk1 kinase was initially identified in yeast as a gene involved in cell cycle control in response to DNA damage caused by ionizing radiation [26]. Diverse types of DNA damaging insults, such as ultraviolet light, ionizing radiation (IR), or DNA-alkylating drugs have been shown to promote Chk1 phosphorylation, resulting in cycle arrest in the G2/M phase [46–49] (Fig. 1). Subsequent studies showed that Chk1 is phosphorylated by an ATM related kinase ATR in response to DNA damage, and is required for the G2/M cell cycle checkpoint induced by ionizing radiation [47]. Specifically, Chk1 phosphorylates and inactivates CDC25 phosphatases, which play a critical role in different phases of the cell cycle including in S phase and mitotic entry [48, 49]. The CDC25B and CDC25C phosphatases dephosphorylate CDK1 at Thr14 and Tyr15 thus promoting activation of the CDK1/cyclin B1 complex [48, 49]. Inhibitory phosphorylation of CDC25B and CDC25C by Chk1 promotes their inactivation and triggers G2/M arrest [46–49] (Figs. 1 and 2). Importantly, Chk1 also phosphorylates and activates Wee1, a kinase that is responsible for inhibitory phosphorylation of CDK1 on Thr14 and Tyr15 and G2/M arrest [50, 51].

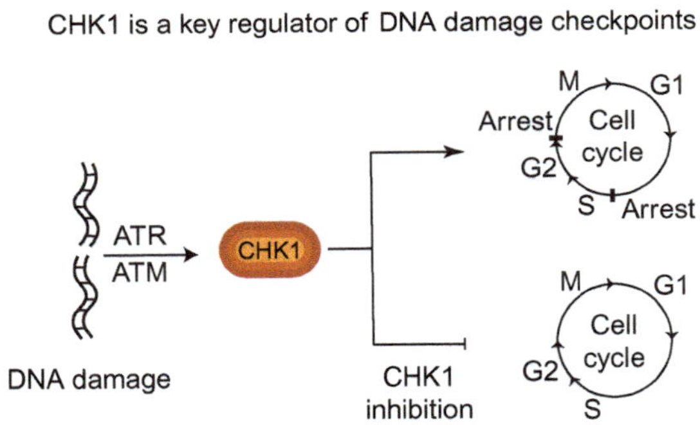

FIG. 1 The role of Chk1 kinase in regulation of cell cycle checkpoints. *Adopted with modifications from Lilly Oncology Pipeline with permissions from Eli Lilly Inc.*

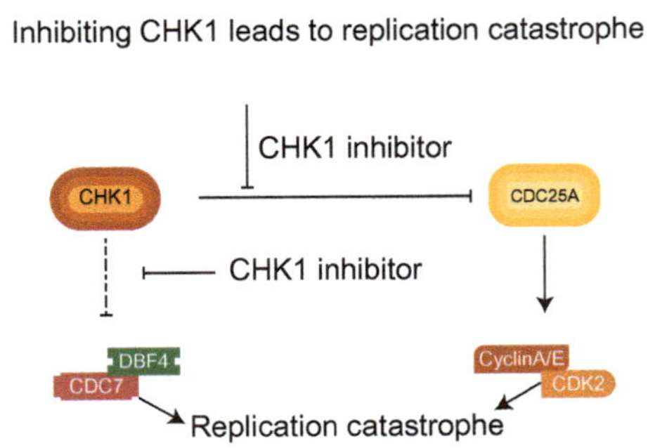

FIG. 2 Consequences of Chk1 inhibition in tumor cells. *Adopted with modifications from Lilly Oncology Pipeline with permissions from Eli Lilly Inc.*

In addition to its role in the G2/M checkpoint, Chk1 has also been shown to be involved in the replication checkpoint (intra-S phase checkpoint), by regulating the turnover of CDC25A, which controls the CDKs responsible for the S phase entry [52]. Furthermore, Chk1 is involved in the stabilization of stalled replication forks [53, 54]. Chk1 is thought to protect stalled replication forks from cleavage by endonucleases [55]. Inhibition of Chk1 in the presence of an antimetabolite may result in the cleavage of the stalled replication forks by endonucleases leading to accumulation of DNA strand breaks, and subsequent cell death [55].

Chk1 activity has also been implicated in homologous recombination repair of the damaged DNA [56, 57]. Specifically, Chk1-mediated phosphorylation of RAD51 appears to be necessary for its recruitment to the sites of single strand DNA in response to DNA damage or inhibition of replication by antimetabolites [56, 57]. Cells with depleted Chk1 are unable to form RAD51 foci in response to DNA damage, and mutation of the Chk1 phosphorylation site in RAD51 causes hypersensitivity to hydroxyurea [56, 57]. In addition, Chk1 phosphorylates the C-terminal domain of BRCA2 which is critical for interaction with RAD51 and subsequent recruitment of RAD51 to nuclear foci in response to a replication arrest, or exposure to DNA damaging agents such as ionizing radiation [58].

Upstream of Chk1 is an ATR kinase, which is activated by single strand DNA damage or by cross-linking agents that generate stalled replication forks [26, 44–49]. ATR is then recruited to the sites of DNA damage by single strand DNA coated with replicative protein A (RPA), where it phosphorylates and activates Chk1 [26, 44–49]. Both ATR and Chk1 knockout mice display early embryonic lethality [59, 60]. Blastocysts from Chk1 null mice show gross nuclear abnormalities, fail to arrest cell cycle in response to DNA replication block and DNA damage, and die of p53-independent apoptosis, indicating the essential role of Chk1 in the G2/M checkpoint in mammalian cells [59]. Embryonic stem cells with genetic deletion of Chk1 have a severe proliferation defect and lack a functional G2/M DNA damage checkpoint in response to ionizing radiation [59]. ATR null blastocysts accumulate chromosomal breaks and subsequently die of caspase-dependent apoptosis presumably secondary to compromised genomic integrity [60].

In contrast, ATM is activated in response to double strand breaks and is recruited to double strand break sites by the MRE11-RAD51-NBS1 complex [44, 45]. ATM subsequently phosphorylates Chk2 on its N-terminal regulatory domain, leading to its activation which involves dimerization and auto-phosphorylation. Activated Chk2 in turn triggers the G1-S checkpoint by directly phosphorylating p53 within its N-terminal transactivation domain. This Chk2-mediated phosphorylation diminishes the interaction between p53 and its negative regulator MDM2, and enhances its interaction with the transcriptional coactivator p300 [44, 45]. The net result is stabilization and activation of p53, which can then promote G1 arrest by inducing transcription of its target gene *CDKN1A*, an inhibitor of cyclin-dependent kinases [44, 45]. Chk2-deficient mice appeared normal, but are resistant to radiation-induced apoptosis, and are deficient in the G1/S cell cycle checkpoint [61]. Transcriptional induction of p53 target genes such as *CDKN1A* in response to ionizing radiation was not observed in Chk2 null cells, indicating that Chk2 is essential for p53-mediated G1 arrest caused by DNA damage [61]. Likewise, fibroblasts from ATM deficient mice displayed an abnormal radiation-induced G1/S checkpoint [62].

Inhibition of Chk1/Chk2 and Wee1 activity in cancer

Several lines of evidence indicate that abrogation of the G1/S and G2/M checkpoints via inhibition of Chk1/Chk2 activity may sensitize tumor cells to DNA damaging agents like chemotherapy and ionizing radiation. Indeed, Chk1 and Chk2 have well-established roles in the induction of intra-S phase, G2/M and G1/S checkpoints respectively in response to genotoxic stress [26, 44, 45, 63] (Figs. 1 and 2). Activation of these checkpoints in response to DNA damage is expected to promote cell survival by allowing DNA repair to occur. Since TP53 mutant cells are deficient in the G1/S checkpoint and are thought to rely on the remaining G2/M checkpoint to repair their DNA, abrogation of the G2/S checkpoint with a Chk1 inhibitor is expected to sensitize TP53 mutant cells to radiation and cisplatin chemotherapy. Additionally, given the important role Chk1 appears to play in the key steps of homologous recombination, including RAD51 loading to single strand DNA, strand invasion, and formation of recombination intermediates, inhibition of Chk1 activity in homologous recombination-deficient tumor cells (for example BRCA1/BRCA2 mutants) may result in premature mitotic entry, leading to cell death by mitotic catastrophe [56, 57, 63]. Therefore, inhibition of either Chk1 or Wee1 in combination with a DNA damaging agent such as cisplatin or ionizing radiation may result in the inappropriate cell cycle progression and cell death (Fig. 2).

To this end, UCN-01 was the first Chk1 inhibitor reported in the literature [64, 65]. It blocks Chk1 activity in vitro and CDC25C phosphorylation in cells in response to DNA damage [65]. Inhibition of Chk1 also caused an increase in late replication origin firing in S-phase arrested cells, and blocked elongation of previously arrested forks, indicating that Chk1 activity is required for both intra-S-phase checkpoint and for maintenance of the replication fork integrity [66]. UCN-01 showed a robust in vivo antitumor activity as a single agent when tested in p53 mutant head and neck squamous cell carcinoma (HNSCC) xenografts in nude mice [67]. The clinical development of this inhibitor was halted due to lack of specificity (off target inhibition of protein kinase C), unusually long half-life, and unpredictable clinical pharmacology due to the binding to alpha1-acid glycoprotein in plasma [68].

A selective Chk1 inhibitor MK-8776 induced DNA double strand breaks when combined with hydroxyurea, possibly due to a collapse of stalled replication forks, and sensitized cancer cells to chemotherapy including cytarabine, gemcitabine, cisplatin, and 5-fuorouracil [69]. This selective inhibitor also demonstrated an improved radiosensitizing effect in a clonogenic assay as compared to UCN-01 (SER of 1.39 vs 1.07 in Hela cells), and exacerbated radiation-induced cell death, likely due to an enhancement of aberrant mitosis [70]. A randomized phase II trial of MK-8776 combined with cytosine arabinoside in relapsed and refractory acute myeloid leukemia did not show a survival benefit despite a robust increase DNA damage in circulating blasts as assessed by gamma-H2AX [71].

A potent Chk1/Chk2 inhibitor AZD7762 sensitized p53 mutant head and neck cancer cells to cisplatin in a synthetically lethal manner, and induced polyploidy and mitotic catastrophe when combined with cisplatin [72]. AZD7762 also inhibited RAD51 focus formation and homologous recombination repair in p53 mutant cells [73]. A Wee1 inhibitor PD0166285 that was identified in a high throughout screen inhibited ionizing radiation-induced phosphorylation of CDK1, abrogated G2/M arrest, and sensitized p53 mutant cells to radiation with a sensitivity enhancement ratio (SER) of 1.23 [74]. Interestingly, restoration of wild type p53 function reversed the effects of PD0166285 Tyr15 phosphorylation, G2/M checkpoint abrogation and radiation induced cell death [75]. A highly specific Wee1 inhibitor AZD1775 sensitized tumor cells to radiation via the induction of replicative stress independent of p53 mutational status [76]. Importantly, an intact homologous recombination repair (BRCA2 wild-type) was required for AZD1775-mediated sensitization of cancer cells to chemoradiation [77]. A triple therapy with Chk1 inhibitor CCT244747, paclitaxel, and radiation had a synergistic effect on elimination of radioresistant p53-deficient head and neck cancer cells, especially when radiation was delivered 6h after Chk1 inhibition in vitro [78]. This triple therapy combination was also efficacious in delaying tumor growth in radioresistant p53-deficient xenograft models [78]. An orally bio-available selective Chk1 inhibitor GNE-900 demonstrated more rapid kinetics of checkpoint abrogation, premature mitotic entry and accelerated cell death in p53 mutant cells as compared p53 wild type cells [79].

Pre-clinical and clinical development of Chk1/Chk2 inhibitor prexasertib in head and neck cancer

Prexasertib (LY2606368) is a novel Chk1/Chk2 inhibitor which had been recently shown to cause DNA double strand breaks in the S-phase cell population and cell death as single agent presumably by causing replication catastrophe (Fig. 3) [80]. Combination of prexasertib with 2Gy dose of ionizing radiation decreased proliferation of head and neck cancer cell lines independent of p53 and HPV status [81]. It also increased cell death as a single agent and potentiated cell killing when combined with ionizing radiation. On the molecular level, prexasertib caused persistent DNA double strand breaks (γ-H2AX), and caspase-3 cleavage in p53 wild-type, null, and mutant cells, as well as in HPV+ cells [81]. Consistent with earlier

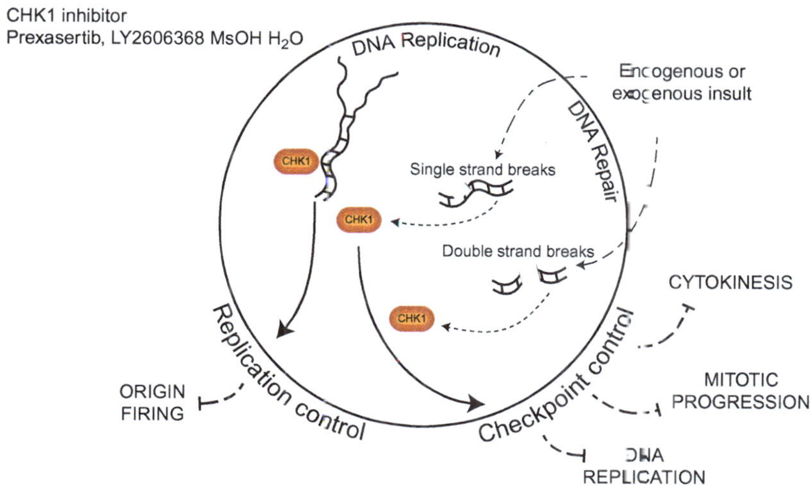

FIG. 3 The role of Chk1 inhibitor Prexasertib in abrogating G2/M and S-phase checkpoint controls in cancer cells. *Adopted with modifications from Lilly Oncology with permissions from Eli Lilly Inc.*

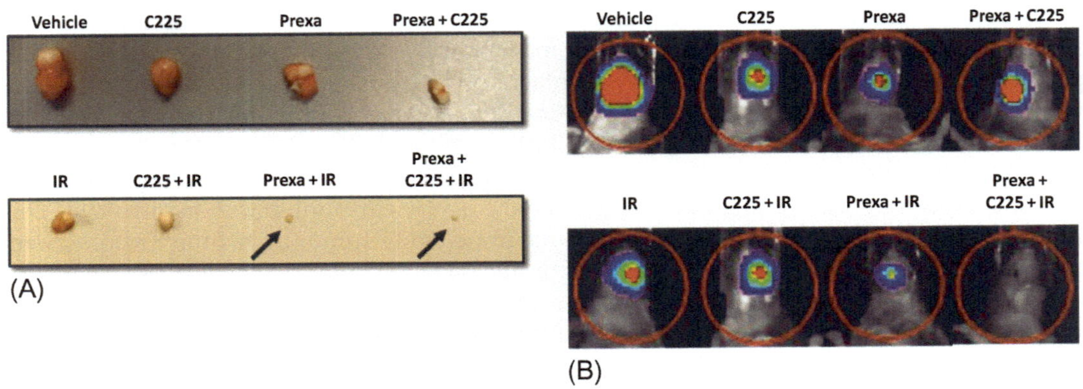

FIG. 4 Chk1 inhibitor Prexasertib delays tumor growth in both (A) HPV negative orthotopic and (B) HPV+ heterotopic xenograft models when combined with Cetuximab and IR. *Adopted from Zeng L, et al. Combining Chk1/2 inhibition with cetuximab and radiation enhances in vitro and in vivo cytotoxicity in head and neck squamous cell carcinoma. Mol Cancer Ther 2017;16(4):591–600 with permissions from AACR.*

reports, prexasertib reduced the proportion of cells in the G1 phase and induced accumulation of cells in the S-phase, possibly reflecting the accumulation of DNA damage during the replicative phase of the cell cycle. Prexasertib was also efficacious in vivo, both as a single agent, and when combined with ionizing radiation, causing a tumor growth delay in mice bearing orthotopic UM-SCC1-luciferase (p53 null) and heterotopic UM-SCC47 (p53 wild-type, HPV+) HNSCC xenograft mice [81] (Fig. 4).

These preclinical results supported further clinical development of prexasertib in head and neck cancer. A phase I dose-escalation study of prexasertib (LY2606368) in patients with solid tumors including squamous cell carcinomas (SCC) revealed dose-limiting hematological toxicities and established the maximum-tolerated dose of 105 mg/m^2 once every 14 days [82]. 33% of patients had a best overall response of stable disease, 6/15 were SCC [82]. Prexasertib monotherapy was further evaluated in patients with advanced SCC including head and neck (HNSCC) [83]. Median progression-free survival was 1.6 months for HNSCC. The rate of complete response, partial response, or stable disease for patients with HNSCC in this phase Ib study was 28%. Given the acceptable safety profile and activity as a single agent in patients with advanced HNSCC, it was recommended for a phase 2 trial. A list of current trials with prexasertib is listed in Table 2.

Combination therapy with prexasertib has also been tested. Cetuximab was shown to promote DNA damage by causing a transient inhibition of non-homologous end joining (NHEJ) and homologous recombination-mediated repair of double strand breaks [18, 19]. Furthermore, by creating a transient DNA repair deficiency, cetuximab was shown to potentiate cytotoxic effects of PARP inhibition [18, 19]. Given the above described effects of cetuximab on DNA repair pathways and the involvement of Chk1 in homologous recombination repair and checkpoint control, it was hypothesized that combined inhibition of Chk1 activity with prexasertib and EGFR signaling with cetuximab would sensitize head and neck cancer cells to radiation [81]. Indeed, a combination of prexasertib with cetuximab and ionizing radiation decreased cell proliferation and potentiated apoptotic cell death presumably by abrogating cetuximab and IR-induced checkpoint activation [81]. When tested in HNSCC xenografts

TABLE 2 Clinical trials with prexasertib

Trial identifier	Detail	Status
NCT01115790	A phase 1 dose escalation study of Chk1 inhibitor LY2606368 in participants with advanced cancer Safety and toxicity of prexasertib in participants with advanced or metastatic cancer, or squamous cell cancer of the head and neck, or squamous cell cancer of any tumor type	Phase 1, completed [82]
NCT01115790	Evaluation of Prexasertib, a Checkpoint Kinase 1 Inhibitor, in a Phase Ib Study of Patients With Squamous Cell Carcinoma	Phase 1, completed [83]
NCT02555644	A Study of Prexasertib (LY2606368) With Chemotherapy and Radiation in Participants With Head and Neck Cancer	Phase 1, recruiting
NCT02124148	A Study of Prexasertib (LY2606368) With Chemotherapy or Targeted Agents in Participants With Advanced Cancer	Phase 1, recruiting
NCT02873975	A Study of LY2606368 (Prexasertib) in Patients With Solid Tumors With Replicative Stress or Homologous Repair Deficiency	Phase 2, recruiting
NCT03057145	Combination Study of Prexasertib and Olaparib in Patients With Advanced Solid Tumors	Phase 1, recruiting
NCT03495323	A Study of Prexasertib (LY2606368), CHK1 Inhibitor, and LY3300054, PD-L1 Inhibitor, in Patients With Advanced Solid Tumors	Phase 1, recruiting

in mice, triple therapy with prexasertib plus cetuximab and radiation induced tumor growth delay of both HPV-negative and HPV-positive xenografts [81] (Fig. 4).

Based on these data, a phase 1b trial of Prexasertib in combined with radiation and either Cisplatin (Part A) or Cetuximab (Part B) as a definitive therapy in non-metastatic previously untreated head and neck cancer patients is currently on-going (NCT02555644, Table 2). A phase 1 clinical trial testing Prexasertib together with other anti-cancer drugs (cisplatin, cetuximab, pemetrexed, fluorouracil or PI3K inhibitor LY3023414) in recurrent advanced or metastatic head and neck cancer is also ongoing (NCT02124148, Table 2).

Resistance mechanisms to CHK1 inhibition

The mechanisms of resistance to Chk1 inhibitors are poorly understood. Early studies with the first Chk1 inhibitor UCN-01 have shown that Chk1 inhibition causes a prolonged activation of the MAPK pathway including MEK1/2 and Erk1/2 phosphorylation [84]. Consistent with this observation, genetic deletion of Chk1 in Drosophila resulted in an increase in MEK activation [85]. The combination of Chk1 and MEK1/2 inhibitors suppressed UNC-01-stimulated MAPK activity, potentiated apoptosis and acted synergistically with radiation in clonogenic survival assays, indicating that a concurrent inhibition of the MEK/ERK pathway may be required to overcome the resistance to Chk1 inhibitors [84]. Furthermore, the PI3K pathway had been recently implicated in the resistance to the Chk1 inhibitor prexasertib [86]. Indeed, an M-molecular subtype of a triple negative breast cancer patient-derived xenografts (PDX) that displayed resistance to prexasertib, also showed a transcriptional upregulation of the PI3K/mTOR pathway [87]. Consistent with this result, a dual inhibition of the Chk1 and PI3K/mTOR pathways showed an additive effect in this PDX model [86]. A recent whole genome screen identified loss of FAM122A, a highly conserved protein that inhibits PP2A, as another potential mechanism of resistance to Chk1 inhibition [88]. In summary, the mechanisms of Chk1 inhibition appear to be multifactorial, and involve distinct molecular pathways (MAPK, PI3K/mTOR). Therefore, multimodality therapy combining several distinct pathway inhibitors with radiation and chemotherapy may be required to overcome the reported resistance.

Future directions

As the clinical trials testing the hypothesis of checkpoint inhibition in head and neck cancer are ongoing, the evaluation for possible synergistic effects of combining checkpoint inhibitors with other targeted molecular therapeutics in different genetic backgrounds warrant further investigation. The concept of synthetic lethality was initially introduced in 1945 by a Drosophila geneticist at Columbia University, the term used to describe genes or parts of chromosomes which are not lethal when taken separately, but become lethal when combined together in a cell or an organism [87]. The concept was subsequently used in cancer research to describe synthetic lethal interactions between BRCA1/2 mutations which result in a deficient HR, and poly(ADP-ribose) polymerase PARP inhibitors [89].

Given the above described loss of the G1/S checkpoint by p53 mutant tumors, and the reliance of these tumors on the remaining intra-S phase and G2/M checkpoints, Chk1 inhibitors were predicted to have synthetic lethal interactions with p53 mutations [90]. Indeed, inhibition of Chk1/2 activity was recently shown to sensitize p53 deficient HNSCC cells to cisplatin via induction of mitotic catastrophe in a synthetically lethal manner [72]. Likewise, Chk1 inhibition produced a synthetic lethality in either p53- or p21-deficient cells exposed to DNA damage [91]. Moreover, a synthetic lethality interaction was also described between the Chk1 upstream kinase ATR and p53 deficiency in tumor cells and adult mice [92]. Interestingly, a synthetic lethality interaction was also described between Chk1 and ATR in cancer cells, indicating that simultaneous inhibition of both Chk1 and ATR may be beneficial when combined with DNA-damaging cancer therapy [92]. Indeed, a combined treatment with the above described Chk1 inhibitor AZD7762 and ATR inhibitor VE-821 resulted in the arrest of replication forks, accumulation of single strand DNA, and synergistic cell death by replication catastrophe, presumably due to an excessive CDK-mediated origin firing in tumor cells [73].

Finally, a synthetic lethality had recently been identified between Chk1 inhibition and loss-of-function mutations of enhancer of zeste homolog 2 (EZH2), an enzymatic component of the polycomb repressive complex 2 (PRC2), which has been implicated in a variety of malignancies including head and neck cancer [93]. Furthermore, the inhibition of EZH2 histone methyltransferase activity with an EZH2 inhibitor, GSK126, sensitized tumor cells to Chk1 inhibition [94]. Intriguingly, a synthetic lethality was recently described between BRCA1 and BRCA2-deficient cells and Chk1 inhibition [95, 96]. Loss of BRCA1 sensitized tumor cells to Chk1 inhibitors in vitro and in vivo, and produced excessive amounts of DNA damage and inability to complete the S-phase, while the loss of BRCA2 sensitized pancreatic cancer cells to Chk1 inhibition depending on TP53 and KRAS status [95, 96].

In summary, recent experimental evidence supports further clinical development of Chk1/2 inhibitor therapies in combination with DNA damaging agents, especially in tumor cells with homologous recombination repair and G1/S checkpoint deficiencies. More work needs to be done to assess whether the combinations of Chk1 inhibitors with other molecular therapies such as ATR, PARP or EZH2 inhibitors in specific tumor genetic backgrounds (BRCA1/2 and TP53 mutations) would synergize with first line DNA damaging chemotherapy agents and ionizing radiation in providing a tangible therapeutic benefit for head and neck cancer patients.

Acknowledgment

Work described in this chapter was supported, in part, by research funding from Eli Lilly.

References

[1] Vigneswaran N, Williams MD. Epidemiologic trends in head and neck cancer and aids in diagnosis. Oral Maxillofac Surg Clin North Am 2014;26(2):123–41.
[2] American Cancer Society. Cancer facts & figures. The Society: Atlanta, GA; 2019. p. 1–71.
[3] Siegel RL, Miller KD, Jemal A. Cancer statistics, 2016. CA Cancer J Clin 2016;66(1):7–30.
[4] Ramroth H, et al. Occupational wood dust exposure and the risk of laryngeal cancer: a population based case-control study in Germany. Am J Ind Med 2008;51(9):648–55.
[5] Peng WJ, Mi J, Jiang YH. Asbestos exposure and laryngeal cancer mortality. Laryngoscope 2016;126(5):1169–74.
[6] Chang ET, Adami HO. The enigmatic epidemiology of nasopharyngeal carcinoma. Cancer Epidemiol Biomark Prev 2006;15(10):1765–77.
[7] Lippman SM, Sudbo J, Hong WK. Oral cancer prevention and the evolution of molecular-targeted drug development. J Clin Oncol 2005;23(2):346–56.
[8] Chan AT, Teo PM, Huang DP. Pathogenesis and treatment of nasopharyngeal carcinoma. Semin Oncol 2004;31(6):794–801.
[9] Osazuwa-Peters N, et al. 40-year incidence trends for oropharyngeal squamous cell carcinoma in the United States. Oral Oncol 2017;74:90–7.
[10] Dok R, Nuyts S. HPV positive head and neck cancers: molecular pathogenesis and evolving treatment strategies. Cancers (Basel) 2016;8(4):41–56.
[11] Rieckmann T, et al. HNSCC cell lines positive for HPV and p16 possess higher cellular radiosensitivity due to an impaired DSB repair capacity. Radiother Oncol 2013;107(2):242–6.
[12] Weaver AN, et al. DNA double strand break repair defect and sensitivity to poly ADP-ribose polymerase (PARP) inhibition in human papillomavirus 16-positive head and neck squamous cell carcinoma. Oncotarget 2015;6(29):26995–7007.
[13] Huang SM, Harari PM. Modulation of radiation response after epidermal growth factor receptor blockade in squamous cell carcinomas: inhibition of damage repair, cell cycle kinetics, and tumor angiogenesis. Clin Cancer Res 2000;6(6):2166–74.
[14] Dittmann K, et al. Radiation-induced epidermal growth factor receptor nuclear import is linked to activation of DNA-dependent protein kinase. J Biol Chem 2005;280(35):31182–9.
[15] Li L, et al. Erlotinib attenuates homologous recombinational repair of chromosomal breaks in human breast cancer cells. Cancer Res 2008;68(22):9141–6.
[16] Prati B, Marangoni B, Boccardo E. Human papillomavirus and genome instability: from productive infection to cancer. Clinics (Sao Paulo) 2018;73(Suppl 1):e539s.
[17] Ang KK, et al. Human papillomavirus and survival of patients with oropharyngeal cancer. N Engl J Med 2010;363(1):24–35.

[18] Nowsheen S, et al. Cetuximab augments cytotoxicity with poly (adp-ribose) polymerase inhibition in head and neck cancer. PLoS ONE 2011;6(8).
[19] Bonner JA, et al. Enhancement of cetuximab-induced radiosensitization by JAK-1 inhibition. BMC Cancer 2015;15:673.
[20] Poeta ML, et al. TP53 mutations and survival in squamous-cell carcinoma of the head and neck. N Engl J Med 2007;357(25):2552–61.
[21] Stransky N, et al. The mutational landscape of head and neck squamous cell carcinoma. Science 2011;333(6046):1157–60.
[22] Pickering CR, et al. Integrative genomic characterization of oral squamous cell carcinoma identifies frequent somatic drivers. Cancer Discov 2013;3(7):770–81.
[23] Zhou G, Liu Z, Myers JN. TP53 mutations in head and neck squamous cell carcinoma and their impact on disease progression and treatment response. J Cell Biochem 2016;117(12):2682–92.
[24] Muller PA, Vousden KH. p53 mutations in cancer. Nat Cell Biol 2013;15(1):2–8.
[25] Hanahan D, Weinberg RA. Hallmarks of cancer: the next generation. Cell 2011;144(5):646–74.
[26] Thompson R, Eastman A. The cancer therapeutic potential of Chk1 inhibitors: how mechanistic studies impact on clinical trial design. Br J Clin Pharmacol 2013;76(3):358–69.
[27] Vincenzi B, et al. The biological properties of cetuximab. Crit Rev Oncol Hematol 2008;68(2):93–106.
[28] Reimers N, et al. Combined analysis of HPV-DNA, p16 and EGFR expression to predict prognosis in oropharyngeal cancer. Int J Cancer 2007;120(8):1731–8.
[29] Kumar B, et al. EGFR, p16, HPV titer, Bcl-xL and p53, sex, and smoking as indicators of response to therapy and survival in oropharyngeal cancer. J Clin Oncol 2008;26(19):3128–37.
[30] Hong A, et al. Relationships between epidermal growth factor receptor expression and human papillomavirus status as markers of prognosis in oropharyngeal cancer. Eur J Cancer 2010;46(11):2088–96.
[31] Kong CS, et al. The relationship between human papillomavirus status and other molecular prognostic markers in head and neck squamous cell carcinomas. Int J Radiat Oncol Biol Phys 2009;74(2):553–61.
[32] Jackson SP, Bartek J. The DNA-damage response in human biology and disease. Nature 2009;461(7267):1071–8.
[33] National Comprehensive Cancer Network. Head and neck cancers. Version 2. 2018. Available from: https://www.nccn.org/professionals/physician_gls/pdf/head-and-neck.pdf; 2018 [cited 06/20/2018].
[34] Adelstein DJ, et al. An intergroup phase III comparison of standard radiation therapy and two schedules of concurrent chemoradiotherapy in patients with unresectable squamous cell head and neck cancer. J Clin Oncol 2003;21(1):92–8.
[35] Calais G, et al. Randomized trial of radiation therapy versus concomitant chemotherapy and radiation therapy for advanced-stage oropharynx carcinoma. J Natl Cancer Inst 1999;91(24):2081–6.
[36] Pignon JP, et al. Chemotherapy added to locoregional treatment for head and neck squamous-cell carcinoma: three meta-analyses of updated individual data. MACH-NC Collaborative Group. Meta-analysis of chemotherapy on head and neck cancer. Lancet 2000;355(9208):949–55.
[37] Pignon JP, et al. Meta-analysis of chemotherapy in head and neck cancer (MACH-NC): an update on 93 randomised trials and 17,346 patients. Radiother Oncol 2009;92(1):4–14.
[38] Blanchard P, et al. Meta-analysis of chemotherapy in head and neck cancer (MACH-NC): a comprehensive analysis by tumour site. Radiother Oncol 2011;100(1):33–40.
[39] Ang KK, et al. Impact of epidermal growth factor receptor expression on survival and pattern of relapse in patients with advanced head and neck carcinoma. Cancer Res 2002;62(24):7350–6.
[40] Bonner JA, et al. Radiotherapy plus cetuximab for squamous-cell carcinoma of the head and neck. N Engl J Med 2006;354(6):567–78.
[41] Bonner JA, et al. Radiotherapy plus cetuximab for locoregionally advanced head and neck cancer: 5-year survival data from a phase 3 randomised trial, and relation between cetuximab-induced rash and survival. Lancet Oncol 2010;11(1):21–8.
[42] Ang KK, et al. Randomized phase III trial of concurrent accelerated radiation plus cisplatin with or without cetuximab for stage III to IV head and neck carcinoma: RTOG 0522. J Clin Oncol 2014;32(27):2940–50.
[43] Gillison ML, et al. Radiotherapy plus cetuximab or cisplatin in human papillomavirus-positive oropharyngeal cancer (NRG oncology RTOG 1016): a randomised, multicentre, non-inferiority trial. Lancet 2018;393:40–50.
[44] Bartek J, Lukas J. Chk1 and Chk2 kinases in checkpoint control and cancer. Cancer Cell 2003;3(5):421–9.
[45] Manic G, et al. Trial watch: targeting ATM-CHK2 and ATR-CHK1 pathways for anticancer therapy. Mol Cell Oncol 2015;2(4).

[46] Walworth N, Davey S, Beach D. Fission yeast chk1 protein kinase links the rad checkpoint pathway to cdc2. Nature 1993;363(6427):368–71.
[47] Liu Q, et al. Chk1 is an essential kinase that is regulated by Atr and required for the G(2)/M DNA damage checkpoint. Genes Dev 2000;14(12):1448–59.
[48] Boutros R, Lobjois V, Ducommun B. CDC25 phosphatases in cancer cells: key players? Good targets? Nat Rev Cancer 2007;7(7):495–507.
[49] Peng CY, et al. Mitotic and G2 checkpoint control: regulation of 14-3-3 protein binding by phosphorylation of Cdc25C on serine-216. Science 1997;277(5331):1501–5.
[50] Lee J, Kumagai A, Dunphy WG. Positive regulation of Wee1 by Chk1 and 14-3-3 proteins. Mol Biol Cell 2001;12(3):551–63.
[51] O'Connell MJ, et al. Chk1 is a wee1 kinase in the G2 DNA damage checkpoint inhibiting cdc2 by Y15 phosphorylation. EMBO J 1997;16(3):545–54.
[52] Sorensen CS, et al. Chk1 regulates the S phase checkpoint by coupling the physiological turnover and ionizing radiation-induced accelerated proteolysis of Cdc25A. Cancer Cell 2003;3(3):247–58.
[53] Scorah J, McGowan CH. Claspin and Chk1 regulate replication fork stability by different mechanisms. Cell Cycle 2009;8(7):1036–43.
[54] Petermann E, Woodcock M, Helleday T. Chk1 promotes replication fork progression by controlling replication initiation. Proc Natl Acad Sci U S A 2010;107(37):16090–5.
[55] Forment JV, et al. Structure-specific DNA endonuclease Mus81/Eme1 generates DNA damage caused by Chk1 inactivation. PLoS ONE 2011;6(8).
[56] Sorensen CS, et al. The cell-cycle checkpoint kinase Chk1 is required for mammalian homologous recombination repair. Nat Cell Biol 2005;7(2):195–201.
[57] Fujinaka Y, et al. ATR-Chk1 signaling pathway and homologous recombinational repair protect cells from 5-fluorouracil cytotoxicity. DNA Repair (Amst) 2012;11(3):247–58.
[58] Bahassi EM, et al. The checkpoint kinases Chk1 and Chk2 regulate the functional associations between hBRCA2 and Rad51 in response to DNA damage. Oncogene 2008;27(28):3977–85.
[59] Takai H, et al. Aberrant cell cycle checkpoint function and early embryonic death in Chk1(-/-) mice. Genes Dev 2000;14(12):1439–47.
[60] Brown EJ, Baltimore D. ATR disruption leads to chromosomal fragmentation and early embryonic lethality. Genes Dev 2000;14(4):397–402.
[61] Takai H, et al. Chk2-deficient mice exhibit radioresistance and defective p53-mediated transcription. EMBO J 2002;21(19):5195–205.
[62] Barlow C, et al. Atm-deficient mice: a paradigm of ataxia telangiectasia. Cell 1996;86(1):159–71.
[63] Qiu Z, Oleinick NL, Zhang J. ATR/CHK1 inhibitors and cancer therapy. Radiother Oncol 2018 126(3):450–64.
[64] Bunch RT, Eastman A. Enhancement of cisplatin-induced cytotoxicity by 7-hydroxystaurosporine (UCN-01), a new G2-checkpoint inhibitor. Clin Cancer Res 1996;2(5):791–7.
[65] Graves PR, et al. The Chk1 protein kinase and the Cdc25C regulatory pathways are targets of the anticancer agent UCN-01. J Biol Chem 2000;275(8):5600–5.
[66] Feijoo C, et al. Activation of mammalian Chk1 during DNA replication arrest: a role for Chk1 in the intra-S phase checkpoint monitoring replication origin firing. J Cell Biol 2001;154(5):913–23.
[67] Patel V, et al. Antitumor activity of UCN-01 in carcinomas of the head and neck is associated with altered expression of cyclin D3 and p27(KIP1). Clin Cancer Res 2002;8(11):3549–60.
[68] Fuse E, et al. Unpredicted clinical pharmacology of UCN-01 caused by specific binding to human alpha1-acid glycoprotein. Cancer Res 1998;58(15):3248–53.
[69] Montano R, et al. Preclinical development of the novel Chk1 inhibitor SCH900776 in combination with DNA-damaging agents and antimetabolites Mol Cancer Ther 2012;11(2):427–38.
[70] Suzuki M, et al. MK-8776, a novel Chk1 inhibitor, exhibits an improved radiosensitizing effect compared to UCN-01 by exacerbating radiation-induced aberrant mitosis. Transl Oncol 2017;10(4):491–500.
[71] Webster JA, et al. Randomized phase II trial of cytosine arabinoside with and without the CHK1 inhibitor MK-8776 in relapsed and refractory acute myeloid leukemia. Leuk Res 2017;61:108–16.
[72] Gadhikar MA, et al. Chk1/2 inhibition overcomes the cisplatin resistance of head and neck cancer cells secondary to the loss of functional p53. Mol Cancer Ther 2013;12(9):1860–73.
[73] Morgan MA, et al. Mechanism of radiosensitization by the Chk1/2 inhibitor AZD7762 involves abrogation of the G2 checkpoint and inhibition of homologous recombinational DNA repair. Cancer Res 2010;70(12):4972–81.

[74] Wang Y, et al. Radiosensitization of p53 mutant cells by PD0166285, a novel G(2) checkpoint abrogator. Cancer Res 2001;61(22):8211–7.
[75] Li J, et al. Wild-type TP53 inhibits G(2)-phase checkpoint abrogation and radiosensitization induced by PD0166285, a WEE1 kinase inhibitor. Radiat Res 2002;157(3):322–30.
[76] Cuneo KC, et al. Wee1 kinase inhibitor AZD1775 radiosensitizes hepatocellular carcinoma regardless of TP53 mutational status through induction of replication stress. Int J Radiat Oncol Biol Phys 2016;95(2):782–90.
[77] Kausar T, et al. Sensitization of pancreatic cancers to gemcitabine chemoradiation by WEE1 kinase inhibition depends on homologous recombination repair. Neoplasia 2015;17(10):757–66.
[78] Barker HE, et al. CHK1 inhibition radiosensitizes head and neck cancers to paclitaxel-based chemoradiotherapy. Mol Cancer Ther 2016;15(9):2042–54.
[79] Blackwood E, et al. Combination drug scheduling defines a "window of opportunity" for chemopotentiation of gemcitabine by an orally bioavailable, selective ChK1 inhibitor, GNE-900. Mol Cancer Ther 2013;12(10):1968–80.
[80] King C, et al. LY2606368 causes replication catastrophe and antitumor effects through CHK1-dependent mechanisms. Mol Cancer Ther 2015;14(9):2004–13.
[81] Zeng L, et al. Combining Chk1/2 inhibition with cetuximab and radiation enhances in vitro and in vivo cytotoxicity in head and neck squamous cell carcinoma. Mol Cancer Ther 2017;16(4):591–600.
[82] Hong D, et al. Phase I study of LY2606368, a checkpoint kinase 1 inhibitor, in patients with advanced cancer. J Clin Oncol 2016;34(15):1764–71.
[83] Hong DS, et al. Evaluation of prexasertib, a checkpoint kinase 1 inhibitor, in a phase Ib study of patients with squamous cell carcinoma. Clin Cancer Res 2018;24(14):3263–72.
[84] McKinstry R, et al. Inhibitors of MEK1/2 interact with UCN-01 to induce apoptosis and reduce colony formation in mammary and prostate carcinoma cells. Cancer Biol Ther 2002;1(3):243–53.
[85] Mogila V, Xia F, Li WX. An intrinsic cell cycle checkpoint pathway mediated by MEK and ERK in *Drosophila*. Dev Cell 2006;11(4):575–82.
[86] Ding Y, Jiangang L, Calley JN, Qian HR, Iversen PW, Ebert PJ, Beckmann RP, Donoho GP, Martinez R, Wu W, Lin AB, Bowden E, Aggarwal A. PI3K/AKT signaling pathway is transcriptionally elevated in prexasertib-resistant TNBC PDX models [abstract]. Proceedings of the American Association for Cancer Research annual meeting 2018, Chicago, IL, Cancer Res 2018;78(13 Suppl):2586.
[87] Dobzhansky T. Genetics of natural populations; recombination and variability in populations of *Drosophila pseudoobscura*. Genetics 1946;31:269–90.
[88] Deraska P, Reavis H, Labe S, D'Andrea A, Kozono D. Whole-genome CRISPR/Cas9 screen of the CHK1 inhibitor prexasertib implicates FAM122A loss as a potential resistance mechanism. Proceedings of the American Association for Cancer Research annual meeting, Chicago, IL, Cancer Res 2018;78(13 Suppl):4287.
[89] Lord CJ, Ashworth A. PARP inhibitors: synthetic lethality in the clinic. Science 2017;355(6330):1152–8.
[90] Gurpinar E, Vousden KH. Hitting cancers' weak spots: vulnerabilities imposed by p53 mutation. Trends Cell Biol 2015;25(8):486–95.
[91] Origanti S, et al. Synthetic lethality of Chk1 inhibition combined with p53 and/or p21 loss during a DNA damage response in normal and tumor cells. Oncogene 2013;32(5):577–88.
[92] Ruzankina Y, et al. Tissue regenerative delays and synthetic lethality in adult mice after combined deletion of Atr and Trp53. Nat Genet 2009;41(10):1144–9.
[93] Leon ET, Rapoz-D'Silva T, Rahman S, Magnussen M, Farah N, Guerra-Assunção JA, Gupta R, Nacheva EP, Herrero J, Linch DC, Mansour MR. Synthetic lethal screening identifies CHK1 inhibition as an exploitable vulnerability in EZH2 deficient T-ALL. 59th annual meeting of the American-Society-of-Hematology (ASH), Atlanta, GA, Blood 2017;130(Suppl 1):1325.
[94] Chang JW, et al. EZH2 is associated with poor prognosis in head-and-neck squamous cell carcinoma via regulating the epithelial-to-mesenchymal transition and chemosensitivity. Oral Oncol 2016;52:66–74.
[95] Paculova H, et al. BRCA1 or CDK12 loss sensitizes cells to CHK1 inhibitors. Tumour Biol 2017;39(10).
[96] Hattori H, et al. Context dependence of checkpoint kinase 1 as a therapeutic target for pancreatic cancers deficient in the BRCA2 tumor suppressor. Mol Cancer Ther 2011;10(4):670–8.

CHAPTER

15

GPCRs in head and neck squamous cell carcinoma

Victoria H. Wu, J. Silvio Gutkind

Department of Pharmacology, UCSD Moores Cancer Center, La Jolla, CA, United States

Abstract

Current standard therapies for head and neck squamous cell carcinoma (HNSCC) include surgery, radiotherapy, and chemotherapy, but therapeutic outcomes remain poor. Targeted therapies have been developed by taking advantage of the aberrant EGFR signaling cascade in HNSCC. These include monoclonal antibodies, such as cetuximab, which target EGFR. However, cetuximab has shown only moderate efficacy in the clinic. Since aberrant GPCR activity has been linked to some of the most prevalent human diseases, including many cancers, they represent promising candidates to target in HNSCC. Here, we present a comprehensive analysis of mutations, differential gene expression, and copy number variations of GPCRs in HNSCC, and also highlight emerging functional impacts of GPCRs in lymph node metastasis, angiogenesis, tumor-promoting oncocrine signaling networks, including EGFR transactivation, and immune evasion.

Abbreviations

A2AR	adenosine 2A receptor
ADAM17	a disintegrin and metalloproteinase 17
AREG	amphiregulin
CNV	copy number variation
COX2	cyclooxygenase 2
DE	differential expression
EGFR	epidermal growth factor receptor
GALR	galanin receptors
GISTIC	genomic identification of significant targets in cancer
GPCR	heterotrimeric G protein coupled receptor
GRM	metabotropic glutamate receptor
HB-EGF	heparin-binding epidermal growth factor like growth factor
HER	human epidermal growth factor
HNSCC	squamous cell carcinomas of the head and neck
HPV	human papillomavirus
IL-6R	interleukin-6 receptor

Improving the Therapeutic Ratio in Head and Neck Cancer
https://doi.org/10.1016/B978-0-12-817868-3.00015-9

MDSC	myeloid-derived suppressor cells
PGE2	prostaglandin E2
RGS	regulators of G protein signaling
TAM	tumor-associated macrophages
TGFα	transforming growth factor-α
TNF	tumor necrosis factor
Tregs	regulatory T cells

Conflicts of interest

JSG is a member of the Scientific Advisory Board of Oncoceutics Inc. and Domain Therapeutics.

Introduction

As more than 90% of neoplastic head and neck lesions of the oral cavity, larynx and pharynx are of squamous cell origin, they are typically referred to collectively as squamous cell carcinomas of the head and neck (HNSCC). The main traditional risk factors for HNSCC are tobacco use and excess alcohol consumption, however human papillomavirus (HPV) infection is also associated with HNSCC, and it is responsible for up to 25% of all cases with its incidence continually increasing [1, 2]. HPV negative HNSCC are characterized by mutations, gene deletion, or epigenetic inactivation of *p16* (*CDKN2A*), frequent mutations in *TP53*, and a large overexpression of EGFR [3, 4], while HPV+ HNSCC cases often do not exhibit *TP53* mutations or *p16* genomic alterations, as the viral oncoproteins E6 and E7 inhibit the tumor suppressive functions of p53 and RB, respectively [5]. Current treatments for HNSCC include surgery, radiotherapy, and chemotherapy, but long-term 5-year survival rates remain poor (40–50%). Despite aggressive treatment plans, the proximity of the head and neck region to vital organs, the limited utility of radiotherapy in advanced stages of the disease, and the toxicity of chemotherapy and available drugs lead to poor survival rates [6]. This is expected to improve in the future with the most recent approval of immunotherapies that restore the patient's own anti-cancer immune response; in June of 2019, the FDA approved pembrolizumab as a first-line treatment for recurrent or metastatic HNSCC following presentation on the final results of the KEYNOTE-048 trial (https://ascopubs.org/doi/abs/10.1200/JCO.2019.37.15_suppl.6000). To date, the main genetic biomarkers defining HNSCC include human papillomavirus (HPV) infection, epidermal growth factor receptor (EGFR) expression, and p53 status [7, 8]. These properties have been correlated with differential responses to chemotherapy, radiation treatment, and overall survival [7, 8]. Though this knowledge has opened the door to new therapeutic strategies, accumulated drug and radiation resistance and lack of a fuller understanding of tumor and immune signaling mechanisms has hindered the ability to achieve more efficacious clinical responses in HNSCC patients. Since GPCRs control a large variety of signaling pathways in normal physiological processes, and are dysregulated in some of the most prevalent human diseases, more than 30% of all FDA-approved drugs target GPCRs [9, 10]. However, their establishment as drug targets in oncology is just beginning to be realized [11, 12]. Emerging studies support a role for GPCRs in the regulation of tumor growth in HNSCC, as well as HNSCC metastasis, angiogenesis, and immune evasion. Clearly, there is a need for better prognostic biomarkers and targets for HNSCC treatment, and the role of GPCRs may provide still unexplored opportunities for intervention in this disease.

GPCR mutations in HNSCC

The contribution of the ~400 GPCR genes in HNSCC is still under evaluation, but their specific expression patterns and presence of somatic mutations suggest a role in HNSCC progression that can now be investigated. Many GPCRs are mutated in HNSCC, among which GPR98 is the most frequently mutated (9%, Table 1A). Though mutations in GPR98, an adhesion receptor, have been well studied in febrile seizures and Usher syndrome, a genetic disease causing combined deafness and blindness, its ligand and physiological functions are currently poorly understood in cancer [13, 14]. The top 20 most frequently mutated GPCRs in HNSCC also include 5 GPCRs from the metabotropic glutamate receptor (GRM) family. These glutamate receptors (GRM1–8) have been shown to be the most predominant players in glutamatergic signaling in multiple cancers [15]. In oral cancer, the role of GRM5 has been studied in the context of tumor cell migration, where stimulation of this receptor promoted cell motility and invasion, including CXCR4-related metastasis [16, 17]. In addition, glutamate, the ligand for these receptors, is largely utilized in the tumor microenvironment as an energy source for massive cell proliferation, which makes the GRM receptor family an intriguing area of study as it is expected to be persistently stimulated by locally produced natural agonist or by activating mutations.

The recent discovery of hot spot mutations in G proteins with oncogenic potential can give insight on the roles of receptors linked to particular G proteins in HNSCC progression. Recent mutational analyses have shown that *GNAS* is the most highly mutated G protein in human cancer (Table 1B) [12]. The oncogenic activity of *GNAS* was first described in pituitary adenomas but since then, it has been found to be mutated in multiple neoplasms, predominantly at the codon 201 hotspot [18, 19]. These mutations lead to activation of adenylate cyclase and constitutive cAMP signaling through the reduction of GTP hydrolysis and adopting persistent activation even with GDP bound [20]. Though the direct functional impact of *GNAS* mutations in HNSCC has not been explored, *GNAS* has been linked to potent inflammatory mediators, particularly through the Gαs-linked prostaglandin receptors. Cyclooxygenase 2 (COX2) is typically overexpressed in HNSCC and COX2-derived prostaglandins, particularly prostaglandin E2 (PGE2), are produced by HNSCC cells and can enhance activation of PI3K and AKT while relieving inhibitory phosphorylation of β-catenin, feeding into Gαs oncogenic signaling [21]. Recent described experimental studies suggest that *GNAS* mutants cause squamous differentiation in the skin [22], and ongoing studies in our laboratory support that *GNAS* can cooperate with *Ras* oncogenes to initiate skin SCC lesions (unpublished observation). Whether this is the case for HNSCC warrants further investigation.

Copy number variations (CNVs) in G proteins and GPCRs

In addition to mutations, aberrant gene expression and copy number variations (CNVs) can also alter downstream signaling of GPCRs and G proteins even without a mutation,

TABLE 1 Frequency of mutations in GPCRs and G proteins in cancer

	Receptor (IUPHAR designation)	Gene	cBIO link	HNSC (507)
Top 20 most frequently mutated GPCRs in HNSCC				
1	ADGRV1	***ADGRV1 (GPR98)***	ADGRV1 (GPR98)	9.07
2	$mGlu_1$ receptor	***GRM1***	GRM1	8.48
3	$mGlu_3$ receptor	***GRM3***	GRM3	6.90
4	ADGRB3	***ADGRB3 (BAI3)***	ADGRB3 (BAI3)	6.51
5	CELSR1	***CELSR1***	CELSR1	6.31
6	CELSR2	***CELSR2***	CELSR2	6.31
7	$mGlu_8$ receptor	***GRM8***	GRM8	6.31
8	ADGRG4	***ADGRG4 (GPR112)***	ADGRG4 (GPR112)	6.11
9	ADGRL3	***ADGRL3 (LPHN3)***	ADGRL3 (LPHN3)	5.13
10	$mGlu_7$ receptor	***GRM7***	GRM7	4.54
11	GPR179	***GPR179***	GPR179	3.95
12	$mGlu_5$ receptor	***GRM5***	GRM5	3.75
13	NPFF2 receptor	***NPFFR2***	NPFFR2	3.75
14	ADGRL4	***ADGRL4 (ELTD1)***	ADGRL4 (ELTD1)	3.55
15	GPR158	***GPR158***	GPR158	3.55
16	GPR149	***GPR149***	GPR149	3.35
17	5-HT_{1A} receptor	***HTR1A***	HTR1A	3.35
18	CELSR3	***CELSR3***	CELSR3	3.16
19	ADGRB2	***ADGRB2 (BAI2)***	ADGRB2 (BAI2)	2.96
20	GPRC6 receptor	***GPRC6A***	GPRC6A	2.96

(A)

	Protein designation	Gene	cBIO link	HNSC (507)
Top 10 most frequently mutated G proteins in HNSCC				
1	$G\alpha_s$	***GNAS***	GNAS	2.96
2	$G\alpha_z$	***GNAZ***	GNAZ	1.38
3	$G\alpha_{15}$	***GNA15***	GNA15	1.18
4	$G\beta_3$	***GNB3***	GNB3	1.18
5	$G\alpha_{t3}$	***GNAT3***	GNAT3	1.18
6	$G\beta_2$	***GNB2***	GNB2	1.18
7	G_{olf}	***GNAL***	GNAL	0.99
8	$G\gamma_{13}$	***GNG13***	GNG13	0.99
9	$G\alpha_{o1}$	***GNAO1***	GNAO1	0.79
10	$G\alpha_{14}$	***GNA14***	GNA14	0.79

(B)

This table was adapted from Nieto Gutierrez A, McDonald PH. GPCRs: emerging anti-cancer drug targets. Cell Signal 2018;41:65–74, showing the top most frequently mutated GPCR (A) and G protein (B) genes.

leading to cancer initiation and progression. During cell division, somatic alterations can be acquired at random and can contribute to tumor growth. CNVs can be analyzed with the algorithm, GISTIC (genomic identification of significant targets in cancer), which evaluates the frequency and amplitude of observed events to identify genes with CNVs that may contribute to tumorigenesis [23].

Among the GPCRs with significant CNVs (q-value < 0.25), copy number deletion in *GALR1* was the most significant variation (Table 2A) [12]. Remarkably, subtypes of the galanin receptors (GALR) have been shown to play a role in HNSCC transformation. Early studies supported that GALR1 signaling leads to inhibition of cell proliferation in GALR1-transfected HNSCC cells in vitro, suggesting that HNSCC tumors may dampen GALR1 signaling to promote proliferation [24–26]. GALR1 signals through a Gαi pathway and ERK1/2 activation leads to inhibition of proliferation, in this case by persistent signaling causing cell cycle arrest [24–26]. In addition, the role of GALR2 in HNSCC cells has been described as anti-proliferative and pro-apoptotic. Numerous mechanisms have been proposed for GALR2-induced apoptosis, including activation of caspase-3-dependent cell death, and upregulation of pro-apoptotic regulatory proteins of the BCL2 family [27–29]. In addition to CNV loss, one epigenetic study demonstrated that the *GALR1* gene is methylated in 52.7% of HNSCC cell lines and that this methylation is correlated with decreased expression [30, 31]. Methylation of GALR2 was also observed in primary HNSCC tumors, and interestingly, 11% of the tumors exhibited hyper-methylation on genes encoding galanin, GALR1 and GALR2. In HNSCC patients, methylation of *GALR1* or *GALR2* can predict poor survival and is also correlated with increased tumor size and decreased disease-free survival, making it a potential prognostic factor in HNSCC [31, 32].

The most significant genomic alterations in G proteins include Gβ and Gγ G protein subunits (*GNG11* and *GNB1*), and this likely reflects the importance of Gβγ-dimers in the signaling pathway for HSNCC tumorigenesis (Table 2B).

Overexpression of GPCRs and G proteins in HNSCC

The Cancer Genome Atlas is comprehensive, cancer genomics database, launched by the National Cancer Institute and includes high-throughput genomic analysis of over 30 tumors and 20,000 patients and information on methylation, large-scale protein expression, gene expression, copy number variation, SNP genotyping, genome wide DNA methylation, microRNA, and exon sequencing profiling [33]. With this database, we have extracted RNA sequencing data from the HNSCC cohort of patients, including primary tumor and normal tissue, and used DESeq to determine differentially expressed genes in tumors versus normal tissue. DESeq is a software package which takes data from high-throughput sequencing assays and tests for differential expression (DE) using negative binomial distribution, producing differential expression data with good statistical power [34].

DE analysis of GPCRs in HNSCC versus normal tissue revealed 35 significantly ($P < .05$) upregulated or downregulated GPCR and G protein genes (Fig. 1). Intriguingly, many of the neuropeptide Y receptors are significantly upregulated (*NPY6R*, *NPY5R*, *NPY1R*). These NPY receptors are activated by neuropeptide Y, present in high amounts in the central and peripheral nervous systems, regulating a variety of key neurological functions. In addition to

TABLE 2 The most significant CNVs of GPCRs and G Proteins in cancer

Gene	q-Value	Deletion or amplification	RNA correlation
(A) Most significantly altered GPCRs in all HNSCC			
GALR1	7.94E−27	DEL	
GPR35	1.28E−25	DEL	0.1901
GPR55	1.28E−25	DEL	0.1132
CXCR1	1.28E−25	DEL	0.137
CXCR2	1.28E−25	DEL	0.1519
GPBAR1	1.28E−25	DEL	0.1952
HTR2B	1.28E−25	DEL	0.1242
NMUR1	1.28E−25	DEL	−0.0233
DRD2	1.89E−21	DEL	0.2581
GNRHR2	7.84E−19	DEL	
CYSLTR2	1.21E−15	DEL	0.0411
HTR2A	1.21E−15	DEL	−0.0181
LPAR6	1.21E−15	DEL	0.207
MLNR	1.21E−15	DEL	0.0622
CALCR	1.40E−10	AMP	0.2817
GPR153	7.08E−07	DEL	0.3515
GPR3	7.08E−07	DEL	−0.1729
TAS1R1	7.08E−07	DEL	0.0134
TAS1R2	7.08E−07	DEL	
TAS1R3	7.08E−07	DEL	
CNR2	7.08E−07	DEL	0.1218
HTR1D	7.08E−07	DEL	0.0031
HTR6	7.08E−07	DEL	0.0693
ADRA2C	0.00046209	DEL	−0.0159
DRD5	0.00046209	DEL	0.1196
GPR78	0.00046209	DEL	
CCKBR	0.00051729	DEL	−0.0308
DRD4	0.00051729	DEL	−0.0518
MRGPRE	0.00051729	DEL	
MRGPRG	0.00051729	DEL	

TABLE 2 The most significant CNVs of GPCRs and G Proteins in cancer—Cont'd

Gene	q-Value	Deletion or amplification	RNA correlation
GPR125	0.00087216	DEL	0.28
LTB4R	0.0025503	DEL	0.3379
LTB4R2	0.0025503	DEL	0.328
BDKRB1	0.0057325	DEL	0.2766
BDKRB2	0.0057325	DEL	0.3199
GPR132	0.0057325	DEL	0.0512
GPR65	0.0057325	DEL	
GPR68	0.0057325	DEL	0.282
AVPR2	0.0095371	AMP	−0.0472
PTGER4	0.018866	AMP	0.1608
CHRM5	0.11567	DEL	0.0764
GPR176	0.11567	DEL	0.0377
CMKLR1	0.22646	DEL	0.0078
LGR5	0.22646	DEL	0.1172
GPR133	0.22646	DEL	0.039
FZD10	0.22646	DEL	0.0355

GPCRs, q-value < 0.25 and |mRNA correlation > 0.33

(B) Most significantly altered G proteins in all HNSCC

Gene	q-Value	Deletion or amplification	RNA correlation
GNG11	1.40E − 10	AMP	
GNGT1	1.40E − 10	AMP	0.3662
GNB1	7.08E − 07	DEL	0.6402

G proteins, q-value < 0.25 and |mRNA correlation| > 0.33

This table was adapted from Nieto Gutierrez A, McDonald PH. GPCRs: emerging anti-cancer drug targets. Cell Signal 2018;41:65–74, showing the most highly significant (GISTIC q-value < 0.25) CNVs in GPCR (A) and G protein (B) genes and the correlation with mRNA expression.

HNSCC, NPY receptors have been shown to be overexpressed in a variety of human cancers, including breast carcinomas, glioblastomas, testicular cancers, and neuroblastomas [35, 36]. As a result, these receptors have become an attractive target for cancer imaging for diagnosis and treatment [37]. Furthermore, in a study determining the methylation status of eight neuropeptide receptor gene promoters in HNSCC, *NPY5R* and *NPY1R* were methylated in 35.1% and 35.9% of patients, respectively. Adding to this, the same study determined that when the NPY2R promoter was methylated, the odds ratio for tumor recurrence and lymph node metastasis were increased [38]. Interestingly, also among the most significantly upregulated genes are *TACR1* and *SST1*, and methylation studies also showed that hypermethylation

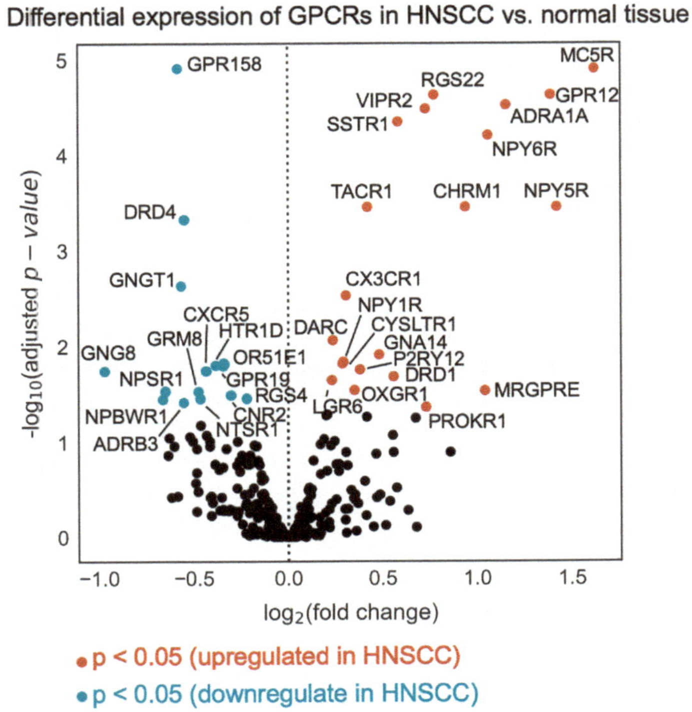

FIG. 1 Differential expression of GPCR and G protein genes in HNSCC. All GPCR and G protein genes were extracted from the UCSC TCGA PanCan Cohort RNA-seq dataset and DESeq was used to assess differential expression of HNSCC tumor versus normal samples. Shown is a volcano plot of significantly ($P < .05$) upregulated *(red)* or downregulated *(blue)* genes.

of both of these genes is associated with significantly reduced disease-free survival and a higher recurrence rate of HNSCC [39]. Altogether, these studies suggest a role for the neuropeptide Y receptors in HNSCC, which is reflected at the transcriptome and epigenetic level.

An interesting and important component of GPCR signaling is the regulators of G protein signaling (RGS), which function as GTPase-accelerating proteins to promote GTP hydrolysis by the Gα subunit. There are over 20 RGS proteins and they participate in G protein signal termination, and they can use their multiple domains to serve as signaling scaffolds; their roles in cancer have been examined in lung and bladder cancers [40]. In HNSCC, *RGS22* is significantly upregulated while *RGS4* is significantly downregulated. Though the specific roles of these RGS proteins in HNSCC have not been examined, one group has investigated 98 SNPs in 17 RGS genes, illuminating a homozygous variant genotype of *RGS2* as being correlated with increased risk of cancer recurrence [41]. Considering that RGSs can dampen the function of G proteins, or bias G protein signaling, we can expect that these changes in RGS expression may have a broad impact on GPCR signaling in HNSCC.

Several G protein genes were also identified in our differential expression analysis, including *GNA14*, *GNG8*, and *GNGT1*. *GNA14* has been shown to promote gastric tumor development following TNFα/TNFR1 signaling, but the roles of *GNG8* and *GNGT1* in HNSCC and other cancers remain to be explored [42]. Although not statistically significant by our stringent criteria, the relevance of overexpression of other G proteins has been examined by in depth studies in collections of HNSCC samples and cell lines. For example, a recent study showed that *GNA13* levels can dictate drug resistance and tumor-initiating phenotypes in HNSCC cells both in vitro and in vivo [43]. *GNA13* is upregulated in many solid tumors and has been shown to affect survival and metastasis in HNSCC patients [43]. Furthermore, the expression of a constitutively activated form of *GNA12* led to cell migration and invasion in oral cancer cell lines, further suggesting that *GNA12* is a major player in lymph node metastasis in HNSCC [44].

Oncocrine signaling of GPCRs through EGFR in HNSCC

In addition to aberrant signaling as a result of mutations, overexpression, and CNVs, the full complexity of GPCR signaling pathways in HNSCC is also reflected in signaling cascades triggered by ligands produced by cancer cells or the variety of cells within the tumor stroma. This autocrine and paracrine signaling processes, termed oncocrine signaling, results in cancer progression, immune evasion, angiogenic signaling, and tumor dissemination as a result of multilayered cross talk between GPCRs and other signaling pathways that together, promote survival and proliferation of cancer cells (Fig. 2).

Perhaps the most extensively studied signaling cascade in HNSCC is the epidermal growth factor receptor (EGFR). The phosphorylation cascade following EGFR activation is mediated by tyrosine kinases working through PI3K-PTEN-AKT, MAPK, ERK, and JAK/STAT pathways [45]. Aberrant EGFR activity leads to cancer cell proliferation, metastatic activity, and angiogenesis and upregulation of EGFR is seen in HNSCC cell lines and also associated with reduced survival in patients [46, 47]. These groundbreaking studies led to the development of Cetuximab, which blocks EGFR activity as the first approved targeted agent for HNSCC [48]. Nevertheless, Cetuximab has only shown modest improvement of efficacy in the clinic. In addition, drug resistance to this agent has mostly been attributed to compensation by growth

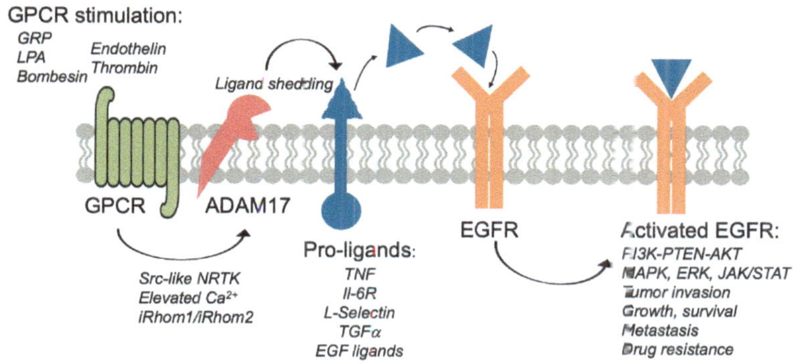

FIG. 2 Oncocrine signaling of GPCRs through EGFR in HNSCC. GPCR ligands bind GPCRs to stimulate the ADAM17 metalloproteinase through Src-like nonreceptor tyrosine kinase and elevation of intracellular calcium, leading to the release of ErbB ligands, and PKC-dependent ligand shedding. ADAM17, whose maturation is controlled by iRhom1 and iRhom2, shed ligands bound to the membrane, and these ligands go on to bind EGFR for downstream signaling leading to the PI3K-PTEN-AKT, MAPK, ERK, and JAK/STAT signaling axes. Altogether, these processes lead to tumor invasion, growth, and survival, as well as metastasis and angiogenesis. *IL-6*, interleukin-6; *TGFα*, tumor growth factor alpha; *TNF*, tumor necrosis factor (see text for details).

promoting signaling pathways independent of EGFR [49]. GPCR signaling represents a prominent mechanism contributing to the activation of the EGFR pathway and drug resistance.

Many GPCR ligands have been shown to contribute to HNSCC carcinogenesis through activation of the EGFR pathway. Ligands like gastrin-releasing peptide (GRP), lysophosphatidic acid (LPA), bombesin, endothelin, and thrombin have all been shown to induce HSNCC growth through EGFR activation [50–54]. These ligands induce autocrine or paracrine release of transforming growth factor-α (TGFα), amphiregulin (AREG), tumor necrosis factor (TNF), interleukin-6 receptor (IL-6R), and L-selectin, which are ligands for EGFR typically present on the cell surface, ultimately contributing to EGFR mediated tumor growth [55, 56]. In HNSCC, the indirect activation of EGFR by GPCRs comes from stimulation of Src, PI3K/AKT, STATs, and MAPK, among which Src has been shown to promote tumor invasion, PI3K/AKT activation leads to growth and survival of tumor cells, and MAPK and STATs both can induce growth [45].

Mature ligands acting on EGFR in an autocrine fashion are the main players for the oncocrine signaling occurring between GPCRs and EGFR in cancer cells. Growing evidence is rendering an important role for A disintegrin and metalloproteinase 17 (ADAM17), an enzyme that is involved in regulating the activation of human epidermal growth factor (HER) family receptors through the shedding or liberation of their ligand precursors [57, 58]. Some notable ligands include TGFα, EGF, AREG, heparin-binding epidermal growth factor like growth factor (HB-EGF), heregulin (HRG), and epiregulin (EPG) [57, 58]. Consequently, aberrant activity of ADAM17 as well as overexpression of HER ligands can lead to enhanced EGFR downstream signaling, thus promoting tumor growth, angiogenesis, and metastasis [57, 58]. Several components of GPCR signaling have been shown to promote ADAM17 activation, including Src-like nonreceptor tyrosine kinases and elevation of intracellular calcium leading to release of ErbB ligands, and PKC-dependent ligand shedding [59–61]. Furthermore, iRhom1 and iRhom2, members of the Rhomboid family of intramembrane serine proteinases, control the maturation and function of ADAM17 [62, 63]. Gene silencing of iRHOM1 with siRNA leads to decreased EGFR activation induced by GRP, suggesting that EGFR function can be modulated indirectly by rhomboids and ADAM17 downstream from GPCR signaling [62]. The impactful role of EGFR in propagating signals from GPCR stimulation suggests that GPCRs bridge the stimuli at the cell surface with prolonged downstream ERK activation through ADAM17 and rhomboid proteins, leading to physiological and pathophysiological responses of relevance to HNSCC biology.

Cell growth promoting by GPCRs

In addition to multiple neuropeptide receptors that activate EGFR indirectly (above), inflammatory mediators acting on GPCRs can promote HNSCC growth. As previously mentioned, overexpression of COX2 is an early event in HNSCC progression. High levels of this enzyme can be detected in pre-malignant HNSCC lesions, and particularly in high-risk oral lesions [64]. COX2 levels can also be enhanced by the exposure of oral keratinocytes to extracts from areca nut, whose quid represents one of the most important HNSCC risk factor in Southeast Asia [65]. Consistently, HNSCC lesions and their derived cell lines release PGE2, which exerts pro-proliferative, angiogenic, and metastatic effects, albeit the specific EP receptors mediating these effects are not well established [66]. These findings provided the rationale for a number of ongoing chemoprevention trials using COX-2 inhibitors in HNSCC [67].

GPCRs and metastasis

A clear role for GPCRs in HNSCC is also exemplified by the high levels of chemokines that are released from squamous carcinoma cells, including CXCL1/GROα and CXCL8/IL-8, and the overexpression of their receptors, CXCR1 and CXCR2, in HNSCC cells [68]. In this case, the release of these chemokines is initiated by the persistent activation of NFκB, which is highly active in this tumor type [69]. These chemokines have a profound effect by creating a permissive microenvironment for tumor growth. The most frequently expressed chemokine receptors in HNSCC cells are CXCR1, CXCR2, CCR7, and CXCR4, and their level of expression often serves as a predictor of local/regional invasion of cervical lymph nodes [68, 70]. Whereas these receptors can transduce proliferative signals in some HNSCC cells, their autocrine activation may help cells acquire a more motile phenotype while promoting the release of matrix-degrading proteases, such as MMP-2 and MMP-9, thereby enhancing the ability of HNSCC cells to invade adjacent tissues, gain access to the blood stream or lymphatic circulation, and promote locoregional metastasis [71].

CCR7 binds the ligands CCL21 and CCL19, and represents one of the best studied chemokine receptors driving cancer metastasis in HNSCC, as well as inhibition of cell death pathways, tumor progression, and therapeutic resistance, particularly due to the activation of the PI3K/AKT signaling cascade [72]. CCL19 binding to CCR7 mediates cell migration and invasion through PI3K, MAPKs, and JAK3 [73–75]. In addition, CCL21 promotes both migration and speed of invasion through ERK1/2 and JNK activation [73]. CCR7 typically guides lymphocytes and dendritic cells to regional lymph nodes, and expression of CCR7 on tumor cells can dictate lymph node metastasis and has emerged as an important prognostic marker for HNSCC, which frequently metastasizes to regional lymph nodes.

The SDF-1/CXCL12-CXCR4 pathway has been well studied in regards to lymph node and distant organ metastasis. SDF-1/CXCL12-induced activation of ERK1/2, AKT/PKB and Src family kinases stimulates CXCR4-expressing HNSCC cells to migrate and proliferate [76]. Additionally, SDF-1/CXCL12 has been shown to be highly expressed in multiple distant organs that are also common sites of metastasis, including lungs, liver, and bone marrow, making it a key player in distant organ metastasis [77].

GPCRs in cancer immunotherapy HNSCC

In the last decade, cancer immunotherapy revolutionized the oncology platform, and numerous studies have elucidated the important roles for immune cells in the tumor microenvironment and their implications for cancer development and therapeutic resistance. A subset (<20%) of HNSCC patients benefit from immunotherapy, and early trials for immune checkpoint inhibitors in recurrent and metastatic HNSCC [78, 79] led the approval of anti-PD-1 checkpoint inhibitors for HNSCC treatment [80]. Tumors employ multiple immune evasion mechanisms, including the recruitment of highly immunosuppressive myeloid-derived suppressor cells (MDSC), regulatory T cells (Tregs), and M2 tumor-associated macrophages (TAM) in order to condition the tumor microenvironment to be immunosuppressive cytokine rich [81, 82]. The receptors that steer immune cell migration toward the chemokine gradients in the tumor microenvironment are dominated by GPCRs. In most cancers, including HNSCC, MDSCs mainly express CCR2, CXCR2, and CXCR4 [83]. Moreover, CXCL8, which binds to the CXCR1 and CXCR2 chemokine receptors, is generally thought to drive immunosuppression and promote angiogenesis and progression and has been found in higher concentration in the serum of HNSCC patients compared to healthy patients [83, 84]. CXCL8 can also directly stimulate the proliferation of HNSCC cells [85]. CXCL8 can also recruit Treg cells, but their recruitment is mostly dictated by expression of CCR4 and CCR10, which bind to CCL28 in the hypoxic tumor microenvironment [86]. Interestingly, blocking CCR4 with an antagonist leads to antitumor immunity in mice and prolonged survival, suggesting that blocking of Treg trafficking could be a promising immunotherapy in HNSCC [87].

Chemokine receptors direct cytotoxic, anti-tumor immune cells as well. More specifically, CD8 T cells and NK cells express CXCR3 and bind to CXCL9 and CXCL10 in order to migrate into tumors. In HNSCC patients, expression of CXCL10, which is part of the interferon-induced signature, is significantly associated with response to radiotherapy and overall survival [88]. The activation of these cytotoxic CD8 T cells is dependent on dendritic cells that present cancer cell antigens to drive effector T cell activation, and these dendritic cells are guided to draining lymph nodes largely by CCR7 [89, 90].

A wide variety of chemokine receptors dictate the infiltration and diversity of immune cells that are guided to the tumor, and this can provide therapeutic opportunities using small molecule inhibitors, negative allosteric modulators, and monoclonal antibodies, which can be utilized to dampen immunosuppressive cell recruitment and disable immune evasive mechanisms. Another mechanism of immune evasion involves a nucleoside, adenosine, which is present in high concentrations in areas of tissue damage and hypoxia. Adenosine is an important pharmacological regulator that is released by the extracellular breakdown of ATP, aided by CD39 and CD73 [91]. Adenosine binds to the GPCR, adenosine 2A receptor (A2AR), and immunostaining of A2AR in HNSCC samples has shown that increased expression correlates with larger tumor size and lymph node metastasis [92]. Mechanistically, A2AR is a Gαs-linked GPCR, and cAMP signaling leads to desensitization of chemokine receptors and reduction in pro-inflammatory cytokine production [93, 94]. A2AR stimulation has also been shown to expand Tregs, and blockade with an A2AR antagonist decreased the growth of spontaneously developed HNSCC tumors in a transgenic mouse model, whereby *TGFBR1* and *PTEN* knockout leads to adenosine pathway activation and spontaneous HNSCC tumors [92, 93]. Clinical trials using A2AR antagonists are underway in patients with advanced non-small cell lung cancer (NSCLC), and may represent an exciting target for HNSCC in the very near future.

Conclusions

GPCRs are the most widely studied receptors, but their roles in HNSCC initiation and progression are not fully understood. We expect that new efforts on the study of GPCRs in cancer, which we refer collectively as "onco-GPCRome," will provide a rationale of the evaluation of the preclinical and clinical benefits of targeting this receptor family in cancer [12]. Specifically, we can envision that disrupting the transactivation of EGFR by multiple GPCRs expressed in each HNSCC lesion may increase the efficiency of EGFR targeting agents, including cetuximab, while blocking PGE2 receptors may decrease HNSCC progression, similar to COX2 inhibitors but with fewer side effects. Many GPCRs can be targeted for the development of new cancer imaging agents for diagnosis and monitoring treatment response. In the new era of cancer immunotherapy, we can also envision that the explosion of knowledge on the multiple roles of GPCRs in cancer immunology may soon be exploited to improve patient response to current and future immune oncology agents. Specifically, with <20% of HNSCC patients responding to recently approved immune checkpoint inhibitors, we expect that the use of antagonist or negative allosteric inhibitors restricting the recruitment of immune suppressive cells (e.g., MDSCs and Tregs) or allosteric modulators of receptors involved in cytotoxic T cells, dendritic cells, and NK cell recruitment and function, may soon provide an opportunity for the development of more effective multimodal precision immunotherapies.

In addition, the emerging view that disruption of GPCR-mediated oncocrine circuits can prevent cancer progression and metastasis may support the possibility of repurposing approved drugs as an adjuvant treatment in HNSCC cancer survivors to prevent cancer recurrence [12]. As such, a widely used β blocker, propranolol, has been shown to improve the prognosis of skin melanoma patients [95] and to limit EGFR inhibitor resistance in lung cancer [96]. Based on these findings, propranolol is under evaluation as an adjuvant treatment in multiple cancer types. We expect that the future analysis of large patient health datasets in HNSCC survivors may provide opportunities for new GPCR-based treatments to delay or prevent tumor relapse.

Overall, GPCRs are widely dysregulated in cancer but underexploited for cancer therapy [12]. Targeting GPCRs and their oncogenic signaling pathways may provide opportunities to limit resistance to cetuximab and other targeted agents in HNSCC, as well as to identify new preventive strategies and more effective combination immunotherapies.

Acknowledgment

This work was supported by NIH grants R33CA225291 (NCI), U54CA209891 (NCI), and U01DE028227 (NIDCR).

References

[1] Auluck A, Hislop G, Bajdik C, Poh C, Zhang L, Rosin M. Trends in oropharyngeal and oral cavity cancer incidence of human papillomavirus (HPV)-related and HPV-unrelated sites in a multicultural population: the British Columbia experience. Cancer 2010;116:2635–44.

[2] Hocking JS, Stein A, Conway EL, Regan D, Grulich A, Law M, Brotherton JM. Head and neck cancer in Australia between 1982 and 2005 show increasing incidence of potentially HPV-associated oropharyngeal cancers. Br J Cancer 2011;104:886–91.

[3] Forastiere A, Koch W, Trotti A, Sidransky D. Head and neck cancer. N Engl J Med 2001;345:1890–900.

[4] Mao L, Hong WK, Papadimitrakopoulou VA. Focus on head and neck cancer. Cancer Cell 2004;5:311–6.

[5] DeFilippis RA, Goodwin EC, Wu L, DiMaio D. Endogenous human papillomavirus E6 and E7 proteins differentially regulate proliferation, senescence, and apoptosis in HeLa cervical carcinoma cells. J Virol 2003;77:1551–63.

[6] Haddad R, Wirth L, Posner M. Emerging drugs for head and neck cancer. Expert Opin Emerg Drugs 2006;11:461–7.
[7] Gillison ML, Chaturvedi AK, Anderson WF, Fakhry C. Epidemiology of human papillomavirus-positive head and neck squamous cell carcinoma. J Clin Oncol 2015;33:3235–42.
[8] Kumar B, Cordell KG, Lee JS, Worden FP, Prince ME, Tran HH, Wolf GT, Urba SG, Chepeha DB, Teknos TN, Eisbruch A, Tsien CI, Taylor JM, D'Silva NJ, Yang K, Kurnit DM, Bauer JA, Bradford CR, Carey TE. EGFR, p16, HPV Titer, Bcl-xL and p53, sex, and smoking as indicators of response to therapy and survival in oropharyngeal cancer. J Clin Oncol 2008;26:3128–37.
[9] Hauser AS, Attwood MM, Rask-Andersen M, Schioth HB, Gloriam DE. Trends in GPCR drug discovery: new agents, targets and indications. Nat Rev Drug Discov 2017;16:829–42.
[10] Santos R, Ursu O, Gaulton A, Bento AP, Donadi RS, Bologa CG, Karlsson A, Al-Lazikani B, Hersey A, Oprea TI, Overington JP. A comprehensive map of molecular drug targets. Nat Rev Drug Discov 2017;16:19–34.
[11] Nieto Gutierrez A, McDonald PH. GPCRs: emerging anti-cancer drug targets. Cell Signal 2018;41:65–74.
[12] Wu V, Yeerna H, Nohata N, Chiou J, Harismendy O, Raimondi F, Inoue A, Russell RB, Tamayo P, Gutkind JS. Illuminating the Onco-GPCRome: novel G protein-coupled receptor-driven oncocrine networks and targets for cancer immunotherapy. J Biol Chem 2019;294(29):11062–86.
[13] Nakayama J, Fu YH, Clark AM, Nakahara S, Hamano K, Iwasaki N, Matsui A, Arinami T, Ptacek LJ. A nonsense mutation of the MASS1 gene in a family with febrile and afebrile seizures. Ann Neurol 2002;52:654–7.
[14] Mathur P, Yang J. Usher syndrome: hearing loss, retinal degeneration and associated abnormalities. Biochim Biophys Acta 2015;1852:406–20.
[15] Yu LJ, Wall BA, Wangari-Talbot J, Chen S. Metabotropic glutamate receptors in cancer. Neuropharmacology 2017;115:193–202.
[16] Kuribayashi N, Uchida D, Kinouchi M, Takamaru N, Tamatani T, Nagai H, Miyamoto Y. The role of metabotropic glutamate receptor 5 on the stromal cell-derived factor-1/CXCR4 system in oral cancer. PLoS ONE 2013;8.
[17] Park SY, Lee SA, Han IH, Yoo BC, Lee SH, Park JY, Cha IH, Kim J, Choi SW. Clinical significance of metabotropic glutamate receptor 5 expression in oral squamous cell carcinoma. Oncol Rep 2007;17:81–7.
[18] Vallar L, Spada A, Giannattasio G. Altered Gs and adenylate cyclase activity in human GH-secreting pituitary adenomas. Nature 1987;330:566–8.
[19] O'Hayre M, Vazquez-Prado J, Kufareva I, Stawiski EW, Handel TM, Seshagiri S, Gutkind JS. The emerging mutational landscape of G proteins and G-protein-coupled receptors in cancer. Nat Rev Cancer 2013;13:412–24.
[20] Hu Q, Shokat KM. Disease-causing mutations in the G protein Gαs subvert the roles of GDP and GTP. Cell 2018;173:1254–64. e11.
[21] Abrahao AC, Castilho RM, Squarize CH, Molinolo AA, dos Santos-Pinto Jr D, Gutkind JS. A role for COX2-derived PGE2 and PGE2-receptor subtypes in head and neck squamous carcinoma cell proliferation. Oral Oncol 2010;46:880–7.
[22] Iglesias-Bartolome R, Torres D, Marone R, Feng X, Martin D, Simaan M, Chen M, Weinstein LS, Taylor SS, Molinolo AA, Gutkind JS. Inactivation of a Gα(s)-PKA tumour suppressor pathway in skin stem cells initiates basal-cell carcinogenesis. Nat Cell Biol 2015;17:793–803.
[23] Mermel CH, Schumacher SE, Hill B, Meyerson ML, Beroukhim R, Getz G. GISTIC2.0 facilitates sensitive and confident localization of the targets of focal somatic copy-number alteration in human cancers. Genome Biol 2011;**12**:R41.
[24] Henson BS, Neubig RR, Jang I, Ogawa T, Zhang Z, Carey TE, D'Silva NJ. Galanin receptor 1 has anti-proliferative effects in oral squamous cell carcinoma. J Biol Chem 2005;280:22564–71.
[25] Takebayashi S, Ogawa T, Jung KY, Muallem A, Mineta H, Fisher SG, Grenman R, Carey TE. Identification of new minimally lost regions on 18q in head and neck squamous cell carcinoma. Cancer Res 2000;60:3397–403.
[26] Kanazawa T, Iwashita T, Kommareddi P, Nair T, Misawa K, Misawa Y, Ueda Y, Tono T, Carey TE. Galanin and galanin receptor type 1 suppress proliferation in squamous carcinoma cells: activation of the extracellular signal regulated kinase pathway and induction of cyclin-dependent kinase inhibitors. Oncogene 2007;26:5762–71.
[27] Kanazawa T, Kommareddi PK, Iwashita T, Kumar B, Misawa K, Misawa Y, Jang I, Nair TS, Iino Y, Carey TE. Galanin receptor subtype 2 suppresses cell proliferation and induces apoptosis in p53 mutant head and neck cancer cells. Clin Cancer Res 2009;15:2222–30.
[28] Kanazawa T, Misawa K, Misawa Y, Maruta M, Uehara T, Kawada K, Nagatomo T, Ichimura K. Galanin receptor 2 utilizes distinct signaling pathways to suppress cell proliferation and induce apoptosis in HNSCC. Mol Med Rep 2014;10:1289–94.

[29] Uehara T, Kanazawa T, Mizukami H, Uchibori R, Tsukahara T, Urabe M, Kume A, Misawa K, Carey TE, Suzuki M, Ichimura K, Ozawa K. Novel anti-tumor mechanism of galanin receptor type 2 in head and neck squamous cell carcinoma cells. Cancer Sci 2014;105:72–80.

[30] Verma M, Srivastava S. Epigenetics in cancer: implications for early detection and prevention. Lancet Oncol 2002;3:755–63.

[31] Misawa K, Ueda Y, Kanazawa T, Misawa Y, Jang I, Brenner JC, Ogawa T, Takebayashi S, Grenman RA, Herman JG, Mineta H, Carey TE. Epigenetic inactivation of galanin receptor 1 in head and neck cancer. Clin Cancer Res 2008;14:7604–13.

[32] Misawa Y, Misawa K, Kanazawa T, Uehara T, Endo S, Mochizuki D, Yamatodani T, Carey TE, Mineta H. Tumor suppressor activity and inactivation of galanin receptor type 2 by aberrant promoter methylation in head and neck cancer. Cancer 2014;120:205–13.

[33] Cancer Genome Atlas Research N, Weinstein JN, Collisson EA, Mills GB, Shaw KR, Ozenberger BA, Ellrott K, Shmulevich I, Sander C, Stuart JM. The cancer genome atlas pan-cancer analysis project. Nat Genet 2013;45:1113–20.

[34] Anders S, Huber W. Differential expression analysis for sequence count data. Genome Biol 2010;11:R106.

[35] Korner M, Reubi JC. NPY receptors in human cancer: a review of current knowledge. Peptides 2007;28:419–25.

[36] Li J, Tian Y, Wu A. Neuropeptide Y receptors: a promising target for cancer imaging and therapy. Regen Biomater 2015;2:215–9.

[37] Khan IU, Beck-Sickinger AG. Targeted tumor diagnosis and therapy with peptide hormones as radiopharmaceuticals. Anti Cancer Agents Med Chem 2008;8:186–99.

[38] Misawa K, Imai A, Mochizuki D, Misawa Y, Endo S, Hosokawa S, Ishikawa R, Mima M, Shinmura K, Kanazawa T, Mineta H. Genes encoding neuropeptide receptors are epigenetic markers in patients with head and neck cancer: a site-specific analysis. Oncotarget 2017;8:76318–28.

[39] Kanazawa T, Misawa K, Misawa Y, Uehara T, Fukushima H, Kusaka G, Maruta M, Carey TE. G-protein-coupled receptors: next generation therapeutic targets in head and neck cancer? Toxins (Basel) 2015;7:2959–84.

[40] Hurst JH, Hooks SB. Regulator of G-protein signaling (RGS) proteins in cancer biology. Biochem Pharmacol 2009;78:1289–97.

[41] Wang J, Lippman SM, Lee JJ, Yang H, Khuri FR, Kim E, Lin J, Chang DW, Lotan R, Hong WK, Wu X. Genetic variations in regulator of G-protein signaling genes as susceptibility loci for second primary tumor/recurrence in head and neck squamous cell carcinoma. Carcinogenesis 2010;31:1755–61.

[42] Oshima H, Ishikawa T, Yoshida GJ, Naoi K, Maeda Y, Naka K, Ju X, Yamada Y, Minamoto T, Mukaida N, Saya H, Oshima M. TNF-α/TNFR1 signaling promotes gastric tumorigenesis through induction of Noxo1 and Gna14 in tumor cells. Oncogene 2014;33:3820–9.

[43] Rasheed SAK, Leong HS, Lakshmanan M, Raju A, Dadlani D, Chong FT, Shannon NB, Rajarethinam R, Skanthakumar T, Tan EY, Hwang JSG, Lim KH, Tan DS, Ceppi P, Wang M, Tergaonkar V, Casey PJ, Iyer NG. GNA13 expression promotes drug resistance and tumor-initiating phenotypes in squamous cell cancers. Oncogene 2018;37:1340–53.

[44] Gan CP, Patel V, Mikelis CM, Zain RB, Molinolo AA, Abraham MT, Teo SH, Abdul Rahman ZA, Gutkind JS, Cheong SC. Heterotrimeric G-protein alpha-12 (Gα12) subunit promotes oral cancer metastasis. Oncotarget 2014;5:9626–40.

[45] Kalyankrishna S, Grandis JR. Epidermal growth factor receptor biology in head and neck cancer. J Clin Oncol 2006;24:2666–72.

[46] Chung CH, Ely K, McGavran L, Varella-Garcia M, Parker J, Parker N, Jarrett C, Carter J, Murphy BA, Netterville J, Burkey BB, Sinard R, Cmelak A, Levy S, Yarbrough WG, Slebos RJ, Hirsch FR. Increased epidermal growth factor receptor gene copy number is associated with poor prognosis in head and neck squamous cell carcinomas. J Clin Oncol 2006;24:4170–6.

[47] Rubin Grandis J, Zeng Q, Tweardy DJ. Retinoic acid normalizes the increased gene transcription rate of TGF-alpha and EGFR in head and neck cancer cell lines. Nat Med 1996;2:237–40.

[48] Bonner JA, Raisch KP, Trummell HQ, Robert F, Meredith RF, Spencer SA, Buchsbaum DJ, Saleh MN, Stackhouse MA, LoBuglio AF, Peters GE, Carroll WR, Waksal HW. Enhanced apoptosis with combination C225/radiation treatment serves as the impetus for clinical investigation in head and neck cancers. J Clin Oncol 2000;18:47S–53S.

[49] Chen LF, Cohen EE, Grandis JR. New strategies in head and neck cancer: understanding resistance to epidermal growth factor receptor inhibitors. Clin Cancer Res 2010;16:2489–95.

[50] Thomas SM, Bhola NE, Zhang Q, Contrucci SC, Wentzel AL, Freilino ML, Gooding WE, Siegfried JM, Chan DC, Grandis JR. Cross-talk between G protein-coupled receptor and epidermal growth factor receptor signaling pathways contributes to growth and invasion of head and neck squamous cell carcinoma. Cancer Res 2006;66:11831–9.
[51] Zhang Q, Bhola NE, Lui VW, Siwak DR, Thomas SM, Gubish CT, Siegfried JM, Mills GB, Shin D, Grandis JR. Antitumor mechanisms of combined gastrin-releasing peptide receptor and epidermal growth factor receptor targeting in head and neck cancer. Mol Cancer Ther 2007;6:1414–24.
[52] Siegfried JM, Krishnamachary N, Gaither Davis A, Gubish C, Hunt JD, Shriver SP. Evidence for autocrine actions of neuromedin B and gastrin-releasing peptide in non-small cell lung cancer. Pulm Pharmacol Ther 1999;12:291–302.
[53] Gschwind A, Prenzel N, Ullrich A. Lysophosphatidic acid-induced squamous cell carcinoma cell proliferation and motility involves epidermal growth factor receptor signal transactivation. Cancer Res 2002;62:6329–36.
[54] Altorki NK, Keresztes RS, Port JL, Libby DM, Korst RJ, Flieder DB, Ferrara CA, Yankelevitz DF, Subbaramaiah K, Pasmantier MW, Dannenberg AJ. Celecoxib, a selective cyclo-oxygenase-2 inhibitor, enhances the response to preoperative paclitaxel and carboplatin in early-stage non-small-cell lung cancer. J Clin Oncol 2003;21:2645–50.
[55] Zhang Q, Thomas SM, Xi S, Smithgall TE, Siegfried JM, Kamens J, Gooding WE, Grandis JR. SRC family kinases mediate epidermal growth factor receptor ligand cleavage, proliferation, and invasion of head and neck cancer cells. Cancer Res 2004;64:6166–73.
[56] Lorenzen I, Lokau J, Korpys Y, Oldefest M, Flynn CM, Kunzel U, Garbers C, Freeman M, Grotzinger J, Dusterhoft S. Control of ADAM17 activity by regulation of its cellular localisation. Sci Rep 2016;6:35067.
[57] Lee DC, Sunnarborg SW, Hinkle CL, Myers TJ, Stevenson MY, Russell WE, Castner BJ, Gerhart MJ, Paxton RJ, Black RA, Chang A, Jackson LF. TACE/ADAM17 processing of EGFR ligands indicates a role as a physiological convertase. Ann N Y Acad Sci 2003;995:22–38.
[58] Zheng X, Jiang F, Katakowski M, Zhang ZG, Lu QE, Chopp M. ADAM17 promotes breast cancer cell malignant phenotype through EGFR-PI3K-AKT activation. Cancer Biol Ther 2009;8:1045–54.
[59] Mifune M, Ohtsu H, Suzuki H, Nakashima H, Brailoiu E, Dun NJ, Frank GD, Inagami T, Higashiyama S, Thomas WG, Eckhart AD, Dempsey PJ, Eguchi S. G protein coupling and second messenger generation are indispensable for metalloprotease-dependent, heparin-binding epidermal growth factor shedding through angiotensin II type-1 receptor. J Biol Chem 2005;280:26592–9.
[60] Seals DF, Courtneidge SA. The ADAMs family of metalloproteases: multidomain proteins with multiple functions. Genes Dev 2003;17:7–30.
[61] Horiuchi K, Le Gall S, Schulte M, Yamaguchi T, Reiss K, Murphy G, Toyama Y, Hartmann D, Saftig P, Blobel CP. Substrate selectivity of epidermal growth factor-receptor ligand sheddases and their regulation by phorbol esters and calcium influx. Mol Biol Cell 2007;18:176–88.
[62] Zou H, Thomas SM, Yan ZW, Grandis JR, Vogt A, Li LY. Human rhomboid family-1 gene RHBDF1 participates in GPCR-mediated transactivation of EGFR growth signals in head and neck squamous cancer cells. FASEB J 2009;23:425–32.
[63] Cavadas M, Oikonomidi I, Gaspar CJ, Burbridge E, Badenes M, Felix I, Bolado A, Hu T, Bileck A, Gerner C, Domingos PM, von Kriegsheim A, Adrain C. Phosphorylation of iRhom2 controls stimulated proteolytic shedding by the metalloprotease ADAM17/TACE. Cell Rep 2017;21:745–57.
[64] Shibata M, Kodani I, Osaki M, Araki K, Adachi H, Ryoke K, Ito H. Cyclo-oxygenase-1 and -2 expression in human oral mucosa, dysplasias and squamous cell carcinomas and their pathological significance. Oral Oncol 2005;41:304–12.
[65] Tang DW, Lin SC, Chang KW, Chi CW, Chang CS, Liu TY. Elevated expression of cyclooxygenase (COX)-2 in oral squamous cell carcinoma—evidence for COX-2 induction by areca quid ingredients in oral keratinocytes. J Oral Pathol Med 2003;32:522–9.
[66] Gallo O, Masini E, Bianchi B, Bruschini L, Paglierani M, Franchi A. Prognostic significance of cyclooxygenase-2 pathway and angiogenesis in head and neck squamous cell carcinoma. Hum Pathol 2002;33:708–14.
[67] Lin DT, Subbaramaiah K, Shah JP, Dannenberg AJ, Boyle JO. Cyclooxygenase-2: a novel molecular target for the prevention and treatment of head and neck cancer. Head Neck 2002;24:792–9.
[68] Wang B, Hendricks DT, Wamunyokoli F, Parker MI. A growth-related oncogene/CXC chemokine receptor 2 autocrine loop contributes to cellular proliferation in esophageal cancer. Cancer Res 2006;66:3071–7.

[69] Ondrey FG, Dong G, Sunwoo J, Chen Z, Wolf JS, Crowl-Bancroft CV, Mukaida N, Van Waes C. Constitutive activation of transcription factors NF-(kappa)B, AP-1, and NF-IL6 in human head and neck squamous cell carcinoma cell lines that express pro-inflammatory and pro-angiogenic cytokines. Mol Carcinog 1999;26:119–29.
[70] Almofti A, Uchida D, Begum NM, Tomizuka Y, Iga H, Yoshida H, Sato M. The clinicopathological significance of the expression of CXCR4 protein in oral squamous cell carcinoma. Int J Oncol 2004;25:65–71.
[71] Samara GJ, Lawrence DM, Chiarelli CJ, Valentino MD, Lyubsky S, Zucker S, Vaday GG. CXCR4-mediated adhesion and MMP-9 secretion in head and neck squamous cell carcinoma. Cancer Lett 2004;214:231–41.
[72] Wang J, Seethala RR, Zhang Q, Gooding W, van Waes C, Hasegawa H, Ferris RL. Autocrine and paracrine chemokine receptor 7 activation in head and neck cancer: implications for therapy. J Natl Cancer Inst 2008;100:502–12.
[73] Liu FY, Safdar J, Li ZN, Fang QG, Zhang X, Xu ZF, Sun CF. CCR7 regulates cell migration and invasion through MAPKs in metastatic squamous cell carcinoma of head and neck. Int J Oncol 2014;45:2502–10.
[74] Muller A, Sonkoly E, Eulert C, Gerber PA, Kubitza R, Schirlau K, Franken-Kunkel P, Poremba C, Snyderman C, Klotz LO, Ruzicka T, Bier H, Zlotnik A, Whiteside TL, Homey B, Hoffmann TK. Chemokine receptors in head and neck cancer: association with metastatic spread and regulation during chemotherapy. Int J Cancer 2006;118:2147–57.
[75] Zhao ZJ, Liu FY, Li P, Ding X, Zong ZH, Sun CF. CCL19-induced chemokine receptor 7 activates the phosphoinositide-3 kinase-mediated invasive pathway through Cdc42 in metastatic squamous cell carcinoma of the head and neck. Oncol Rep 2011;25:729–37.
[76] Uchida D, Begum NM, Almofti A, Nakashiro K, Kawamata H, Tateishi Y, Hamakawa H, Yoshida H, Sato M. Possible role of stromal-cell-derived factor-1/CXCR4 signaling on lymph node metastasis of oral squamous cell carcinoma. Exp Cell Res 2003;290:289–302.
[77] Zlotnik A, Burkhardt AM, Homey B. Homeostatic chemokine receptors and organ-specific metastasis. Nat Rev Immunol 2011;11:597–606.
[78] Mehra R, Seiwert TY, Gupta S, Weiss J, Gluck I, Eder JP, Burtness B, Tahara M, Keam B, Kang H, Muro K, Geva R, Chung HC, Lin CC, Aurora-Garg D, Ray A, Pathiraja K, Cheng J, Chow LQM, Haddad R. Efficacy and safety of pembrolizumab in recurrent/metastatic head and neck squamous cell carcinoma: pooled analyses after long-term follow-up in KEYNOTE-012. Br J Cancer 2018;119:153–9.
[79] Ferris RL, Blumenschein Jr G, Fayette J, Guigay J, Colevas AD, Licitra L, Harrington K, Kasper S, Vokes EE, Even C, Worden F, Saba NF, Iglesias Docampo LC, Haddad R, Rordorf T, Kiyota N, Tahara M, Monga M, Lynch M, Geese WJ, Kopit J, Shaw JW, Gillison ML. Nivolumab for recurrent squamous-cell carcinoma of the head and neck. N Engl J Med 2016;375:1856–67.
[80] Moskovitz J, Moy J, Ferris RL. Immunotherapy for head and neck squamous cell carcinoma. Curr Oncol Rep 2018;20:22.
[81] Veglia F, Perego M, Gabrilovich D. Myeloid-derived suppressor cells coming of age. Nat Immunol 2018;19:108–19.
[82] Devaud C, John LB, Westwood JA, Darcy PK, Kershaw MH. Immune modulation of the tumor microenvironment for enhancing cancer immunotherapy. Oncoimmunology 2013;2.
[83] Kumar V, Patel S, Tcyganov E, Gabrilovich DI. The nature of myeloid-derived suppressor cells in the tumor microenvironment. Trends Immunol 2016;37:208–20.
[84] Trellakis S, Bruderek K, Dumitru CA, Gholaman H, Gu X, Bankfalvi A, Scherag A, Hutte J, Dominas N, Lehnerdt GF, Hoffmann TK, Lang S, Brandau S. Polymorphonuclear granulocytes in human head and neck cancer: enhanced inflammatory activity, modulation by cancer cells and expansion in advanced disease. Int J Cancer 2011;129:2183–93.
[85] Chan LP, Wang LF, Chiang FY, Lee KW, Kuo PL, Liang CH. IL-8 promotes HNSCC progression on CXCR1/2-meidated NOD1/RIP2 signaling pathway. Oncotarget 2016;7:61820–31.
[86] Facciabene A, Peng X, Hagemann IS, Balint K, Barchetti A, Wang LP, Gimotty PA, Gilks CB, Lal P, Zhang L, Coukos G. Tumour hypoxia promotes tolerance and angiogenesis via CCL28 and T(reg) cells. Nature 2011;475:226–30.
[87] Sun W, Li WJ, Wei FQ, Wong TS, Lei WB, Zhu XL, Li J, Wen WP. Blockade of MCP-1/CCR4 signaling-induced recruitment of activated regulatory cells evokes an antitumor immune response in head and neck squamous cell carcinoma. Oncotarget 2016;7:37714–27.
[88] Rentoft M, Coates PJ, Loljung L, Wilms T, Laurell G, Nylander K. Expression of CXCL10 is associated with response to radiotherapy and overall survival in squamous cell carcinoma of the tongue. Tumour Biol 2014;35:4191–8.

[89] Chen DS, Mellman I. Oncology meets immunology: the cancer-immunity cycle. Immunity 2013;39:1–10.
[90] Randolph GJ, Angeli V, Swartz MA. Dendritic-cell trafficking to lymph nodes through lymphatic vessels. Nat Rev Immunol 2005;5:617–28.
[91] Young A, Mittal D, Stagg J, Smyth MJ. Targeting cancer-derived adenosine: new therapeutic approaches. Cancer Discov 2014;4:879–88.
[92] Ma SR, Deng WW, Liu JF, Mao L, Yu GT, Bu LL, Kulkarni AB, Zhang WF, Sun ZJ. Blockade of adenosine A2A receptor enhances CD8(+) T cells response and decreases regulatory T cells in head and neck squamous cell carcinoma. Mol Cancer 2017;16:99.
[93] Ohta A, Gorelik E, Prasad SJ, Ronchese F, Lukashev D, Wong MK, Huang X, Caldwell S, Liu K, Smith P, Chen JF, Jackson EK, Apasov S, Abrams S, Sitkovsky M. A2A adenosine receptor protects tumors from antitumor T cells. Proc Natl Acad Sci U S A 2006;103:13132–7.
[94] Hatfield SM, Sitkovsky M. Oxygenation to improve cancer vaccines, adoptive cell transfer and blockade of immunological negative regulators. Oncoimmunology 2015;4.
[95] De Giorgi V, Grazzini M, Benemei S, Marchionni N, Botteri E, Pennacchioli E, Geppetti P, Gandini S. Propranolol for off-label treatment of patients with melanoma: results from a cohort study. JAMA Oncol 2018;4.
[96] Nilsson MB, Sun H, Diao L, Tong P, Liu D, Li L, Fan Y, Poteete A, Lim SO, Howells K, Haddad V, Gomez D, Tran H, Pena GA, Sequist LV, Yang JC, Wang J, Kim ES, Herbst R, Lee JJ, Hong WK, Wistuba I, Hung MC, Sood AK, Heymach JV. Stress hormones promote EGFR inhibitor resistance in NSCLC: implications for combinations with beta-blockers. Sci Transl Med 2017;9(415):eaao4307.

CHAPTER

16

Clinical trial design in head and neck cancer-challenges and opportunities

Malini M. Patel[a,b], *Justine Yang Bruce*[c], *Randall J. Kimple*[d]

[a]Division of Medical Oncology, Department of Medicine, Robert Wood Johnson Medical School, New Brunswick, NJ, United States [b]Rutgers Cancer Institute of New Jersey, New Brunswick, NJ, United States [c]Division of Hematology and Oncology, Department of Medicine, University of Wisconsin Carbone Cancer Center, University of Wisconsin School of Medicine and Public Health, University of Wisconsin, Madison, WI, United States [d]Department of Human Oncology, University of Wisconsin Carbone Cancer Center, University of Wisconsin School of Medicine and Public Health, University of Wisconsin, Madison, WI, United States

Abstract

Over the last few decades, the treatment of head and neck cancer has evolved in a multidisciplinary approach. While small gains in outcomes have been produced, much more work needs to be done to have achieved significant gains in survival and quality of life. This highlights the importance of carrying out well designed clinical trials. Several accrual hurdles have been identified over the years, including relative low incidences of head and neck cancer, clinician and patient-driven factors, and standardization of available resources. From a trial implementation perspective, appropriate site selection, meaningful endpoints, and trial schemas also contribute to the success of a study. This chapter will discuss the current challenges in head and neck trials implementation and the opportunities available to ensure the execution of meaningful research

Abbreviations

HHAC	historically high accruing centers
HLAC	historically low accruing centers
HNC	head and neck cancer
HNCIG	head and neck cancer intergroup
HPV	human papilloma virus
HRQOL	health related quality of life
LED	laryngo-pharyngeal dysfunction free survival
NPC	nasopharyngeal cancer
OPC	oropharyngeal carcinoma
ORR	overall response rate
OS	overall survival
PFS	progression free survival

Improving the Therapeutic Ratio in Head and Neck Cancer
https://doi.org/10.1016/B978-0-12-817868-3.00016-0

Conflict of interest

No potential conflicts of interest were disclosed.

The treatment of head and neck cancer (HNC) evolved into a multidisciplinary approach over the last three decades. Treatment options include single modality surgical resection or radiotherapy for early stage disease, combined modality treatment with definitive concurrent chemoradiation, surgery followed by adjuvant radiotherapy with or without chemotherapy, induction chemotherapy followed by radiation or chemoradiation for locally advanced disease, and finally chemotherapy or immunotherapy for metastatic disease. Some have pointed to an improvement in survival of head and neck cancer patients over the last 20 years, but this is likely due to an increased incidence of human papillomavirus related cancers which carry an improved prognosis rather than to improved outcomes with current treatments [1]. In spite of the adoption of multidisciplinary treatments, survival outcomes for patients have remained stagnant, highlighting the importance of conducting well designed clinical trials.

Accrual challenges

One of the biggest hurdles in clinical trial design for head and neck malignancies has undoubtedly been poor accrual. Up to a quarter of global head and neck trials have closed early due to insufficient accrual numbers [2]. This poses a significant problem as lack of accruals leads to premature closures, waste of financial and human resources, and lengthy enrollment periods. Various contributing factors may explain barriers to enrollment.

The overall low incidence of head and neck cancer likely contributes to poor accrual rates, by effectively having a smaller pool of patients per clinical trial slot. In 2018, there were approximately 65K new cases of oral cavity, pharyngeal, and laryngeal malignancies in the United States. This is in contrast to approximately 269K and 234K cases of breast cancer and lung cancer, respectively [3]. The impact of the relative rarity of HNC on meeting accrual goals can be highlighted in the example of nasopharyngeal carcinoma (NPC). Endemic to Southeast Asia, this region typically is successful in enrolling patients on NPC trials. In contrast, the landmark North America Intergroup 0099 trial, a collaborative effort by SWOG, ECOG, and RTOG, required 6 years to enroll 193 patients [4].

Continued collaborative efforts are paramount in ensuring the success of clinical trial completion. In an era of limited financial resources, decreased funding and increasing regulatory requirements, trials must not only enroll enough patients to meet statistical power but should be done in a reasonable amount of time to avoid losing relevance to the research question being examined. One third of federally funded clinical trials failed to meet minimum accrual goals. These problematic trends led the National Cancer Institute (NCI) to establish rules for flagging studies for closure if they fail to reach minimal enrollment targets at pre-specified timepoints. Other international cooperative groups such as the EORTC have also implemented similar policies for early closure of slow accruing studies [5].

International collaborations, though in theory seem logical to overcome accrual issues, carry a unique set of difficulties. These include working with multiple and redundant review processes, delays in protocol and amendment changes, oversight in data management regulation, and procurement and distribution of drugs. The Head and Neck Cancer Intergroup (HNCIG) was established in 2015 to tackle some of these obstacles. The established mission is to "promote and conduct high quality clinical trials and meta-analyses to improve outcomes for patients with head and neck tumors." Full members include the Canadian Cancer Trials Group (CCTG), Cancer Trials Ireland, Danish Head and Neck Cancer Group, ECOG-ACRIN, EORTC, Fudan University (China), Gruppo Oncologico del Nord Ovest, GORTEC, Hellenic Cooperative Oncology Group, Hong Kong Nasopharyngeal Carcinoma Study group, IAG KDT, Japanese Clinical Oncology group, MACH-NC Collaborative group, National Cancer Research Infrastructure, National Cancer Center, Nederlandse Werkgroep Hoofd-Hals Tumoren, NRG Oncology, Tata Memorial Center (India), Taiwan Cooperative Oncology Group, and trans-Tasman Radiation Oncology Group [6]. While there are many logistical barriers from a regulatory and financial standpoint in establishing a HNCIG trial, this collaboration serves as an excellent first step which will hopefully mirror the successes of other already productive international cooperative groups such as the Gynecological Cancer Intergroup or the Breast International Group.

Patient driven variables must also be considered when examining the low accrual to HNC trials. A British study evaluating barriers to recruitment for surgical trials identified patient preferences for one arm over the other and aversion to randomization as the most common reasons for refusal of participation. Other notable reasons included excessively complex patient information and protocol demands such as extra traveling and appointments [7]. Additionally, HNC patients tend to have more comorbidities, may have secondary tobacco related cancers, and are perceived to be less fit, all which may exclude patients from participation in trials [2].

Many clinical trials for head and neck cancer patients involve the combination of chemotherapy with radiotherapy. This imposes additional burdens on patients who often have to travel 5 days/week for 6–7 weeks to receive their radiotherapy at the enrolling institution. When compared to chemotherapy focused trials which often require travel once every 2–3 weeks, it is not surprising that many patients decline to participate in these trials.

Clinician driven factors also exist. A Canadian study anonymously polled 207 physicians and identified extra paperwork, additional clinic visits, extended follow up of trial patients and patient education needs as deterrents to clinical trial participation. Providers were also queried on strategies to increase motivation for participation- commonly encountered responses included alert systems to draw attention to new trials, regular local meetings to review protocols, screening logs to aid in accrual, and academic or work load credit for patient accruals and clinical trial participation [8].

Taking this all together, innovative strategies must be undertaken to improve accrual. The lack of awareness of clinical trials in HNC must be rectified to boost recruitment efforts. A more vigilant effort must be made to keep databases such as clinicaltrials.gov current to provide accurate information of available studies to patients and clinicians. Print materials, web-based videos, social media, continuing medical education events, and clinical talks have all been invaluable to disseminating awareness as well.

Site selection

Appropriate site selection must also be considered to ensure success of a study. As the care of head and neck oncology patients moved to a multidisciplinary model, the need for technical and clinical skill increases. In an effort to increase accrual, complex designed trials requiring a great deal of supportive care resources and clinical expertise are typically offered to a large number of sites. This leads to some sites enrolling few if any patients, and perhaps compromising the standardization of care that is received on protocol. In the TROG 02.02 trial, patients were randomized to chemoradiation alone versus chemoradiation plus tirapazamine. While the trial was ultimately negative, an important observation came to light: adherence to planned radiation therapy was a strong predictor for overall survival [9]. Furthermore, the probability of being non-compliant to radiotherapy directly correlated to the number of patients a site enrolled. Centers enrolling more than 20 patients had a higher 2years overall survival versus site that enrolled fewer than 5 patients (70% vs 50%; $P<.001$) [10].

In a retrospective study from a high-volume center, it was found that the single most important factor in choosing a radiotherapy location was the residing distance from the high-volume center, with patients living in the same county choosing a high-volume center over a low volume center. This is not surprising as radiation therapy requires daily treatment for 5–7weeks in most instances. Radiation treatment at a high-volume center was associated with improvements in local regional control, disease free survival and overall survival compared to HNC patients who were evaluated at a high-volume center but ultimately received their radiotherapy at a low volume center. The authors postulated this may be related to institutional variations with tumor coverage and radiotherapy treatment delivery [11]. Another retrospective study examined the institutional experience on overall survival of patients with stage III/IV HNC treated on RTOG 0129. The authors used historical trial accruals as the surrogate for institutional experience, classifying sites into historically low accruing centers (HLAC) or historically high accruing centers (HHACs). Patients at HLACs were more likely to have radiotherapy protocol deviations and a worse overall survival. This decrease in survival was present after adjustment of prognostic factors and radiotherapy compliance, suggesting the potential of other unstudied variables such as utilization of tumor boards, number of treating colleagues, presence of a training program, and ancillary support such like speech and swallow therapists, nutritional support and specialized nursing care [12].

The effect on survival based on institutional experience must be considered when designing future clinical trials and determining the optimal sites for study initiation. At the very least, trials should stratify patients according to the volume seen at a site—this may give future insight of the real-world applicability of specific interventions. Limiting the complexity of studies (i.e., restricting to non-adaptive designs, less intense treatment regimens) may allow for lower volume centers to offer trials and enroll patients without compromising outcomes. Alternatively, large multidisciplinary trials requiring radiation or surgical delivery could be restricted to high volume centers, creating a more complete exploration of the research questions at hand. This strategy may limit access to trials and further halt the pace of study completion. The optimal strategy remains debated.

Changing landscape of tumor biology

Initial approaches to HNC viewed the disease as one disease, with a similar natural history regardless of primary site or associated epidemiological risk factors. Early clinical trials in HNC allowed for broad inclusion criteria, resulting in heterogeneous study populations. This ultimately made it difficult to arrive at clear conclusions.

In the INT 0099 trial, untreated patients with nasopharyngeal carcinoma were randomized to radiotherapy alone (control arm) versus concurrent radiotherapy with high dose cisplatin followed by three cycles of adjuvant chemotherapy (experimental arm) [4].

Impressively, the trial included most of the SWOG, ECOG, and RTOG institutions. All histological subtypes were included, with a quarter of the patients being WHO type I. This was later a criticism as WHO type I patients were quite distinct in regards to tumor biology from the WHO type II/III patients seen in endemic areas. The study was terminated early after a significant difference in survival was seen in 147 assessed patients. With analysis for only a little over half the projected goal of patients, one can argue that there were just as many questions as answers left on the table, such as whether there were specific subgroups that only benefited from chemotherapy (advanced nodal stage or tumor stage, differentiated vs undifferentiated histological subtypes) [13].

Early clinical trials also failed to account for biological heterogeneity in oropharyngeal carcinoma. In 2007, the WHO officially recognized Human papillomavirus (HPV) as a viral etiology of oropharyngeal carcinoma. Several groups showed that this subset of OPC had better outcomes irrespective of stage or extent of nodal disease. In RTOG 0129, the authors conclude that there was no difference in survival between standard or accelerated fractionation radiotherapy with cisplatin, however HPV positive OPC had a significantly better 3-year OS compared to HPV negative OPC (82% vs 57%; $P<.001$) [14]. HPV positive patients were younger, typically had a better performance status, and are more likely to be non-smokers or light smokers. These findings established the rationale to stratify patients based on HPV status in future clinical trials. Recognizing that HPV positive OPC is a distinct biological entity, contemporary trials have emerged that are dedicated to this favorable subset. Most trials aim at de-escalation strategies for good risk HPV positive patients (see Chapter 3), with the ultimate goal of reducing long term side effects without compromising oncological outcome. Alternatively, HPV negative oropharyngeal carcinomas may require treatment intensification strategies.

Applicability of a trial

Caution must be exercised by designing a trial with a treatment algorithm that will be difficult to implement in "real world" practice. Using the INT 0099 example, only 55% of trial patients received all three cycles of adjuvant cisplatin and 5-FU and only 63% received all the concurrent chemotherapy doses with radiation [4]. A similarly conducted study examined the efficacy of the INT 0099 protocol in Japanese patients and saw that even with use of lower doses of chemotherapy, only 63% and 38% completed concurrent chemoradiation and adjuvant chemotherapy, respectively [15]. Despite this, clinicians will cite standard of care to include three cycles of adjuvant chemotherapy.

Contemporary clinical trial designs

The approach to cancer biology has shifted from considering tumors based on their primary site and stage to approaching them as those with a heterogeneous collection of molecular alterations. Rather than continuing to treat all patients with a particular stage of cancer from a particular site, we rely more on next generation sequencing to get an in-depth analysis of potential oncogenic drivers that can be manipulated for an anti-tumor effect. As we move to a precision medicine model in oncological care, our trial design must also reflect this new paradigm of therapeutic intent.

Basket trials allow for tumors from several primary tumor sites or histology and rely on the presence of a specific alteration to determine eligibility. These trials are typically phase II and may include a single drug or multiple drugs. All patients with a common actionable mutation are pooled together for analysis. Umbrella trials enroll patients with a particular primary tumor to receive one of several drugs depending on their genomic alterations present. As the development of small molecules that target a signaling pathway increase, these newer trial designs allow for more rapid study of efficacy [16]. As discussed throughout this book, the development of resistance to single-agent targeted therapy remains a significant problem. The importance of pre-clinical research to identify approaches to overcome this resistance cannot be over-estimated.

Basket or umbrella trials are ideal approaches to study rare cancer types, such as salivary gland carcinomas, where there are a paucity of data and approved systemic therapies. The EORTC has initiated the first umbrella trial design for recurrent/metastatic HNC where patients are randomized to standard of care, targeted therapies or immunotherapies based on their genomic profile. The study at present is using four study arms and testing three different targets (Fig. 1).

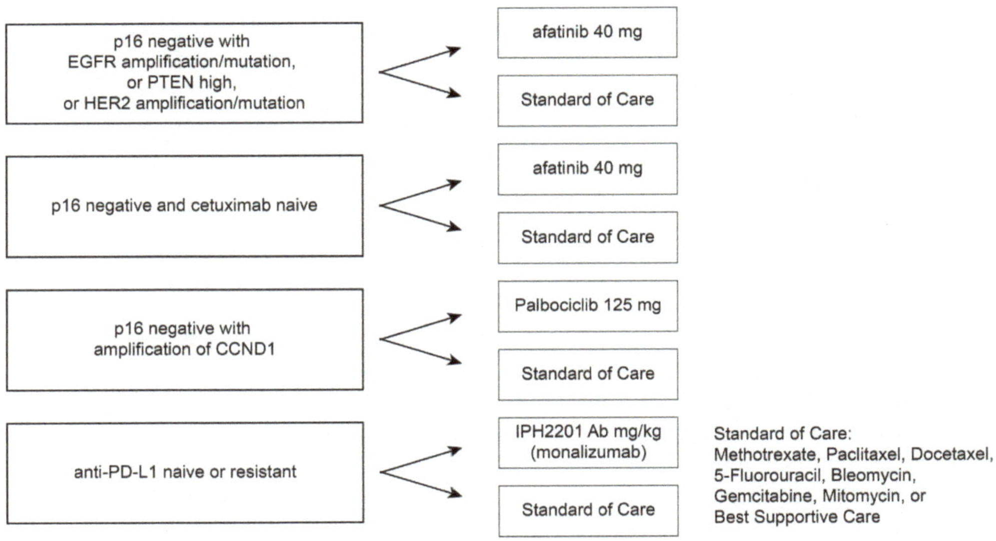

FIG. 1 Schema for EORTC 1559 HNCG: UPSTREAM study, a biomarker driven study which will assign treatment based on potential biomarkers and molecular alterations identified in the central platform.

The US Food and Drug Administration defines adaptive clinical trials as a "study that includes a prospectively planned opportunity for modification of one or more specified aspects of the study design and hypotheses based on the analysis of data from the subjects in the study." These modifications could include early stopping rules for inferior or superior treatment arms, adjustment of sample sizes, population enrichment, adding treatment arms or subgroups, or altering treatment assignment probabilities.

Sample size reassessments use interim analysis data during the trial to determine if the sample size needs to be increased to have sufficient power. Sample sizes can also be reduced, for example in the case of a particular intervention that appears to be futile. Response adaptive randomization alters the allocation ratio when interim analysis results are favorable, ensuring subsequent enrolled patients are more likely to be assigned to that treatment arm. Adaptive enrichment modifies eligibility criteria, so that if interim analysis shows a particular subgroup is more likely to respond, subsequent eligibility and enrollment will be restricted or favored for that subgroup [17]. These approaches, may make clinicians more comfortable in offering enrollment on clinical trials to patients as they are more likely to benefit from successful trials and less likely to be enrolled on futile study arms. Unfortunately, these same approaches can be confusing for patients to understand and may interfere with enrollment.

Predictive biomarkers offer information about the effect of a therapeutic intervention. In many instances, study investigators may not have strong enough pre-clinical evidence on the treatment benefit for biomarker positive tumors. Many trials design options exist for these scenarios. An enrichment design involves randomizing only the biomarker positive patients to the novel therapy versus standard therapy. This is useful when examining the treatment effect only in the population enriched subgroup for which there is prior evidence that the novel treatment is beneficial, but prospective validation is necessary. In this case, there is prior evidence that it will not be beneficial for the biomarker negative subpopulation to receive the novel therapy, providing rationale for excluding a subset access to a particular drug. Enrichment designed trials allow for efficient means to study the experimental treatment effect in the biomarker positive subgroup by using smaller sample sizes and this ultimately leads to cost savings. Additionally, interim monitoring allows for adjusting the trial so that the most promising subpopulation receives an intervention. For example, if in a phase II/III study, a particular intervention is futile in the biomarker negative subgroup, the phase III eligibility can be limited to the biomarker positive patients [17].

The NRG-HN 001 randomized phase II/III study aims at individualizing treatment for locally advanced NPC based on plasma EBV status. In the first portion of the trial, patients with positive pre-treatment EBV DNA are treated with definitive chemoradiation. Post chemoradiation, EBV DNA is measured and those with undetectable plasma EBV DNA are randomized to observation versus standard of care cisplatin and 5-FU $\times$ 3 cycles. Those with detectable EBV DNA are randomized to cisplatin/5FU versus gemcitabine/paclitaxel to determine if the latter regimen may be more effective than standard cisplatin/5FU. Innovative trial designs such as this should continue to be encouraged to allow for more personalized treatment approaches.

Due to the use of surgery in the treatment of head and neck cancer patients, window of opportunity trials offers a valuable approach in which to investigate novel therapeutics. Traditionally, novel agents are studied in the metastatic setting, but this presents many challenges to drug development. Often, metastatic patients have acquired several mechanisms of drug resistance that may render a drug inactive, causing investigators to prematurely presume the futility of a potentially beneficial agent. Additionally, challenges arise for meaningful translational research in identifying predictive biomarkers as this may necessitate repeat biopsies which poses an ethical dilemma in the palliative setting. Window of opportunity trials enroll patients prior to surgery and involve the delivery of a short course of therapy (Table 1). The duration of treatment is typically only a few days to a few weeks to avoid delays in delivery of curative therapy. Nevertheless, the duration of treatment must take into the account the time needed for a drug to reach a pharmacokinetic steady state

TABLE 1 Ongoing window-of-opportunity trials in head and neck cancer

NCT number	Title	Interventions	Phases
NCT03737968	Window of Opportunity Trial of Durvalumab (MEDI4736) or Durvalumab/Tremelimumab as Neoadjuvant Chemotherapy to Identify Immune Dynamics in Surgically Resectable Head and Neck Cancer Patients.	Durvalumab +/− Tremelimumab	Phase 2
NCT03769311	Cetuximab in Head and Neck Cancer Patients	Cetuximab	Phase 2
NCT03510390	Metformin in Head and Neck Squamous Cell Carcinoma: Effect on Tissue Oxygenation	Metformin	Not applicable
NCT03028766	WEE1 Inhibitor With Cisplatin and Radiotherapy: A Trial in Head and Neck Cancer	AZD1775 + Cisplatin	Phase 1
NCT02002182	ADXS 11-001 Vaccination Prior to Robotic Surgery, HPV-Positive Oropharyngeal Cancer	ADXS11-001 (ADXS-HPV)	Phase 2
NCT03854032	Nivolumab and BMS986205 in Treating Patients With Stage II–IV Squamous Cell Cancer of the Head and Neck	Nivolumab + IDO1 Inhibitor BMS-986205	Phase 2
NCT03795610	Window of Opportunity Study of IPI-549 in Patients With Locally Advanced HPV+ and HPV− Head and Neck Squamous Cell Carcinoma	IPI-549	Phase 2
NCT02769520	Efficacy Study of Pembrolizumab in Relapsed, Locally Recurrent Squamous Cell Cancer of the Head and Neck	Pembrolizumab	Phase 2
NCT03238365	Window of Opportunity Trial of Nivolumab and Tadalafil in Patients With Squamous Cell Carcinoma of the Head and Neck	Nivolumab + Tadalafil	Early phase 1
NCT03138070	A Window of Opportunity Study to Assess the Modulation of Biomarkers in Head and Neck Squamous Cell Cancer (HNSCC) By Preoperative Treatment With BYL719	BYL719	Not applicable
NCT03618654	Durvalumab With or Without Metformin in Treating Participants With Head and Neck Squamous Cell Carcinoma	Metformin + Durvalumab	Phase 1
NCT03575598	Sitravatinib (MGCD516) and Nivolumab in Oral Cavity Cancer Window Opportunity Study	Sitravatinib + Nivolumab	Early phase 1
NCT02882308	Preoperative Administration of Olaparib With Cisplatin or With Durvalumab or Alone or no Tratment in Patients Who Are Candidates for Surgery of Carcinoma of the Head and Neck	Olaparib +/− Cisplatin or +/− Durvalumab	Phase 2

[18]. Responses can be quickly assessed by imaging, pathology, or clinical features to provide an early endpoint of efficacy.

MK-3475-689, a randomized, international phase III trial is examining pembrolizumab prior to surgery and in combination with standard of care adjuvant therapy in stage III/IVA locoregionally advanced HNSCC. In the trial schema, those in the experimental arm receive two doses of pembrolizumab, 3 weeks apart, prior to planned resection, with one of the primary endpoints being major pathological response. These trials can be relatively easy to accrue to as the number of study visits required for a patient is relatively small, and it provides a cancer-directed treatment in the interval between diagnosis and when the patient can be scheduled into the operating room. Importantly, these studies also provide an opportunity to study biomarkers of response or resistance and can help investigators prioritize treatments to bring forward to larger study cohorts.

Choosing appropriate endpoints

The gold standard for a primary endpoint in clinical trial design remains overall survival. It leaves little room for interpretation as it's clearly defined as the time from randomization to time of death. While the endpoint itself is not influenced by compliant follow up and response assessments, this endpoint can often be confounded by non-cancer related mortality and the type and lines of post progression therapy received. This is an especially relevant point for the head and neck population who often have frequent competing comorbidities for death.

As research focuses shift on delivery effective care with limiting short and long-term toxicity, health related quality of life (HRQOL) serves as an important endpoint in HNC trials. These can be measured in a variety of validated assessment tools such as EORTC QLQ-C30, QLQ-H&N35, MDADI, and the EQ-5D-5L questionnaires [19].

Nonetheless, non-traditional alternative endpoints are helpful in the context of particular trials designs. One such example is in organ preservation trials. The Veterans Affairs Laryngeal Cancer Study Group (VALCSG) performed a pivotal trial where the aim was to identify an improvement in laryngectomy free survival. In this trial, patients were randomized to upfront laryngectomy versus induction chemotherapy followed by radiotherapy, with salvage laryngectomy in non-responders. While the overall survival in both groups was similar, close to two thirds of patients were able to avoid a laryngectomy. While endpoints like this in the correct clinical trial context provide important information, caution should be exercised in the inappropriate creation of artificial endpoints. A laryngectomy free survival may not necessarily equate to a meaningful clinical benefit. A laryngo-pharyngeal dysfunction free survival (LED) may be a more appropriate measure, which documents death, local relapse, need for total/partial laryngectomy, or presence of a tracheostomy or feeding tube for >2 years as measures of failure [13].

Objective response rate is an endpoint that is reported in clinical trials but its relevance and correlation to overall survival remains debatable. Two examples highlight this point. In induction chemotherapy trials, overall responses of up to 70% are encountered. However, despite high overall response rates to induction chemotherapy, when given in a sequential approach with radiation +/− chemotherapy, there is no survival benefit seen [20, 21]. In the case of induction chemotherapy, a relatively high overall response rate fails to yield a survival benefit. Alternatively, relatively low overall response rates should not reflexively create an assumption that an intervention does not have a clinically meaningful benefit. In KEYNOTE-012, the overall response rates (ORR) to pembrolizumab was modest at 18%, however >70% of patients had durable responses lasting at least 12 months [22]. The use of ORR does not capture patients with a durable stable response or differentiate outcomes based on the depth of response or time to next treatment.

Progression-free survival (PFS) is also commonly used as a surrogate for overall survival and HRQOL. There are several limitations to using PFS, especially in immune-oncology trials, which make up a large portion of contemporary HNC studies. It does not reliably predict outcomes in immunotherapy trials where there has been OS benefit but no difference in PFS. For example, in the phase III randomized CHECKMATE 141 study, the PFS difference between the nivolumab and chemotherapy arms was not statistically different, despite a statistical difference in overall survival [23]. Additionally, PFS, does not capture pseudoprogression, a well-known phenomenon seen with immune-oncology drugs.

Several other endpoints are included in HNC studies as secondary objectives of study. Local control and locoregional control are frequently reported in trials investigating surgery or radiotherapy treatments. An important issue to point out when using one of these secondary endpoints is that they do not provide information on organ function which may be relevant to patient quality of life. This is particularly important when determining the optimal local strategy (i.e., surgery vs radiation therapy) in the management of locally advanced head and neck cancer. Distant control is commonly reported to study the effect of systemic treatment, however these must be analyzed in the context of overall survival. Induction chemotherapy trials have suggested improved distant control, but this has failed to result in an overall survival benefit [20].

Moving forward, the design of HNC clinical trials must not only include traditional endpoints such as overall survival but incorporate HRQOL measures and patient reported outcomes (PRO) as means to determine the future treatment paradigms. In additional to survival, long term late toxic effects, overall function and quality of life, and the psychological impact from treatments, are all important data points that can be captured in clinical trial design.

Acknowledgments

This project was supported in part by the Specialized Program of Research Excellence (SPORE) program, through the NIH National Institute for Dental and Craniofacial Research (NIDCR) and National Cancer Institute (NCI), grant P50DE026787. The content is solely the responsibility of the authors and does not necessarily represent the official views of the NIH.

References

[1] Pulte D, Brenner H. Changes in survival in head and neck cancers in the late 20th and early 21st century: a period analysis. Oncologist 2010;15(9):994–1001.

[2] Haddad RI, Chan AT, Vermorken JB. Barriers to clinical trial recruitment in head and neck cancer. Oral Oncol 2015;51(3):203–11.

[3] American Cancer Society. Cancer facts & figures 2018. Atlanta: American Cancer Society; 2018. 2018Contract.

[4] Al-Sarraf M, LeBlanc M, Giri PG, Fu KK, Cooper J, Vuong T, Forastiere AA, Adams G, Sakr WA, Schuller DE, Ensley JF. Chemoradiotherapy versus radiotherapy in patients with advanced nasopharyngeal cancer: phase III randomized intergroup study 0099. J Clin Oncol 1998;16(4):1310–7.

[5] Cooperative Group Guidelines for the Development, Conduct, and Analysis of Clinical Trials with International Collaborating Institutions. National cancer institute—cancer therapy evaluation program. Available from: https://ctep.cancer.gov/branches/ctmb/clinicaltrials/docs/nci_clin_intl_guidelines.pdf; 2007.

[6] Le QT, Welch JJ, Vermorken JB, Rischin D, Mehanna H. Formation of an international intergroup to coordinate clinical trials in head and neck cancers: HNCIG. Oral Oncol 2017;71:180–3.

[7] Kaur G, Hutchison I, Mehanna H, Williamson P, Shaw R, Tudur Smith C. Barriers to recruitment for surgical trials in head and neck oncology: a survey of trial investigators. BMJ Open 2013;3(4):e002625.

[8] Mahmud A, Zalay O, Springer A, Arts K, Eisenhauer E. Barriers to participation in clinical trials: a physician survey. Curr Oncol 2018;25(2):119–25.

[9] Rischin D, Peters LJ, O'Sullivan B, Giralt J, Fisher R, Yuen K, Trotti A, Bernier J, Bourhis J, Ringash J, Henke M, Kenny L. Tirapazamine, cisplatin, and radiation versus cisplatin and radiation for advanced squamous cell carcinoma of the head and neck (TROG 02.02, HeadSTART): a phase III trial of the Trans-Tasman Radiation Oncology Group. J Clin Oncol 2010;28(18):2989–95.

[10] Peters LJ, O'Sullivan B, Giralt J, Fitzgerald TJ, Trotti A, Bernier J, Bourhis J, Yuen K, Fisher R, Rischin D. Critical impact of radiotherapy protocol compliance and quality in the treatment of advanced head and neck cancer: results from TROG 02.02. J Clin Oncol 2010;28(18):2996–3001.

[11] Naghavi AO, Echevarria MI, Strom TJ, Abuodeh YA, Venkat PS, Ahmed KA, Demetriou S, Frakes JM, Kim Y, Kish JA, Russell JS, Otto KJ, Chung CH, Harrison LB, Trotti A, Caudell JJ. Patient choice for high-volume center radiation impacts head and neck cancer outcome. Cancer Med 2018;7(10):4964–79.

[12] Wuthrick EJ, Zhang Q, Machtay M, Rosenthal DI, Nguyen-Tan PF, Fortin A, Silverman CL, Raben A, Kim HE, Horwitz EM, Read NE, Harris J, Wu Q, Le QT, Gillison ML. Institutional clinical trial accrual volume and survival of patients with head and neck cancer. J Clin Oncol 2015;33(2):156–64.

[13] Adelstein DJ. Clinical trial design in head and neck cancer: what has the oncologist learned? Lancet Oncol 2012;13(7):e318–23.

[14] Ang KK, Harris J, Wheeler R, Weber R, Rosenthal DI, Nguyen-Tan PF, Westra WH, Chung CH, Jordan RC, Lu C, Kim H, Axelrod R, Silverman CC, Redmond KP, Gillison ML. Human papillomavirus and survival of patients with oropharyngeal cancer. N Engl J Med 2010;363(1):24–35.

[15] Demizu Y, Sasaki R, Soejima T, Maruta T, Okamoto Y, Yamada K, Yoden E, Ejima Y, Ota Y, Ishida H, Nibu K, Sugimura K. Efficacy and feasibility of cisplatin-based concurrent chemoradiotherapy for nasopharyngeal carcinoma. Jpn J Clin Oncol 2006;36(10):620–5.

[16] Simon R. Genomic alteration-driven clinical trial designs in oncology. Ann Intern Med 2016;165(4):270–8.

[17] Thorlund K, Haggstrom J, Park JJ, Mills EJ. Key design considerations for adaptive clinical trials: a primer for clinicians. BMJ 2018;360:k698.

[18] Schmitz S, Duhoux F, Machiels JP. Window of opportunity studies: do they fulfil our expectations? Cancer Treat Rev 2016;43:50–7.

[19] Hezel M, von Usslar K, Kurzweg T, Lorincz BB, Knecht R. Endpoints and cutpoints in head and neck oncology trials: methodical background, challenges, current practice and perspectives. Eur Arch Otorhinolaryngol 2016;273(4):837–44.

[20] Cohen EE, Karrison TG, Kocherginsky M, Mueller J, Egan R, Huang CH, Brockstein BE, Agulnik MB, Mittal BB, Yunus F, Samant S, Raez LE, Mehra R, Kumar P, Ondrey F, Marchand P, Braegas B, Seiwert TY, Villaflor VM, Haraf DJ, Vokes EE. Phase III randomized trial of induction chemotherapy in patients with N2 or N3 locally advanced head and neck cancer. J Clin Oncol 2014;32(25):2735–43.

[21] Haddad R, O'Neill A, Rabinowits G, Tishler R, Khuri F, Adkins D, Clark J, Sarlis N, Lorch J, Beitler JJ, Limaye S, Riley S, Posner M. Induction chemotherapy followed by concurrent chemoradiotherapy (sequential chemoradiotherapy) versus concurrent chemoradiotherapy alone in locally advanced head and neck cancer (PARADIGM): a randomised phase 3 trial. Lancet Oncol 2013;14(3):257–64.

[22] Seiwert TY, Burtness B, Mehra R, Weiss J, Berger R, Eder JP, Heath K, McClanahan T, Lunceford J, Gause C, Cheng JD, Chow LQ. Safety and clinical activity of pembrolizumab for treatment of recurrent or metastatic squamous cell carcinoma of the head and neck (KEYNOTE-012): an open-label, multicentre, phase 1b trial. Lancet Oncol 2016;17(7):956–65.

[23] Ferris RL, Blumenschein Jr G, Fayette J, Guigay J, Colevas AD, Licitra L, Harrington K, Kasper S, Vokes EE, Even C, Worden F, Saba NF, Iglesias Docampo LC, Haddad R, Rordorf T, Kiyota N, Tahara M, Monga M, Lynch M, Geese WJ, Kopit J, Shaw JW, Gillison ML. Nivolumab for recurrent squamous-cell carcinoma of the head and neck. N Engl J Med 2016;375(19):1856–67.

Index

Note: Page numbers followed by *f* indicate figures and *t* indicate tables.

FSC
www.fsc.org
MIX
Papier | Fördert gute Waldnutzung
FSC® C083411

Kolibri 360 GmbH
Ferdinand-Jühlke-Straße 7
99095 Erfurt, Deutschland
produktsicherheit@kolibri360.de